T0413532

Barile's Clinical Toxicology

Principles and Mechanisms,
Third Edition

Barile's Clinical Toxicology

Principles and Mechanisms, Third Edition

Frank A. Barile, MS, PhD

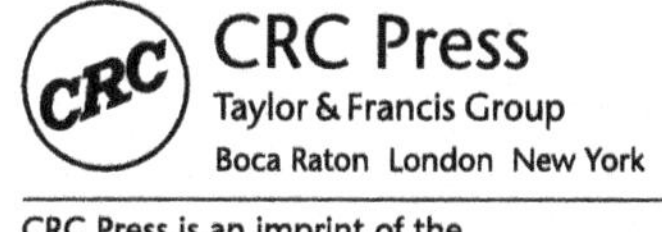

CRC Press is an imprint of the
Taylor & Francis Group, an informa business

CRC Press
Taylor & Francis Group
6000 Broken Sound Parkway NW, Suite 300
Boca Raton, FL 33487-2742

Printed on acid-free paper

International Standard Book Number-13: 978-1-4987-6530-5 (Hardback)

Visit the Taylor & Francis Web site at
http://www.taylorandfrancis.com

and the CRC Press Web site at
http://www.crcpress.com

Pauline

through all the edits

Contents

SECTION II Toxicity of Therapeutic Agents

SECTION III Toxicity of Nontherapeutic Agents

Preface

The third edition of the original text, now *Barile's Clinical Toxicology: Principles and Mechanisms*, represents a journey into a rapidly changing field that traditionally did not present fast-moving developments. Unlike other evolving technological fields such as computer engineering and mobile devices, toxicology has not been a prime mover in the applied sciences. In comparison with other biomedical applications, advances in clinical toxicology are also not as rapid as those encountered within genetic, stem cell, or immunologic therapeutics. However, the cross talk among these areas involving clinical applications of toxicology has enabled the latter to express a variety of new approaches and knowledge within the last 10 years, advances that reflect the problems and issues encountered by society. The topics reflect the increasing issues seen in substance abuse and treatment, environmental contamination, adverse reactions encountered with traditional cancer chemotherapeutic agents, and chemical and biological threats to public health. In particular, as mentioned in the Preface to the 2nd Edition, clinical toxicology has shifted consistently toward applied realms; the basic science of poisons, however, in this author's opinion, remains with the traditional understanding of the basic science of toxicology, notwithstanding the causes, mechanisms, and pathology.

This third edition presents the field of clinical toxicology as a thought-provoking application, especially when viewed in the context of social media and the entertainment industry as part of law enforcement and crime-solving theater. To students in the health and biomedical sciences and to established health professionals, the arena of medical toxicology is inspiring. The reviews, criticisms, and suggestions garnered over the last 10 years, since the advent of the first edition, have enabled the production of an updated, applied, and mechanistic text. The second edition contained more detailed emergency protocols, new and revised therapies, extensive insight into mechanisms, and recent references. Each chapter had been revised, while two new chapters were included (Introduction to Chemical- and Drug-Receptor Interactions [Chapter 11] and a presentation of Toxicogenomics [Chapter 12]). These new chapters are also included in the third edition of Part I, Introduction to Basic Principles of Toxicology, which should be useful for understanding the basic principles of toxicology by undergraduate and graduate students. In addition, some chapters, such as Cardiovascular Toxicology (Chapter 20), Alcohols and Aldehydes (Chapter 24), Metals (Chapter 26), and Reproductive and Developmental Toxicology (Chapter 32), have been completely rewritten and updated with biochemical formulas, chemical structures, toxicological mechanisms, and treatment principles and protocols.

The third edition highlights new fields of toxic exposure, particularly as related to controlled substances. Since the original investigations into the mechanisms of opioid substance use, the situation has significantly worsened, prompting regulators to relabel the field as an "epidemic." In addition, developments in U.S. federal and state regulations concerning marijuana use have resulted in the development of unforeseen and unpredictable social and toxicological consequences. Thus, specific antidotes, therapeutic interventions, and general supportive measures addressing the old—yet new—problems are inserted as important guidelines of treatment. Consequently, the 3rd edition necessitated at least 120 additional pages from the 2nd.

In general, and as with previous versions, the third edition examines the complex interactions associated with clinical toxicological events as a result of therapeutic drug administration or deliberate or inadvertent chemical and drug exposure. Special emphasis is placed on signs and symptoms of syndromes and pathology caused by chemical exposure and administration of clinical drugs. The source, pharmacological and toxicological mechanisms of action, toxicokinetics, medicinal chemistry, clinical management of toxicity, and detection and identification of the drug or chemical in body fluids are discussed. Proprietary names are inserted where applicable, and particularly when publicly available. Historically well-known but obsolete drugs are mentioned only as a matter of reference. Other contemporary issues in clinical toxicology, including the various means of possible exposure to therapeutic and nontherapeutic agents, an overview of protocols for managing various toxic ingestions, and the antidotes and treatments associated with their pathology, are conveyed.

The objective of this edition is to instill a greater respect among health professionals and students for clinical toxicology as they strive to understand the serious consequences of exposure to drugs and the multitude of chemicals. As societies become more technologically adept at many levels, their thirst for information about chemical exposure increases. Accordingly, students, health professionals, and emergency personnel develop greater awareness and appreciation of the costs of drug administration and of exposure to nontherapeutic compounds, of the risk from biological threats, and of the adverse effects of herbal products. As a generation becomes more knowledgeable, its preparedness is sturdy, and its response to the threat is more formidable.

Acknowledgments

My appreciation is extended to several individuals and colleagues, whose edits, comments, and opinions have greatly contributed to the final version of the third edition of *Barile's Clinical Toxicology: Principles and Mechanisms.* Among them, I am appreciative of Ms. Ann Yang's dedication with reference retrieval and accurate evaluations; Mr. Anirudh Chintalapati for his insightful and informative discussions on the subject of clinical toxicology and contributions to the chapters; and my past graduate students, Angela Aliberti, Sanket Gadhia, Neha Chavan and Devin O'Brien for their constant vision with purpose, motivation, technical expertise, dependability, individual strength, and forceful forward momentum. I am impressed with the focused motivation and keen interest in the field of clinical toxicology of my young visiting scholar, Nicholas Cordero, whose bright cerebral glow illuminated the entire lab. My gratitude is also extended to my graduate classes in the Department of Pharmaceutical Sciences in toxicology and the College of Pharmacy, whose interest and inquisitiveness stimulated a series of thoughtful and unique class discussions. In addition, Ms. Luzmery Benitez dedicated invaluable administrative assistance toward the completion of this text.

I also acknowledge the concern and commitment from the editors at CRC Press, particularly publisher Barbara Norwitz (retired), managing editor Stephen Zollo, editorial assistant Ms. Laura Piedrahita, and R. Sweda of Deanta Global Publishing. Their patience and focus were of great support in the completion of this project.

Finally, I am continuously indebted to my wife Pauline, whose encouragement, inspiration, and reassurance have assiduously sustained my efforts, for believing in my abilities, in the force of my dedication, and in the value of my contributions.

Author

Frank A. Barile, PhD, is Professor in the Toxicology Division and past chairman of the Department of Pharmaceutical Sciences, St. John's University College of Pharmacy and Health Sciences, New York.

Dr. Barile received his BS in Pharmacy, MS in Pharmacology, and PhD in Toxicology at St. John's University. After doing a postdoctoral fellowship in Pulmonary Pediatrics at the Albert Einstein College of Medicine, Bronx, NY, he moved to the Department of Pathology, Columbia University—St. Luke's Roosevelt Hospital, NY, as a research associate. In these positions, he investigated the role of pulmonary toxicants in collagen metabolism in cultured lung cells. In 1984, he was appointed assistant professor in the Department of Health Sciences at City University of NY. Sixteen years later, he rejoined St. John's University in the Department of Pharmaceutical Sciences in the College of Pharmacy.

Dr. Barile holds memberships in several professional associations, including the U.S. Society of Toxicology, the American Association of University Professors, the American Association for the Advancement of Science, the American Society of Hospital Pharmacists, New York City Pharmacists Society, New York Academy of Sciences, and New York State Council of Health System Pharmacists. He is past president of the *In Vitro* and Alternative Methods Specialty Section of the U.S. Society of Toxicology and a former member of the Scientific Advisory Committee for Alternative Toxicological Methods, the National Institute of Environmental Health Sciences, U.S. National Institutes of Health (NIH). He is former editor of *Toxicology in Vitro* and *Journal of Pharmacological & Toxicological Methods*, published by Elsevier Ltd.

Dr. Barile is the recipient of Public Health Service research grants from the National Institute of General Medical Sciences, the National Institutes of Health, and grants from private foundations dedicated to alternatives to animal testing. He has authored approximately 100 original research manuscripts, review articles, research abstracts, and conference proceedings in peer-reviewed toxicology and biomedical journals. He has also published several books and related chapters in the field. He contributed original *in vitro* toxicology data to the international Multicenter Evaluation for In Vitro Cytotoxicity program. He lectures regularly to pharmacy and toxicology undergraduate and graduate students in clinical and basic pharmaceutical and toxicological sciences and was awarded "Professor of the Year" for the College of Pharmacy by the St. John's University Student Government Association (2003).

Dr. Barile has served on several U.S. government advisory committees, including: Toxicology Assessment Peer Review Committee, U.S. Environmental Protection Agency; National Institutes of Health Review Panel, Study Section Reviewer (2013); Scientific Advisory Committee on Alternative Toxicological Methods (SACATM, 2005–2009); National Toxicology Program (NTP) Interagency Center for the Evaluation of Alternative Toxicological Methods, Support Contract Reviewer (2014); U.S. Food and Drug Administration (U.S. FDA), National Center for Toxicology Research, Systems Biology Subcommittee (2016); National Institute for Occupational

Safety and Health (NIOSH) and Oak Ridge Associated Universities, SK Profiles Review Group (2014); U.S. Food and Drug Administration advisory committee on alternative toxicological methods (2013); and has served as vice president of the Dermal Toxicology Specialty Section for the Society of Toxicology (2013–2014).

Dr. Barile received the Faculty Recognition Award, American Association of University Professors Faculty Association Award, St. John's University, in 2003, 2004–2005, 2013–2014, and 2015–2016 and received the prestigious Public Health Service Medallion from the director of the National Institute of Environmental Health Sciences, Dr. Linda Birnbaum, for contributions to the Scientific Advisory Committee for Alternative Toxicological Methods (2009).

Dr. Barile continues original research on the cytotoxic effects of therapeutic drugs, environmental chemicals, and controlled substances on cultured mammalian embryonic stem cells and human induced pluripotent stem cells and has dedicated his professional life to the advancement of *in vitro* alternative methods to reduce, replace and refine animal toxicology testing.

Contributors

Anirudh J. Chintalapati
Department of Pharmaceutical Sciences
St. John's University College of Pharmacy and Health Sciences
Queens, New York

Zacharoula Konsoula
Syneos Health
Washington, D.C.

Section I

Introduction to Basic Toxicological Principles

1 Introduction

1.1 INTRODUCTION

As in other biomedical fields, the evolution of toxicology has progressed logarithmically. Prehistoric methods of identifying substances by trial and error involved tasting, applying, cooking with, inhaling, or accidental encounter with natural products in the environment, after which the effects could be observed and monitored. This type of exploration required careful observation, recording, and follow-up of the effects.

Historically, the "science of poisons" developed from an applied subject matter to its current unique discipline. It is conceivable that early hominids identified edible plants and animals by primitive experimentation, the dire consequences of which were quickly memorized. As early human civilizations began to understand the fields of agriculture and tool-making, they learned about the usefulness and risks associated with a variety of substances. Metals, such as lead, arsenic, and mercury, were quick to find a role in pottery, as tonics, as miracle cures, or as ingredients of culinary recipes. The aromas emanating from the burning, boiling, or cooking of herbs and plants probably led to an understanding of other possible uses. For instance, the strength derived from the cardiotonic effects of ingesting *Digitalis purpurea*, or from the experience of the religious revelations aided by the hallucinogenic properties of the ergot-rich plants, was at best only mysteriously appreciated by the "medicine men" of the time. Although apparently medieval in its approach, the methodology of toxicological analysis, systematic observation, and practical experimentation would eventually lead to more sophisticated standards and applications.

As societies progressed through the industrial age, more adventurous scholars attempted to decipher the chemical basis of natural compounds. However, since the basic sciences of chemistry and biology were poorly understood, scientists and clinicians would be forced to wait several centuries before the biochemical mechanisms of poisonous interactions could be understood. In fact, it was not until the fifteenth century that the physician Paracelsus, one of the early "Fathers of Toxicology," promoted that all matter was composed of three "primary bodies" (sulfur, salt, and mercury) and that "all substances are poisons, there is none which is not a poison … the right dose differentiates a poison from a remedy" (Paracelsus, 1493–1541). Yet, toxicology, as a field of scholarly endeavor, would need to be tolerant until the mid- to late 1600s to be conceptually accepted as a discipline.

Fast-forward to the latter half of the twentieth century—once again, the field of toxicology derived impetus from analytical chemists, who sought to understand the action of xenobiotics by isolating and identifying the components. Until these developments crystallized, most traditional toxicologists were trained in classical quantitative chemical analysis as well as organic and inorganic chemistry. Eventually, the field would flourish into a variety of specialties, which today are well accepted as part of the biomedical sciences. Currently, clinical toxicology (also known as *pharmacotoxicology*) enjoys its own particular designation, the subject of which is advanced in the subsequent chapters.

The early investigators had a plan—a systematic approach, as far as they recognized, that would sustain human existence, maintain and advance health status, promote the health of family and community, treat the incomprehensible traumas encountered, and prevent the maladies of the day. Today, the objectives are similar, and the plan for the field is indistinguishable from prior times. The methodologies and understanding, however, have advanced significantly, whereas the objectives are immutable.

1.2 BASIC DEFINITIONS

1.2.1 Toxicology

The classic definition of toxicology has traditionally been understood as the study of xenobiotics, or simply as the science of poisons—that is, the interaction of exogenous agents with mammalian physiological compartments. For purposes of organizing the nomenclature, chemicals, compounds, or drugs are often referred to as *agents*. And since such agents have the ability to induce undesirable effects, they are usually referred to as *toxins*. Yet, most agents were justifiably initially used for benevolent purposes. The desire to seek remedies eventually led to the discoveries that therapeutic drugs also did harm. Thus, this text uses terms such as *agent*, *chemical*, *drug*, or *toxin* interchangeably, depending on the application. Consequently, toxicology involves internal and external physiological exposure to toxins and their interactions with the body's compartments. With time, the nomenclature has evolved to include many chemically unrelated classes of agents. What transforms a chemical into a toxin depends more on the length of time of exposure (duration), the dose (or concentration) of the chemical, and the route of exposure as well as the chemical structure, product formulation, or intended use of the material. As a result, almost any chemical has the potential for toxicity and thus, falls under the broad definition of toxicology.

1.2.2 Clinical Toxicology

Traditionally, clinical toxicology was regarded as the specific discipline of the broader field of toxicology concerned with unforeseen consequences of agents whose intent is to treat, ameliorate, modify, or prevent disease states, or the effect of drugs that, at one time, were intended to be used as such. These compounds would fall under the classification of therapeutic agents (Section II). A more liberal definition of clinical toxicology involves not only the toxic effects of therapeutic agents but also those chemicals whose intention is not therapeutic. Humans and animals are exposed to such substances through environmental pathways (i.e., gases) or as components of industrial advances (metals). A relatively new application of drug use is as the result of societal behavior (alcohol and substance abuse), as chemical by-products of industrial development (gases, hydrocarbons, and radiation), or as essential components of urban, suburban, or agricultural technologies (pesticides, insecticides, and herbicides). Thus, these chemicals are classified as nontherapeutic agents but are in fact associated with a variety of well-known clinical signs and symptoms, which warrant discussion as part of a clinical toxicology text.

1.3 TYPES OF TOXICOLOGY

1.3.1 General Toxicology

General toxicology involves the broad application of toxins and their interaction with biological systems. As with any scientific, biomedical, or clinical discipline, however, the term *general toxicology* has lost its popularity and has been replaced by more specialized fields of study. With the development of advanced methodologies in biotechnology, the requirement for increased training in the field has made it necessary to accommodate the discipline with an expanding body of specialties. Thus, a variety of adjectives evolved to further define the particular disciplines within the toxicological framework.

1.3.2 Mechanistic Toxicology

Mechanistic toxicology refers to the identification of the cause of toxicity of a chemical at the cellular or tissue level. This may involve the interaction of a toxic chemical with cellular biochemical pathways, interference with protein and enzyme reactions within the cell, or interface of the molecule with a membrane or cytoplasmic receptor. In any of these events, the classification of toxicity of a chemical, therefore, may be expressed in terms of its mechanism of toxicity. A similar expression, mechanism of action, is universally applied in the study of pharmacology. Thus, mechanistic toxicology seeks to answer the biochemical, physiological, or organic basis of a chemical's effect on eukaryotic systems.

1.3.3 Regulatory Toxicology

Regulatory toxicology refers to the administrative codes associated with the availability of potential toxins encountered in society. Regulatory toxicology thus refers to a systematic approach that defines, directs, and dictates the rate at which an individual may encounter a synthetic or naturally occurring toxin. Based on the scientific basis of evaluation of the toxic characteristics of a chemical, government administrative agencies establish guidelines for the chemical, drug, or molecule and regulate its maintenance in the environment or within the therapeutic market. The guidelines are generally promulgated by agencies whose jurisdiction and regulations are established by federal, state, and local authorities, although most governments adjudicate laws and regulations through federal authority.

1.3.4 Descriptive Toxicology

Descriptive toxicology is a comprehensive attempt at explaining the toxic agents and their applications. Descriptive toxicology developed principally as a method for bridging the vacuum between the science and the public's understanding of the field, especially when it became necessary for nonscientific sectors to comprehend the importance of toxicology. For instance, the study of metals in the environment (metal toxicology) has become a popular description for toxicologists interested in examining the role of heavy or trace metals in the environment. Clinical toxicology may also be considered as a descriptive category that involves the analysis of drugs or chemicals whose exposure is associated with pathological consequences, therapeutic intervention, known signs and symptoms, and public health involvement. Other descriptive areas in the field are defined in Table 1.1. More recently, research areas have intensified in toxicology and have blossomed into several broad descriptive fields, including, but not limited to, the study of apoptosis, receptor-mediated signal transduction, gene expression, proteomics, oxidative stress, and toxicogenomics.

TABLE 1.1
Definition of Descriptive Toxicology

Type of Toxicology	Definition
General	Applies to the broad application of toxins and their interaction with biological system
Mechanistic	Involves the interaction of a chemical with cellular biochemical and physiological pathways
Regulatory	Establishes guidelines that are generally promulgated by agencies whose jurisdiction and authority are established by federal, state, and local authorities
Genetic	Incorporates molecular biology principles in applications of forensic sciences, such as DNA testing, as legal evidence in court proceedings
Occupational	Examines hazards associated with toxic exposure in the workplace
In vitro	Refers to the development of cell culture techniques as alternatives to animal toxicity testing
Analytical	Laboratory methods associated with identification and chemical characterization of a chemical
Immuno-	Effects of chemicals or drugs on the cells and organs of the mammalian immune system, particularly the interactions with the cellular components of immune pathways
Neuro-	Effects of chemicals or drugs on the cells and organs of the mammalian neural system, particularly the interactions with the cellular components of the neuroendocrine pathways
Developmental	Effects of chemicals or drugs on reproductive systems, growth and progress of reproductive organs, and embryonic growth

1.4 TYPES OF TOXICOLOGIST

Table 1.2 summarizes different specialties of toxicology and the practitioners trained for those appropriate fields. In general, many toxicologists have roles that overlap within the disciplines. For instance, clinical toxicologists may also be involved in understanding the mechanisms of toxic agents, whether from a clinical application or a basic research perspective.

1.4.1 Forensic Toxicologist

From its inception in the medieval era to its maturity as a distinct and separate discipline in the 1950s, toxicology was taught primarily as an applied science. More recently, toxicology has evolved from the applications of analytical and clinical chemists, whose job definition was the chemical identification and analysis of body fluids. Thus, the first modern toxicologists were chemists with specialized training in inorganic separation methods, including chromatographic techniques. Later, analytical chemists employed thin-layer and gas chromatography. As the instrumentation evolved, high-performance liquid chromatography was incorporated into the analytical methods to isolate minute quantities of compounds from complex mixtures of toxicological importance. Forensic toxicologists integrated these techniques to identify compounds from mixtures of sometimes unrelated poisons as a result of incidental or deliberate exposure. Initially, forensic sciences profited from the application of the principles of chemical separation methods for the identification of controlled substances in body fluids. Later, forensic toxicologists applied biological principles of antigen–antibody interaction for paternity testing. By using the principles of blood grouping and "exclusion" of the potential outcomes of paternal contribution to offspring phenotype, it was feasible to eliminate the possibility of a male as the father of a child. Antigen–antibody interactions also became the basis for enzyme-linked immunosorbent assays (ELISA), currently used for specific and sensitive identification of drugs in biological fluids. Radioimmunoassays (RIAs) use similar antigen–antibody reactions while incorporating radiolabeled ligands as indicators. DNA separation and sequencing techniques have now almost totally replaced traditional paternity exclusion testing. These methods are also the basis for inclusion or exclusion of evidence in criminal and civil cases.

TABLE 1.2
Descriptive Areas of Toxicology and Toxicologists

Types of Toxicology	Types of Toxicologists
General	Academic
Mechanistic	Research
Regulatory	Clinical
Analytical	Forensic
Descriptive	Regulatory

1.4.2 Clinical Toxicologist

Clinical toxicologists have evolved and branched away from their forensic counterparts. The clinical toxicologist is interested in the identification, diagnosis, and treatment of a condition, pathology, or disease resulting from environmental, therapeutic, or illicit exposure to chemicals or drugs. Exposure is commonly understood to include individual risk of contact with a toxin but can further be defined to include population risk.*

1.4.3 Research Toxicologist

In the academic or industrial arenas, the research toxicologist examines the broad issues in toxicology in the laboratory. Academic concerns include any of the public health areas where progress in understanding toxicological sciences is necessary. This includes the elucidation of mechanistic, clinical, or descriptive toxicological theories. In the pharmaceutical industry, research toxicologists are needed to conduct phase I trials—that is, preclinical testing of pharmaceuticals before clinical evaluation and marketing. Preclinical testing involves the toxicity testing of candidate compounds, which are chemically and biochemically screened as potentially useful therapeutic drugs. The toxicity testing procedures include both *in vitro* and animal protocols.

1.4.4 Regulatory Toxicologist

Regulatory toxicologists are employed primarily in government administrative agencies. Their role is to sanction, approve, and monitor the use of chemicals by enforcing rules and guidelines. The guiding principles are promulgated through laws enacted by appropriate federal, state, and local jurisdictions, which grant the regulatory agency its authority. Thus, through these regulations, the agency determines who is accountable and responsible for manufacturing, procurement, distribution, marketing, and ultimately, release and dispensing of potentially toxic compounds to the public.

REFERENCES

SUGGESTED READINGS

Abraham, J., The science and politics of medicines control. *Drug Saf.* 26, 135, 2003.

Bateman, D.N., EJCP and clinical toxicology: the first 40 years. *Eur. J. Clin. Pharmacol.* 64, 127, 2008.

Blaauboer, B.J., The contribution of in vitro toxicity data in hazard and risk assessment: current limitations and future perspectives. *Toxicol. Lett.* 180, 81, 2008.

Bois, F.Y., Applications of population approaches in toxicology. *Toxicol. Lett.* 120, 385, 2001.

Chartier, M., Morency, L.P., Zylber, M.I., and Najmanovich, R.J., Large-scale detection of drug off-targets: hypotheses for drug repurposing and understanding side-effects. *BMC Pharmacol. Toxicol.* 18, 18, 2017.

* Risks to groups of persons from exposure to radiation, pollutants, and chemical or biological toxins that would necessitate diagnosis and treatment are classified as population risks.

Downs, J.W. and Hakkinen, P.J., What online toxicology resources are available at no cost from the (US) National Library of Medicine to assist practicing OEM physicians? *J. Occup. Environ. Med.* 57, e85, 2015.

Eaton, D.L. and Gilbert, S.G., Chapter 2. Principles of toxicology, in *Casarett and Doull's Toxicology: The Basic Science of Poisons*, Klassen, C.D. (Ed.), 8th ed., McGraw-Hill, New York, 2013.

Greenberg, G., Internet resources for occupational and environmental health professionals. *Toxicology* 178, 263, 2002.

Huggins, J., Morris, J., and Peterson, C.E., Distance learning and toxicology: new horizons for Paracelsus. *Toxicol. Appl. Pharmacol.* 207, 735, 2005.

Kaiser, J., Toxicology. Tying genetics to the risk of environmental diseases. *Science* 300, 563, 2003.

Meyer, O., Testing and assessment strategies, including alternative and new approaches. *Toxicol. Lett.* 140, 21, 2003.

Tennant, R.W., The National Center for Toxicogenomics: using new technologies to inform mechanistic toxicology. *Environ. Health Perspect.* 110, A8, 2002.

Turner, P.V. Haschek, W.M., Bolon, B., Diegel, K., Hayes, M.A., McEwen, B., Sargeant, A.M., Scudamore, C.L., Stalker, M., von Beust, B., Wancket, L.M., Commentary: the role of the toxicologic pathologist in academia. *Vet. Pathol.* 52, 7, 2015.

REVIEW ARTICLES

Dayan, A.D., Development immunotoxicology and risk assessment. Conclusions and future directions: a personal view from the session chair. *Hum. Exp. Toxicol.* 21, 569, 2002.

Decristofaro, M.F. and Daniels, K.K., Toxicogenomics in biomarker discovery. *Methods Mol. Biol.* 460, 185, 2008.

Fowler, J. and Galli, C.L., EUROTOX's view regarding the role and training of certified European registered toxicologists (ERT). *Toxicol. Lett.* 168, 192, 2007.

Harris, S.B. and Fan, A.M., Hot topics in toxicology. *Int. J. Toxicol.* 21, 383, 2002.

Johnson, D.E. and Wolfgang, G.H., Assessing the potential toxicity of new pharmaceuticals. *Curr. Top. Med. Chem.* 1, 233, 2001.

Kane-Gill, S.L. Kane-Gill, S.L., Dasta, J.F., Buckley, M.S., Devabhakthuni, S., Liu, M., Cohen, H., George, E.L., Pohlman, A.S., Agarwal, S., Henneman, E.A., Bejian, S.M., Berenholtz, S.M., Pepin, J.L., Scanlon, M.C., Smith, B.S. Clinical practice guideline: safe medication use in the ICU. *Crit. Care Med.* 45, e877, 2017.

Lewis, R.W., Billington, R., Debryune, E., Gamer, A., Lang, B., and Carpanini, F., Recognition of adverse and non-adverse effects in toxicity studies. *Toxicol. Pathol.* 30, 66, 2002.

Proudfoot, A.T., Good, A.M., and Bateman, D.N., Clinical toxicology in Edinburgh, two centuries of progress. *Clin. Toxicol. (Phila)* 51, 509, 2013.

Schrenk, D., Regulatory toxicology: objectives and tasks defined by the working group of the German society of experimental and clinical pharmacology and toxicology. *Toxicol. Lett.* 126, 167, 2002.

Turkez, H., Arslan, M.E., and Ozdemir, O., Genotoxicity testing: progress and prospects for the next decade. *Expert Opin. Drug Metab. Toxicol.* 13, 1089, 2017.

Wilks, M.F. Wilks, M.F., Blaauboer, B.J., Schulte-Hermann, R., Wallace, H.M., Galli, C.L., Haag-Grönlund, M., Matović, V., Teixeira, J.P., Zilliacus, J., Basaran, N., Bonefeld-Jørgensen, E.C., Bourrinet, P., Brueller, W., Claude, N., Miranda, J.P., Gundert-Remy, U., Håkansson, H., Kovatsi, L., Liesivuori, J., Lindeman, B., Lison, D., Leconte, I., Martínez-López, E., Murias, M., Michel, C., Scheepers, P.T.J., Stanley, L., Tsatsakis, A. The European Registered Toxicologist (ERT): current status and prospects for advancement. *Toxicol. Lett.* 259, 151, 2016.

2 Risk Assessment and Regulatory Toxicology

2.1 RISK ASSESSMENT

2.1.1 Introduction

Risk assessment is a complex process aimed at calculating, evaluating, and predicting the potential for harm to a defined community from exposure to chemicals and their by-products. The tolerance of chemicals in the public domain is initially determined by the need for such chemicals, by the availability of alternative methods for obtaining the desired qualities of the chemicals, and by the economic impact of the presence or absence of chemical agents on the economic status of the region. For example, over the last few decades, the amounts and frequency of use of pesticides and herbicides have increased exponentially, especially in the agricultural industry. Unequivocally, the benefits of these chemicals have resulted in greater yields and quality of agricultural products as a consequence of reduced encroachment by insects and wild-type plant growth. Indeed, greater availability of food products and more sustainable produce has been realized. The balance between safe and effective use of such chemicals and the risk to the population constantly undergoes reevaluation to determine whether pesticides or herbicides are associated with developmental, carcinogenic, or endocrine abnormalities.

2.1.2 Hazard Identification/Risk Assessment (HIRA)

HIRA in toxicology involves the analysis of a continuum of information that ultimately leads to identifying, labeling, and predicting the relative risk of a toxic event resulting from exposure to a chemical insult. This analytical method identifies hazards that merit special attention. For prediction purposes, emergency planning and mitigation (circumstantial) planning essentially rely on this regulatory guide. Ultimately, emergency measures are derived from the summary of the process, what actions must be planned for, and what resources are likely to be incorporated into the measures. Table 2.1 outlines the types of hazards associated with potential toxicological consequences from a regulatory viewpoint. The table also places microbiological, chemical, and radiologic contact in the context of other potential hazardous exposures.

2.1.3 Dose–Response Evaluation

Dose–response evaluation or dose–response assessment is a gauge of the relationship between exposure to a particular dose or doses of toxicant and the relative incidence

TABLE 2.1
Types and Examples of Hazards with Toxicological Consequences

Hazard Categories	Types	Some Examples
Geologic and natural	Weather	Hurricanes, tornadoes, storms
	Hydrologic	Floods, droughts
	Wildfire	Urban and wildland fires
	Geologic	Earthquakes, landslides, volcanic
Public health	Microbiological	Viral, bacterial, fungal, parasitic epidemics
	Chemical	See text for description
Terrorist threats	Microbiological	Endemic and epidemic risk to populations,
	Chemical	land, resources and environment
	Radiological	
Workplace	Hospital	Nosocomial infections
	Chemical	Benzene
	Energy source	Electricity, radiation

of adverse effects on the exposed population in question. Waters and Jackson (2008) describe the contents of several databases that use dose–response evaluation in hazard identification and risk assessment. Graphic and quantitative analysis, lowest effective dose, and acute, subchronic, and chronic data derived from a variety of test systems based on thousands of chemicals are used to support hazard classification of potential toxicants. The databases include the EPA Gene-Tox Database, the EPA/IARC Genetic Activity Profile (GAP) Database, and the Toxicological Activity Profile (TAP) prototype database. Many of these informational electronic systems provide primarily either bibliographic information or summary factual data from toxicological studies; however, relatively few contain adequate information to allow the application of dose–response models. Woodall (2008) describes the Exposure-Response database (ERDB), which allows the entry of detailed toxicological results from experimental studies in support of dose–response assessment. The model thus permits a high level of automation in the performance of various types of dose–response analyses.

2.1.4 Exposure Assessment and Assessment Modeling

An important component of the risk assessment process is exposure assessment, which measures the intensity, frequency, duration, route, and site distribution of a chemical agent to an exposed population. The populations that lie in the conduit of potential risk for exposure include not only human and other domestic mammalian targets but also environmental entities, such as fish, plants, and wildlife colonies. Also, common predictable pathways are not limited to obvious exposure routes, such as the oral, dermal, inhalation, and parenteral routes. Other environmental corridors that offer more subtle exposure routes include groundwater, surface and subsurface soil and water, contaminated food and potable water, and airborne contaminants.

The U.S. Environmental Protection Agency (EPA) uses fundamental exposure assessment programs and assessment modeling that:

1. Identify exposure pathways
2. Characterize exposure settings
3. Quantify the amount of exposure

To this effect, the Exposure Factors Program (U.S. EPA) addresses issues related to exposure factors (see EPA website in References: Review Articles) and uses this data to assess exposure to contaminants in the environment. The *Exposure Factors Handbook* compiled by the EPA offers a summary of the available statistical data on various factors used in assessing human exposure. These factors (Table 2.2), along with relevant data, are provided for exposure assessors who need to obtain data on standard factors to calculate human exposure to toxic chemicals.

Figures 2.1 through 2.4 illustrate modeling scenarios that track movement of contaminants through environmental paradigms. For instance, groundwater models quantify the movement of subsurface water and provide inputs to subsurface contaminant transport models (Figure 2.1). The simulation provides insight into groundwater and contaminant behavior and allows quantitative assessments for environmental decision-making. Surface water models (Figure 2.2) simulate contaminant movement and concentration in lakes, streams, estuaries, and marine environments. This model allows regulators to understand how exposure to contaminants affects aquatic environments. Similarly, models that emulate possible contamination in the food chain can aid in tracing contaminated aquatic and terrestrial environments, which typically result in the bioaccumulation of chemicals within all trophic levels of an ecosystem, and in estimating the chemical impacts on exposed plant, prokaryotic, and eukaryotic kingdoms (Figure 2.3). Multimedia and AQUATOX models track contaminants that travel through the atmosphere, soil, surface water, cohabitant (Figure 2.4), and aquatic systems (Figure 2.5). An advantage of the multimedia computer simulation is its ability to monitor and quantify the impacts of exposure as the contaminants navigate concurrently and sequentially throughout these environments (Figure 2.4).

TABLE 2.2
Various Factors Used in Assessing Human Exposure as Provided by the U.S. EPA

Sources	Activities
Water	Drinking water consumption
Soil	Soil ingestion or intake rates
Inhalation rates	Air pollution, domestic and environmental exhaust
Dermal factors	Including skin area and soil adherence factors
Oral consumption	Consumption of fruits and vegetables, fish, meats, dairy products, homegrown foods, breast milk intake
Human activity factors	Consumer product use, residential characteristics

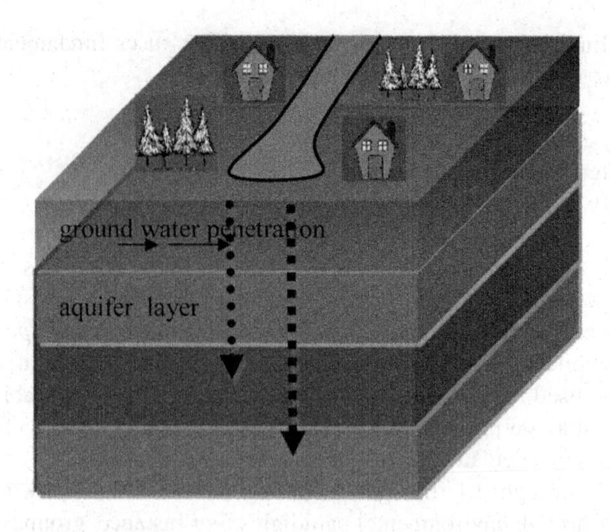

FIGURE 2.1 Ground water model.

(From U.S. EPA website, public access, www.epa.gov/ceampubl, last viewed March 2019.)

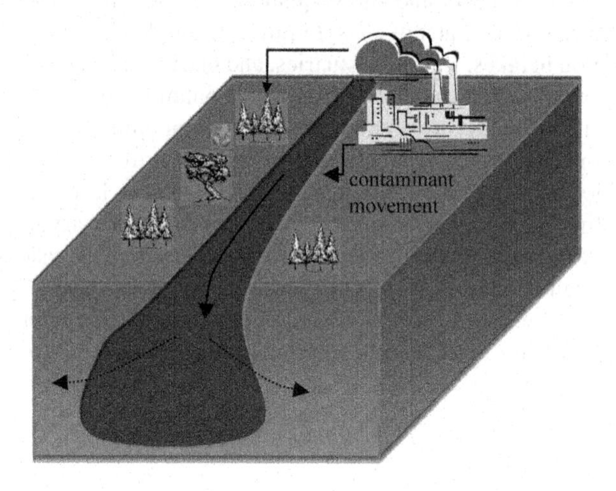

FIGURE 2.2 Surface water model.

(From U.S. EPA website, public access, www.epa.gov/ceampubl, last viewed March 2019.)

AQUATOX is an applied simulation model for use by environmental analysts that predicts the fate of various pollutants and their effects on the ecosystem, including fish, invertebrates, and aquatic plants (Figure 2.5).

2.1.5 Risk Characterization

The National Academy of Sciences' National Research Council defines risk characterization as the process of estimating the incidence of a health effect under the various conditions of human exposure. Risk characterization is initiated through the

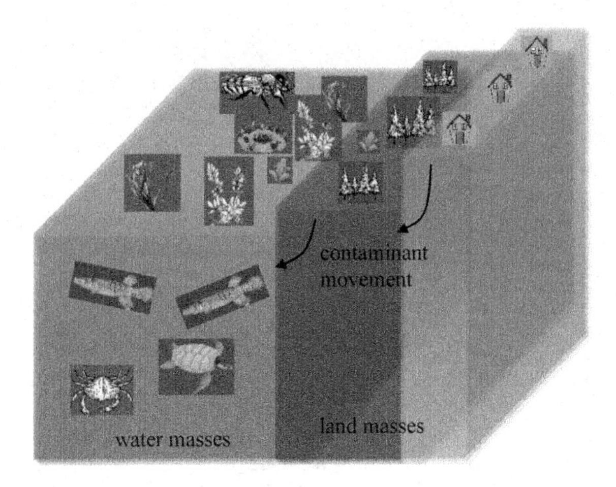

FIGURE 2.3 Food chain model.

(From U.S. EPA website, public access, www.epa.gov/ceampubl, last viewed March 2019.)

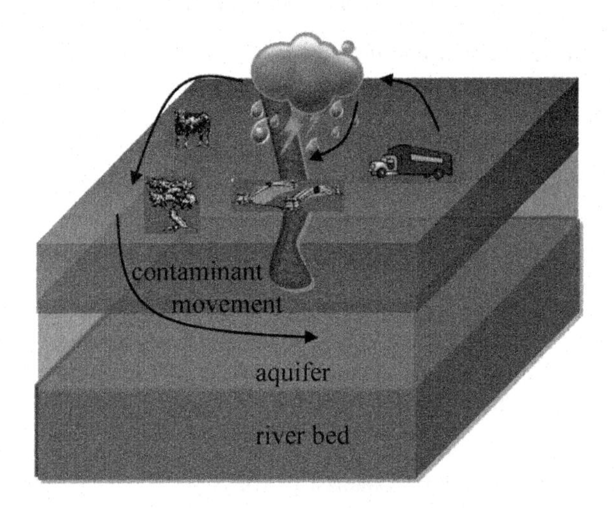

FIGURE 2.4 Multimedia model.

(From U.S. EPA website, public access, www.epa.gov/ceampubl, last viewed March 2019.)

combination of estimates following exposure assessment and dose–response assessment. Thus, the summary effects of the uncertainties in making these assessments often preview the magnitude of human risk, including attendant uncertainties.

The description of risk characterization has since evolved through the EPA's management of risk assessment and the decision-making process. For instance, the EPA describes risk characterization as the risk associated with the particular exposures in the situation being considered for regulation. Presentation and display of information, as well as exposure times potency (unit risk), nature and weight of evidence, uncertainty of the component parts, and distribution of population risk, should be

FIGURE 2.5 AQUATOX model.

(Adapted from U.S. EPA website, www.epa.gov/ceampubl, last viewed March 2019.)

included in the final decision-making process. Risk characterization translates the strengths and weaknesses of the assessment by purposefully synthesizing risk considerations into an incorporated, reasonable analysis. Thus, risk characterization is the final, integrative step of risk assessment and conveys the risk assessor's judgment as to the nature and existence of human health or ecological risks.

2.2 REGULATORY TOXICOLOGY

In the United States, several administrative government agencies are responsible for the application, establishment, and enforcement of rules, regulations, and directives associated with safe and effective chemical and drug use. A summary of the regulatory agencies and acts under their jurisdiction is presented in Table 2.3.

2.2.1 Nuclear Regulatory Commission (NRC)

Headed by a five-member commission, the NRC is an independent agency established by the Energy Reorganization Act of 1974 to regulate civilian use of nuclear materials.* The NRC's primary mission is to protect the public health and safety, and the environment, from the effects of radiation from nuclear reactors, materials, and waste facilities. In addition, it monitors national defense and promotes security from radiological threat. NRC carries out its mission through the enactment

* The Atomic Energy Act of 1954 established the single agency known as the Atomic Energy Commission. It was responsible for the development and production of nuclear weapons and for both the development and safety regulation of civilian uses of nuclear materials. The Act of 1974 split these functions, assigning to the Department of Energy the responsibility for the development and production of nuclear weapons and other energy-related work and delegating to the NRC the regulatory work noted here.

TABLE 2.3
Summary of Regulatory Agencies and Acts under Their Jurisdiction

Agency	Act	Regulatory Responsibility
NRC	Energy Reorganization Act of 1974	Established the NRC; development and production of nuclear weapons, promotion of nuclear power, and other energy-related work; provides protections for employees who raise nuclear safety concerns (later amendment)
EPA	Clean Air Act	Air quality, air pollution
	Clean Water Act	Water pollution, waste treatment management, toxic pollutants
	CERCLA	"Superfund"; cleanup of hazardous substances released in air, land, water
	FIFRA	Safety and regulation of pesticides
	RCRA	Generation, transportation, and disposal of hazardous waste
	Safe Drinking Water Act	Standards for drinking water; establishes MCLs
	Toxic Substances Control Act	Production, processing, importation, testing, and use of potentially hazardous chemicals; testing for HVHE and HPV chemicals
FDA	Federal Food, Drug, and Cosmetic Act	Safety and/or efficacy of food and color additives, medical devices, premarketing drug approval, cosmetics
DEA	Controlled Substances Act	Manufacture, distribution, dispensing, registration, handling of narcotics, stimulants, depressants, hallucinogens, and anabolic steroids; monitors chemicals used in the illicit production of controlled substances
CPSC	Consumer Product Safety Act	Safety standards for consumer products
	FHSA	Toxic, corrosive, radioactive, combustible, or pressurized labeling requirements for hazardous substances
	PPPA	Packaging of hazardous household products
OSHA	OSHA	Sets occupational safety and health standards, and toxic chemical exposure limits, for working conditions

CERCLA: Comprehensive Environmental Response, Compensation, and Liability Act; FIFRA: Federal Insecticide, Fungicide, and Rodenticide Act; FHSA: Federal Hazardous Substances Act; HPV: high production volume chemicals; HVHE: high volume-high exposure chemicals; MCLs: maximum contaminant levels for drinking water; OSHA: Occupational Safety and Health Act; PPPA: Poison Prevention Packaging Act; RCRA: Resource Conservation and Recovery Act.

of policymaking, protection of workers from radiation hazards, development of standards, inspection of facilities, investigation of cases, research and development, licensing in the procurement, storage, and use of radioactive materials, and adjudication.

2.2.2 Environmental Protection Agency (EPA)

Headed by a presidentially appointed administrator, the EPA's mission is to protect human health and safeguard the natural environment. The EPA employs over 18,000 engineers, scientists, environmental protection specialists, and staff in program offices, regional offices, and laboratories.

The EPA's mandate is guided by federal laws protecting human health and the environment. It oversees natural resources, human health, economic growth, energy, transportation, agriculture, industry, and international trade in establishing environmental policy. The EPA provides leadership in environmental science, research, education, and assessment efforts, and works closely with other federal agencies, state and local governments, and Indian tribes to develop and enforce regulations. The EPA sets national standards and delegates responsibility to states and tribes for issuing permits, enforcing compliance, and issuing sanctions. It monitors and enforces a variety of voluntary pollution prevention programs and energy conservation efforts, particularly with industrial concerns.

2.2.3 The Food and Drug Administration (FDA)

At the beginning of the twentieth century, revelations about filth in the Chicago stockyards shocked the nation into awareness that, in an industrial economy, protection against unsafe products is beyond any individual's means. The U.S. Congress responded to Upton Sinclair's best-selling book *The Jungle* by passing the Food and Drug Act of 1906, which prohibited interstate commerce of misbranded and adulterated food and drugs. Enforcement of the law was entrusted to the U.S. Department of Agriculture's Bureau of Chemistry, which later became the U.S. Food and Drug Administration (U.S. FDA). The Act was the first of more than 200 laws, some of which are discussed in this section as well as in Sections 2.2.4.–2.2.6. The regulations constitute a comprehensive and effective network of public health and consumer protections. The FDA's mission, therefore, is to promote and protect public health by ensuring the safety and efficacy of products in the market and monitoring them for continued safety. Overall, the FDA regulates $1 trillion worth of products a year.

The FDA is under the jurisdiction of the Department of Health and Human Services (HHS) and consists of nine centers: Center for Biologics Evaluation and Research (CBER), Center for Devices and Radiological Health (CDRH), Center for Drug Evaluation and Research (CDER), Center for Food Safety and Applied Nutrition (CFSAN), Center for Veterinary Medicine (CVM), National Center for Toxicological Research (NCTR), Office of the Commissioner (OC), and the Office of Regulatory Affairs (ORA).

The Code of Federal Regulations (CFR) is a codification of the rules published in the Federal Register by executive departments and agencies of the federal government. The CFR is divided into 50 titles, which represent broad areas subject to federal regulation, with environmental regulations contained mainly in title 40. Products regulated by the FDA through the CFR include food and food additives (except for meat, poultry, and some egg products); medical and surgical devices; therapeutic drugs; biological products (including blood, vaccines, and tissues for transplantation); animal drugs and feed; and radiation-emitting consumer and medical products. Furthermore, the CFR acts to prevent the willful contamination of all regulated products and improve the availability of medications to prevent or treat injuries caused by biological, chemical, or nuclear agents.

The Federal Food, Drug, and Cosmetic Act (FD&C) of 1938 (see Table 2.3) was passed after a legally marketed toxic elixir killed 107 children and adults. The FD&C Act authorized the FDA to demand evidence of safety for new drugs, issue standards for food, and conduct factory inspections.

The Kefauver–Harris Amendments of 1962 were inspired by the thalidomide tragedy in Europe. The Amendments strengthened the rules for drug safety and required manufacturers to prove drug effectiveness.*

The Medical Device Amendments of 1976 followed a U.S. Senate finding that faulty medical devices had caused 10,000 injuries, including 731 deaths. The law established guidelines for the safety and effectiveness of new medical devices.

2.2.4 Drug Enforcement Administration (DEA)

The DEA is under the policy guidance of the Secretary of State and the HHS and is headed by an administrator. Its mission is to enforce U.S. controlled substances laws and regulations and to prosecute individuals and organizations involved in the growing, manufacture, or distribution of controlled substances appearing in or destined for illicit traffic. It also recommends and supports nonenforcement programs aimed at reducing the availability of illicit controlled substances on the domestic and international markets. The DEA's primary responsibilities include investigation and preparation for the prosecution of major violators of controlled substance laws operating at interstate and international levels; prosecution of criminals and drug gangs; management of a national drug intelligence program in cooperation with federal, state, local, and foreign officials to collect, analyze, and disseminate strategic and operational drug intelligence information; seizure and forfeiture of assets derived from, traceable to, or intended to be used for illicit drug trafficking; enforcement of the provisions of the Controlled Substances Act (see next paragraph) as they pertain to the manufacture, distribution, and dispensing of legally produced controlled substances; and coordination and cooperation with federal, state, and local law enforcement officials on mutual drug enforcement efforts, including the reduction of the availability of illicit abuse-type drugs, crop eradication, and training of foreign officials.

The Controlled Substances Act (CSA), Title II of the Comprehensive Drug Abuse Prevention and Control Act of 1970, is a consolidation of numerous laws. It regulates the manufacture and distribution of narcotics, stimulants, depressants, hallucinogens, anabolic steroids, and chemicals used in the illicit sale or production of controlled substances. The CSA places all regulated substances into one of five schedules based on the substance's therapeutic importance, toxicity, and potential for abuse or addiction. Table 2.4 describes the federal schedules and some examples of drugs within these classifications. Proceedings to add, delete, or change the schedule of a drug or other substance may be initiated by the DEA or the HHS. Any other interested party

* Prompted by the FDA's vigilance and monitoring of clinical cases, thalidomide was prevented from entering the U.S. market.

TABLE 2.4
Controlled Substances Schedules and Description

Schedule	Description of Regulated Substances	Some Examples of Listed Substances
I	Drug or substance has a high potential for abuse, has no currently accepted medical use in treatment in the United States, and there is a lack of accepted safety for its use under medical supervision.	Benzylmorphine, etorphine, heroin, dimethyltryptamine (DMT), marijuana, lysergic acid diethylamide (LSD), mescaline, peyote, psilocybin, cocaine, tetrahydrocannabinols (THC)
II	The drug or substance has a high potential for abuse, has a currently accepted medical use in	Fentanyl, levorphanol, methadone, opium and derivatives
IIN*	treatment in the United States with or without severe restrictions, and may lead to severe psychological or physical dependence.	Amphetamine, methamphetamine, methylphenidate
III	The drug or substance has a potential for abuse less than the substances in Schedules I and II,	Codeine not >90 mg per dosage unit, buprenorphine
IIIN[a]	has a currently accepted medical use in treatment in the United States, and may lead to moderate or low physical dependence or high psychological dependence.	Ketamine, phendimetrazine, benzphetamine, anabolic steroids, testosterone
IV	The drug or substance has a low potential for abuse relative to those drugs or substances in Schedule III, has a currently accepted medical use in treatment in the United States, and may lead to limited physical dependence or psychological dependence.	Barbital, chloral hydrate, meprobamate, phenobarbital
V	The drug or substance has a low potential for abuse, has a currently accepted medical use in treatment in the United States, and may lead to limited physical dependence or psychological dependence.	Codeine not >200 mg/100 ml, diphenoxylate not >2.5 mg and atropine sulfate not >25 μg per dosage unit (Lomotil®),

[a] N: nonnarcotic drugs of respective schedule.

may also petition the DEA, including the manufacturer of a drug, a medical society or association, a pharmacy association, a public interest group concerned with drug abuse, a state or local government agency, or individuals.

The CSA also creates a closed system of distribution for those authorized to handle controlled substances. The cornerstone of this system is the registration of all those authorized by the DEA to handle controlled substances. All individuals and firms that are registered are required to maintain complete and accurate inventories and records of all transactions involving controlled substances as well as security for the storage of controlled substances.

2.2.5 Consumer Products Safety Commission (CPSC)

The CPSC is an independent federal regulatory agency created to protect the public from unreasonable risks of injuries and deaths associated with some 15,000 types of consumer products. Its mission is to inform the public about product hazards through local and national media coverage, publication of booklets, and product alerts; through a website, a telephone hotline, the National Injury Information Clearinghouse, and the CPSC's Public Information Center; and through responses to Freedom of Information Act (FOIA) requests. CPSC was created by Congress in 1972 under the Consumer Product Safety Act.*

The CPSC fulfills its mission through the development of voluntary standards with industry and by issuing and enforcing mandatory standards or banning consumer products if no feasible standard adequately protects the public. The CPSC obtains recall of products or arranges for their repair. In addition, it conducts research on potential product hazards, informs and educates consumers, and responds to consumer inquiries. Some of the product guidelines under the CPSC jurisdiction are listed in Table 2.5.

The Poison Prevention Packaging Act (PPPA) of 1970 is administered by the CPSC. The PPPA requires child-resistant packaging of hazardous household products. There have been significant declines in reported deaths from ingestions of toxic household substances by children since the inception of this Act. However, it is estimated that more than 1 million calls to poison control centers are still registered following unintentional exposure to medicines and household chemicals by children under 5 years of age. Of these, more than 85,000 children

TABLE 2.5
Examples of Consumer Products with Established Safety and Monitoring Guidelines under CPSC Jurisdiction

All-Terrain Vehicles Safety	Home Heating Equipment
Art and crafts safety	Household products safety
Bicycle safety	Indoor air quality
Child safety	Older consumers safety
Children's furniture	Outdoor power equipment safety
Clothing safety	Playground safety
Consumer product safety review	Poison prevention
Crib safety and SIDS[a] reduction	Pool and spa safety
Electrical safety	Public use products
Fire safety	Recreational and sports safety
General information	Reports
Holiday safety	Toy safety

[a] SIDS: sudden infant death syndrome.

* The CPSC is headed by three presidentially nominated commissioners (one of whom is chairman) and confirmed by the Senate for staggered 7 year terms. The three commissioners set policy for CPSC. The CPSC is not part of any other department or agency in the federal government. The congressional affairs, equal employment and minority enterprise, general counsel, inspector general, secretary, and executive director report directly to the chairman.

are examined in emergency departments (ED), resulting in almost 50 deaths of children each year.*

Some of the reasons accounting for continuing ingestions are availability of non-child-resistant packaging, on request, for prescription medication; availability of one non-child-resistant size of over-the-counter medications; inadequate quality control by manufacturers leading to defective child-resistant closures; misuse of child-resistant packaging in the home (leaving the cap off or unsecured, transferring the contents to a non-child-resistant package); and violations by health professionals. Consequently, the CPSC has designed a textbook to educate health professionals, particularly pharmacists and physicians, about the child-resistant packaging program. It is intended to be incorporated into medical and pharmacy school curricula to bring greater awareness of their legal responsibilities. Some of the substances covered by the PPPA regulations are listed in Tables 2.6 and 2.7.

2.2.6 Occupational Safety and Health Administrations (OSHA)

Created by the Occupational Safety and Health Act of 1970, OSHA ensures safe and healthful working conditions for the working public. It accomplishes this task through inspectors and staff personnel. OSHA authorizes the enforcement of the standards developed under the Act in cooperation with 26 states. In addition, it provides for research, information, education, and training in the field of occupational safety and health.

The agency fulfills its mandate by setting standards established by regulations. For example, the *OSHA Lead Standards for General Industry and Construction* require employers to provide biological monitoring for workers exposed to airborne lead above the action level. Monitoring must be provided for lead and zinc protoporphyrin (or free

TABLE 2.6
Therapeutic Substances Covered by the PPPA of 1970

Therapeutic Category	Examples of Drugs and Substances
Analgesics	Aspirin, acetaminophen, methyl salicylate
Controlled substances	Opioids, S/H, stimulants
Vitamins and dietary supplements	Iron-containing drugs and other dietary supplements
Local anesthetics	Lidocaine, dibucaine, and minoxidil
Nonprescription NSAIDs	Ibuprofen, naproxen, and ketoprofen
Nonprescription antihistamines	Diphenhydramine (Benadryl®)
Nonprescription antidiarrheal products	Loperamide (Imodium®)
General	Human oral prescription drugs (with some exceptions and exemptions)

NSAIDs: nonsteroidal anti-inflammatory drugs; S/H: sedative/hypnotics.

* According to the National Electronic Injury Surveillance System (a CPSC database of emergency room visits) and the American Association of Poison Control Centers (2003).

TABLE 2.7
Non-therapeutic Substances or Additives Covered by the PPPA of 1970

Nontherapeutic Category	Description[a]
Furniture polish	Non-emulsion-type liquid furniture polishes containing ≥10% of mineral seal oil and/or other petroleum distillates and having a viscosity <100 Saybolt universal seconds at 100 °F
Methyl salicylate	Liquid preparations containing >5% by weight of methyl salicylate
Sodium and/or potassium hydroxide	Household substances in dry forms such as granules, powder, and flakes, ≥10% by weight of free or chemically non-neutralized sodium and/or potassium hydroxide
Kindling preparations	Illuminating preparations that contain ≥10% by weight of petroleum distillates and have a viscosity of less than 100 Saybolt universal seconds at 100 °F
Methyl alcohol	Household substances in liquid form containing ≥4% by weight of methyl alcohol
Turpentine	Household substances in liquid form containing ≥10% by weight of turpentine
Sulfuric acid	Household substances containing ≥10% by weight of sulfuric acid, except such substances in wet-cell storage batteries

[a] Contents must be packaged in accordance with the provisions of 16 CFR 1700.15—poison prevention packaging standards.

erythrocyte protoporphyrin) in blood (see Chapter 24, Metals, for description). The employer is required to have these analyses performed by a laboratory that meets accuracy requirements specified by OSHA. The *OSHA List of Laboratories Approved for Blood Lead Analysis* is designed to provide a source to locate laboratories determined by OSHA to meet the requirements of the accuracy provisions of the lead standards. Laboratories voluntarily provide proficiency test data to OSHA for evaluation.

REFERENCES

SUGGESTED READINGS

Nicoll, A. and Murray, V., Health protection—a strategy and a national agency. *Public Health* 116, 129, 2002.

Patterson, J., Hakkinen, P.J., and Wullenweber, A.E., Human health risk assessment: selected Internet and world wide web resources. *Toxicology* 173, 123, 2002.

REVIEW ARTICLES

Accomasso, L., Cristallini, C., and Giachino, C., Risk assessment and risk minimization in nanomedicine: a need for predictive, alternative, and 3Rs strategies. *Front. Pharmacol.* 9, 228, 2018.

Acquavella, J., Doe, J., Tomenson, J., Chester, G., Cowell, J., and Bloemen, L., Epidemiologic studies of occupational pesticide exposure and cancer: regulatory risk assessments and biologic plausibility. *Ann. Epidemiol.* 13, 1, 2003.

Agathokleous, E., Kitao, M., and Calabrese, E.J., Environmental hormesis and its fundamental biological basis: rewriting the history of toxicology. *Environ. Res.* 165, 274, 2018.

Barlow, S.M. Greig, J.B., Bridges, J.W., Carere, A., Carpy, A.J., Galli, C.L., Kleiner, J., Knudsen, I., Koeter, H.B., Levy, L.S., Madsen, C., Mayer, S., Narbonne, J.F., Pfannkuch, F., Prodanchuk, M.G., Smith, M.R., Steinberg, P. Hazard identification by methods of animal-based toxicology. *Food Chem. Toxicol.* 40, 145, 2002.

Buchanan, J.R., Burka, L.T., and Melnick, R.L., Purpose and guidelines for toxicokinetic studies within the National Toxicology Program. *Environ. Health Perspect.* 105, 468, 1997.

Carere, A. and Benigni, R., Strategies and governmental regulations. *Teratog. Carcinog. Mutagen.* 10, 199, 1990.

Casciano, D.A., FDA: a science-based agency. *FDA Consum.* 36, 40, 2002.

Clewell, H.J. 3rd, Andersen, M.E., and Barton, H.A., A consistent approach for the application of pharmacokinetic modeling in cancer and noncancer risk assessment. *Environ. Health Perspect.* 110, 85, 2002.

Conolly, R.B., The use of biologically based modeling in risk assessment. *Toxicology* 181, 275, 2002.

Eason, C. and O'Halloran, K., Biomarkers in toxicology versus ecological risk assessment. *Toxicology* 181, 517, 2002.

Holsapple, M.P. and Wallace, K.B., Dose response considerations in risk assessment—an overview of recent ILSI activities. *Toxicol. Lett.* 180, 85, 2008.

Indans, I., The use and interpretation of *in vitro* data in regulatory toxicology: cosmetics, toiletries and household products. *Toxicol. Lett.* 127, 177, 2002.

Kowalski, L., Denne, J., Dyer, R., Garrahan, K., and Wentsel, R.S., Overview of EPA Superfund human health research program. *Int. J. Hyg. Environ. Health* 205, 143, 2002.

Lathers, C.M., Risk assessment in regulatory policy making for human and veterinary public health. *J. Clin. Pharmacol.* 42, 846, 2002.

Liebsch, M. and Spielmann, H., Currently available *in vitro* methods used in the regulatory toxicology. *Toxicol. Lett.* 127, 127, 2002.

Lorber, M., Indirect exposure assessment at the United States Environmental Protection Agency. *Toxicol. Ind. Health* 17, 145, 2001.

Lu, F.C. and Sielken, R.L. Jr., Assessment of safety/risk of chemicals: inception and evolution of the ADI and dose-response modeling procedures. *Toxicol. Lett.* 59, 5, 1991.

Munro, I.C., Renwick, A.G., and Danielewska-Nikiel, B., The threshold of toxicological concern (TTC) in risk assessment. *Toxicol. Lett.* 180, 151, 2008.

Ritter, L. and Arbuckle, T.E., Can exposure characterization explain concurrence or discordance between toxicology and epidemiology? *Toxicol. Sci.* 97, 241, 2007.

Temple, R., Policy developments in regulatory approval. *Stat. Med.* 21, 2939, 2002.

U.S. Environmental Protection Agency, *Risk Characterization Handbook*, Science Policy Council, Publication No. EPA 100-B-00-002 (2003). www.epa.gov/OSA/spc/pdfs/rchandbk.pdf, last viewed March 2019.

U.S. Environmental Protection Agency, *Distribution and Support Policy for Exposure Assessment Models.* www.epa.gov/ebtpages/enviriskassessmentexposureassessment.html, last viewed March 2019.

Walker, V.R., Boyles, A.L., Pelch, K.E., Holmgren, S.D., Shapiro, A.J., Blystone, C.R., Devito, M.J., Newbold, R.R., Blain, R., Hartman, P., Thayer, K.A., Rooney, A.A. Human and animal evidence of potential transgenerational inheritance of health effects: an evidence map and state-of-the-science evaluation. *Environ. Int.* 14, 115, 2018.

Warheit, D.B., Hazard and risk assessment strategies for nanoparticle exposures: how far have we come in the past 10 years? *F1000Res* 7, 376, 2018.

Waters, M. and Jackson, M., Databases applicable to quantitative hazard/risk assessment—towards a predictive systems toxicology. *Toxicol. Appl. Pharmacol.* 233, 34, 2008.

Wexler, P., Online toxicology resources in support of risk assessment from the U.S. National Library of Medicine. *Toxicol. Appl. Pharmacol.* 233, 68, 2008.

Woodall, G.M., An exposure-response database for detailed toxicity data. *Toxicol. Appl. Pharmacol.* 233, 14, 2008.

3 Therapeutic Monitoring of Adverse Drug Reactions (ADRs)

3.1 ADVERSE DRUG REACTIONS (ADRS) IN CLINICAL PRACTICE

The Institute of Medicine of the National Academy of Sciences estimated in 2006 that medical errors harm as many as 1.5 million people per year in the United States. The extra medical costs of treating drug-related injuries occurring in hospitals alone conservatively amount to $3.5 billion a year, and this estimate does not take into account lost wages and productivity or additional health-care costs. A 2002 study of 368 hospitals blamed overworked nursing staff, doctors' handwriting, and computer-entry errors for drug mistakes as the most common reasons for the errors. In essence, more than 80% of adults in the United States take at least one medication,* and almost a third take at least five different medications. On average, a hospital patient is subject to at least one medication error per day, with considerable variation in error rates across facilities. The few existing studies of the costs associated with medication errors are limited to the health-care costs incurred by preventable injuries. Thus, errors can occur with any therapeutic or supplemental drug product at any point in the curative or restorative handling or administration of the medications and in any hospital, long-term, or community setting. The frequency of medication errors and preventable medication-related injuries is of sufficient concern to justify the promulgation of studies, analyses, regulations, and programs by The Joint Commission and Institute of Medicine as mandated by U.S. Senate and Congressional committees.

3.1.1 The Joint Commission (TJC)

Established in 1951, TJC is an independent, not-for-profit organization that sets accreditation standards for improving the quality and safety of care provided by health-care organizations. TJC evaluates and accredits more than 15,000 health-care organizations and programs in the United States and evaluates an organization's compliance with these standards and certification requirements. Among the sentinel events† that TJC investigates are the growing concerns surrounding medication safety.

* Includes prescription or over-the-counter drugs, vitamins, minerals, herbal supplements, and alternative medications. The term *drug safety and quality* does not include known risks associated with the medication itself, such as purity or component or ingredient integrity.

† A sentinel event is an unexpected risk or occurrence involving death or serious injury and usually requires immediate investigation and response.

3.1.2 Growing Medication Safety Concerns

Chartered by the U.S. Congress in 1970 as a component of the National Academy of Sciences, the Institute of Medicine (IOM) serves as a scientific and independent national advisor for improving the public health. The advice promoted by IOM is projected as unbiased, evidence-based information concerning health and science policy to regulatory agencies, health-care professionals, and the general public. In support of this mission, the Medicare Modernization Act of 2003 (Section 107(c)) mandated that the IOM develop a full understanding of drug safety and quality issues through an evidence-based review of literature, case studies, and analysis. Table 3.1 lists the objectives of the Act and the methods of analysis that the IOM developed to address the concerns surrounding growing medication errors.

3.1.3 National Patient Safety Goals (NPSG)

In the process of promoting patient safety, TJC established 16 national patient safety goals (NPSGs, 2009) to promote specific improvements in patient safety with respect to medication and health-care protection. TJC defines drug* safety as those issues relating to the safe, effective, appropriate, and efficient use of medications.

TABLE 3.1
Institute of Medicine (IOM) Objectives and Methods Concerning Growing Medication Errors

Objectives	Methods
Prioritize resources for national quality improvement efforts and influence national health-care policy concerning medication errors	Estimate the incidence, severity, and costs of medication errors
Evaluate alternative approaches to reducing medication errors	Analyze factors to improve efficacy, cost-effectiveness, appropriateness, feasibility, and institutional barriers to reduce medication errors
Achieve short-term and long-term drug safety goals	Elucidate the goals and expected results and support the business case for them
Identify critical success factors and key controls for achieving success	Provide guidance to consumers, health-care providers, payers, and other stakeholders
Assess opportunities and impediments to broaden national implementation of medication error reductions	Provide guidance to policymakers and government agencies to promote a national agenda for medication error reduction
Develop an applied research agenda to evaluate the health and cost impacts of alternative interventions	Assess collaborative public and private strategies for implementing the research agenda through AHRQ and other government agencies.

AHRQ: Agency for Healthcare Research and Quality.

* *Drug* includes prescription therapies and biologics excluding blood and blood products and tissues for transplantation or grafting.

The sphere of activity of the pharmaceutical industry incorporates five components or subsystems of the medication use system:

1. selecting and procuring the drug by the pharmacy;
2. prescribing and selecting the drug for the patient;
3. preparing and dispensing;
4. administering the drug; and
5. monitoring the patient for effect.

Today, as the potency, effectiveness, and specificity of clinical drugs improve, the toxicity associated with them also increases. As a result, the prevention and treatment of clinical toxicity is not limited to fortuitous encounters with random chemicals but may take on a more subtle form of exposure. This exposure, commonly referred to as *pharmacotoxicology*, is related to continuous and chronic use of prescription as well as nonprescription, *over-the-counter* (OTC) medications for the treatment or prevention of pathologic conditions. Thus, NPSGs, as established by TJC, and their derived medication safety concerns aim to minimize the toxicity resulting from therapeutic agents. The TJC identifies and describes the 16 goals in the 2008 national patient safety goals hospital program manual along with applicable rationales and implementation expectations (The Joint Commission, 2018). In addition, Table 3.2 lists a number of medication safety concerns from which implementation expectations* (IEs) are developed. The IEs are used as part of a scoring system, are based on frequency of expectations, and are used to monitor patient compliance, to set clinical standards of performance, and to establish clinical assessment surveys as guides for hospital accreditation requirements. The IEs are then scaled and tabulated and are discussed as part of a universal protocol for addressing patient safety (Table 3.3).

3.2 FACTORS THAT CONTRIBUTE TO ADVERSE DRUG REACTIONS (ADRS)

Factors that influence the pharmacotoxicology of therapeutic drugs are outlined in Table 3.4. These aspects of ADRs supplement the NPSG set by The Joint Commission and are described in the following sections.

3.2.1 Inadequate Monitoring of Prescribed Drugs

Inadequate monitoring of prescribed drugs requires coordination between laboratory therapeutic monitoring of medication blood concentrations and the intended therapeutic goals. It also involves attention to the appearance of adverse reactions that may occur during the course of treatment. This practice is instituted by the prescriber, and its progress is necessarily continued by other health-care professionals who are responsible for the patient. Patients must also be vigilant as far as they can be relied on to be aware of and report any subjective untoward reactions. However,

* IEs are factors that are scored and are based on frequency of expectations that contribute to the quantification of adverse drug reactions.

TABLE 3.2
Abbreviated List of Medication Safety Concerns Derived from National Patient Safety Goals Implemented by TJC

Medication Safety Concerns	Description	Examples of Applications Involved in Errors	Common Risk Factors
High-alert medications	Drugs associated with significant sentinel events and high-risk ADRs	Insulin, opioids, conc. KCl, conc. KPhos, conc. NaCl, IV anticoagulants	Improper programming of pump rates; dose and concentrations; improper storage
General assessment of medication risks	Communication, identification, reporting, labeling, checking	All formularies, especially commonly used medications	Reduce the risk of errors associated with health-care providers
"Do not use" abbreviations	A standardized list of abbreviations, acronyms, and symbols that are not to be used within an institution	Abbreviations on handwritten or preprinted forms	Associated with dose, concentrations, directions, units
Look-alike sound-alike medications (LASA)	Pairs of drug names that look or sound alike, leading to confusion and error	See TJC website for list of frequently encountered LASA medication pairs	Periodic review of formularies, reporting habits, institutional CE
Improve safety with anticoagulant medications	Errors associated with initiation and maintenance protocols; compliance, and dietary and drug interactions	Unfractionated and low–molecular weight heparins, warfarin	Narrow TI, neonatal errors; errors of omission, prescribing, dose
Medication reconciliation	Process for comparing individual's outpatient and inpatient list of medications	Health-care team-oriented approach incorporating communication with patient and family	Avoidance of errors of omission, duplication, dosing, interactions; include intake/discharge counseling

ADR: adverse drug reaction; CE: continuing education; IV: intravenous; conc. KCl: concentrated potassium chloride; KPhos: potassium phosphate; NaCl: sodium chloride; TI: therapeutic index; TJC: The Joint Commission.

this requires that sufficient medical information is explained and discussed with the patient concerning the medication and its influence on the condition. Thus, the health-care team approach to monitoring the progress of the patient's treatment is regarded as the cornerstone of effective therapy.

The practice of monitoring and coordinating health care, however, sometimes requires more resources than is practical. When resources are scarce, or their implementation is not feasible, serious drug toxicity can result. For instance, many medication errors occur when patients transition from home to the hospital, to the emergency room, between units of the hospital, or back home from these care settings.

TABLE 3.3
Institute for Safe Medication Practices (ISMP) Reported List of Drugs Prone to Medication Errors and Strategies to Overcome

Drug Class	Strategies
Narcotics/Opioids	Consider patient-specific factors in prescribing; assess opioid tolerance and other comorbidities; limit strengths of available drugs in automated dispensing cabinet.
Antibiotics	Ensure appropriate diagnosis; avoid wrong dosage.
Antipsychotics	Avoid wrong dosage.
Antithrombotics	Familiarize with novel oral anticoagulants and other injectable anticoagulants, implementing mandatory pharmacy renal function evaluation for patients on new oral anticoagulants, and mandatory discharge counseling for patients.
Insulins	Use U-500 syringes or U-500 pens rather than U-100/tuberculin syringes.

TABLE 3.4
Other Factors That Influence Pharmacotoxicology of Therapeutic Drugs

Inadequate monitoring of prescribed drugs
Improper adherence to directions
Inadequate patient compliance
Over-prescribing of medications
Drug–drug interactions
Drug–disease interactions
Drug–nutrition interactions
Allergic reactions
Inadequate attention to medication warnings
Unsupervised and/or illicit use of online prescription drug use

The complications arise when patients or their caretakers are unaware of the possible negative interactions between drugs that they have been using and those that are newly prescribed as a result of emergency or admission to the hospital. To avoid, predict, obviate, or prevent ADRs resulting in this situation, some institutions have initiated MedRecon®, a new automated decision support feature in the patient's electronic health records that consistently reconciles new medication orders with previous or currently used drugs (Porcelli, et al., 2010). The system relies on integrated automatic prompts that ensure the medication history present in the electronic medical record is carefully considered at all points of care, from admission to the hospital, within the units and through discharge. The prompts alert the operator to the entire medication history and force the prescriber to consider the continuation, discontinuation, or substitution of drugs. In addition, the hospital pharmacist reviews the medication history on patient discharge as an added alert feature. Since the adoption of MedRecon® in New York City hospitals, medication reconciliation rates have improved from 34% to 100%, while discrepancies have dropped to 1.4% from 20%.

3.2.2 Improper Adherence to Prescribed Directions

Improper adherence to prescribed directions is a common cause of drug toxicity. Elderly patients often have trouble reading the small print on medication bottles, often leading to improper dosage administration. Patients also may not adequately follow-up on treatment protocols. Frequently, they either miss scheduled appointments to discuss the development of any untoward effects or fail to report warning signs. Often, patients take the liberty of adjusting dosage regimens without proper consultation.

3.2.3 Over-Prescribing and Overuse of Medications

In 2019, 4.3 billion retail prescription drugs will be filled at U.S. pharmacies, amounting to 19 annual per capita prescriptions. On average, adults aged 65 or over account for 32 annual per capita prescription drugs filled at pharmacies (H.J. Kaiser Family Foundation). It is not unreasonable to conclude, therefore, that health-care practitioners in the United States, and around the world, have a tendency to over-prescribe medications. Today, the public relies heavily on scientific breakthroughs and developments, especially in the field of drug development. This is a justifiable assumption, since biotechnology, and its drug development product, has increased life expectancy. Consequently, compromising social behaviors, such as poor nutritional habits and lack of daily routine exercise programs, are often substituted by increased drug use.

3.2.4 Drug–Drug and Drug–Disease Interactions

Further complicating the preponderance and availability of drugs, drug interactions from both prescription and OTC medications can compromise therapy. Accordingly, drug–drug and drug–disease interactions are often difficult to identify, monitor, and adjust. The question then arises of whether the untoward or unanticipated effect is due to the combination of medications or to exacerbation of the condition. Corrective measures such as adjustment of dosage, removal of one or more medications, or substitution of one of the components in the treatment schedule, or further tests, may be necessary.

3.2.5 Allergic Reactions

Any drug preparation has the potential to elicit an allergic reaction. Even a careful patient medication history may not reveal an allergic tendency. Immediate hypersensitivity reactions occurring within minutes or hours of consumption are often easiest to detect. This is because the patient is more alert to the possibility of a cause–effect response soon after one or two doses are ingested. A patient, however, may not associate the effects of an administered dose with an allergic reaction occurring 72 hours later, typical of a *delayed hypersensitivity reaction*. Thus, the patient may not arrive at the critical connection between these events. Careful monitoring of drug toxicity, therefore, is essential to differentiate a true hypersensitivity reaction, immediate or delayed, from any incidental or pathologic complication.*

3.2.6 Medication Warnings

Inadequate attention to medication warnings, whether buried in the small print of package inserts or on the back of labels of OTC products, results in significant toxicity. Weight-loss products containing sympathetic stimulants are notorious for causing cardiovascular complications in patients with or without heart disease. As described in subsequent chapters, excessive use of other OTC products can cause cumulative toxicities or further complicate normal physiological function. Such products include aspirin, acetaminophen, antihistamines, and products used for wakefulness, sleep, cough, colds, and energy.

3.2.7 Medication Errors

Medication errors by health professionals account for a large proportion of unanticipated, untoward toxicity. As noted earlier, with the increased availability of and reliance on drug treatment, the possibility of medication errors is amplified. Table 3.5 lists some common sources of errors by health-care professionals. Recently, in an effort to reduce the high rate of medical errors, the U.S. Food and Drug Administration (FDA) announced that it would require bar codes on all medications so that hospitals could use scanners to make sure patients get the correct dose of the right drug. The new requirement is one of several steps the agency has taken to reduce medical errors.

3.2.8 Adverse Drug Reactions (ADRS)

ADR is a ubiquitous term, which describes any undesirable effect of medication administration in the course of therapeutic intervention, treatment, or prevention of disease. The FDA defines ADR as "a reaction that is noxious, unintended, and occurs at doses normally used in man for the prophylaxis, diagnosis, or therapy of disease." The term *ADR* has come to replace the common usage of *side effect* often associated with a drug, although the latter is still popular with the public. An ADR may or may not be anticipated depending on the patient and the condition in question. It is estimated that 15 to 30% of all hospitalized patients have a drug reaction. In addition, one out of five injuries or deaths of hospitalized patients and 5% of all admissions to hospitals are ascribed to ADRs. Adverse reactions double the length of a hospital stay. Overall, over 2 million ADRs occur annually, of which 350,000

* It is interesting to note that during the course of obtaining medication histories, health-care professionals often do not assess *how* the patient has concluded that they may have an allergic tendency. Subjective negative reactions to a drug in the past, which may not be associated with an allergy toward the medication, may preclude the inclusion of the drug as part of future treatment for that patient.

TABLE 3.5
Common Sources of Errors by Health-Care Professionals

Prescriber	Pharmacist	Nurse/PA
Confusion with similar-sounding drug names (LASA)	Illegible prescription order	Illegible prescription order
Prescribing wrong dosage	Lack of follow-up on questionable medication or dosage	Lack of follow-up on questionable medication
Transcribing drug for an unintended patient	Confusion with similar-sounding drug names	Confusion with similar-sounding drug names
Inadequate familiarity with the product	Dispensing wrong medication or wrong dosage	Administering wrong medication or wrong dosage
Inadequate attention to age, sex, and weight of patient	Dispensing medication to the wrong patient	Administering medication to the wrong patient
Misdiagnosis	Lack of attention to follow-up signs and symptoms	Lack of attention to follow-up signs and symptoms
Failure to set a therapeutic endpoint	Lack of attention to drug interactions	Inadequate attention to time of administration
Inadequate counseling	Inadequate counseling	Inadequate counseling

LASA: look-alike sound-alike medications; PA: physician's assistant.

occur in nursing homes. ADR is the fourth leading cause of death in the United States (100,000 deaths) at a cost of $136 billion yearly.

Each medication possesses several descriptive terms, which detail any and all toxicities that have been encountered in preclinical and clinical use. Table 3.6 lists the categories of untoward effects and their definitions, collectively known as ADRs, including a separate category referred to as Adverse Reactions, which are understood for all FDA-approved drugs.

TABLE 3.6
Categories of Untoward Effects

Categories of Untoward Effects	Description
Warnings and precautions	Specify conditions in which use of the drug may be hazardous and establish parameters to monitor during therapy
Drug interactions	Outline a brief summary of documented, clinically significant drug–drug, drug–laboratory test, and drug–food interactions
Contraindications	Specify those conditions in which the drug should not be used
Adverse reactions	Describe the undesirable results of drug administration in addition to or extension of the desired therapeutic effect (side effect)
Overdosage	Describe the administration of a quantity of drug that is greater than required for a therapeutic effect

3.3 TREATMENT OF ADRS AND POISONING IN PATIENTS

3.3.1 History

Among the first treatment modalities used in the presence of unknown, unsuspected toxicity were the induction of emesis and the use of activated charcoal. Early Greek and Roman civilizations recognized the importance of administering charcoal to victims of poisoning as well as for treating anthrax and epilepsy. Early American folk medicine called for the use of common refined charcoal, obtained from the bark of black cherry trees or peach stones and ground to a powder, mixed with common chalk, and administered for the treatment of stomach distress due to unknown ingestion. Bread burned to charcoal, dissolved in cold water, was used as a remedy for dysentery. The induction of emesis was also suggested in early civilizations for reversing the effects of ingested poisons. The use of ipecac was described by South American Indians and first mentioned by Jesuit Friars in 1601. Brazilian, Colombian, and Panamanian ipecac were introduced into Europe in 1672 and well established in medicine by the turn of the century. Today, syrup of ipecac is readily available and should be a part of all household medicine chests.

3.3.2 Poison Control Centers (PCCs)

The first specialized medical units devoted to treating poisoned patients were started in Denmark and Hungary, while poison information services began in the Netherlands in the 1940s. PCCs started in the 1950s in Chicago and were well established by the 1970s. The role of the PCC has evolved significantly in the last 50 years to provide public and professional toxicological services associated with emergency management and prevention. Table 3.7 lists some of the aims of the PCC.

PCCs are staffed by a board certified medical toxicologist and supportive personnel. Board certification is obtained by passing the certification examination after approval of the required minimum experience. The American Board of Applied Toxicology (ABAT), the American Association of Poison Control Centers (AAPCC), the American Academy of Clinical Toxicology (AACT), and the American Board of Medical Subspecialties (for physicians) are positioned to administer and maintain standards and board certification in medical toxicology.

TABLE 3.7
Role of the Poison Control Center

Provide product ingredient information
Provide information on treatment of poisoned patients
Supply direct information to patients
Provide diagnostic and treatment information to health-care professionals
Institute education programs for health-care professionals
Sponsor poison prevention activities

3.3.3 Clinical Management of ADRs

The clinical management of ADRs presents situations that require careful examination by health-care professionals or trained emergency department (ED) personnel. In addition, ADRs can be detected through insightful questioning of patients by the pharmacist, since very often, they are the first health professionals readily available. Thus, a team-oriented approach, whether in an outpatient or inpatient setting, is critical to ascertaining the nature of potential toxic reactions to therapeutic drugs.

In general, acute, sudden reactions to drugs are usually the result of allergic tendencies, contraindications, or overdosage. Even one therapeutic dose of a medication can result in a serious ADR. These reactions, such as the appearance of a sudden rash, ostensibly may be of considerable concern to the patient and should be treated without delay. These reactions, however, do not always require an emergency room visit.

Alternatively, ADRs associated with chronic administration of medications are more subtle and insidious. Chronic toxicity occurs as a result of prolonged accumulation, the effects of which will develop into organ-specific toxicity. The adverse reactions in this case may be difficult to differentiate from a coexisting pathologic condition. For example, chronic low-grade digoxin toxicity often mimics symptoms of congestive heart failure, the precise condition it is designed to treat (see Chapter 21).

Overall, the hallmark of effective management of ADRs includes an approach that is not all that different from the "medical ABCs"—obtaining a detailed patient history, performing a physical examination, arriving at a differential diagnosis, requesting supplementary laboratory tests, reaching a final diagnosis, prescribing of therapy, and follow-up and evaluation of condition.

Included in a general medical history are questions pertaining to the use of drugs (prescription, OTC, or illicit), alcohol, tobacco, and any dietary supplements (vitamins or herbal products). In addition, the time and method of administration should be ascertained.

3.3.4 Clinical Management of Toxicologic Emergencies

The clinical management of a patient presenting with apparent poisoning involves a stepwise approach to securing effective treatment of the suspected toxicologic emergency. Whether the perceived emergency situation is in the hospital emergency room or in the field, such as at home or on the street, it is important to begin management of the situation in a systematic manner. This includes 1. stabilization of the patient; 2. clinical evaluation (including, if possible, history of the events leading to the emergency), physical examination, and laboratory and/or radiological tests; 3. prevention of further absorption, exposure, or distribution; 4. enhancement of elimination of the suspected toxin; 5. administration of an antidote; and 6. supportive care and follow-up.

1. *Stabilization of the patient* involves a general assessment of the situation, the environment, and the overall appearance of the patient, and maintenance of vital signs. Removal of the victim from an obvious source of contamination, such as from

fumes, gas, or spilled liquid, is of primary concern. This is followed by maintenance of the ABCs of clinical management; that is, maintaining **a**irway, **b**reathing, and **c**irculation, which are crucial to survival. This includes monitoring of blood pressure and heart rate, ensuring that respirations are adequate, checking the status of the pupils, and determination of skin temperature, color, and turgor. There may be wide variability in the initial signs and symptoms, especially in the first few minutes post exposure. Stabilization includes not only the return of vital signs to normal but also normal rhythm.

2. *Clinical evaluation* includes documentation of the history of the events leading to the emergency. This is accomplished more easily in the emergency room, or on phone intake, if the patient is conscious and able to respond. Determination of the substance ingested can also be obtained by inspection of the area adjacent to the victim and questioning of any potential witnesses. Time of exposure is critical and can influence the course of treatment and the prognosis. If time permits, calls to a local pharmacy or physician's office may contribute some valuable information as to the etiologic agent. Location of the victim, neighborhood, and activity in the vicinity may also divulge some hints as to the nature of the ingested agent.

Physical examination involves identification of a constellation of clinical signs and symptoms that, together, are likely associated with exposure to certain classes of toxic agents. Also known as the identification of the toxic syndrome or *toxidrome*, this compilation of observations allows the initiation of treatment and progress toward follow-up care and support. Table 3.8 describes five of the most common toxidromes and their clinical signs and symptoms that suggest a particular toxic agent, or category of agents, is responsible. Commonly observed features are listed first, followed by peripheral or more severe signs and symptoms. The agents and the mechanisms responsible for the toxidromes are discussed in subsequent chapters. It is important to note that the clinical picture becomes complicated when multiple drugs or chemicals are involved. The manifestations of one drug may mask or prevent the identification of other chemicals.

TABLE 3.8
Common Toxidromes and Their Clinical Features

Common Toxidromes	Features
Anticholinergic	Dry mucous membranes, flushed skin, urinary retention, decreased bowel sounds, altered mental status, dilated pupils, cycloplegia
Sympathomimetic	Psychomotor and physical agitation, hypertension, tachycardia, hyperpyrexia, diaphoresis, dilated pupils, tremors, seizures (if severe)
Cholinergic	SLUDGE (sialorrhea, lacrimation, urination, diaphoresis, gastric emptying), BBB (bradycardia, bronchorrhea, bronchospasms), muscle weakness, intractable seizures
Opioid	Central nervous system depression, miosis, respiratory depression, bradycardia, hypotension, coma
Benzodiazepine	Mild sedation, unresponsive or comatose with stable vital signs; transient hypotension, respiratory depression

Although *stat** tests are available for most clinical drugs, laboratory tests and radiological examinations have limited turnaround time for identification of the cause of toxicity. Consequently, substance analysis of biological specimens is generally performed for confirmation of differential diagnosis and follow-up treatment. Initializing emergency measures should not depend on obtaining laboratory results. Qualitative and quantitative analysis of biological specimens, however, may be necessary for optimal patient management for specific drugs, including: acetaminophen, aspirin, digoxin, iron, lead, lithium, and theophylline. Drug screens aid in diagnosis, and sodium (Na^+), chloride (Cl^-), and bicarbonate ions (HCO_3^-), as well as glucose, contribute to determination of the serum osmolarity. High serum concentrations of drugs, enough to cause unstable clinical conditions, alter the osmolarity and create a gap between 1. the osmoles measured by laboratory analysis and 2. the calculated osmoles. This *anion gap* is a nonspecific, yet instrumental, diagnostic manipulation and can be of assistance in monitoring the progression of treatment. The anion gap is calculated according to the following formula:

$$\left[Na^+ - \left(Cl^- + HCO_3^-\right)\right]$$

where the concentrations of ions are in milliequivalents per liter. A normal value is less than 12. An elevated anion gap suggests metabolic acidosis. The most common chemical agents that contribute to an elevated anion gap include *a*lcohol, *t*oluene, *m*ethanol, *p*araldehyde, *i*ron, *l*actic acid, *e*thylene glycol, and *s*alicylates. Other conditions suggested by an elevated anion gap include *d*iabetic acidosis and *u*remia. Paraldehyde, an hypnotic drug, was also frequently encountered but is no longer available (the acronym AT MUD PILES, which includes paraldehyde, may still be of some mnemonic use).

Another nonspecific diagnostic aid used in monitoring the progression of toxicity is the *osmolar gap*. The following formula is used to calculate the osmolar gap:

$$\left(2 \times Na^+\right) + \left([\text{glucose}]/18\right) + \left(\text{BUN}/2.8\right)$$

where glucose and blood urea nitrogen (BUN) concentrations are in milligrams per deciliter. The osmolar gap is determined by subtracting the calculated value from the measured osmolarity. The normal osmolar gap is less than 10 mOsm. An increase in the gap above 10 suggests poisoning with ethanol, ethylene glycol, isopropanol, or methanol. It is important to note that depending on the method used to determine the osmolar gap, a normal value does not rule out poisoning with ethylene glycol or methanol.

3. *Prevention of any further absorption or exposure* to a toxic agent initially involves removing the patient from the environment, especially in the presence of gaseous fumes or corrosive liquids. In the event of dermal exposure to a liquid, removal of the contaminated clothes and thorough rinsing with water are important steps. Rinsing the exposed area with soap and water are of great benefit for acid and phenol burns.

---

* Quick, immediate, or on emergency order.

Limiting the exposure to oral intoxication of an agent should be pursued immediately after a known ingestion. In the home or ED, several methods can be employed to limit intestinal absorption, enhance bowel evacuation, or promote emesis. Activated charcoal is the best method to diminish intestinal absorption. A 1 to 2 g/kg dose, as a slurry, is given orally or through a large-bore (36–40 French) nasogastric tube. A dose of 1000–2000 m^2 surface area per gram of activated charcoal can effectively absorb 50% of an orally ingested chemical 1 hour later. The material binds high–molecular weight organics, by noncovalent forces, more effectively than low–molecular weight inorganic molecules. Activated charcoal, however, is not effective for metals, such as lead or iron, hydrocarbons, acids, or alkalis.

Historically, PCCs recommended the use of ipecac syrup prior to referral to the ED, especially if the victim was at home, to start the gastric emptying process as early as possible. Follow-up telephone calls by the PCC were then part of the protocol to gauge patient progress and outcome. The use of ipecac-induced emesis in the management of poisoned patients as a gastric emptying technique has declined significantly, however. Despite its reliable ability to induce vomiting, the treatment modality has shown limited effectiveness in preventing drug absorption, particularly when given more than 30–90 minutes following ingestion of a toxic substance. In addition, there are potentially significant contraindications and adverse effects associated with the use of the emetic. Table 3.9 outlines the conclusions of the AAPCC panel regarding the use of ipecac syrup and lists the rare situations in which the benefit-to-risk ratio in its administration is acceptable. According to the panel guidelines, only in such circumstances is the administration of ipecac syrup advisable in response to a specific recommendation from a PCC, ED physician, or other qualified medical personnel. In addition, the routine stocking of ipecac in all households with young children is not recommended, although the consensus on which households might benefit from having ipecac available is still debatable. Instead, individual practitioners and PCCs are best able to determine the circumstances when ipecac syrup should be routinely available in the household, and when there are contraindications for its use (Table 3.10).

Gastric lavage, or "pumping the stomach," has limited value, since it has not been demonstrated to be of much clinical benefit. It is most effective when administered through an orogastric tube within 1 hour of ingestion. In addition, the procedure

TABLE 3.9
Acceptable Situations Where Ipecac Administration Is Potentially Beneficial

No apparent contraindication to the use of ipecac syrup
No apparent or known risk of serious toxicity to the victim with the administration of ipecac
No alternative therapy available or no other treatment known to be effective to decrease gastrointestinal absorption (e.g., activated charcoal)
Possibility that there will be a delay of greater than 1 hour before the patient will arrive at an ED
Probability that ipecac syrup can be administered within 30–90 minutes of ingestion of the toxin
Likelihood that ipecac syrup administration will not adversely affect more definitive treatment that might be provided at an ED

TABLE 3.10
Contraindications to the Routine Administration of Ipecac

May delay the administration or reduce the effectiveness of activated charcoal, oral antidotes, or whole bowel irrigation
Should not be administered to a patient who has a decreased level or loss of consciousness
Should not be administered to a patient who has ingested a corrosive substance (e.g., caustic acid or alkali) or hydrocarbon with high aspiration potential
Contraindicated in children under 6 years of age

is uncomfortable and, like syrup of ipecac, exposes the patient to the possibility of tracheal aspiration of stomach contents.

4. *Enhancement of elimination* of suspected chemical agents or drugs is accomplished using whole bowel irrigation. In adults, oral administration of polyethylene glycol (Golytely®, Colyte®), at a rate of 2 L/h, can flush ingested toxic agents through the bowel. Administration of the preparation is continued for 4–5 hours or until the bowel effluents are clear. The method is useful for enhanced elimination of sustained-release preparation of capsules or tablets, cellophane packets of street heroin or cocaine, and agents not effectively absorbed with charcoal.

Hemodialysis has historical and sometimes empirical value in enhancing elimination. The hemodialysis machine pumps the patient's blood through a dialysis membrane, the purpose of which is to decrease the drug's volume of distribution. Ideally, the compound should be of low molecular weight, higher water solubility, and low protein binding capacity. The procedure appears to be useful when other measures have failed, especially in the treatment of amphetamine, antibiotic, boric acid, chloral hydrate, lead, potassium, salicylate, and strychnine poisoning.

Alkalinization or acidification of urine, although based on valid chemical pharmacokinetic principles of ion trapping and acid–base reactions, is not clinically recommended. Practically, the concept is not effective, and it may aggravate or complicate the removal of agents that interfere with acid–base balance.

5. *Administration of an antidote.* As noted earlier, adherence to the ABC principles and good supportive care are the hallmarks of treatment of the poisoned patient. Once the agent responsible is suspected or identified, the administration of an antidote may be necessary. Table 3.11 organizes a variety of toxins according to their classification and available antidotes. Although only a small number of antidotes are available, many of these agents can completely reverse the toxicologic consequences of poisoning. Also, specific antidotes are discussed further under the individual chapter headings (chelating agents for metal poisoning are reviewed separately in Chapter 24, Metals). However, it should be noted that antidotes are associated with their own adverse reactions and toxicity. In addition, the effectiveness of antidotes is compromised in the presence of overdose from multiple agents.

6. *Supportive care and maintenance.* After the primary goal of stabilization of vital signs is achieved, supportive care and maintenance are essential. This may require several more hours in the ED before release or transfer to an intensive care unit (ICU). Only a few drugs and chemicals have a tendency for delayed toxicity,

including iron tablets or elixir, salicylates, acetaminophen, opioids, sedative-hypnotics (barbiturates and benzodiazepines), and paraquat. In addition, salicylate and iron toxicity, as well as lead poisoning, may exhibit multiple phases of acute toxicity, interspersed with stages of remission of signs and symptoms. Monitoring in an ICU is essential to ensure complete recovery. In the event of suspected deliberate intoxication, ingestion, or administration, psychiatric assessment, forensic analysis, and police investigation are warranted.

3.4 DRUG IDENTIFICATION AND METHODS OF DETECTION

The analysis of drugs in biological fluids, in tissues, or in mixtures with other compounds is a complex science. Several areas of toxicology rely on the identification of drugs and chemicals for monitoring of toxic effects.

Forensic toxicology laboratories have the responsibility of identifying drug ingredients predominantly of unknown origin and separating them from additives and contaminants. The assortment of drug specimens that confronts the criminalist presents a formidable task for identifying illicit drugs. The first step in detecting and identifying a suspected chemical agent is the application of a screening test. The results of this procedure determine the next test, or series of tests, to confirm the presence of the specific compound. Initial analytical methods range from simple color tests and chromatography screens to more sophisticated gas chromatographic-mass spectrometry detection. In addition, extraction techniques are sometimes necessary to separate the suspected active constituent from the complex nature of the specimen.

Table 3.12 outlines the approaches, chemical substances, and methods employed in the forensic toxicology laboratory for screening and identifying common drugs of abuse. Details of the procedures and analytical methods for specific drugs and chemicals are found in References and Suggested Readings. Available methods for identification of classes of chemical substances in biological specimens are also included in each of the respective chapters.

Clinical toxicology laboratories are involved in the detection of drugs in suspected cases of drug abuse or overdosage or for monitoring of drugs in sports. Drug abuse tests are generally noninvasive and employ mostly screening tests for the more common illicit compounds. Urine is used for large-scale testing because of its ease of acquisition and its noninvasive nature. Analysis for unauthorized use of drugs is performed in a variety of settings. Testing of drug abusers in drug rehabilitation clinics and hospitals is done to monitor the effectiveness of the rehabilitation program for the patient. Monitoring of drug use in the workplace, in schools, and in individuals involved in critical jobs (such as police and pilots) is a relatively recent phenomenon. Testing is required or is mandatory to protect employees, employers, students, and the public from the harmful consequences of illicit drug use by individuals responsible for public safety. The identification of drugs in drug abuse cases and monitoring of suspected drug use in sports is generally accomplished by rapid screening of urine samples.

Monitoring of drugs in cases of overdosage falls within the realm of the clinical (and sometimes forensic) toxicology laboratory. The major reason for analysis

TABLE 3.11
Classification of Toxins and Their Specific Antidotes

Classification of Toxins	Examples of Specific Toxic Agents	Antidote
Alcohols	Ethylene glycol	Ethanol, fomepizol, pyridoxine
	Methanol	Ethanol, fomepizol, folic acid, leucovorin
Analgesics	Acetaminophen	N-acetylcysteine
	Aspirin	Sodium bicarbonate, ipecac
Anticholinergics	Cholinergic blockers	Physostigmine
	Tricyclic antidepressants	Sodium bicarbonate
Anticoagulants	Heparin	Protamine
	Warfarin	Vitamin K_1 (phytonadione)
Arthropod bites and stings	Black widow spider bite	*Latrodectus* antivenom
	Brown recluse spider bite	*Loxosceles* antivenom
	Rattlesnake bite	*Crotalidae* antivenom
	Scorpion sting	Antivenin
Benzodiazepines	Diazepam, alprazolam	Flumazenil
Cardiovascular drugs	Digitalis glycosides	Digoxin immune FAB
	β-blockers	Glucagon
	Calcium channel blockers	Calcium, glucagon
Gases	Chlorine gas	Sodium bicarbonate[a]
	Hydrogen sulfide gas	Sodium nitrite
	Carbon monoxide	100% oxygen
	Cyanide	Amyl nitrite, sodium nitrite, sodium thiosulfate
Infectious agents	Clostridium botulinum *(botulism)*	*Botulinum* antitoxin
Metals	Arsenic	BAL
	Copper	D-penicillamine
	Mercury	BAL, DMSA
	Iron	Deferoxamine
	Lead	Calcium-disodium-EDTA, dimercaprol, DMSA, BAL, D-penicillamine
Methemoglobinemia-inducing agents	Nitrites, nitrates	Methylene blue
Opiates	Narcotic analgesics and heroin	Naloxone, naltrexone, nalmefene
Pesticides	Organophosphate insecticides	Atropine followed by pralidoxime
	Carbamate insecticides	Atropine

[a] For treatment of accompanying metabolic acidosis.

BAL: British anti-lewisite; DMSA: 2,3-dimercaptosuccinic acid; EDTA: ethylenediamine tetraacetate; FAB: fragment antigen-binding.

TABLE 3.12
Forensic Toxicology Laboratory Initial Testing (Screening) Methods Based on Specimen Availability

Specimen Available	Chemical Substances	Method
Blood or urine	Volatiles (alcohols, acetone); gases (CO)	HS/GC
	Opiates, benzylecgonine	RIA
Urine only	Amphetamines, cannabinoids, PCP, methadone, barbiturates	EI
	Salicylates, APAP	CT
	Other basic drugs[a]	GC/MS or GC/NP
Blood only or tissue only	Volatiles	HS/GC
	Opiates, barbiturates benzylecgonine	RIA
	Salicylates (blood only)	CT
	Other basic drugs*	GC/MS or GC/NP

[a] See Table 3.13.
APAP: acetaminophen; CO: carbon monoxide; CT: color test; EI: enzyme immunoassay; GC: gas chromatography; HPLC: high-performance liquid chromatography; HS: headspace; MS: mass spectrometry; NP = nitrogen-phosphorus detector; PCP: phencyclidine; RIA: radio immunoassay; SP: spectrophotometry.

of fluids in overdose cases is to establish the source and extent of chemical toxicity and to apply an appropriate treatment program. Samples obtained in overdose cases (intentional or accidental) are usually more amenable to preparation and analysis than samples supplied to the forensic toxicology laboratory. Based on the signs and symptoms of the overdose (coma, respiratory depression, hallucinations, or cardiac arrest), the range of possible chemical suspects is also narrower. Screening methods and confirmation tests are available for proper management of poisoning cases.

As mentioned earlier, *therapeutic drug monitoring* is necessary for clinical scrutiny of ADRs. It is also important for managing the appropriate therapeutic response to potent medications and preventing inadequate or toxic responses. Therapeutic drug plasma levels are requested to adjust the dose for optimal effect. A variety of more sophisticated analytical techniques have been developed for monitoring therapeutic drugs, including high-performance liquid chromatography (HPLC), gas chromatography (GC), and bioassays (enzyme immunoassay [EI], radioimmunoassay [RIA]).

Table 3.13 outlines some of the drugs and the methods used in routine toxicology and therapeutic screens. Basic drugs encompass the majority of compounds detected in toxicology laboratories. The extraction of active constituents from biological specimens is a necessary prerequisite for the analysis of drugs by GC/nitrogen-phosphorus detector (NP). Acidic or basic drugs are extracted from biological fluids or tissue homogenates by adjusting the medium pH to acidic or basic, respectively. The compounds are then extracted and back-extracted into appropriate inorganic and organic solvents after further adjustment of pH. Calibration curves, negative and

TABLE 3.13
Drugs Tested in Routine Toxicology and Therapeutic Screens Based on Their Chemical Class

Chemical Class	Chemical Substances	Method
Volatiles and gases	Ethanol, acetone	HS/GC
	CO	SP, CT
Acidic and Neutral drugs	APAP, salicylates	CT, HPLC
	Barbiturates	EI, RIA, HPLC
	Anticonvulsants (phenytoin, carbamazepine); digoxin	HPLC, RIA
Basic drugs	Amitriptyline, amoxapine, chlorpheniramine, diphenhydramine, doxepin, doxylamine, ethylbenzylecgonine, fluoxetine, flurazepam, haloperidol, imipramine, ketamine, lidocaine, meperidine, methorphan, propoxyphene, verapamil	GC, GC/MS
	Benzylecgonine, cocaine, codeine, morphine	EI, RIA, GC/MS
	Chlordiazepoxide, diazepam, methadone, PCP	EI, GC, GC/MS
	Chlorpromazine, thioridazine	CT, GC, GC/MS
	Caffeine, theophylline	GC, GC/MS, HPLC
Hormones and miscellaneous	Anterior pituitary hormones, thyroid hormones, specific binding proteins, antibiotics, antineoplastics	RIA

APAP: acetaminophen; CO: carbon monoxide; CT: color test; EI: enzyme immunoassay; GC: gas chromatography; HPLC: high-performance liquid chromatography; HS: headspace; MS: mass spectrometry; PCP: phencyclidine; RIA: radio immunoassay; SP: spectrophotometry.

positive controls, and internal and external standards are part of the routine procedures. Some examples of drugs that are not *routinely* tested in toxicology screens include highly polar drugs (antibiotics, diuretics, and lithium); highly volatile compounds (nitrous oxide and hydrocarbons); and highly nonpolar compounds (steroids, digoxin, and plant derivatives). Other toxic anions and alkaloids that are difficult to detect, or not routinely screened, are gamma-hydroxybutyrate (GHB), cyanide, LSD, and the ergotamines. Specific details of the techniques are discussed in referenced suggested readings. Analytical methods for specific drugs and chemicals are also located under the headings for the classes of compounds in subsequent chapters.

REFERENCES

SUGGESTED READINGS

Brunton, L., Chabner, B.A., and Knollman, B., *Goodman & Gilman's The Pharmacological Basis of Therapeutics*, 12th ed., McGraw-Hill, New York, 2010.

de Jonge, M.E., Huitema, A.D., Schellens, J.H., Rodenhuis, S., and Beijnen, J.H., Individualized cancer chemotherapy: strategies and performance of prospective studies on therapeutic drug monitoring with dose adaptation: a review. *Clin. Pharmacokinet.* 44, 147, 2005.

H.J. Kaiser Family Foundation. www.kff.org, last viewed March 2019.

Meyer, C., *American Folk Medicine*, Plume Books, New American Library, New York, 1973.

Mowry, J.B., Spyker, D.A., Brooks, D.E., McMillan, N., and Schauben, J.L., 2014 Annual Report of the American Association of Poison Control Centers' National Poison Data System (NPDS): 32nd Annual Report. *Clin. Toxicol. (Phila)* 53, 962, 2015.

The Joint Commission, 2018 National Patient Safety Goals Hospital Program Manual Chapter. www.jointcommission.org/standards_information/npsgs.aspx, last accessed March 2019.

The National Academy of Sciences, Institute of Medicine, *Preventing Medication Errors: Quality Chasm Series*. www.nap.edu/read/11623/chapter/1, last accessed March 2019.

Touw, D.J., Neef, C., Thomson, A.H., and Vinks, A.A., Cost-effectiveness of therapeutic drug monitoring: a systematic review. *Ther. Drug Monit.* 27, 10, 2005.

REVIEW ARTICLES

Bowden, C.A. and Krenzelok, E.P., Clinical applications of commonly used contemporary antidotes. A US perspective. *Drug Saf.* 16, 9, 1997.

Dawson, A.H. and Whyte, I.M., Evidence in clinical toxicology: the role of therapeutic drug monitoring. *Ther. Drug Monit.* 24, 159, 2002.

Epstein, F.B. and Hassan, M., Therapeutic drug levels and toxicology screen. *Emerg. Med. Clin. North Am.* 4, 367, 1986.

Fountain, J.S. and Beasley, D.M., Activated charcoal supercedes ipecac as gastric decontaminant, *N. Z. Med. J.* 111, 402, 1998.

Guernsey, B.G., Ingrim, N.B., Hokanson, J.A., Doutré, W.H., Bryant, S.G., Blair, C.W., and Galvan, E., Pharmacists' dispensing accuracy in a high-volume outpatient pharmacy service: focus on risk management, *Drug Intell. Clin. Pharm.* 17, 742, 1983.

Henry, J.A. and Hoffman, J.R., Continuing controversy on gut decontamination. *Lancet* 352, 420, 1998.

Jackson, W.A., Antidotes. *Trends Pharmacol. Sci.* 23, 96, 2002.

Jones, A., Recent advances in the management of poisoning. *Ther. Drug Monit.* 24, 150, 2002.

Korb-Savoldelli, V., Boussadi, A., Durieux, P., and Sabatier, B., Prevalence of computerized physician order entry systems-related medication prescription errors: a systematic review. *Int. J. Med. Inform.* 111, 112, 2018.

Krenzelok, E.P., New developments in the therapy of intoxications. *Toxicol. Lett.* 127, 299, 2002.

Krenzelok, E.P., Ipecac syrup-induced emesis ... no evidence of benefit. *Clin. Toxicol. (Phila)* 43, 11, 2005.

Larmené-Beld, K.H.M., Alting, E.K. and Taxis, K., A systematic literature review on strategies to avoid look-alike errors of labels. *Eur. J. Clin. Pharmacol.* 74, 985, 2018.

Larsen, L.C. and Cummings, D.M., Oral poisonings: guidelines for initial evaluation and treatment. *Am. Fam. Physician* 57, 85, 1998.

Lheureux, P., Askenasi, R., and Paciorkowski, F., Gastrointestinal decontamination in acute toxic ingestions. *Acta. Gastroenterol. Belg.* 61, 461, 1998.

Limaye, N. and Jaguste, V., Risk-based monitoring (RBM) implementation: challenges and potential solutions. *Ther. Innov. Regul. Sci.* 2168479018769284, 2018. doi:10.1177/2168479018769284.

Mahtani, K.R., Jefferson, T., Heneghan, C., Nunan, D., and Aronson, J.K., What is a "complex systematic review"? Criteria, definition, and examples. *BMJ Evid. Based Med.* 23, 127, 2018.

Manoguerra, A.S. and Cobaugh, D.J., Guidelines for the Management of Poisoning Consensus Panel, Guideline on the use of ipecac syrup in the out-of-hospital management of ingested poisons. *Clin. Toxicol. (Phila)* 43, 1, 2005.

Neville, R.G., Robertson, F., Livingstone, S., and Crombie, I.K., A classification of prescription errors. *J. R. Coll. Gen. Pract.* 39, 110, 1989.

Porcelli, P.J., Waitman, L.R., Brown, S.H., Review of Medication Reconciliation Issues and Experiences with Clinical Staff and Information Systems, *Appl Clin Inform.* 1, 442, 2010.

Rubenstein, H. and Bresnitz, E.A., The utility of Poison Control Center data for assessing toxic occupational exposures among young workers. *J. Occup. Environ. Med.* 43, 463, 2001.

Schaper, A. and Ebbecke, M., Intox, detox, antidotes—evidence based diagnosis and treatment of acute intoxications. *Eur. J. Intern. Med.* 45, 66, 2017.

Seifert, S.A. and Jacobitz, K., Pharmacy prescription dispensing errors reported to a regional poison control center. *J. Toxicol. Clin. Toxicol.* 40, 919, 2002.

Vernon, D.D. and Gleich, M.C., Poisoning and drug overdose. *Crit. Care Clin.* 13, 647, 1997.

Warner, A., Setting standards of practice in therapeutic drug monitoring and clinical toxicology: a North American view. *Ther. Drug Monit.* 22, 93, 2000.

4 Classification of Toxins in Humans

4.1 INTRODUCTION

The classification of chemical agents is a daunting task considering the vast number and complexity of chemical compounds in the public domain. With the variety of chemicals and drugs available, and their varied pharmacological and toxicological effects, an agent may find itself listed in several different categories. Even compounds with similar structures or pharmacological actions may be alternatively grouped according to their toxicological activity or physical state.

Recently, the European Union established regulations addressing the assessment of all relevant industrial chemicals concerning the registration, evaluation, authorization, and restriction of chemicals (REACH). The REACH program directs industrial concerns with the responsibility for more efficient risk assessment. In particular, the guidance document promotes an integrated and updated weight-of-evidence approach for gathering information from different sources with various degrees of uncertainty. It sets minimum data requirements for use and draws conclusions on the toxicity of a substance based on known regulatory endpoints—*c*lassification and *l*abeling (C&L), *p*ersistent, *b*ioaccumulative, and *t*oxic (PBT) assessment, and *p*redicted *no*-*e*ffect *c*oncentrations (PNEC) derivation.

The following outline of the classification systems is generally accepted for understanding the complex nature of toxins. The mechanisms of toxicity are discussed in subsequent chapters.

4.2 TARGET ORGAN CLASSIFICATION

4.2.1 Agents Affecting the Hematopoietic System

Hematotoxicology is the effect of therapeutic and nontherapeutic chemicals on circulating blood components, coagulation processes, and blood-forming tissues (Bloom, 1993. Understanding the toxicity borne by these physiologic systems requires a working knowledge of the hematopoietic process and its cellular constituents—that is, the process involving the formation of blood and its components. Figure 4.1 illustrates the process of hematopoiesis, including the development and ontogeny of erythrocytes (red blood cells) and leukocytes (white blood cells). Chemicals that affect hematopoiesis, therefore, may interfere with the normal physiologic function of any of its cells and proteins, including erythrocytes and their oxygen-carrying capacity residing in hemoglobin; thrombocytes (platelets) and their role in fibrin clot formation; and basophils, neutrophils, and eosinophils and their antimicrobial activities. Chemicals that interfere with white blood cell components and their formation are

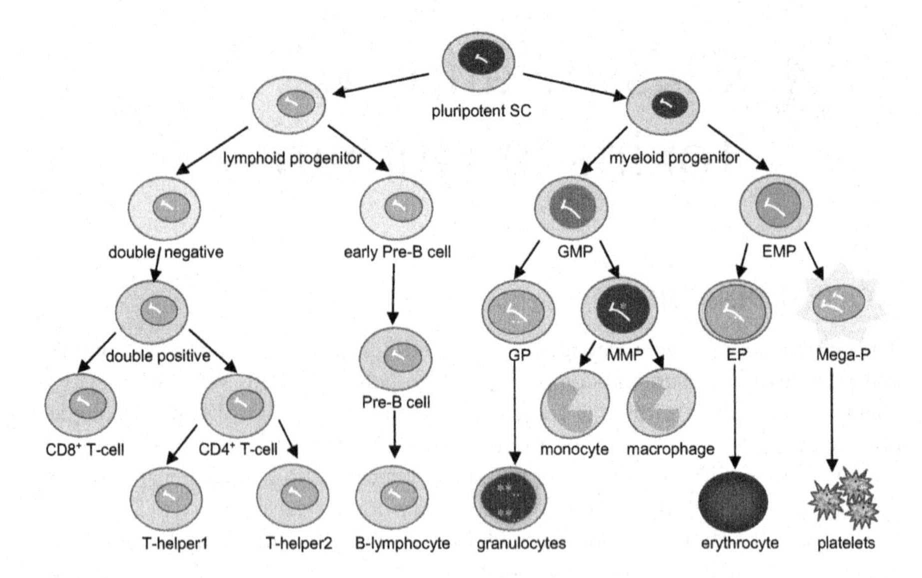

FIGURE 4.1 Hematopoiesis.

generally classified as leukemogenic agents and are discussed individually in later chapters.

Figure 4.2 illustrates the intrinsic and extrinsic pathways of coagulation, incorporating the tissue factors and phospholipids associated with the coagulation cascade. Anticoagulants, such as the clinical drug and pesticide warfarin, interfere with the synthesis of coagulation proteins and factors. A detailed explanation of the mechanism of warfarin and other anticoagulants is outlined in Chapter 28, Pesticides.

4.2.2 Immunotoxic Agents

The immune system, and the mechanisms by which the body maintains homeostasis and protection against microorganisms, is often the target of chemical agents and therapeutic drugs. Tables 4.1 and 4.2 organize the immune system according to the concept of *innate immunity*—that is, the immunological protection afforded without the production of antibodies—and *acquired immunity*—the adaptive response induced as a result of exposure to antigen. Familiarity with the role of the immune system allows an understanding of the mechanisms by which compounds, especially a number of therapeutic immunosuppressive drugs, produce toxicity.

4.2.3 Hepatotoxic Agents

The liver is a target organ for many chemicals and therapeutic drugs, predominantly as a result of its high metabolic capacity, extensive circulation, and sizeable surface area. For instance, the hepatotoxic effects of vinyl chloride and carbon tetrachloride have been known for decades to be a consequence of industrial exposure. The seriousness of hepatotoxicity of the type-2 antidiabetic drug Rezulin® (troglitazone) was

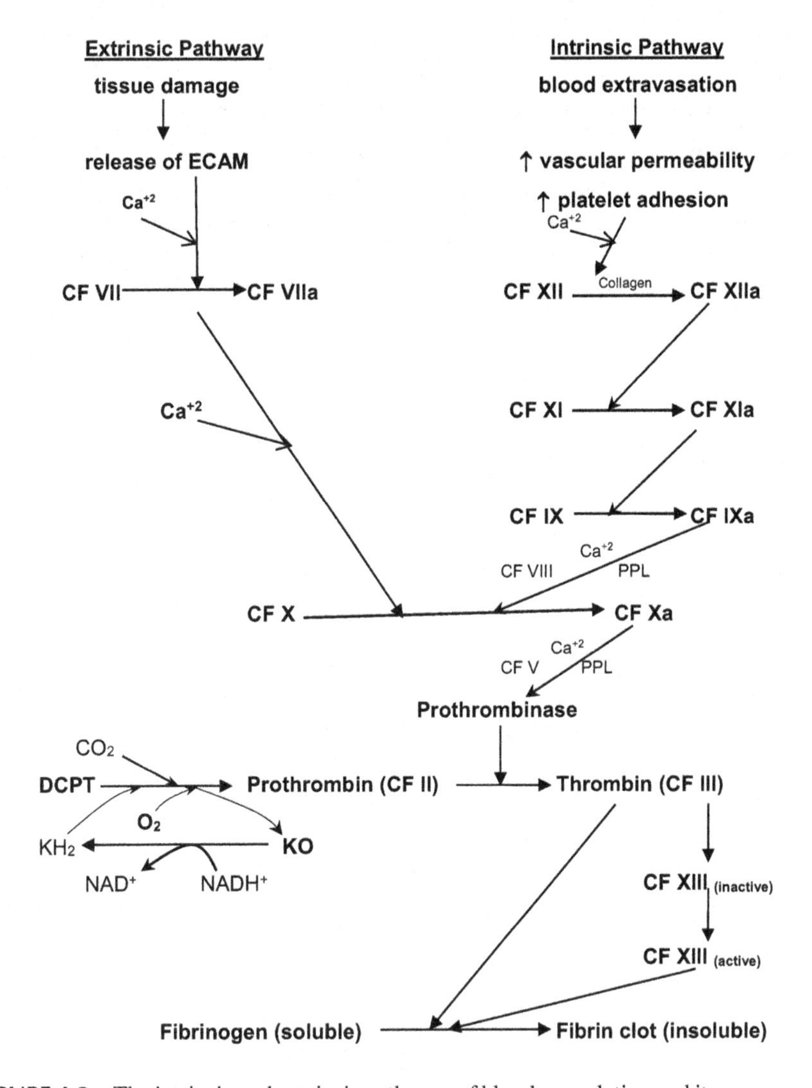

FIGURE 4.2 The intrinsic and extrinsic pathways of blood coagulation and its components.

identified only after the drug had been prescribed to over 800,000 patients for about 6 years, which prompted its withdrawal from the prescription drug market. Table 4.3 lists the physiological functions and biochemical pathways in the liver, representative chemicals that interfere with these functions, and the resulting injury associated with these pathways.

4.2.4 Nephrotoxic Agents

It is not surprising that the kidneys bear a significant burden from the assault of xenobiotic exposure, since they are ultimately involved with the maintenance of

TABLE 4.1
Characteristics and Properties of Innate Immunity

Innate Immunity	Characteristics and Properties
Susceptibility	Result of genetic influences
Epithelial barriers	Desquamation, desiccation, fatty acid secretion, mucosal enzymes
Lymphoid tissue	Bone marrow, thymus, spleen, lymphatics; BALT, GALT, MALT
Inflammatory response	PMN, APP, cytokines and vasoactive mediators; calor, rubor, tumor, dolor[a]
Complement	Complement proteins and cell lysis
Phagocytosis	Pinocytosis, endocytosis, opsonization, degranulation; cells include granulocytes and the monocyte/macrophage system
Nonspecific extracellular killing	NK cells, cytotoxic T-cells

APP: acute phase proteins; BALT: bronchial-associated lymphoid tissue; GALT: gut-associated lymphoid tissue; MALT: mucosa-associated lymphoid tissue; NK: natural killer; PMN: polymorphonuclear cells.
[a] Fever, redness, swelling, pain (respectively).

TABLE 4.2
Characteristics and Properties of Acquired Immunity

Acquired Immunity	Subdivision	Characteristics and Properties
Active immunity	Cell-mediated immunity	T-lymphocytes: allogeneic response, cytokine production, tumor cell cytotoxicity, immunologic memory
	Humoral immunity	B-lymphocytes: antigen–antibody response, cytokine production, complement activation, lipopolysaccharide, immunologic memory, vaccination
Passive immunity	Passive immunization: oral, parenteral administration of antiserum	Administration of antitoxin, antivenom; colostrums; placental transfer of antibodies; bone marrow transplant

homeostasis. The functions of the kidney in maintaining this role include monitoring of blood electrolyte composition and concentration, elimination of metabolic waste, regulation of extracellular fluid volume, maintenance of blood pressure, and adjustment of acid–base balance. Metabolically, the kidneys synthesize and secrete important hormones, such as renin and erythropoietin, necessary for regulation of blood pressure and hematopoiesis, respectively. Metabolically, they convert vitamin D_3 to the active 1,25-dihyroxy metabolite. Figure 4.3 illustrates the anatomical arrangement of the nephron, the functional kidney cell unit. Direct or indirect exposure to nephrotoxic agents may result in ultrastructural damage to any of the principal components of the nephron. Consequently, the pathologic response to a toxic insult is not

TABLE 4.3
Functions of the Liver and Characteristics of Chemical Agents Associated with Hepatotoxicity

Physiologic Liver Function	Associated Pathways	Representative Drugs or Toxins	Resulting Damage
Biochemical homeostasis	Glucose storage and synthesis	CCL_4, ethanol, insulin	Hypoglycemia
	Cholesterol synthesis and uptake		Hypercholesterolemia, fatty liver
Protein synthesis	Clotting factors, albumin, lipoproteins (LDL, VLDL)	Iron, metals, aspirin	Hemorrhage, hypoalbuminemia, fatty liver
Bioactivation and detoxification	P450 isozymes	CCL_4, acetaminophen	Increased sensitivity to drug interactions, drug overdose
	Glutathione (GSH) levels	Acetaminophen	
	Oxygen-dependent bioactivation	Allyl alcohol, ethanol, iron	
Bile formation and secretion	Bilirubin and cholesterol synthesis; uptake of lipids and vitamins	Amoxicillin	Bile duct damage, malnutrition, gallstones, steatorrhea
Hepatic sinusoids	Exchange of LMW substances between hepatocytes and sinusoid	Metals, anabolic steroids, cyclophosphamide	Jaundice, gallstones, neurotoxicity, interference with secondary male sex characteristics
Direct hepatocyte toxicity	Inflammation, wound healing and repair	Acetaminophen, arsenic, ethanol, vinyl chloride	Cell death, fibrosis, cirrhosis
	Tumors	Aflatoxin, androgens	Hepatocellular carcinoma

LDL: low-density lipoproteins; LMW: low–molecular weight; VLDL: very low-density lipoproteins.

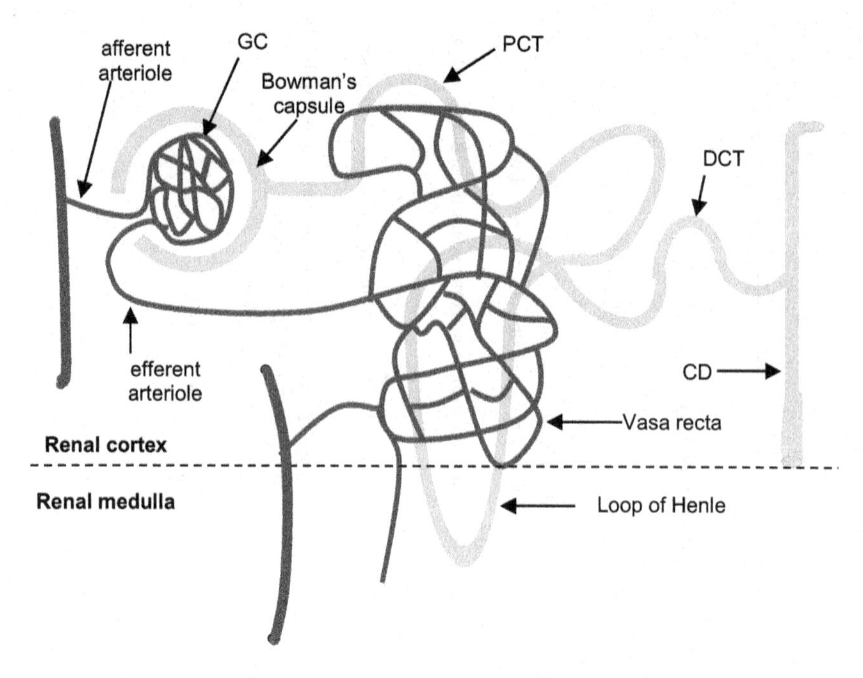

FIGURE 4.3 Schematic diagram of the nephron, the functional cell unit of the kidney.

necessarily different for various chemicals affecting alternate parts of the nephron. For example, acute renal failure (ARF) is the most common manifestation of toxic renal exposure, resulting in an abrupt decline in glomerular filtration rate (GFR). The anatomical sites of chemically induced ARF, and some of the drugs and chemicals associated with injury, include:

1. *prerenal* (i.e., prior to entering the glomerulus): diuretics, vasodilator antihypertensive agents, angiotensin-converting enzyme inhibitors;
2. *vasoconstriction of afferent/efferent arterioles*: cyclosporin, nonsteroidal anti-inflammatory agents (NSAIDs);
3. *crystalluria*: sulfonamides, acyclovir, ethylene glycol;
4. *proximal and distal tubular toxicity*: aminoglycosides, vancomycin;
5. *glomerular capillary endothelial injury*: cocaine, quinine, conjugated estrogens; and
6. *glomerular injury*: gold compounds, NSAIDs, penicillamine.

4.2.5 Pulmonary Toxic Agents

There are numerous lung diseases caused by direct exposure to inhalation of toxic airborne products, traditionally referred to as *inhalation toxicology*. These agents have historically been classified according to occupational diseases because of their association with inhalation and the workplace. Agents that produce pathological changes to the pulmonary system, however, whether airborne or blood-borne, may also be classified according to their pulmonary target organ toxicity. Today, the differences in organizing chemicals according to their target organ or route of exposure are more subtle, given the ubiquitous nature of environmental pollution. Tables 4.4 and 4.5 outline some acute and chronic diseases, respectively, associated with direct inhalation or blood exposure to select toxic compounds. Hallmark

TABLE 4.4
Acute Pulmonary Responses and Associated Select Toxic Agents

Common Pathology	Characteristic Syndrome	Toxic Agents
Bronchitis	Bronchial inflammatory response	Arsenic, chlorine, chromium (VI)
Hard metal disease	Hyperplasia and metaplasia of bronchial epithelium	Titanium and tungsten carbides
Silicosis	Acute silicosis	Silica
Pulmonary edema	Exudative lung injury with thickening of alveolar-capillary wall	Histamine, paraquat, phosgene, beryllium (berylliosis), nitrogen oxide, nickel
Bronchoconstriction	Decrease in airway diameter and increase in airflow resistance	Histamine, cholinergic drugs, prostaglandins, leukotrienes, β-adrenergic inhibitors
Acute pulmonary fibrosis	Increased deposition of collagen in alveolar interstitium	Paraquat, bleomycin

TABLE 4.5
Chronic Pulmonary Responses and Associated Select Toxic Agents

Common Pathology	Characteristic Syndrome	Toxic Agents
Chronic pulmonary fibrosis	Increased deposition of collagen in alveolar interstitium (resembles ARDS)	Paraquat, bleomycin, aluminum dust, beryllium, BCNU, cyclophosphamide, kaolin (kaolinosis), ozone, talc
Emphysema	Abnormal enlargement of airspaces distal to terminal bronchioles, destruction of alveolar walls, without fibrosis	Cadmium oxide, tobacco smoking, air pollution
Asthma	Narrowing of bronchi due to direct or indirect inhalation of toxicants, increased airway reactivity	Isocyanates, tobacco smoking, air pollution, ozone
Carcinoma	Bronchogenic, adenocarcinoma, squamous cell carcinoma	Asbestos fibers (asbestosis), tobacco smoking, metallic dusts, mustard and radon gases, formaldehyde, chromium, arsenic, perchlorethylene
Pulmonary lipidosis	Intracellular macrophages containing abnormal phospholipid complexes	Amiodarone, chlorphentermine

ARDS: adult respiratory distress syndrome; BCNU: 1,3-bis-(2-chloroethyl)-1-nitrosourea.

features that characterize the syndromes are also presented. Some agents lack systematic classification because of the unusual nature of the pulmonary pathology. Such compounds include monocrotaline, a naturally occurring plant alkaloid found in herbal teas, grains, and honey, which produces endothelial hyperplasia with pulmonary arterial hypertension. Naphthalene, a principal component of tar and petroleum distillates, produces extensive bronchial epithelium necrosis. Oxygen toxicity results in bronchial metaplasia and alveolar destruction (see Chapter 23, Gases). Many more industrial chemicals have been associated with pulmonary toxicity, as evidenced from epidemiological data and from studies conducted with experimental animals.

4.2.6 Agents Affecting the Nervous System

Elucidating the mechanisms by which neurotoxicants exert their clinical effects has been instrumental in understanding many concepts of neurophysiology. These mechanisms have prompted breakthroughs, in part, through the development of psychoactive drugs aimed at the same target receptors in common with neurotoxic compounds. Figure 4.4 illustrates a representative diagram of the neuron and some of its components. Examination of the components of the neuron makes it possible to envision the types of damage a neurotoxic agent is capable of eliciting. Some pathologic effects include neuronopathies, axonopathies, myelinopathies, and interference with neurotransmission. Table 4.6 classifies clinically important neurotoxicants according to these major categories and their sites of action. It is important to note that although these compounds appear to have anatomical site specificity, their action may uniformly affect any aspect of the central, autonomic, and peripheral nervous systems, thus leading to a variety of complications and effects.

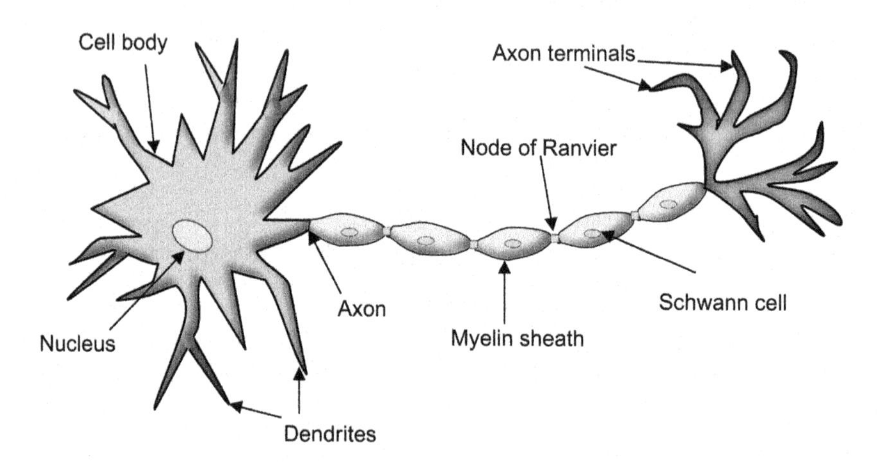

FIGURE 4.4 Schematic diagram of the neuron, the functional cell unit of the nervous system.

TABLE 4.6
Mechanisms of Neurotoxicity and Associated Select Toxic Agents

Mechanism of Neurotoxicity	Representative Toxic Agents	Anatomical Sites of Toxic Action
Neuropathy	Doxorubicin, methyl mercury, catecholamines, MPTP	Dorsal root ganglion, autonomic ganglia, visual cortex, cerebellum; midbrain, thalamic, hypothalamic centers
Axonopathy	Carbon disulfide, acrylamide, chloroquine, lithium, organophosphates	Central and peripheral axons, ascending and descending spinal cord axons
Myelinopathy	Hexachlorophene, lead	Central and peripheral myelinated axons
Interference with neurotransmission	Nicotine, cocaine, amphetamines	Central cognitive and motor centers; associative and integrative pathways; dopaminergic, serotonergic, cholinergic systems

MPTP: 1-methyl-4-phenyl-1,2,3,6-tetrahydropyridine (causes Parkinson-like syndrome).

4.2.7 Agents Affecting the Cardiovascular (CV) System

Subsequent to significant breakthroughs in the treatment of cancers in the last 30 years, CV disease is now the primary cause of mortality among adults in the United States. Simultaneously, a major effort to develop drugs to treat heart disease has resulted in a rapid turnover of these drugs and the establishment of new categories of CV agents. Therefore, adverse drug reactions of therapeutic CV drugs, as well as those of other unrelated chemicals and xenobiotics, have been documented to provoke significant toxic effects on the heart and circulatory system. Toxicity generally occurs as a result of direct and indirect actions on the myocardium, the mechanisms of which are summarized in Tables 4.7 and 4.8 for some commonly encountered agents. The toxicity of therapeutic drugs, administered with the intention of treating a heart condition, is not necessarily different from adverse drug reactions of compounds not intended to treat heart conditions. Thus, the mechanisms of toxicity are seemingly related.

4.2.8 Dermatotoxic Agents

The skin boasts the largest surface area of any organ and has traditionally been understood as the first line of defense against microbial and xenobiotic insult. It is reasonable to anticipate, therefore, that this organ is often a target for environmental toxicants. In addition, since many drugs are administered percutaneously, the skin distributes chemicals in a process known as *transdermal drug delivery*. Consequently, the most common manifestations of dermatotoxic agents include contact dermatitis and to a lesser extent, phototoxicity. Table 4.9 summarizes the most common agents responsible for skin toxicity and their mechanisms.

TABLE 4.7
Mechanisms of Cardiovascular Toxicity Associated with Select CV and Non-CV Therapeutic Agents

Drug Classification	Representative Agents	Cardiotoxic Effects	Mechanism of Cardiotoxicity
	Agents Used in the Treatment of CV Disease		
Antiarrhythmic: Class I	Disopyramide, phenytoin, procainamide	↓ conduction velocity	Na^+ channel inhibition
Antiarrhythmic: Class II	Propranolol, metoprolol, atenolol	Heart block	β-adrenergic receptor inhibition
Antiarrhythmic: Class III	Amiodarone, quinidine	↑ duration of AP	K^+ channel inhibition
Antiarrhythmic: Class IV	Diltiazem, verapamil	↓ AV conduction	Ca^+ channel inhibition
Cardiac glycosides	Digoxin	↓ AV conduction	Na^+, K^+-ATPase inhibition
Catecholamines	Epinephrine, isoproterenol, norepinephrine	Tachycardia	β_1-adrenergic receptor stimulation
	Non-CV Therapeutic Agents		
Antibacterial: aminoglycosides	Gentamicin, kanamycin, streptomycin, tobramycin	Negative inotropy	↓ Ca^+ influx
Antibacterial: fluoroquinolones	Ciprofloxacin, levofloxacin, moxifloxacin	↑ AP duration	K^+ channel inhibition
Antibacterial: macrolides	Azithromycin, clarithromycin, erythromycin	↑ AP duration	K^+ channel inhibition
Antifungal	Amphotericin B	Negative inotropy	Ca^+ channel inhibition
Antihistamines	Astemizole, terfenadine	↑ AP duration	K^+ channel inhibition
Antineoplastic: anthracyclines	Daunorubicin, doxorubicin	Cardiomyopathy	Oxidative stress, apoptosis
Antiviral: nucleoside analog RTI	Stavudine, zidovudine	Cardiomyopathy	Mitochondrial injury
Appetite suppressants	Amphetamines, fenfluramine, phentermine	Tachycardia	Nonselective β-adrenergic receptor stimulation
Bronchodilators	Albuterol, metaproterenol, salmeterol, terbutaline	Tachycardia	Nonselective β-adrenergic receptor stimulation
General anesthetics	Halothane, enflurane	Negative inotropic effect	Ca^{+2} channel inhibition
Local anesthetics	Cocaine	Myocardial ischemia/infarction	Sympathomimetic stimulant
Methylxanthines	Theophylline	Tachycardia	PDE inhibition

(*Continued*)

TABLE 4.7 (CONTINUED)
Mechanisms of Cardiovascular Toxicity Associated with Select CV and Non-CV Therapeutic Agents

Drug Classification	Representative Agents	Cardiotoxic Effects	Mechanism of Cardiotoxicity
Nasal decongestants	Ephedrine and ephedra alkaloids, phenylephrine, PPA, pseudoephedrine	Tachycardia	Nonselective β-adrenergic receptor stimulation
Phenothiazines	Chlorpromazine, thioridazine	Tachycardia, arrhythmia, ↓ AV conduction	Anticholinergic effects
SSRIs	Fluoxetine, paroxetine, risperidone	Bradycardia	Ca^+ channel inhibition
Thyroid hormones	Thyroxine, triiodothyronine	Tachycardia	Altered Ca^+ homeostasis
Tricyclic antidepressants	Amitriptyline, desipramine, imipramine,	ST segment elevation, QT interval prolongation	Na^+, K^+, Ca^+ channel inhibition

AP = action potential; CV = cardiovascular; PDE = phosphodiesterase; PPA = phenylpropranolamine; RTIs = reverse transcriptase inhibitors; SSRIs = selective serotonin reuptake inhibitors.

Modified from K.S. Ramos (2001).

TABLE 4.8
Mechanisms of Cardiovascular Toxicity–Associated Select Nontherapeutic Agents

Drug Classification	Representative Agents	Cardiotoxic Effects	Mechanism of Cardiotoxicity
	Nontherapeutic Agents with CV Toxicity		
Alcohols	Ethanol	↓ conductivity	Oxidative stress
Bacterial endotoxins	Gram-negative bacteria	Bacterial endocarditis	Cardiogenic shock
Halogenated hydrocarbons	Carbon tetrachloride, chloroform, triochloroethane	Altered conduction	↑ adrenergic sensitivity
Heavy metals	Cadmium, lead	Negative inotropy	Altered Ca^{+} homeostasis
Ketones	Acetone	Altered conduction	↑ adrenergic sensitivity
Solvents	Toluene, xylene	Altered conduction	↑ adrenergic sensitivity

TABLE 4.9
Responses of the Skin to Dermatotoxic Agents

Reaction	Common Chemicals or Toxic Agents Associated with Dermatotoxicity
Allergic contact dermatitis	Topical antibiotics, antiseptics; preservatives, cosmetics; plant resins, leather products, industrial solvents, cleaning products, and paints; jewelry and metals
Carcinogenic changes	UV light, ionizing radiation, PAH, arsenic
Chemical burn	Inorganic and organic acids or bases; ammonia; liquids or concentrated vapors of halides, oxides; phenol, phosphorus; organic solvents
Irritant dermatitis	When applied topically, any chemical or drug at low dose has this potential
Photoallergy	Topical salicylic acid derivatives; hexachlorophene; camphor, menthol, phenol as additives; wood and leather oils
Phototoxicity	Furocoumarins (psoralen derivatives); PAH; tetracyclines; sulfonamides; eosin dyes
	Pigmentary disturbances
Hyperpigmentation	Anthracene; hydroquinones, metals, psoralens; UV light; amiodarone, chloroquine, tetracyclines
Hypopigmentation	BHT, hydroquinones, phenolic compounds
Urticaria	Antibacterial antibiotics; food preservatives (BHT, benzoic acid) and additives (CMC, pectin); plants, weeds; latex (natural rubber gloves); lectins (grains, nuts, fruit), pectin (fruit), and a variety of meats, poultry, fish, vegetables, grains, and spices

BHT: butylhydroxytoluene; PAH: polycyclic aromatic hydrocarbons (coal tar, anthracene derivatives).

Among the more common dermal reactions, contact dermatitis and photosensitivity predominate. Contact dermatitis reactions include

1. *irritant dermatitis*—a nonimmune reaction of the skin as a result of toxic exposure, characterized by eczematous itching or thickened eruptions;
2. *chemical burn*—coagulative necrosis resulting from extreme corrosivity from a reactive chemical, characterized by potent skin corrosivity, exothermic reaction, blistering, scar formation, and systemic absorption;
3. *allergic contact dermatitis*—a type IV, delayed hypersensitivity reaction, possessing dose-dependent intensity and requiring initial sensitization, adaptation, and subsequent re-challenge; the reaction ranges from mild irritation, eczematous reactions, eruptions, and flaring to full systemic immune manifestations.

Photosensitivity involves

1. *Phototoxicity*—an abnormal sensitivity to ultraviolet and visible light produced as a result of systemic or topical exposure; it is characterized by red, blistering eruptions as well as hyperpigmentation and thickening;
2. *Photoallergy*—a type IV delayed hypersensitivity reaction.

Other dermatotoxic responses include *chloracne*, a disfiguring form of acne characterized by comedones,* papules, pustules, or cysts, not limited to the facial region; *pigmentary disturbances*, represented as hyper- or hypo-pigmentation; *urticaria*, commonly known as *hives*; and *carcinogenic changes*.

4.2.9 Agents Affecting the Reproductive System

In the last 50 years, the effects of environmental and occupational exposure to industrial chemicals have posed alarming risks to human health and safety. Like other endocrine organs, the male and female gonads act to maintain the reproductive integrity of the species. Thus, chemicals and drugs currently identified as environmental estrogens, whether or not they are structurally similar to the hormone, have the potential for endocrine disruption—that is, the ability of chemicals, drugs, and metals to alter steroidogenesis. Although the etiology of their adverse effects is unclear, their mechanisms influence maternal and paternal dynamics. Tables 4.10 and 4.11 outline several classes of drugs and chemicals (endocrine disruptors) that alter male and female reproductive systems, respectively. Their target mechanisms and possible adverse effects are noted.

4.2.10 Agents Affecting the Endocrine System

Endocrine glands coordinate activities that maintain homeostasis throughout the body. Through their hormonal secretions, and in cooperation with the nervous system, these specialized organs regulate the action of their target cells, increasing or decreasing their respective activities. Consequently, many toxicants have the potential for disrupting

* A comedone is a dilated hair follicle infundibulum filled with keratin, squamae, and sebum; it is the primary lesion of acne vulgaris.

TABLE 4.10
Representative Drugs and Chemicals that Affect the Male Reproductive System

Classification	Representative Agents	Target Cells	Adverse Effects
Alkylating agents	Busulfan, chlorambucil, cyclophosphamide, nitrogen mustard	Gonadogenesis	Testicular development
Androgen derivative	Danocrine	Steroidogenic enzymes	Testicular atrophy
Chlorphenoxy herbicide	TCDD	Spermatogenesis	Testicular development
Diuretic	Spironolactone	Steroidogenic enzymes	Testicular atrophy
Heavy metal	Cadmium	Sertoli cell	Testicular toxicity
	Lead	Spermatocytes	Spermatocidal, infertility
Plasticizer	Phthalate esters	Sertoli cell	Testicular toxicity

TCDD: 2,3,4,8-tetrachlorodibenzo-p-dioxin.

TABLE 4.11
Representative Drugs and Chemicals That Affect the Female Reproductive System

Classification	Representative Agents	Target Cells	Adverse Effects
Alkylating agents	Busulfan, chlorambucil, cyclophosphamide, nitrogen mustard	Gonadogenesis	Ovarian development
Androgen derivative	Danocrine	Steroidogenic enzymes	Disrupts gonadotropin cycles
Steroids and derivatives	Estrogens, progestins	Hypothalamic–pituitary–ovarian axis	Disrupts gonadotropin action

hormonal pathways, either by site-specific toxicity at the endocrine level or by interference with the feedback mechanisms. A list of the endocrine organs, the potential toxicants, and their effects is summarized in Table 4.12. Some of the toxicants exert their effects by causing proliferative changes, resulting in hyperplastic and neoplastic transformation, while other agents inhibit steroid hormone synthesis and secretion.

4.3 CLASSIFICATION ACCORDING TO USE IN THE PUBLIC DOMAIN

4.3.1 Insecticides, Herbicides, Fungicides, and Rodenticides (Pesticides)

The U.S. Environmental Protection Agency (EPA) defines pesticides as substances or a mixture of substances intended for preventing, destroying, repelling,

TABLE 4.12
Target Endocrine Organs and Potential Toxicants

Organ	Representative Toxicants[a]	Target Cell Mechanism	Adverse Effects
Adrenal cortex	Acrylonitrile	Necrosis	Adrenal atrophy
	Anilines	Lipidosis	
	Corticosteroids	Proliferative lesions	
Adrenal medulla	Atenolol, excess dietary calcium intake, nicotine, reserpine, vitamin D_3	Proliferative lesions	Adrenal hypersecretion and atrophy
Parathyroid gland	Aluminum	Inhibits PTH secretion	Hypocalcemia
	Ozone	Hyperplasia	Parathyroid atrophy
Pituitary	Calcitonin, estrogens	Hyperplasia, neoplasia	Pituitary tumors, hypersecretion
Thyroid gland	Thiouracil, sulfonamides	Proliferative changes	Thyroid tumors
	Thiocyanate, thiourea, methimazole, PTU	Inhibit thyroid hormone synthesis	Hypertrophy
	Excess iodide, lithium	Inhibit thyroid hormone secretion	Hypertrophy and goiter

[a] Most of the evidence is from studies performed with laboratory animals.
PTH: parathyroid hormone; PTU: propylthiouracil.

or mitigating any pest. In general, pesticides are classified according to the biological target or organism killed. Four major classes of pesticides include insecticides, herbicides, fungicides, and rodenticides (Tables 4.13 and 4.14). Because of the physiological and biochemical similarities between target species and mammalian organisms, there is an inherent toxicity associated with pesticides in the latter. In addition, within each classification, compounds are identified according to mechanism of action, chemical structure, or synthetic source. For instance, although there are many fungicide categories, fungicidal toxicity in humans is mostly of low order. Similarly, fumigants range from carbon tetrachloride to ethylene oxide and are used to kill insects, roundworms, and fungi in soil, stored grain, fruits, and vegetables. Their toxicity, however, is limited to occasional occupational exposure.

4.3.2 Food and Color Additives

Direct food or color additives are intentionally incorporated in food and food processing for the purpose of changing, enhancing, or masking color. They are also used for a variety of functionalities, ranging from anticaking agents to stabilizers, thickeners, and texturizers. This area is encompassed within the field of food toxicology, and the reader is referred to any of the review articles concerning food ingredients and contaminants.

TABLE 4.13
Classification of Insecticides

Classification	Representative Agents	Mechanism of Toxicity in Mammalian Species
Organophosphorus esters	Ethyl parathion, malathion, mevinphos, sarin, soman, tabun, TEPP	Acetylcholinesterase inhibition
Organochlorine compounds	Aldrin, DDT, chlordane, dieldrin, endrin, lindane, methiochlor, methoxychlor	Selective neurotoxicity
Carbamate esters	Aldicarb, carbofuran, carbaryl (Sevin), propoxur (Baygon)	Reversible acetylcholinesterase inhibition
Pyrethroid esters	Cismethrin, pyrethrin I	Topical inflammatory reaction
Avermectins	Avermectin B1a, ivermectin	Interfere with neuronal transmission
Botanical derivatives	Nicotine	Autonomic NS, neuromuscular stimulant
	Rotenoids	Inhibit neuronal transmission

DDT: dichlorodiphenyltrichloroethane; NS: nervous system; TEPP: tetraethyl pyrophosphate.

TABLE 4.14
Classification of Herbicides and Rodenticides

Classification	Representative Agents	Mechanism of Toxicity in Mammalian Species
	Herbicides	
Chlorphenoxy compounds	2,4-D; 2,4,5-T; TCDD (dioxin)	Neuromuscular, carcinogenic, teratogenic[a]
Bipyridyl derivatives	Diquat, paraquat	Single electron reduction, free radical generation
Chloroacetanilides	Alachlor, metolachlor	Probable human carcinogens
Phosphonomethyl amino acids	Glufosinate, glyphosate	Metabolic enzyme inhibition
	Rodenticides	
Anticoagulants	Warfarin	Inhibits vitamin K-dependent newly synthesized clotting factors
α-naphthyl thiourea (ANTU)	ANTU	Interferes with microsomal monooxygenases
Miscellaneous	Zinc phosphide	Cellular necrosis
	Fluoroacetic acid	Interferes with tricarboxylic acid cycle
	Red squill	Digitalis-like activity
	Phosphorus	Inhibits carbohydrate, lipid, and protein metabolism
	Thallium	Chelates sulfhydryl-containing enzymes

[a] Mechanism is still debatable.

2,4-D: 2,4-dichlorophenoxyacetic acid; 2,4,5-T: 2,4,5-trichlorophenoxyacetic acid; TCDD: 2,3,7,8-tetrachlorodibenzo-p-dioxin.

4.3.3 Therapeutic Drugs

The toxicological classification of therapeutic agents follows their pharmacological mechanisms of action or their principal target organ of toxicity. Section II of this book addresses the clinical toxicology of therapeutic drugs as an extension of their adverse reactions as well as their direct effect resulting from excessive dose or overdosage.

4.3.4 By-Products of Combustion

This broad range of seemingly unrelated chemical substances represents a toxicological classification not unlike the toxic combustion products (TCP, Chapter 26.4). They are generally released with smoke following the burning of wood, plastics, metals, or any solid or liquid chemical. The by-products are released in gaseous form—the most notorious of these are classified and described separately under different categories. They include the gases ammonia, cyanide, carbon monoxide, aldehydes, hydrocarbons, sulfur oxides, hydrogen sulfide, and arsine, and fumes of metallic oxidation. Signs and symptoms of exposure to toxic by-products are related to upper and lower respiratory tract irritation. The pathology is distinct from asphyxiation due to smoke inhalation or oxygen deprivation from toxic concentrations of simple gases.

4.4 CLASSIFICATION ACCORDING TO SOURCE

4.4.1 Botanical

Contact dermatitis caused by poison ivy is a well-characterized syndrome of acute inflammation. Today, as the name implies, many herbal supplements are classified according to their plant species, implying that their origin is botanical. Their importance in maintaining health has also been related to their natural or organic roots. The toxicity of these agents, however, is poorly understood. A detailed discussion of herbal supplements and related products is addressed in Chapter 24, Herbal Remedies.

4.4.2 Environmental

As a result of industrialization, many chemicals are associated and classified according to their continuous presence in the environment—that is, water, land, and soil. The phenomenon is not limited to Western developed nations but is also a rising problem among developing Asian and African countries.

Environmental toxicology is a distinct discipline encompassing the areas of air pollution and ecotoxicology. A discussion of air pollution necessarily includes outdoor and indoor air pollution, the presence of atmospheric sulfuric acid, airborne particulate matter, interaction of photochemicals with the environment, and chemicals found in smog. Ecotoxicology is that branch of environmental toxicology that investigates the effects of environmental chemicals on the ecosystems in question. The reader is referred to the review articles for further discussion.

4.5 CLASSIFICATION ACCORDING TO EFFECTS

4.5.1 Pathologic

Chemicals are also classified according to their toxic effects on mammalian systems, resulting in a discernable pathology. When the toxin produces a disease state, the resulting descriptive pathology is defined by the toxicity based on noticeable, detectable, or measurable signs and symptoms. This classification thus incorporates reactions that aid in identification of the chemical responsible for the pathology and foster diagnosis of the condition, which can facilitate the subsequent treatment. Examples of reactions that act as a gauge for the condition and present intrinsic therapeutic monitoring include chemical allergies, idiosyncratic reactions, hypersensitivity reactions, and local or systemic responses. Effects of toxic substances and therapeutic classification are discussed further in Chapter 6.

4.5.2 Teratogenic, Mutagenic, and Carcinogenic

Our appreciation of the mechanisms underlying chemical carcinogenesis and teratogenesis became a reality only after it was recognized that chemicals could be systematically classified according to their ability to induce molecular and cellular changes in mammalian organisms. This was critically important for the understanding of carcinogenic risk assessment as well as the development of mechanistically based therapeutic strategies for the prediction and management of human cancers. This category, as with most of the topics noted earlier, is described in subsequent chapters in detail.

4.6 CLASSIFICATION ACCORDING TO PHYSICAL STATE

An apparently simple classification scheme involves categorizing chemicals according to their physical state. The importance of such classification for clinical toxicology lies in the knowledge of the physical nature of the substance and its contribution to determining the availability of the compound, the route of exposure, and the onset of action.

4.6.1 Solids

Most therapeutic drugs are available as solid powders in the form of tablets or capsules. This deliberate drug design is intended for facility of administration, ease of packaging, and convenience. As a result, most clinical toxicity results from oral ingestion of solid dosage forms. Solids, however, are subject to the peculiarities of gastrointestinal absorption, pH barriers, and solubility—factors that ultimately influence toxicity.

4.6.2 Liquids

For *nontherapeutic* categories, compounds in their liquid state are readily accessible. Their presence in the environment, food, drinking water, or household products allows immediate oral or dermal exposure. The availability of liquid dosage forms of *therapeutic* agents is intended for pediatric oral use, for injectables, and for those special situations where liquid ingestion is more convenient, such as when it is easier for a patient to swallow a liquid dose of medication. As with solid dosage forms, liquids are also subject to bioavailability on oral ingestion.

4.6.3 Gases

The gaseous state of a chemical represents the fastest route of exposure outside intravenous injection. Inhaled therapeutic or nontherapeutic agents readily permeate the upper and lower respiratory tracts and transport across alveolar walls, gaining access to the capillary circulation. In addition, immediate absorption also occurs with penetration of the mucous membranes before diffusion of the gas to the lower respiratory tract. Both routes of exposure bypass most gastrointestinal absorption and first-pass (hepatobiliary) detoxification phenomena. Consequently, exposure to toxic gases should be approached in all instances as a clinical emergency.

4.7 CLASSIFICATION ACCORDING TO BIOCHEMICAL PROPERTIES

4.7.1 Chemical Structure

Drugs and chemicals are initially identified, synthesized, and screened for activity according to their chemical classification. Because drug and chemical development depends on the availability of parent compounds, the synthesis of structurally related compounds is an important process for the industry. Thus, understanding the chemical structure of a compound allows reasonable prediction of many of its anticipated and adverse effects. Some examples of agents that are categorized according to their molecular structures include benzodiazepines (sedative-hypnotics), imidazolines (tranquilizers), fluoroquinolones (antibiotics), organophosphorus insecticides, and heavy metals.

4.7.2 Mechanism of Action or Toxicity

Similarly, it is convenient to organize drugs and chemicals according to their physiologic or biochemical target. Examples include β-adrenergic agonists, methemoglobin-producing agents, and acetylcholinesterase inhibitors.

REFERENCES

SUGGESTED READINGS

Kang, Y.J., Chapter 18. Toxic responses of the heart and vascular system, in *Casarett and Doull's Toxicology: The Basic Science of Poisons*, Klaassen, C.D. (Ed.), 8th ed., McGraw-Hill, New York, 2013.

REVIEW ARTICLES

Ahlers, J., Stock, F., and Werschkun, B., Integrated testing and intelligent assessment—new challenges under REACH. *Environ. Sci. Pollut. Res. Int.* 15, 565, 2008.

Autrup, H. , Principles of pharmacology and toxicology also govern effects of chemicals on the endocrine system. *Toxicol. Sci.* 146, 11, 2015.

Blaauboer, B.J., Boobis, A.R., Bradford, B., Cockburn, A., Constable, A., Daneshian, M., Edwards, G., Garthoff, J.A., Jeffery, B., Krul, C., and Schuermans, J. Considering new methodologies in strategies for safety assessment of foods and food ingredients. *Food Chem. Toxicol.* 91, 19, 2016.

Collins, A.R., Annangi, B., Rubio, L., Marcos, R., Dorn, M., Merker, C., Estrela-Lopis, I., Cimpan, M.R., Ibrahim, M., Cimpan, E., Ostermann, M., Sauter, A., Yamani, N.E., Shaposhnikov, S., Chevillard, S., Paget, V., Grall, R., Delic, J., de-Cerio, F.G., Suarez-Merino, B., Fessard, V., Hogeveen, K.N., Fjellsbø, L.M., Pran, E.R., Brzicova, T., Topinka, J., Silva, M.J., Leite, P.E., Ribeiro, A.R., Granjeiro, J.M., Grafström, R., Prina-Mello, A., and Dusinska, M. High throughput toxicity screening and intracellular detection of nanomaterials. *Wiley Interdiscip. Rev. Nanomed. Nanobiotechnol.* 9,1, 2017.

Enoch, S.J., Hewitt, M., Cronin, M.T., Azam, S., and Madden, J.C., Classification of chemicals according to mechanism of aquatic toxicity: an evaluation of the implementation of the Verhaar scheme in Toxtree. *Chemosphere* 73, 243, 2008.

Feron, V.J., Cassee, F.R., Groten, J.P., van Vliet, P.W., and van Zorge, J.A., International issues on human health effects of exposure to chemical mixtures, *Environ. Health Perspect.* 110, 893, 2002.

Fonger, G.C., Hakkinen, P., Jordan, S., and Publicker, S., The National Library of Medicine's (NLM) Hazardous Substances Data Bank (HSDB): background, recent enhancements and future plans. *Toxicology* 325, 209, 2014.

Gobas, F.A.P.C., Mayer, P., Parkerton, T.F., Burgess, R.M., van de Meent, D., and Gouin, T., A chemical activity approach to exposure and risk assessment of chemicals: focus articles are part of a regular series intended to sharpen understanding of current and emerging topics of interest to the scientific community. *Environ. Toxicol. Chem.* 37, 1235, 2018.

Hattan, D.G. and Kahl, L.S., Current developments in food additive toxicology in the USA. *Toxicology* 181, 417, 2002.

Heinemeyer, G., Concepts of exposure analysis for consumer risk assessment. *Exp. Toxicol. Pathol.* 60, 207, 2008.

Kirkland, D.J., Testing strategies in mutagenicity and genetic toxicology: an appraisal of the guidelines of the European Scientific Committee for Cosmetics and Non-Food Products for the evaluation of hair dyes. *Mutat. Res.* 588, 88, 2005.

Leung, M.C.K., Procter, A.C., Goldstone, J.V., Foox, J., DeSalle, R., Mattingly, C.J., Siddall, M.E., and Timme-Laragy, A.R., Applying evolutionary genetics to developmental toxicology and risk assessment. *Reprod. Toxicol.* 69, 174, 2017.

Louekari, K., Sihvonen, K., Kuittinen, M., and Sømnes, V., *In vitro* tests within the REACH information strategies. *Altern. Lab. Anim.* 34, 377, 2006.

McCarty, L.S., Issues at the interface between ecology and toxicology. *Toxicology*, 181, 497, 2002.

Marano, K.M., Liu, C., Fuller, W., and Gentry, P.R., Quantitative risk assessment of tobacco products: a potentially useful component of substantial equivalence evaluations. *Regul. Toxicol. Pharmacol.* 95, 371, 2018.

Oliveira, P.A., Colaço, A., Chaves, R., Guedes-Pinto, H., De-La-Cruz P, L.F., and Lopes, C., Chemical carcinogenesis. *An. Acad. Bras. Cienc.* 79, 593, 2007.

Pohl, H.R., Luukinen, B., and Holler, J.S., Health effects classification and its role in the derivation of minimal risk levels: reproductive and endocrine effects. *Regul. Toxicol. Pharmacol.* 42, 209, 2005.

Ramos, K.S., Toxic response of the heart and vascular systems, in: *The Basic Science of Poisons*, 6th edition, Klaassen, C.D., Ed., McGraw-Hill, New York, 2001, Chapter 18.

Reichard, J.F., Maier, M.A., Naumann, B.D., Pecquet, A.M., Pfister, T., Sandhu, R., Sargent, E.V., Streeter, A.J., and Weideman, P.A. ., Toxicokinetic and toxicodynamic considerations when deriving health-based exposure limits for pharmaceuticals. *Regul. Toxicol. Pharmacol.* 79 Suppl 1, S67, 2016.

Sauer, U.G., Hill, E.H., Curren, R.D., Raabe, H.A., Kolle, S.N., Teubner, W., Mehling, A., and Landsiedel, R. Local tolerance testing under REACH: accepted non-animal methods are not on equal footing with animal tests. *Altern. Lab. Anim.* 44, 281, 2016.

Spycher, S., Netzeva, T.I., Worth, A.P., and Escher, B.I., Mode of action-based classification and prediction of activity of uncouplers for the screening of chemical inventories. *SAR QSAR Environ. Res.* 19, 433, 2008.

Stewart, A.G. and Carter, J., Towards the development of a multidisciplinary understanding of the effects of toxic chemical mixtures on health. *Environ. Geochem. Health* 31, 239, 2009.

Strickland, J., Clippinger, A.J., Brown, J., Allen, D., Jacobs, A., Matheson, J., Lowit, A., Reinke, E.N., Johnson, M.S., Quinn, M.J. Jr., Mattie, D., Fitzpatrick, S.C., Ahir, S., Kleinstreuer, N., and Casey, W. Status of acute systemic toxicity testing requirements and data uses by U.S. regulatory agencies. *Regul. Toxicol. Pharmacol.* 94, 183, 2018.

Watanabe, H., Kobayashi, K., Kato, Y., Oda, S., Abe, R., Tatarazako, N., and Iguchi, T. Transcriptome profiling in crustaceans as a tool for ecotoxicogenomics: *Daphnia magna* DNA microarray. *Cell Biol. Toxicol.* 24, 641, 2008.

Waters, M., and Jackson, M., Databases applicable to quantitative hazard/risk assessment—towards a predictive systems toxicology. *Toxicol. Appl. Pharmacol.* 233, 34, 2008.

Wexler, P. and Phillips, S., Tools for clinical toxicology on the World Wide Web: review and scenario. *J. Toxicol. Clin. Toxicol.* 40, 893, 2002.

Williams, E.S., Panko, J., and Paustenbach, D.J., The European Union's REACH regulation: a review of its history and requirements. *Crit. Rev. Toxicol.* 39, 553, 2009.

5 Exposure

5.1 INTRODUCTION

Given the proper circumstances, any chemical has the potential for toxicity—that is, the same dose of a chemical or drug may be harmless if limited to oral exposure but toxic if inhaled or administered parenterally. Thus, the route and site of exposure have a significant influence on determining the toxicity of a substance. More frequently, a therapeutic dose for an adult may be toxic for an infant or child. Similarly, a substance may not exert adverse effects until a critical threshold is achieved. Thus, to induce toxicity, it is necessary for the chemical to accumulate in a physiological compartment at a concentration sufficiently high to reach the threshold value. Finally, repeated administration over a specific period of time also determines the potential for toxicity. This chapter details the circumstances for exposure and dosage of a drug that favor or deter the potential for clinical toxicity.

5.2 ROUTE OF EXPOSURE

5.2.1 Oral

Oral administration of drugs and toxins is by far the most popular route of exposure. Oral administration involves the presence of several physiological barriers, which must be penetrated or circumvented if an adequate blood concentration of the compound is to be achieved. The mucosal layers of the oral cavity, pharynx, and esophagus consist of stratified squamous epithelium, which serves to protect the upper gastrointestinal (GI) lining from the effects of contact with physical and chemical agents. Simple columnar epithelium lines the stomach and intestinal tracts, which function in digestion, secretion, and some absorption. Immediately underlying the epithelium is the lamina propria, a mucosal layer rich in blood vessels and nerves. Mucosa-associated lymphoid tissue (MALT) is layered within this level, where prominent lymphatic nodules sustain the presence of phagocytic macrophages and granulocytes. Salivary and intestinal glands contribute to the digestive process by secreting saliva and digestive juices. The submucosa, muscularis, and serosa complete the strata that form the anatomical envelope of the GI tract. Enteroendocrine and exocrine cells in the GI tract secrete hormones and in the stomach, secrete acid and gastric lipase. Tables 5.1 and 5.2 illustrate select toxic substances and routes of exposure, including dermal and ocular routes, and their suspected clinical effects.

The primary function of the stomach is the mechanical and chemical digestion of food. Absorption is secondary. Consequently, several factors influence the transit and stability of a drug in the stomach, thereby influencing gastric emptying time (GET). The presence of food delays absorption and dilutes the contents of the stomach, thus reducing subsequent drug transit. An increase in the relative pH of the stomach causes a negative feedback inhibition of stomach churning and motility,

TABLE 5.1
Common Routes of Toxic Exposure with Select Toxicants and Their Effects

Route of Exposure	Toxicants	Toxic Effects
Dermal	Sarin, carbon tetrachloride, hexane	Irritation, inflammation, corrosion; hepatotoxicity, neurotoxicity, depending on rate of absorption
Ocular	Acids: sulfuric acid, picric acid	Acid burns including irritation, inflammation, corrosion
	Bases: ammonia, sodium hydroxide	Alkali burns: as with acids, alkali burns produce similar effects such as irritation, inflammation, corrosion
	Solvents: ethanol, acetone, methanol	Corneal and optic nerve damage
	Miscellaneous: naphthalene, thallium	Cataracts, retinal damage, optic nerve damage
Oral	Pesticides and other organic/inorganic toxicants	GI insult, multi-organ toxicity following systemic absorption
Inhalation	See Table 5.2	See Table 5.2
Parenteral	Desomorphine ("krokodil") abuse	Gangrene, phlebitis, thrombosis, pneumonia

TABLE 5.2
Toxicological Classification of Respiratory Toxicants

Class	Description	Representative Substances
Asphyxiants	Gases that reduce tissue oxygen levels	Nitrogen, propane, methane, carbon dioxide, carbon monoxide
Irritants	Irritation and constriction of airways leading to pulmonary edema and infection	Ammonia, hydrogen chloride, hydrogen fluoride
Necrosis producers	Cellular necrosis and edema	Ozone, nitrogen dioxide
Fibrosis producers	Induce fibrotic tissue, causing airway obstruction and decreased lung capacity	Silicates, asbestos, beryllium
Allergens	Induce allergic response	Sulfur dioxide, isocyanates
Carcinogens	Induce lung cancer	Cigarette smoke, asbestos, arsenic

which also results in delayed gastric emptying. Any factor that slows stomach motility will increase the amount of time in the stomach, prolonging the GET. Thus, the longer the GET, the greater the duration of a chemical within the stomach, and the more susceptible it is to gastric enzyme degradation and acid hydrolysis. In addition, prolonged GET delays passage to, and subsequent absorption in, the intestinal tract.

5.2.2 Intranasal

Intranasal insufflation is a popular method for therapeutic administration of corticosteroids and sympathomimetic amines and for the illicit use of drugs of abuse, such as cocaine and opioids. In the former, the dosage form is usually aerosolized in a metered nasal inhaler. In the latter case, the crude illicit drugs are inhaled through the nares ("snorted") as fine or coarse powders. In either case, absorption is rapid due to the extensive network of capillaries in the lamina propria of the mucosal lining within the nasopharynx. Thus, the absorption rate rivals that of pulmonary inhalation.

5.2.3 Inhalation

The vast surface area of the upper and lower respiratory tracts allows wide and immediate distribution of inhaled powders, particulates, aerosols, and gases. Figure 5.1 illustrates the thin alveolar wall that separates airborne particulates from access to capillary membranes. Once a drug is ventilated to the alveoli, it is transported across the alveolar epithelial lining to the capillaries, where rapid absorption is inevitable.

5.2.4 Parenteral

Parenteral administration includes epidermal, intradermal, transdermal, subcutaneous (S.C.), intramuscular (I.M.), and intravenous (I.V.) injections. Parenteral routes, in general, are subject to minimal initial enzymatic degradation or chemical neutralization, thus bypassing the hindrances associated with passage through epithelial barriers. Figure 5.2 illustrates the layers of the skin: the epidermis, the outermost layer, and the underlying dermis and subcutaneous (hypodermis) layers. Epidermal and upper dermal injections have the poorest absorption capabilities of the parenteral routes, primarily because of limited circulation. The deeper dermal and subcutaneous layers provide entrance to a richer supply of venules and arterioles. An S.C. injection may form a depot within the residing adipose tissue with subsequent leakage into the systemic circulation, or the majority of the injection can enter the arterioles and venules. I.M. injections ensure access to

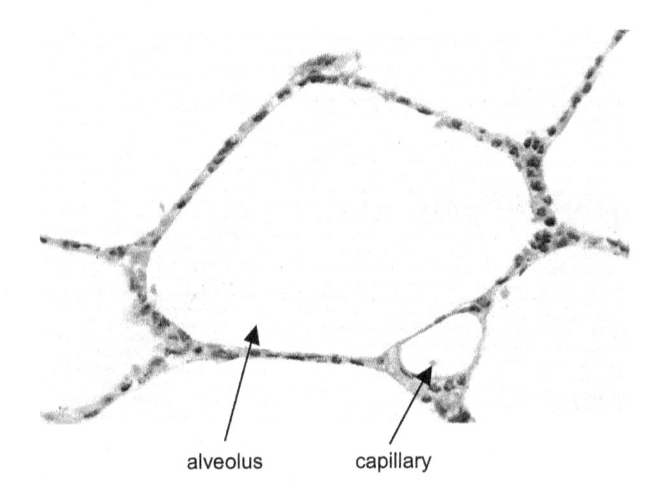

FIGURE 5.1 Diagram of pulmonary alveolus.

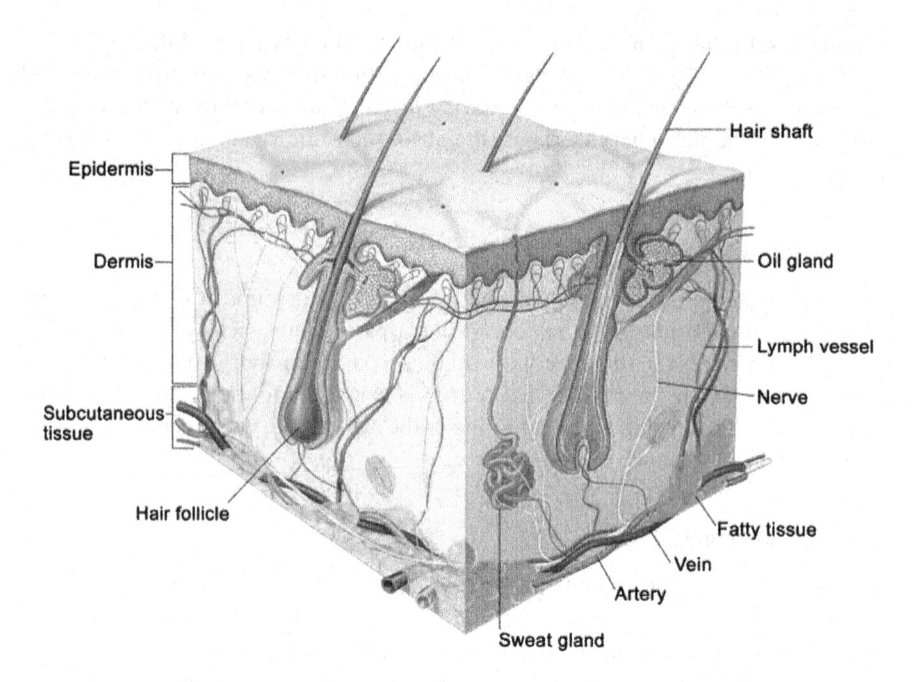

FIGURE 5.2 Epidermis, dermis, and underlying (subcutaneous) layers of the skin. (From PDQ Cancer Information Summaries, U.S. National Center for Biotechnology Information (NCBI), National Institutes of Health (NIH) Terese Winslow, U.S. Govt. Public Domain, 2008)

a more extensive vascular network within skeletal muscle, accounting for more rapid exposure than S.C. injections. I.V. injection is prompt and the most rapid method of chemical exposure, since access to the circulation is direct and immediate. For instance, when ingested orally, a single 40 mg long-acting tablet of oxycodone HCl produces a blood concentration for sufficient analgesia over a 24 hour period. When the same tablet is crushed and injected I.V. by illicit drug users, a naïve victim risks respiratory depression and arrest from the same single tablet. Consequently, the route of exposure contributes as much to toxicity as the dose.

5.3 DURATION AND FREQUENCY

In clinical toxicology, as in pharmacology, duration and frequency depend on the chemical as much as on the route. Thus, for the most part, acute and chronic exposures are relative terms intended for comparative purposes.

5.3.1 Acute Exposure

In general, any exposure less than 24 h may be regarded as *acute*. Exposure to most toxic gases requires less than 24 h for toxicity (carbon monoxide or hydrogen cyanide). In clinical toxicology, however, 72 h may still be an acute exposure, such as in continuous low-dose exposure to acetaminophen in children. In addition, a single I.V. injection of a chemical is certainly classified as an acute exposure. *Subacute* exposure generally refers to repeated exposure to a chemical for greater than 72 h but less than 1 month.

5.3.2 Chronic Exposure

Chronic exposure is any relative time period for which continuous, or repeated intermittent, exposure is required beyond the acute phase for the same chemical to induce a toxic response. Similarly, *subchronic* is also understood to involve a time period between acute and chronic. The traditional timeline of subchronic exposure is understood to be a period of 1 to 3 months. In the realm of pharmacologic toxicology, however, a subchronic exposure may sometimes include repeated exposure for a period greater than 3 months—that is, relative to the acute or chronic administration of the same or similar classes of chemicals. Thus, the terms are flexible adaptations to define the onset of chemical intoxication. In addition, there is considerable overlap in judgment when assigning labels to exposure periods.

5.3.3 Single- or Repeated-Dose Exposure

Frequency of administration considers a single or repeated administration of a dose of the drug or toxin during the exposure period. Although not exclusively, single-dose exposures are essentially associated with acute duration and are the most frequent and convenient methods of experimental drug administration. Most complications and adverse drug reactions (ADRs), however, are usually associated with *continuous* or *intermittent* repeated administration of a dose of a drug beyond the accepted acute frequency, within the time period defined as acute or chronic. Examples of *continuous* repeated exposure parallel therapeutic administration protocols such as I.V. infusions, delayed-release oral or parenteral dosage forms, administration of volatile drugs using an uninterrupted inhalation apparatus, or transdermal cutaneous delivery systems. *Intermittent* repeated exposure protocols follow programmed hourly, daily, or some other regular frequency of administration. Nevertheless, *intermittent* exposure still maintains consistency; it may be alternating, but it should not be sporadic or irregular. Both continuous and intermittent schemas establish greater potential for adverse effects.

5.4 ACCUMULATION

Dose, duration, frequency, and route of exposure contribute to chemical toxicity in part through accumulation of the compound in physiological compartments. A normal dosage schedule is determined according to a chemical's half-life ($t_{½}$) in plasma and its intended response—that is, the time required for plasma levels to decrease to one-half of the measured or estimated concentration. Thus, if the frequency of administration exceeds the $t_{½}$ of the chemical, its concentration in a compartment is likely to increase beyond the desirable level. Accumulation results from overloading of a drug within this compartment.

5.4.1 According to Physiological Compartment

The body as a unit is considered a one-compartment model. Ideally, a drug capable of distributing uniformly throughout the body would maintain steady-state levels for the exposure period. Blood levels* of the compound may also decrease uniformly, assuming a constant rate of elimination. The body, however, is not a homogeneous chamber. A chemical, once absorbed, can distribute and/or bind to one or more of the many physiological sites. The distribution of the chemical depends largely on its physicochemical characteristics (see Chapter 11, Toxicokinetics, for a detailed description of absorption, distribution, metabolism, and elimination). The compartments include whole blood, serum and serum proteins, plasma and plasma proteins, adipose tissue, interstitial and extracellular fluids, alveolar air space, and bone marrow. In addition, any tissue or organ may preferentially accumulate a chemical, thus acting as a discrete compartment. For instance, many therapeutic drugs, such as warfarin (a vitamin K antagonist anticoagulant), nonspecifically bind to circulating plasma proteins, resulting in apparently lower blood concentrations than predicted. Heavy metals preferentially accumulate in adipose tissue. Consequently, in either case, their toxicity may be experienced for prolonged periods of time, as the compounds are slowly released from this compartment for years after exposure has ceased. Accumulation, therefore, can be predicted based on a chemical's apparent volume of distribution (V_d), which is estimated as the total dose of drug in the body divided by the concentration of drug in the plasma, for a given time period. In general, the greater the V_d, the greater the potential for accumulation in some physiological compartment.†

5.4.2 According to Chemical Properties

Accumulation is also determined by the chemical's properties, including structure, degree of ionization, solubility, and its interaction within the physiological compartment. This phenomenon is guided by the chemical's predominant state of existence in a physiological fluid—that is, its persistence in the fluid compartment as an ionic or nonionic species. In general, at physiological pH, lipid-soluble compounds will preferentially remain in their nonionic state, preferring to bind to, and accumulate in, membranes of tissues and organs. Conversely, water-soluble compounds remain as ionic species at the pH of the blood. Thus, because they are less prone to tissue binding, the ions are readily available for renal secretion and elimination (see Chapter 11, Toxicokinetics, for a complete discussion).

5.4.3 According to Other Biological Factors

As noted earlier, a drug's duration of action, the frequency of administration, the route of administration, and its half-life, interact in a connected way prior to accumulation in a physiologic compartment. In fact, the interaction may occur well before

---

* The terms *blood levels* and *plasma levels* are often used interchangeably, although blood and plasma are distinct anatomical compartments.

† A standard measure for the accumulated internal dose of a chemical is the *body burden*, which refers to the amount of chemical stored in one or several physiological compartments or in the body as a whole.

the compartment is reached and is governed by universal forces, such as membrane permeability, that determine the quantity of compound arriving in the biological sector. Thus, other biological factors that influence the chemical's ultimate destination include

1. *Absorption and reabsorption through tissues, organs, or cell membranes*: this vast area of toxicokinetic factors is discussed in detail in Chapter 10, Toxicokinetics. Nevertheless, absorption, reabsorption, and distribution through biological membranes are essentially governed by ionic and nonionic forces of chemical and solvent (i.e., physiological compartment) interactions.
2. *Cardiac output and blood flow*: by definition, cardiac output (CO) is equal to heart rate (HR) multiplied by stroke volume (SV).* Any factor that reduces HR or SV decreases CO, thus slowing the amount of blood perfusion through vital organs. Toxicologically, a decrease in CO also decreases absorption of chemicals through membranes, slows distribution to physiological compartments, and reduces renal perfusion. Consequently, compromised CO diminishes the normal toxicokinetic influences on a chemical.
3. *Peripheral resistance*: as with a reduction in CO, an increase in peripheral resistance slows blood perfusion. Arteriosclerosis, atherosclerosis, vasoconstriction, or loss of arterial or capillary elasticity—all contribute to a decrease in blood circulation through the peripheral systemic circulation, thus leading to reduced distribution of a chemical into target physiologic compartments.
4. *Sympathetic innervation*: sympathetic stimulation is partly driven by α-adrenergic stimulation—that is, a receptor-mediated sympathetic component of the autonomic nervous system. The presence of these receptors in the periphery and their activation are responsible for, among other physiologic activities, peripheral vasoconstriction. The net effect is decreased fluid perfusion in the extremities, particularly the mucous membranes, ears, nose, throat, and buccal cavity. As with the other factors listed, sympathetic stimulation results in reduced kinetic transport of chemicals through biological membranes and diminished compartmental accumulation.
5. *Percent body fat*: increase in percentage body fat above normal (e.g., greater than that seen in a 70 kg male, or 20–25% body fat) results in greater accumulation of nonionic agents in fat-laden compartments and a greater propensity for distribution. This is particularly problematic for compounds that distribute appreciably to lipid tissue, such as metals, pesticides, and organic solvents.
6. *Renal perfusion*: in the therapeutic realm, determination of renal function is of the utmost importance. Any reduction in renal perfusion, decrease in renal output, or increase in glomerular or tubular resistance necessarily results in accumulation of the parent drug as well as toxic metabolites. Accordingly, depressed renal function has significant deleterious consequences with regard to chemical accumulation.

---

* HR is the number of beats per minute, while the SV equals the volume of blood, in milliliters, per beat.

REFERENCES

SUGGESTED READINGS

Eaton, D.L. and Klassen, C.D., Principles of Toxicology, in *Casarett and Doull's Toxicology: The Basic Science of Poisons*, Klaassen, C.D. (Ed.), 8th ed., McGraw-Hill, New York, 2013.

Lou, I., Wambaugh, J.F., Lau, C., Hanson, R.G., Lindstrom, A.B., Strynar, M.J., Zehr, R.D., Setzer, R.W., and Barton, H.A., Modeling single and repeated dose pharmacokinetics of PFOA in mice. *Toxicol. Sci.* 107, 331, 2009.

Maroni, M., Fanetti, A.C., and Metruccio, F., Risk assessment and management of occupational exposure to pesticides in agriculture. *Med. Lav.* 97, 430, 2006.

Zhang, L., Pfister, M., and Meibohm, B., Concepts and challenges in quantitative pharmacology and model-based drug development. *AAPS J.* 10, 552, 2008.

REVIEW ARTICLES

Albertini, R. Bird, M., Doerrer, N., Needham, L., Robison, S., Sheldon, L., and Zenick, H., The use of biomonitoring data in exposure and human health risk assessments. *Environ. Health Perspect.* 114, 1755, 2006.

Efroymson, R.A. and Murphy, D.L., Ecological risk assessment of multimedia hazardous air pollutants: estimating exposure and effects. *Sci. Total Environ.* 274, 219, 2001.

Furtaw, E.J. Jr., An overview of human exposure modeling activities at the USEPA's National Exposure Research Laboratory. *Toxicol. Ind. Health* 17, 302, 2001.

Graham, S.E. and McCurdy, T., Developing meaningful cohorts for human exposure models. *J. Expo. Anal. Environ. Epidemiol.* 14, 23, 2004.

Iavicoli, I., Carelli, G., and Marinaccio, A., Dose-response relationships in human experimental exposure to solvents. *Dose Response* 4, 155, 2006.

Isaacs, K., McCurdy, T., Glen, G., Nysewander, M., Errickson, A., Forbes, S., Graham, S., McCurdy, L., Smith, L., Tulve, N., Vallero, D., Statistical properties of longitudinal time-activity data for use in human exposure modeling. *J. Expo. Sci. Environ. Epidemiol.* 23, 328, 2013.

Lentz, T.J. Dotson, G.S., Williams, P.R., Maier, A., Gadagbui, B., Pandalai, S.P., Lamba, A., Hearl, F., and Mumtaz, M., Aggregate exposure and cumulative risk assessment—integrating occupational and non-occupational risk factors. *J. Occup. Environ. Hyg.* 12, S112, 2015.

Marant Micallef, C. Shield, K.D., Baldi, I., Charbotel, B., Fervers, B., Gilg Soit Ilg, A., Guénel, P., Olsson, A., Rushton, L., Hutchings, S.J., Straif, K., and Soerjomataram, I., Occupational exposures and cancer: a review of agents and relative risk estimates. *Occup. Environ. Med.* 75, 604, 2018.

Maxim, L., A systematic review of methods of uncertainty analysis and their applications in the assessment of chemical exposures, effects, and risks. *Int. J. Environ. Health Res.* 25, 522, 2015.

McCurdy, T., Glen, G., Smith, L., and Lakkadi, Y., The national exposure research laboratory's consolidated human activity database. *J. Expo. Anal. Environ. Epidemiol.* 10, 566, 2000.

Moya, J. and Phillips, L., Overview of the use of the U.S. EPA exposure factors handbook. *Int. J. Hyg. Environ. Health* 205, 155, 2002.

Paustenbach, D. and Galbraith, D., Biomonitoring and biomarkers: exposure assessment will never be the same. *Environ. Health Perspect.* 114, 1143, 2006.

Phillips, L. and Moya, J., The evolution of EPA's Exposure Factors Handbook and its future as an exposure assessment resource. *J. Expo. Sci. Environ. Epidemiol.* 23, 13, 2013.

Strickland J. Clippinger, A.J., Brown, J., Allen, D., Jacobs, A., Matheson, J., Lowit, A., Reinke, E.N., Johnson, M.S., Quinn, M.J. Jr., Mattie, D., Fitzpatrick, S.C., Ahir, S., Kleinstreuer, N., and Casey, W., Status of acute systemic toxicity testing requirements and data uses by U.S. regulatory agencies. *Regul. Toxicol. Pharmacol.* 94, 183, 2018.

6 Effects

6.1 GENERAL CLASSIFICATION

6.1.1 Introduction to Principles of Immunology

As noted in Chapter 4 (Classification), chemicals and drugs can be categorized, in part, according to their toxic effects. The prediction of the toxicity of a chemical lies in the ability to understand its potential effects based on factors not necessarily related to physicochemical properties. This chapter explores a variety of local and systemic reactions elicited by chemical exposure and the physiological and immunological basis of those effects.

To understand some general principles of immunology, Tables 6.1 through 6.3 introduce and define several important immunological terms and concepts. Table 6.1 lists immunological definitions for understanding the principles of immunology, while Table 6.2 outlines the components of natural immunity—that is, innate, naturally occurring immune reactions resulting in the development of physiologic defenses against foreign agents. Thus, natural immunity involves the development of immunological defense and extracellular killing without the production of specific CD-cells* or antibodies (Table 6.3). Table 6.4 presents other components of natural immunity—that is, the immunological cellular defenses that also do not require specific T- or B-cell adaptation, since these cells do not acquire T-cell receptors (TCR). Table 6.5 introduces the features of adaptive immunity—that is, acquired immunity developed as a result of exposure to a specific antigen and resulting in the expression of cell surface markers, TCR, and the development of immunologic memory. Active immunity and passive immunity are subtypes of the acquired response (Table 4.2 also summarizes the features of acquired immunity). Furthermore, active immunity involves two types of responses, humoral and cell-mediated, while passive immunity is composed of the administration of synthetic antibodies, antivenoms, and antitoxins and therapeutic passive immunization. The antibodies are produced in commercial laboratories in mammalian species using good laboratory practice procedures (GLP). Administration of this preparation bypasses the production of antibodies in the ultimate host.

Although not comprehensive, the tables and definitions are useful for interpreting the events that occur in immunotoxicology. They also are particularly helpful to appreciate the basic interactions of drugs and the immune system and supplement the forthcoming information involving toxicokinetic interactions.

6.1.2 Chemical Allergies

The four types of immunological hypersensitivity (allergic) reactions include type I antibody-mediated reactions; type II antibody-mediated cytotoxic reactions; type III

* CD: cluster of differentiation, used to distinguish subsets of T-cells that have acquired T-cell receptors.

TABLE 6.1
Immunological Definitions for the Interpretation of Principles of Immunology

Term	Definition
Immunology	Study of the science of immunity, induced sensitivity, allergy
Immunity	Resistance to infectious agents, foreign particles, toxins, cells, cancer; concerned with the immune state
Immunogen	Any substance capable of stimulating an immune response
Antigen	Any substance capable of binding to components of the immune system or eliciting a response from cells in contact with immune cells; implies the induction of an immune response, immunologic sensitivity, and a latent time period
Antibody	Immunoglobulins, plasma proteins, formed by activated B-lymphocytes as a result of exposure to antigen; react with specificity to that antigen
Serology	*In vitro* study of antigen–antibody reactions; involves the science of immunity and *in vivo* development of antibodies against specific antigens

TABLE 6.2
Components of Natural Immunity: Natural Defenses

Term	Definition
Susceptibility	Immunologically, genetic influence that renders an individual sensitive to an immune response, infectious disease, or other immune-associated pathology
Epithelial barriers	Anatomical or physiological tissues, organs, or membranes that provide chemical, physical, or immunological barriers against antigens, infectious agents, chemicals, immunogens; examples of anatomical barriers: dermis, mucous membranes; examples of physiological barriers: epithelial desquamation, desiccation, fatty acid secretion, mucosal enzymes
Inflammation	Naturally occurring process of immunologically based reactions usually triggered by membrane compromise of affected blood vessels, resulting in a series of chemical and cellular reactions in adjacent tissues; inflammatory process induced by physical, chemical, or biological agents
Cytokines	Cellular secretory immunological peptides or hormones capable of regulating the inflammatory response; physiological responses include production of fever, allergic reactions, constriction of pulmonary smooth muscle, arterial vasodilation, endothelial cell leakage, antiviral activity, chemoattraction (immune cell activation)
Chemical mediators of inflammation	Long-chain, unsaturated, arachidonic acid derivatives, formed as a result of lipid peroxidation of cellular membranes; stimulated by foreign physical, chemical, or biological agents; capable of inducing an inflammatory response; e.g., prostaglandins (PG), leukotrienes (LT), thromboxanes (TBX)
Complement	Group of about 30 distinct cytolytic serum proteins, the cascade reaction of which is usually activated by antibodies and initiated in response to infectious agents

TABLE 6.3
Cells of Natural Immunity: Extracellular Killing

Cell Type	Description
Leukocytes	White blood cells: 1. *Granular cells* (granulocytes): eosinophil, basophil, neutrophil; 2. *Agranular cells*: lymphocyte, plasma cell, monocyte, dendritic cell
Granulocytes	Multilobed WBCs; migratory; staining due to numerous cytoplasmic granules; particularly active in acute inflammation
Eosinophils	2–5% of peripheral blood granulocytes, bi-lobed; stain yellow-red with routine blood smear; particularly elevated in parasitic (helminth) infections; express F_c receptors for IgE antibodies
Basophils/mast cells	0.5–1% of peripheral blood granulocytes; stain blue with routine blood smear; high concentration of histamine-containing granules; increase in response to allergic, type IV hypersensitivity reactions; tissue form known as *mast cells*
Neutrophils	60–70% of granulocytes; stain violet with routine blood smear (Giemsa stain); phagocytic; hydrolase- and neuraminidase-containing vesicles; express complement and F_c receptors
NK cells	Natural killer cells of lymphocyte origin; early appearance in inflammatory process, do not require antigen specificity[a]; do not acquire TCR
T-cells	Thymus-derived lymphocytes; large, nonphagocytic, granular; acquire TCR and antigen specificity; express $CD4^+$ (T-helper) and CD8+ (T-suppressor) surface markers
B-cells	Bone marrow ("bursa")–derived lymphocytes; differentiate to plasma cells
Plasma cells	Derived from B-lymphocytes; acquire BCR, enabling them to synthesize immunoglobulins
Monocytes/ macrophages	Large, mononuclear, phagocytic, circulating, APC cells of monocytic origin; are secretory and migratory; differentiate to alveolar macrophages, Kupffer cells, splenic macrophages, microglia (CNS)
Dendritic cells	Phagocytic, monocytic/lymphocytic origin; vanguards of portals of microbial entry (e.g., skin, lungs, GIT)
PMN	Polymorphonuclear leukocytes; neutrophilic granulocyte; secretory capability toward bacteria; involved in "early response" in infections
Platelets	Irregularly shaped, myeloid-thrombocytic origin; contain granules, microtubules, and vacuoles but no nucleus; involved in clotting as part of inflammatory–wound-healing process

APC: antigen-presenting cells; BCR: B-cell receptor; CNS: central nervous system; GIT: gastrointestinal tract; TCR: T-cell receptor; WBC: white blood cells.
[a] Most cells of innate immunity are not antigen specific.

immune complex reactions; and type IV delayed-type hypersensitivity (cell—mediated immunity).

Type I antibody-mediated reactions occur as three phases. The initial phase, the *sensitization phase*, is triggered by contact with a previously unrecognized antigen. This reaction entails binding of the antigen to immunoglobulin E (IgE) present on the surface of mast cells and basophils. A second phase, the *activation phase*, follows after an additional dermal or mucosal challenge with the same antigen. This

TABLE 6.4
Components of Natural Immunity: Immunological Cellular Defenses

Term	Definition
Phagocytosis	Engulfment and degradation of particulate matter; includes pinocytosis (membrane invagination), endocytosis (receptor-mediated internalization), and recognition of cell surface receptors
Chemotaxis	Movement of cells in a direction influenced by a chemical stimulus, especially through chemical mediators of inflammation
Opsonization	Specialized phagocytosis, stimulated by *opsonins*—antibodies that interact with the surface of bacteria, responsible for temporary immobilization of the organism in preparation for engulfment by phagocytic cells
Ingestion	Phagocytic process involving active cellular amoeboid movement
Degranulation	Release of the contents of vesicles from granulocytes, particularly lysosomal enzymes or vasoactive amines, ultimately for targeting digestion of ingested material

TABLE 6.5
Components of Adaptive (Acquired) Immunity

Term	Definition
Antigen-specific T-cells	Thymus-derived lymphocytes that have acquired TCR and antigen specificity; express $CD4^{+}$ (T-helper) and CD8+ (T-suppressor) cell surface markers
T-secretor cells	T-cells capable of secreting cytokines (soluble mediators of immunity)
T-memory cells	Associated with immunologic memory and cell transformation and proliferation
Antigen-specific B-cells	Bone marrow ("bursa")–derived lymphocytes that have acquired BCR and antigen specificity; differentiate to plasma cells; secrete antigen-specific antibodies
Immunoglobulins	Antibodies, plasma proteins, produced by plasma cells in response to exposure to antigen; react with specificity to that antigen. *Function*: neutralization of toxins, facilitation of phagocytosis, immobilization of microorganisms, agglutination and precipitation reactions, activation of complement, placental transfer
MHC molecules	Major histocompatibility complex molecules, group of linked loci present on normal or viral-infected cells (MHC-I) or on immunological cells (such as APC, MHC-II); facilitate binding to their respective antigen through T- or B-cell receptors; genes express cell surface histocompatibility antigens; principal determinant of tissue type and transplant compatibility

APC: antigen-presenting cell; BCR: B-cell receptor; MHC: major histocompatibility complex; TCR: T-cell receptor.

phase is characterized by degranulation of mast cells and basophils with a subsequent release of histamine and other soluble mediators. The third stage, the *effector phase*, is characterized by accumulation of preformed and newly synthesized chemical mediators that precipitate local and systemic effects. Degranulation of neutrophils and eosinophils completes the late-phase cellular response.

Antigens involved in type I reactions are generally airborne pollens, including mold spores and ragweed, as well as food ingredients. Ambient factors, such as heat and cold, drugs (opioids and antibiotics), and metals (silver and gold), precipitate chemical allergies of the type I nature. Because of their small molecular weight, the majority of drugs and chemicals, as single entities, generally circulate undetected by immune surveillance systems. Consequently, to initiate the sensitization phase of an antigenic response, chemicals are immunologically handled as haptens.* Some examples of type I hypersensitivity syndromes are described in Table 6.6, and typical effects of chemical allergies are listed in Table 6.7. The effects of Type I chemical allergies are usually acute and appear quickly, as compared with other Types II-IV hypersensitivity reactions.

TABLE 6.6
Examples of Type I Hypersensitivity Syndromes, Causes, Effects, and Signs and Symptoms

Syndrome	Causes	Effects	Signs and Symptoms
Allergic rhinitis (hay fever)	Pollen, mold spores	↑ capillary permeability in nasal and frontal sinuses and mucosal membranes, ↑ vasodilation	Congestion, sneezing, headache, watery eyes
Asthma	Chemicals, environmental, behavioral	Chronic obstructive reaction of LRT involving airway hyperactivity and cytokine release	Dyspnea, edema resulting from mucous hypersecretion, bronchoconstriction, airway inflammation
Atopic dermatitis (allergic dermatitis)	Localized exposure to drugs and chemicals	Initial local mast cell release of cytokines followed by activation of neutrophils and eosinophils	Local erythematous reaction
Food allergies	Lectins and proteins present in nuts, eggs, shellfish, dairy products	↑ capillary permeability, ↑ vasodilation, ↑ smooth muscle contraction	Congestion, sneezing, headache, watery eyes, nausea, vomiting

LRT: lower respiratory tract.

* Haptens are small–molecular weight chemical entities (<1000 kD) that nonspecifically bind to larger circulating polypeptides or glycoproteins. The binding of the chemical induces a conformational change in the tertiary or quaternary structure of the polypeptide. Consequently, the larger molecule is no longer recognized as "self," rendering it susceptible to immune attack. Under these conditions, the chemical has acquired antigenic potential.

TABLE 6.7
Descriptive Inflammatory Effects of Chemical Allergies

Effect	Description
Bullous	Large dermal blisters
Erythema	Redness or inflammation of skin due to dilation and congestion of superficial capillaries (e.g., sunburn)
Flare	Reddish, diffuse blushing of the skin
Hyperemia	Increased blood flow to tissue or organ
Induration	Raised, hardened, thickened skin lesion
Macule	Flat red spot on skin due to increased blood flow
Papule	Raised red spot on skin due to increased blood flow and antibody localization
Petechial	Small, pinpoint dermal hemorrhagic spots
Pruritus	Itching
Purulent	Suppuration; production of pus containing necrotic tissue, bacteria, and inflammatory cells
Urticaria	Pruritic skin eruption characterized by wheal formation
Vesicular	Small dermal blisters
Wheal	Raised skin lesion due to accumulation of interstitial fluid

Type II antibody-mediated cytotoxic reactions differ from type I in the nature of the antigen, the cytotoxic character of the antigen–antibody reaction, and the type of antibody formed (IgM or IgG). In general, antibodies are formed against target antigens that are *altered cell membrane determinants*. Examples of type II reactions include complement-mediated reactions (CM), antibody-dependent cell-mediated cytotoxicity (ADCC), antibody-mediated cellular dysfunction (AMCD), transfusion reactions, Rh incompatibility reactions, autoimmune reactions, and drug-induced reactions, the last of which are of greater interest in clinical toxicology.

As with type I reactions, drug-induced type II cytotoxicity requires that the agent behaves as a hapten. The chemical binds to the target cell membrane and proceeds to operate as an altered cell membrane determinant. This determinant changes the conformational appearance of a component of the cell membrane, not unlike the effect of a hapten. Thus, it induces a series of responses that terminate in antibody induction. The determinant attracts a variety of immune surveillance reactions, including complement-mediated cytotoxic reaction, recruitment of granulocytes, or deposition of immune complexes within the cell membrane. Examples of drugs that traditionally induce type II reactions, and their pathologic effects, are noted in Table 6.8.

Type III immune complex reactions are localized responses mediated by antigen–antibody immune complexes. Type III reactions are stimulated by microorganisms and involve activation of complement. Systemic (serum sickness) and localized (Arthus reaction) immune complex disease, infection-associated immune complex disease (rheumatic fever), and occupational diseases (opportunistic pulmonary fungal infections) induce complement–antibody–antigen complexes. These complexes then trigger release of cytokines and recruitment of granulocytes, resulting in increased vascular permeability and tissue necrosis.

TABLE 6.8
Examples of Drug-Induced Type II Antibody-Mediated Cytotoxic Reactions

Drug	Classification	Pathologic Effect
Chloramphenicol	Antibiotic (bacterial)	Produces agranulocytosis through binding to WBCs
Chlorpropamide	Antidiabetic	Cholestatic jaundice
Erythromycin estolate	Antibiotic (bacterial)	Cholestatic jaundice
Estrogen/progesterone	Oral contraceptives	Induce SLE syndrome through the production of antinuclear antibodies; cholestatic jaundice
Gentamicin, carbenicillin	Antibiotics (bacterial)	Nephrotoxic immune complex reaction
Hydralazine	Antihypertensive	Stimulates SLE syndrome through the production of antinuclear antibodies
Isoniazid	Antibiotic (*Mycobacteria* sp.)	Induces cell-to-cell immune complex reactions
Methyldopa	Antihypertensive	Produces a positive DAT in 15 to 20% of patients and AHA in 1%
NSAIDS	Analgesic/anti-inflammatory	Generate cell–cell immune complex reactions
Penicillins, cephalosporins	Antibiotics (bacterial)	Induce hemolysis by binding to RBCs
Phenacetin	Analgesic	Produces hemolytic anemia through binding to RBCs
Phenytoin	Antiepileptic	Induces SLE syndrome through the production of antinuclear antibodies
Procainamide	Antiarrhythmic	Stimulates SLE syndrome through the production of antinuclear antibodies
Quinidine	Antiarrhythmic	Provokes cell-to-cell immune complex reactions
Testosterone	Anabolic steroids	Cholestatic jaundice

AHA: autoimmune hemolytic anemia; DAT: direct antiglobulin test; NSAIDS: nonsteroidal anti-inflammatory drugs; RBCs: red blood cells; SLE: systemic lupus erythematosus; WBCs: white blood cells.

Type IV (delayed-type) hypersensitivity cell-mediated immunity involves antigen-specific T-cell activation. The reaction starts with an intradermal or mucosal challenge (*sensitization stage*). $CD4^+$ T-cells then recognize MHC-II (major histocompatibility class-II) antigens on antigen-presenting cells (such as Langerhans cells) and differentiate to T_H1 cells. This sensitization stage requires prolonged local contact with the agent, usually for at least 2 weeks. A subsequent repeat challenge stage induces differentiated T_H1 (memory) cells to release cytokines, further stimulating the attraction of phagocytic monocytes and granulocytes. The release of lysosomal enzymes from the phagocytes results in local tissue necrosis. Contact hypersensitivity resulting from prolonged exposure to plant resins and jewelry, for example, is caused by the lipophilicity of the chemical in oily skin secretions, thus acting as a hapten.

6.1.3 Idiosyncratic Reactions

Idiosyncratic reactions are abnormal responses to drugs or chemicals, generally resulting from uncommon genetic predisposition. An exaggerated response to the skeletal muscle relaxant properties of succinylcholine, a depolarizing neuromuscular blocker, classifies as a typical idiosyncratic reaction. In some patients, a congenital deficiency in plasma cholinesterase results in a reduction in the rate of succinylcholine deactivation. As succinylcholine accumulates, respiration fails to return to normal during the postoperative period. Similarly, the cardiotoxic action of cocaine is exaggerated in cases of congenital deficiency of plasma esterases, which are necessary for metabolism of the drug. A paucity of circulating enzymes allows an uncontrolled, sympathetically mediated tachycardia, vasoconstriction, and subsequent heart failure in naïve or habitual cocaine users.

6.1.4 Immediate versus Delayed Effects

In contrast to immune hypersensitivity reactions, some chemical effects are immediate or delayed, depending on the mechanism of toxicity. The acute effects of sedative-hypnotics are of immediate consequence; an overdose raises the risk of death from respiratory depression. The effects of carcinogens, however, may not be demonstrated for generations. An important example of this has been demonstrated with the link between diethylstilbestrol (DES) administration to child-bearing women in the 1950s and the subsequent development of clear cell adenocarcinoma vaginal cancers in the offspring.

6.1.5 Reversible versus Irreversible Reactions

In general, the effects of most drugs or chemicals are reversible until a critical point is reached—that is, when vital function is compromised, or a teratogenic or carcinogenic event develops. In fact, the carcinogenic effect of chemicals, such as those substances present in tobacco smoke, may be delayed for decades until irreversible cellular transformation occurs. Reversibility of a chemical effect may be enacted through the administration of antagonists, by enhancement of metabolism or elimination, by delaying absorption, by intervening with another toxicological procedure that decreases toxic blood concentrations, or by terminating the exposure.

6.1.6 Local versus Systemic Effects

As discussed in Chapter 5 (Exposure), local or systemic effects of a compound depend on the site of exposure. The integument (skin) and lungs are frequent targets of chemical exposure, since these organs are the first site of contact with environmental chemicals. Oral exposure requires absorption and distribution of the agent prior to the development of systemic effects. Hypersensitivity reactions types I and IV are precipitated by local activation of immune responses following a sensitization phase, while drug-induced type II reactions are elicited through oral or parenteral administration.

6.1.7 Target Therapeutic Effects

It is important to remember that desirable therapeutic, or undesirable adverse, effects of pharmaceutical agents have the capacity to prevail over normal physiology. Thus, the action of a drug on its intended target will largely alter the function of that target. Under physiological conditions, the target action may be more predictable, particularly in response to a therapeutic dose (e.g., administration of oral contraceptives or vitamins). Nevertheless, these agents may still produce adverse reactions by the very nature of their ability to alter physiological processes. In addition, since most therapeutic drugs are administered in the presence of, or for the treatment of, a pathological condition, the predictability and consequences of drug administration are not as reliable. In fact, under these circumstances, several factors contribute to the various possible outcomes that may occur in the presence of comorbid surroundings. For instance, the following situations require modification of the therapeutic regime:

1. The toxicokinetics or toxicodynamics of the compound may be altered (as with a drug that exhibits primary renal elimination in the presence of compromised kidney function);
2. The response of the diseased organ, tissue, or system may be distorted in response to drug treatment (as with neurologic toxicities demonstrated with phenothiazine derivatives); and
3. The toxicity of a compound may mimic the very condition for which the drug is used (as with the toxic effects of digitalis glycosides for use in congestive heart failure).

In any event, many other examples of these situations will present themselves in subsequent chapters with further elucidation on the mechanisms of toxic events.

6.2 CHEMICAL INTERACTIONS

6.2.1 Potentiation

Potentiation of toxicity occurs when the toxic effect of one chemical is enhanced in the presence of a toxicologically unrelated agent. The situation can be described numerically as "0 + 2 > 2," whereby a relatively nontoxic chemical alone has little or no effect ("0") on a target organ but may enhance the toxicity of another co-administered chemical ("2"). The hepatotoxicity of carbon tetrachloride, for instance, is greatly increased in the presence of isopropanol.

6.2.2 Additive

Two or more chemicals whose combined effects are equal to the sum of the individual effects are described as having additive interactions. Such is the case with the additive effects of the combination of sedative-hypnotics and ethanol (drowsiness and respiratory depression). Numerically, this is summarized as "2 + 2 = 4."

6.2.3 Synergistic

By definition, a synergistic effect is indistinguishable from potentiation except that in some references, both chemicals must have cytotoxic activity. Numerically, synergism occurs when the sum of the effects of two chemicals is greater than the individual effects, such as is experienced with the combination of ethanol and antihistamines ("1 + 2 > 3"). Often, synergism and potentiation are used synonymously.

6.2.4 Antagonistic

The opposing actions of two or more chemical agents, not necessarily administered simultaneously, are considered to be antagonistic interactions. Different types of antagonism include *functional antagonism*—the opposing physiological effects of chemicals, such as with central nervous system stimulants versus depressants; *chemical antagonism*—drugs or chemicals that bind to, inactivate, or neutralize target compounds, such as with the action of chelators in metal poisoning; *dispositional antagonism*—interference of one agent with the absorption, distribution, metabolism, or excretion (ADME) of another (examples of agents that interfere with absorption, metabolism, and excretion include activated charcoal, phenobarbital, and diuretics, respectively); and *receptor antagonism*—refers to the occupation of pharmacological receptors by competitive or noncompetitive agents, such as the use of tamoxifen in the prevention of estrogen-induced breast cancer.

REFERENCES

SUGGESTED READINGS

Baken, K.A., Vandebriel, R.J., Pennings, J.L., Kleinjans, J.C., and van Loveren, H., Toxicogenomics in the assessment of immunotoxicity. *Methods* 41, 132, 2007.

Burns-Naas, L.A., Hastings, K.L., Ladics, G.S., Makris, S.L., Parker, G.A., and Holsapple, M.P., What's so special about the developing immune system? *Int. J. Toxicol.* 27, 223, 2008.

Colosio, C., Birindelli, S., Corsini, E., Galli, C.L., and Maroni, M., Low level exposure to chemicals and immune system. *Toxicol. Appl. Pharmacol.* 207, 320, 2005.

Descotes, J., Methods of evaluating immunotoxicity. *Expert Opin. Drug Metab. Toxicol.* 2, 249, 2006.

Dietert, R.R., Developmental immunotoxicity (DIT) in drug safety testing: matching DIT testing to adverse outcomes and childhood disease risk. *Curr. Drug Saf.* 3, 216, 2008.

Hammes, B. and Laitman, C.J., Diethylstilbestrol (DES) update: recommendations for the identification and management of DES-exposed individuals. *J. Midwifery Women's Health* 48, 19, 2003.

Hostynek, J.J., Sensitization to nickel: etiology, epidemiology, immune reactions, prevention, and therapy. *Rev. Environ. Health* 21, 253, 2006.

Selgrade, M.K., Immunotoxicity: the risk is real. *Toxicol. Sci.* 100, 328, 2007.

REVIEW ARICLES

Arthur, P.G., Grounds, M.D., and Shavlakadze, T., Oxidative stress as a therapeutic target during muscle wasting: considering the complex interactions. *Curr. Opin. Clin. Nutr. Metab. Care* 11, 408, 2008.

Bolt, H.M. and Kiesswetter, E., Is multiple chemical sensitivity a clinically defined entity? *Toxicol. Lett.* 128, 99, 2002.

Bradberry, S.M. and Vale, J.A., Multiple-dose activated charcoal: a review of relevant clinical studies. *J. Toxicol. Clin. Toxicol.* 33, 407, 1995.

Buckley, N.A., Poisoning and epidemiology: "toxicoepidemiology". *Clin. Exp. Pharmacol. Physiol.* 25, 195, 1998.

Damstra, T., Potential effects of certain persistent organic pollutants and endocrine disrupting chemicals on the health of children. *J. Toxicol. Clin. Toxicol.* 40, 457, 2002.

Foster, J.R., The functions of cytokines and their uses in toxicology. *Int. J. Exp. Pathol.* 82, 171, 2001.

Fowler, S. and Zhang, H., In vitro evaluation of reversible and irreversible cytochrome P450 inhibition: current status on methodologies and their utility for predicting drug-drug interactions. *AAPS J.*, 10, 410, 2008.

Haddad, J.J., On the mechanisms and putative pathways involving neuroimmune interactions. *Biochem. Biophys. Res. Commun.* 370, 531, 2008.

Hastings, K.L., Implications of the new FDA/CDER immunotoxicology guidance for drugs. *Int. Immunopharmacol.* 2, 1613, 2002.

Isbister, G.K., Data collection in clinical toxinology: debunking myths and developing diagnostic algorithms, *J. Toxicol. Clin. Toxicol.* 40, 231, 2002.

Kimber, I., Gerberick, G.F., and Basketter, D.A., Thresholds in contact sensitization: theoretical and practical considerations. *Food Chem. Toxicol.* 37, 553, 1999.

Maurer, T., Skin as a target organ of immunotoxicity reactions. *Dev. Toxicol. Environ. Sci.* 12, 147, 1986.

Tanaka, E., Toxicological interactions between alcohol and benzodiazepines. *J. Toxicol. Clin. Toxicol.* 40, 69, 2002.

Zhou, H.X., Rivas, G., and Minton, A.P. Macromolecular crowding and confinement: biochemical, biophysical, and potential physiological consequences. *Annu. Rev. Biophys.* 37, 375, 2008.

7 Dose–Response

7.1 GENERAL ASSUMPTIONS

7.1.1 Types of Dose–Response Relationships

The discussion of effects of chemicals as a result of exposure to a particular dose (Chapter 6) necessarily must be followed by a discussion of the path by which that dose elicits that response. This relationship has traditionally been known as the *dose–response* relationship.* The result of exposure to the dose can be any measurable, quantifiable, or observable indicator. The response depends on the quantity of chemical exposure or administration within a given time period. Two types of dose–response relationships exist, depending on the number of subjects and doses tested. The *graded dose–response* describes the relationship of an individual test subject or system to increasing and/or continuous doses of a chemical. Figure 7.1 illustrates the effect of increasing doses of several chemicals on cell proliferation *in vitro*. The concentration of the chemical is inversely proportional to the number of surviving cells in the cell culture system.

Alternatively, the *quantal dose–response* is determined by the distribution of responses to increasing doses in a population of test subjects or systems. This relationship is generally classified as an "all-or-none effect," where the test system or organisms are quantified as either "responders" or "nonresponders." A typical quantal dose–response curve is illustrated in Figure 7.2 by the LD_{50} (lethal dose 50%) distribution. The LD_{50} is a statistically calculated dose of a chemical that causes death in 50% of the animals tested. The doses administered are also continuous, or at different levels, and the response is generally mortality, gross injury, tumor formation, or some other criterion by which a standard deviation or "cut-off" value can be determined. In fact, other decisive factors, such as therapeutic dose or toxic dose, can be determined using quantal dose–response curves, from which are derived the ED_{50} (effective dose 50%) and the TD_{50} (toxic dose 50%), respectively.

Graded and quantal curves are generated based on several assumptions. The time period at which the response is measured is chosen empirically or selected according to accepted toxicological practices. For instance, empirical time determinations may be established using a suspected toxic or lethal dose of a substance, and the response is determined over several hours or days. This time period is then set for all determinations of the LD_{50} or TD_{50} for that time period.† The frequency of administration is assumed to be a single dose administered at the start of the time period when the test subjects are acclimated to the environment.

---

* In certain fields of toxicology, particularly mechanistic and *in vitro* systems, the term *dose–response* is more accurately referred to as *concentration–effect*. This change in terminology makes note of the specific effect on a measurable parameter that corresponds to a precise plasma concentration.

† This time period is commonly set at 24 h for LD_{50} determinations.

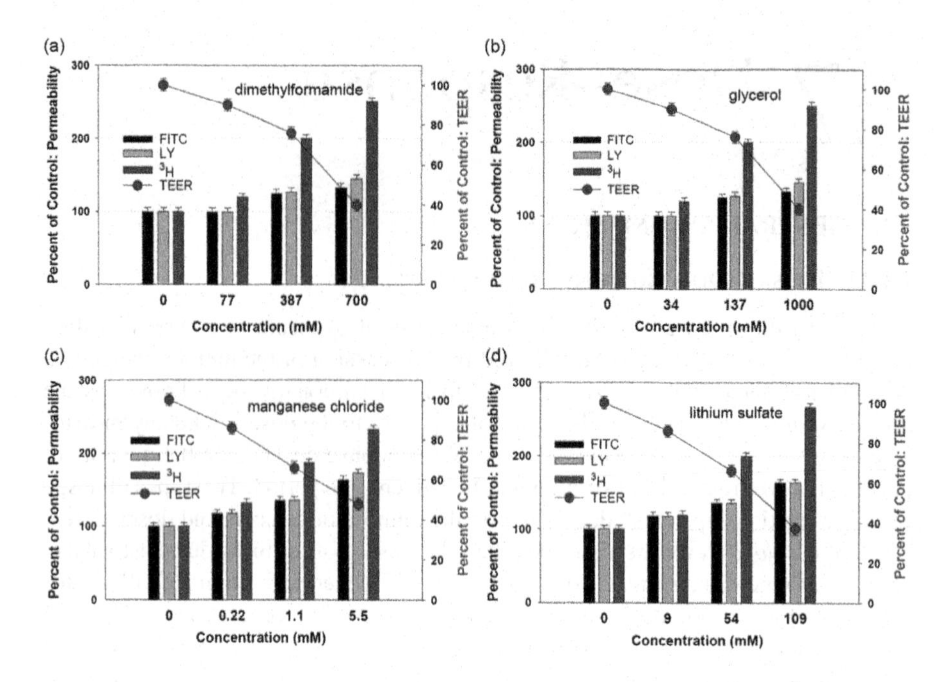

FIGURE 7.1 Graded concentration-effect plots for (a) dimethylformamdie, (b) glycerol, (c) manganese chloride, and (d) lithium sulfate, on PP in confluent monolayers of mouse embryonic stem cells. Scale for % of control for PP markers (bars) are on left axis; for TEER measurements (solid line), scale is on right axis. All control values are set at 100%. FITC = fluorescein isothiocyanate-dextran, LY = lucifer yellow, 3H = [3H]-D-mannitol, PP = paracellular permeability, TEER = transmonolayer specific electrical resistance. (from Calabro, A.R., Konsoula, R., Barile, F.A. 2008)

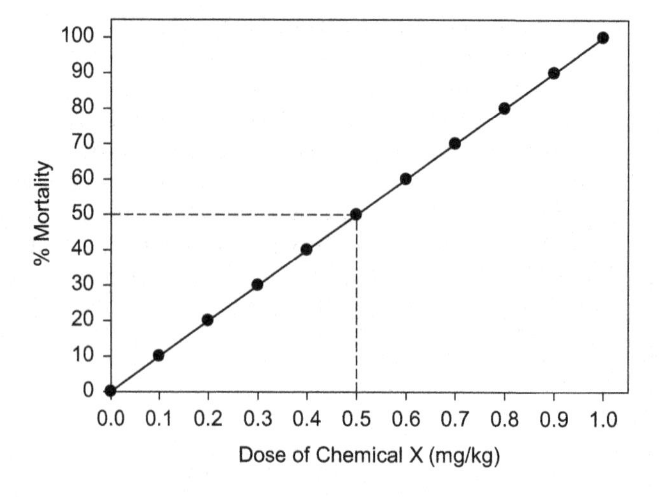

FIGURE 7.2 Quantal dose–response curve showing the experimental derivation and graphic estimation of the LD_{50}.

7.1.2 Concentration–Effect and Presence at the Receptor Site

It is assumed that at the conclusion of an experiment, the measured or observed effect is, in fact, due to the presence of the chemical. The establishment of this causal relationship is critical if valid conclusions are to be drawn from the dose–response curve. It is also presumed that the chemical in question is present at the receptor site or molecular target responsible for the effect. Support for this assumption follows from measurement of concentrations at the cellular or organ level. In fact, as the concentration of a chemical in the affected compartment increases, the degree of response must increase proportionately if this assumption is valid. For this reason, some references have suggested the use of the term *concentration–effect curve* as an alternative label to dose–response. The former is purported to be a more straightforward description of the causal relationship and more accurately reflects the parameters measured.

7.1.3 Criteria for Measurement

Except for lethality, the selection of a measurable or observable endpoint is crucial. A desirable biomarker may be one that accurately reflects the presence of the chemical at the molecular site or suggests that the toxic effect originates from the target organ. The selection of a measurable endpoint thus depends on the suspected mechanism of toxicity, if known, or on empirical determinations based on the chemical formula. Some biomarkers are also subjective, such as reliance on histological grading or calculation of the degree of anesthesia, pain, motor activity, or behavioral change. Thus, the standards for quantifying the endpoint are determined and established prior to the experimental setup.

7.2 LD_{50} (LETHAL DOSE 50%)

7.2.1 Definition

As noted earlier, the LD_{50} is a statistically calculated dose of a chemical that causes death in 50% of the animals tested, based on the objective observation of lethality. This "all-or-none" effect uses lethality as an absolute, unequivocal measurement. The usefulness of the test provides a screening method for toxicity evaluation, which is particularly useful for new, unclassified substances. The determination, however, is antiquated, requires large numbers of animals, does not provide information regarding mechanistic effects or target organ, and does not suggest complementary or selective pathways of toxicity. It is also limited by the route and duration of exposure. Consequently, its routine use in drug testing has become the subject of continuous debate and regulatory review in toxicity testing.

7.2.2 Experimental Protocol

A predetermined number of animals, at least 10 animals per dose and 10 doses per chemical, are selected. Groups of animals are subjected to increasing single doses of the test substance (the dose is calculated as milligrams of substance per kilogram of body weight [mg/kg]). The number of animals that expire within the group at the end of a predetermined time period (usually 24 to 96 hours) is converted to a percentage of the total animals exposed to the chemical. A dose–response relationship is constructed, and the 50% extrapolation is computed from the curve (see Figure 7.2).

Alternatively, the testing of chemicals for biological effects sometimes entails that the response to a chemical is normally distributed—that is, that the highest number of respondents are gathered in the middle dosage range. Figure 7.3 represents a normal frequency distribution achieved with increasing dosage of a chemical versus the cumulative percentage mortality. The bars represent the percentage of animals that died at each dose minus the percentage that died at the next lower dose. As is shown by the normal (Gaussian) distribution, the lowest percentage of animals died at the lowest and highest doses, accounted for by biological variation (hypersensitive vs. resistant animals, respectively).

7.2.3 Factors That Influence the LD_{50}

Several factors influence the reliability of the LD_{50} determination, including the selection of species, the route of administration, and the time of day of exposure and observation. In general, the reproducibility of an LD_{50} relies on adherence to the same criteria in each trial experiment. The same species must be of the same age, sex, strain, weight, and breeder. The route of administration, as well as the time of administration, is critical. Also, the animal care maintenance should be similar in each run, with attention to light and dark cycles, feeding, and waste disposal schedules. Variability in LD_{50} is largely due to changes in parameters between experiments.

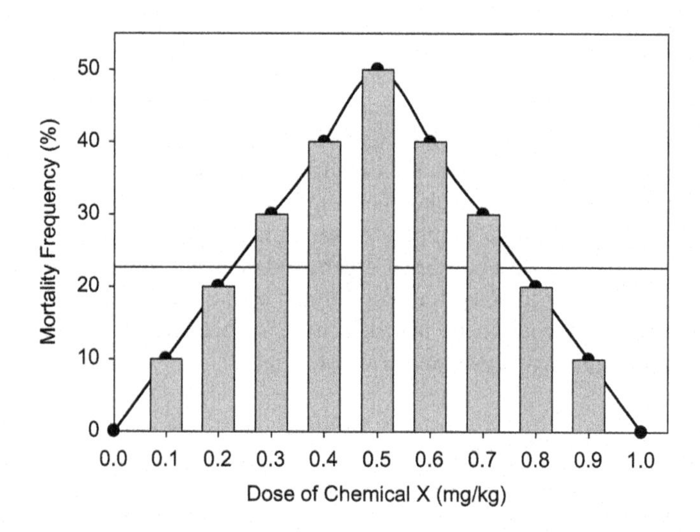

FIGURE 7.3 Normal frequency distribution of mortality frequency (%) versus dose.

7.3 ED_{50} (EFFECTIVE DOSE 50%), TD_{50} (TOXIC DOSE 50%), AND TI (THERAPEUTIC INDEX)

7.3.1 RELATIONSHIP TO LD_{50}

The ED_{50} is analogous to the LD_{50} with the important exception that the former uses an endpoint designed to determine a desirable effect of the test substance occurring in 50% of the animals tested. Such desirable effects might include sedation with a sedative-hypnotic drug or analgesia with an opioid derivative. Similarly, the TD_{50} is calculated based on the measurement of a toxic nonlethal endpoint, such as respiratory depression. The relationship between the lethal (or toxic) dose and the effective dose is calculated and expressed as the therapeutic index (TI):

$$TI = LD_{50} \left(\text{or } TD_{50}\right)/ED_{50}$$

The formula is generally accepted as an approximation of the relative safety of a chemical and is used as a guide for therapeutic monitoring. The TI is illustrated in Figure 7.4 as a graphic representation of the relationship between the LD_{50} and the ED_{50}.

7.3.2 ASSUMPTIONS USING THE TI

The derivation of the TI assumes the measurement of median doses in the experimental protocol.* The median dose is calculated according to the desired effect observed in 50% of the animals tested and represents a quantitative, but relative,

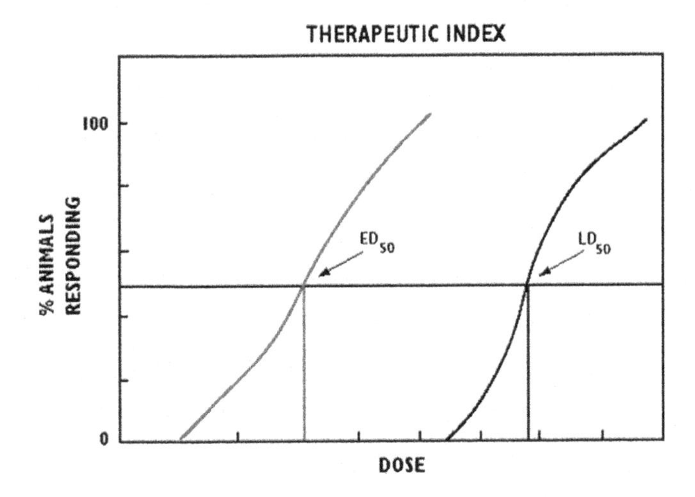

FIGURE 7.4 Graph of therapeutic index (TI). The TI is used to determine the difference between an effective dose and a toxic (or lethal) dose. The TI increases as the difference between the effective (therapeutic) dose and the toxic (or lethal) dose is amplified. The calculation allows an estimation of the potential usefulness of the agent as a therapeutic tool.

---

* The median is the middle value in a set of measurements. Like the mean, but not identical to it, it is a measure of central tendency.

estimation of the margin of safety of a chemical. This information may be valuable in establishing guidelines for estimating potential human risk. As with the determination of the LD_{50}, the TI does not provide information on the slopes of the curves (as with Figure 7.2), mechanism of toxicity, or pharmacological action.

7.4 IC_{50} (INHIBITORY CONCENTRATION 50%)

7.4.1 DEFINITION

The concentration necessary to inhibit 50% of a measured response in an *in vitro* system is known as the IC_{50} (similarly with IC_{75} and IC_{25}, which are sometimes computed as "cut-off" values). The number is calculated based on the slope and linearity of a typical concentration–effect curve (Figure 7.5). The symmetry of the curve is mathematically estimated from *regression analysis* of the plot, "percentage of control" versus "log of concentration." The values for "percentage of control" are derived by converting the absolute values of the measured response to a fraction of the control value—that is, responses measured in the group without chemical treatment. Using the control value as the 100% level, all subsequent groups are transformed into relative percentage values. This manipulation has the advantage of allowing comparison of results between different experiments and different laboratories, even when the absolute values are not identical.

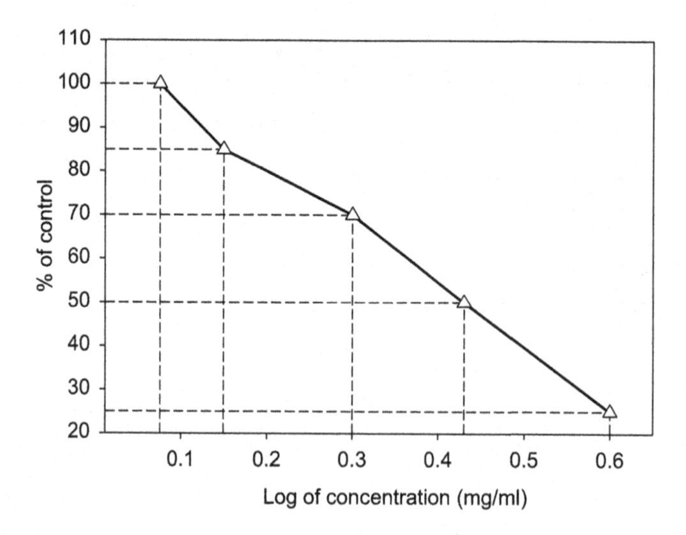

FIGURE 7.5 Graded dose–response (concentration–effect) curve for xylene on cultured human lung cells for 72 hours. The measured parameter (% of control) is cell proliferation. The diagram illustrates extrapolation of the IC_{50} (0.43 mg/ml) as well as the IC_{75} (0.6 mg/ml), IC_{30} (0.3 mg/ml), and IC_{15} (0.15 mg/ml). The latter represent "cut-off" values corresponding to 25%, 70%, and 85% of control, respectively. (From Yang et al., Toxicol. In Vitro, 16, 33, 2002.)

7.4.2 Experimental Determination

IC_{50} determinations are traditionally performed using cell culture methodology or other *in vitro* methods. Manipulation of cells in culture, whether with continuous or finite cell lines or primary cultures, requires attention to several steps employed in the screening assay. For example, a typical procedure begins with inoculation of cells in multiwell plates at a rate of 10^4 cells per square centimeter. Contact-inhibited cultures are grown to stationary phase in appropriate complete media. Monolayers are then incubated with increasing concentrations of sterile, soluble solutions of each chemical for a predetermined period of time. Incubation criteria for mammalian cells generally also dictate 37 °C in a gaseous, humidified atmosphere defined by the requirements of the medium. Prior to the end of the incubation period, indicators, dyes, fixatives, or reactive substrates are added as needed and allowed to incubate. The cultures are then processed, and the reaction product is quantified according to the protocol.

7.4.3 For *In Vitro* Systems

The determination of IC_{50} for *in vitro* systems is analogous to the calculation of the LD_{50} or TD_{50} in animal testing. Unlike the LD_{50} or TD_{50}, however, which represents a median lethal or toxic dose, the IC_{50} allows comparison of the *concentrations* of chemicals necessary to inhibit any measurable parameter, such as cell proliferation, protein synthesis, or DNA synthesis. Thus, it becomes a valuable tool for estimating toxic effects. Using the IC_{50} for estimating toxicity, international validation programs have now successfully proposed the use of *in vitro* methods to supplement or replace animal toxicity testing programs. In addition, *in vitro* methods are more cost-efficient than animal experimentation. It is conceivable, therefore, that animal toxicity testing in the foreseeable future will be augmented with the incorporation of data obtained from *in vitro* systems. Such testing may prove to be more predictive for assessing human clinical toxicity than current animal testing protocols.

REFERENCES

SUGGESTED READINGS

Balls, M., ECVAM and the promotion of international co-operation in the development, validation and acceptance of replacement alternative test methods. *ALTEX* 16, 280, 1999.

Calabro, A.R., Konsoula, R., Barile, F.A. Evaluation of in vitro cytotoxicity and paracellular permeability of intact monolayers with mouse embryonic stem cells. *Toxicology in Vitro* 22, 1273, 2008.

Clemedson, C., Barile, F.A., Chesne, C.G., Cottin, M., Curren, R., Ekwall, B., Ferro, M., Gomez-Lechón, M.J., Imai, K., Janus, J., Kemp, R.B., Kerszman, G., Kjellstrand, P., Lavrijsen, K., Logemann, P., McFarlane-Abdulla, E., Roguet, R., Segner, H., Thuvander, A., Walum, E., and Ekwall, B., MEIC evaluation of acute systemic toxicity. Part VII. Prediction of human toxicity by results from testing of the first 30 reference chemicals with 27 further in vitro assays. *Altern. Lab. Anim. (ATLA)* 28, 161, 2000.

Eaton, D.L. and Klassen, C.D., Principles of toxicology, in *Casarett and Doull's Toxicology: The Basic Science of Poisons*, Klaassen, C.D. (Ed.), 7th ed., McGraw-Hill, New York, 2006, Chapter 2.

Goldberg, A.M., Development of alternative methods. *J. Toxicol. Sci.* 20, 443, 1995.

Neervannan, S., Preclinical formulations for discovery and toxicology: physicochemical challenges. *Expert Opin. Drug Metab. Toxicol.* 2, 715, 2006.

Waters, M.D., Selkirk, J.K., and Olden, K., The impact of new technologies on human population studies. *Mutat. Res.* 544, 349, 2003.

Weltje, L., Integrating evolutionary genetics and ecotoxicology: on the correspondence between reaction norms and concentration-response curves. *Ecotoxicology* 12, 523, 2003.

Yang, A., Cardona, D.L., and Barile, F.A., *In vitro* cytotoxicity testing with fluorescence-based assays in cultured human lung and dermal cells. *Cell Biol. Toxicol.* 18, 97, 2002.

Yang, A., Cardona, D.L., and Barile, F.A., Subacute cytotoxicity testing with cultured human lung cells. *Toxicol. In Vitro* 16, 33, 2002.

REVIEW ARTICLES

Burden, N., Chapman, K., Sewell, F., and Robinson, V., Pioneering better science through the 3Rs: an introduction to the national centre for the replacement, refinement, and reduction of animals in research (NC3Rs). *J. Am. Assoc. Lab. Anim. Sci.* 54, 198, 2015.

Crettaz, P., Pennington, D., Rhomberg, L., Brand, K., and Jolliet, O., Assessing human health response in life cycle assessment using ED10s and DALYs: part 1—Cancer effects. *Risk Anal.* 22, 931, 2002.

Desprez, B. et al., A strategy for systemic toxicity assessment based on non-animal approaches: The Cosmetics Europe Long Range Science Strategy programme. *Toxicol. In Vitro* 50, 137, 2018.

Edler, L. and Kopp-Schneider, A., Statistical models for low dose exposure. *Mutat. Res.* 405, 227, 1998.

Flecknell, P., Replacement, reduction and refinement. *ALTEX* 19, 73, 2002.

Franco, N.H., Sandøe, P., and Olsson, I.A.S., Researchers' attitudes to the 3Rs—an upturned hierarchy? *PLoS One* 13, e0200895, 2018.

Gaylor, D.W. and Kodell, R.L., Dose-response trend tests for tumorigenesis adjusted for differences in survival and body weight across doses. *Toxicol. Sci.* 59, 219, 2001.

Green, S., Goldberg, A., and Zurlo, J., TestSmart—high production volume chemicals: an approach to implementing alternatives into regulatory toxicology. *Toxicol. Sci.* 63, 6, 2001.

Hakkinen, P.J. and Green, D.K. Alternatives to animal testing: information resources via the Internet and World Wide Web. *Toxicology* 173, 3, 2002.

Hanson, M.L. and Solomon, K.R., New technique for estimating thresholds of toxicity in ecological risk assessment. *Environ. Sci. Technol.*, 36, 3257, 2002.

Hilmer SN., ADME-tox issues for the elderly. *Expert Opin. Drug Metab. Toxicol.* 4, 1321, 2008.

Hothorn, L.A., Selected biostatistical aspects of the validation of in vitro toxicological assays. *Altern. Lab. Anim. (ATLA)* 30, 93, 2002.

Maracy, M. and Dunn, G., Estimating dose-response effects in psychological treatment trials: the role of instrumental variables. *Stat. Methods Med. Res.* 20, 191, 2008.

Meador, J.P., McCarty, L.S., Escher, B.I., and Adams, W.J., 10th anniversary critical review: the tissue-residue approach for toxicity assessment: concepts, issues, application, and recommendations. *J. Environ. Monit.* 10, 1486, 2008.

Pennington, D., Crettaz, P., Tauxe, A., Rhomberg, L., Brand, K., and Jolliet, O., Assessing human health response in life cycle assessment using ED10s and DALYs: part 2—non-cancer effects. *Risk Anal.* 22, 947, 2002.

Taylor, K.A. and Emerson, M. Refinement of a mouse cardiovascular model: development, application and dissemination. *F1000Res* 7, 593, 2018.

Worth, A.P. and Balls, M., The importance of the prediction model in the validation of alternative tests. *Altern. Lab. Anim. (ATLA)* 29, 135, 2001.

8 Descriptive Animal Toxicity Tests

8.1 CORRELATION WITH HUMAN EXPOSURE

8.1.1 Human Risk Assessment

The information derived from descriptive animal toxicology testing is useful in determining the potential toxicity of a compound to humans. The objective of these tests is to identify toxic chemicals at an early stage of chemical development, especially if the substance is already commercially available. Together with *in vitro* tests, animal tests are applied as biological markers of chemically induced risks, whether synthetic or naturally occurring. These validated batteries of tests are used by regulatory agencies to screen for or predict human toxic effects in an attempt to establish a significant frame of reference for monitoring environmental chemical threats, therapeutic adverse reactions, and commercial toxicity.

8.1.2 Predictive Toxicology and Extrapolation to Human Toxicity

The ability to predict toxicity in humans with a responsible level of safety is probably the most trying conclusion of animal testing. Correlation of the results of toxicity testing described in animals with human exposure requires careful consideration of the parameters of the animal tests, including the selection of species with similar physiology. The variety of tests available to reach valid conclusions with a particular chemical must be systematically chosen according to the guidelines of the regulatory agencies.

8.2 SPECIES DIFFERENTIATION

8.2.1 Selection of a Suitable Animal Species

Among the mammalian species available to toxicologists for testing chemicals, the rodent is the most useful and convenient. Notwithstanding the ostensible difference in size and a few minor anatomical distinctions, rodent physiology is identical to human. The rodent, therefore, is a suitable animal species that mimics the human. In addition, the species must be able to reflect diverse toxicological challenges, including behavioral, neurological, and cognitive reactions to chemical insult. Thus, rodents are the first species of consideration in general toxicity testing. Other animal species are selected based on the sensitivity of the animal to particular classes of chemicals or due to particular organ similarity. For instance, because of their dermal sensitivity, guinea pigs are used for the testing of local irritants. Rabbits have traditionally been employed for the testing of ocular irritants, as their eyes are anatomically and physiologically similar to those of humans. Larger species, such as dogs, are useful subjects of specific toxicity testing involving cardiovascular and pulmonary agents. The use of primates for testing the behavioral, immunological, microbiological, and neurological response to chemicals is controversial but is often necessary for interpretation of human risk assessment.

8.2.2 Cost-Effectiveness

One of the most important criteria for selecting an appropriate animal species is cost. Animal testing is more costly than *in vitro* methods. The budget for animal testing must account for

1. Procurement of animals, housing, and maintenance within the animal care facility;
2. Daily animal requirements for food, water, and bedding;
3. Training and employment of animal care staff;
4. Veterinarian on-call or on-staff; and
5. Adherence to proper procedures for removal and disposal of specimens and waste.

8.2.3 Institutional Animal Care and Use Committee (IACUC)

The U.S. Public Health Service (PHS) Policy on Humane Care and Use of Laboratory Animals was promulgated in 1986 to implement the Health Research Extension Act of 1985 (Public Law 99–158), "Animals in Research." The Office of Laboratory Animal Welfare (OLAW) at the National Institutes of Health (NIH), which has responsibility for the general administration and coordination of policy on behalf of the PHS, provides specific guidance, instruction, and materials to institutions on the use and care of vertebrate animals in testing, research, and training. This policy requires institutions to establish and maintain proper measures to ensure the appropriate care and use of all animals involved in research, research training, and biological testing activities conducted or supported by the PHS. In addition, any activity involving animals must provide a written assurance acceptable to the PHS setting forth compliance with this policy. Assurances are submitted to OLAW at the NIH and are evaluated by OLAW to determine the adequacy of the institution's proposed program for the care and use of animals in PHS-conducted or supported activities. In summary, the program assurance includes

1. A list of every branch and major component of the institution;
2. The lines of authority and responsibility for administering the program;
3. The qualifications, authority, and responsibility of the veterinarian(s) who will participate in the program;
4. The membership list of the IACUC committee established in accordance with the requirements;
5. The procedures that the IACUC will follow to fulfill the requirements;
6. The health program for personnel who work in laboratory animal facilities or have frequent contact with animals;

7. A synopsis of training or instruction in the humane practice of animal care and use. Training or instruction in research or testing methods that minimize the number of animals required to obtain valid results and minimize animal distress must be offered to scientists, animal technicians, and other personnel involved in animal care, treatment, or use;
8. The gross square footage of each animal facility, the species housed and the average daily inventory of animals in each facility, and any other pertinent information requested by OLAW.

Other functions of the IACUC include periodic review of the institution's program, inspection of facilities, preparation of reports, and submission of suggestions and recommendations on improving the program.

8.3 DESCRIPTIVE TESTS

8.3.1 Required LD_{50} and Two Routes

In accordance with Food and Drug Administration (FDA) regulations establishing current drug testing protocols using animal experiments, the LD_{50} is required for a substance using at least two routes of exposure, usually oral and parenteral routes. Depending on the nature of the substance, the route can be modified to include testing for inhalation, local, or other selective exposure. In addition, the animals are also monitored for gross appearance, behavior, morbidity, and time of onset of any signs and symptoms. The upper limit dose is 2 g/kg of body weight by any route.

8.3.2 Chronic and Subchronic Exposure

Besides acute 24 h exposures, several exposure periods may be incorporated into the protocols. Subchronic exposure usually involves a 90 day exposure period with or without repeated doses. Chronic exposure studies are more difficult, cumbersome, and expensive to conduct. These experiments are reserved for carcinogenicity testing or accumulation studies. The rat is the animal of choice in chronic experiments. Hematology, clinical chemistry, and histological analysis also become important parameters to monitor in chronic programs.

8.3.3 Types of Tests

Routine oral or parenteral 24 h studies in a rodent species are performed along with other types of tests depending on the chemical substance. Local toxicity testing for dermal toxicity or sensitization is conducted with chemicals destined for local use. Cosmetics, nasal or pulmonary inhalation products, toiletries, therapeutic or cosmetic creams, ointments, and lotions are tested for local hypersensitivity reactions. As noted earlier, the guinea pig is the species of choice for dermal testing because of skin sensitivity and anatomical resemblance to the human.

Other animal testing techniques are listed in Table 8.1 and include methods for the determination of ocular irritation, local irritation, teratogenicity, developmental

TABLE 8.1
Other Types of Descriptive Animal Techniques and Their Potential for Use in Toxicity Testing Protocols

Methods	Descriptive Protocol	Preferred Species
Dermal testing	Local irritancy	Guinea pig
Draize method	Ocular irritancy	Rabbit
Fertility and reproductive tests	Developmental toxicity	Rabbit and rodent litters
Immunotoxicity studies	Toxicity related to immune function	Rodent
Mammalian mutagenesis	Altered foci induction*; skin neoplasm, pulmonary neoplasm, and breast cancer induction	Rodent
Mammalian teratology, whole embryo culture	Teratogenicity	Rabbit, rodent
Perinatal and postnatal studies	Developmental toxicity	Rabbit, rodent
Toxicokinetic studies	Influence on A,D,M,E**	Rodent and larger mammals

* In rodent liver.
** A,D,M,E: absorption, distribution, metabolism, elimination.

toxicity, perinatal and postnatal toxicity, mutagenicity, immunotoxicity, and toxicokinetics. Among these tests, the determination of potential ocular toxicity from chemicals and solvents has been traditionally evaluated in rabbits using the Draize method. This controversial test measures eye irritancy and has been the standard method for six decades. Recently, its replacement by *in vitro* tests has been recommended.

REFERENCES

SUGGESTED READINGS

Baumans, V., Science-based assessment of animal welfare: laboratory animals. *Rev. Sci. Tech.* 24, 503, 2005.

Bayne, K.A., Environmental enrichment of nonhuman primates, dogs and rabbits used in toxicology studies. *Toxicol. Pathol.* 31, 132, 2003.

Collins, J.G., Postapproval monitoring and the IACUC. *ILAR J.* 49, 388, 2008.

Everitt, J.I. and Berridge, B.R., The role of the IACUC in the design and conduct of animal experiments that contribute to translational success. *ILAR J.* 58, 129, 2017.

Ford, K.A., Refinement, reduction, and replacement of animal toxicity tests by computational methods. *ILAR J.* 57, 226, 2016.

Garcia, K. and Purcell, J., Which protocol deviations require preapproval? IACUC should develop policies. *Lab. Anim. (N.Y.)* 37, 244, 2008.

Jacobson-Kram, D., Commentary: regulatory toxicology and the critical path: predicting long-term outcomes from short-term studies. *Vet. Pathol.* 45, 707, 2008.

Klein, H.J. and Bayne, K.A., Establishing a culture of care, conscience, and responsibility: addressing the improvement of scientific discovery and animal welfare through science-based performance standards. *ILAR J.* 48, 3, 2007.

Knight, A., Systematic reviews of animal experiments demonstrate poor contributions toward human healthcare. *Rev. Recent Clin. Trials* 3, 89, 2008.
Mazzatorta, P., Estevez, M.D., Coulet, M., and Schilter, B., Modeling oral rat chronic toxicity. *J. Chem. Inf. Model.* 48, 1949, 2008.
Zeiger, E., Reflections on a career and on the history of genetic toxicity testing in the National Toxicology Program. *Mutat. Res.* 773, 282, 2017.

REVIEW ARICLES

Barrow, P.C., Reproductive toxicology studies and immunotherapeutics. *Toxicology* 185, 205, 2003.
Bucher, J.R., The National Toxicology Program rodent bioassay: designs, interpretations, and scientific contributions. *Ann. N. Y. Acad. Sci.* 982, 198, 2002.
Gipson, C., Holt, M.A., and Brown, P., Deciding which animals to use. A word from USDA, FDA and OLAW. *Lab. Anim. (N.Y.)*. 37, 295, 2008.
Grimm, H., Olsson, I.A.S., and Sandøe, P., Harm-benefit analysis—what is the added value? A review of alternative strategies for weighing harms and benefits as part of the assessment of animal research. *Lab. Anim.* 1, 23677218783004, 2018.
Hampshire, V.A., Regulatory issues surrounding the use of companion animals in clinical investigations, trials, and studies. *ILAR J.* 44, 191, 2003.
Haywood, J.R. and Greene, M., Avoiding an overzealous approach: a perspective on regulatory burden. *ILAR J.* 49, 426, 2008.
James, M.L., IACUC training: from new-member orientation to continuing education. *Lab. Anim. (N.Y.)* 31, 26, 2002.
Jenkins, E.S., Broadhead, C., and Combes, R.D., The implications of microarray technology for animal use in scientific research. *Altern. Lab. Anim. (ATLA)* 30, 459, 2002.
Meyer, O., Testing and assessment strategies, including alternative and new approaches. *Toxicol. Lett.* 140, 21, 2003.
Petrini, A., and Wilson, D., Philosophy, policy and procedures of the World Organisation for Animal Health for the development of standards in animal welfare. *Rev. Sci. Tech.* 24, 665, 2005.
Pitts, M., A guide to the new ARENA/OLAW IACUC guidebook. *Lab. Anim. (N.Y.)* 31, 40, 2002.
Rogers, J.V. and McDougal, J.N., Improved method for *in vitro* assessment of dermal toxicity for volatile organic chemicals. *Toxicol. Lett.* 135, 125, 2002.
Silverman, J., Animal breeding and research protocols—the missing link. *Lab. Anim. (N.Y.)* 31, 19, 2002.
Spielmann, H., Validation and regulatory acceptance of new carcinogenicity tests. *Toxicol. Pathol.* 31, 54, 2003.
Steneck, N.H., Role of the institutional animal care and use committee in monitoring research. *Ethics Behav.* 7, 173, 1997.
Waters, M. and Jackson, M., Databases applicable to quantitative hazard/risk assessment—towards a predictive systems toxicology. *Toxicol. Appl. Pharmacol.* 233, 34, 2008.
Whitney, R.A., Animal care and use committees: history and current policies in the United States. *Lab. Anim. Sci.* 37, 18, 1987.
Zhu, H. Bouhifd, M, Donley, E., Egnash, L., Kleinstreuer, N., Kroese, E.D., Liu, Z., Luecht, T., Palmer, J., Pamies, D., Shen, J., Strauss, V., Wu, S., and Hartung, T., Supporting read-across using biological data. *ALTEX* 33, 167, 2016. doi:10.14573/altex.1601252.

9 *In Vitro* Alternatives to Animal Toxicity

9.1 *IN VITRO* METHODS

9.1.1 Cell Culture Methods

Techniques used to culture animal and human cells and tissues on plastic surfaces or in suspension have contributed significantly to the development of toxicology testing. Cell culture techniques are fundamental for understanding mechanistic toxicology and have been used with increasing frequency in all biomedical disciplines. In particular, the mechanisms underlying clinical toxicity have been elucidated in part with the development of *in vitro* methods. Additionally, the necessity to clarify the toxic effects of chemicals and drugs that are developed and marketed at rapid rates has provoked the need for fast, simple, valid, and alternative methods to animal toxicology testing.

The use of cell culture techniques in toxicological investigations is referred to as *in vitro cytotoxicology* or *in vitro toxicology*. The latter term also includes noncellular test systems, such as isolated organelles or high-throughput (HTP) microarrays. The reasons for the increasing interest in and success of these techniques in clinical and research toxicology testing are a consequence of the considerable advancements realized since the advent of the technology:

1. Since the growth of the first mammalian cells in capillary glass tubes was described, cell culture technology has been extensively refined;
2. The *mechanism* of toxicity of chemicals in humans and animals has been elucidated primarily through the use of *in vitro* toxicological methods;
3. The necessity to determine the toxic effects of pharmaceuticals and industrial chemicals, which are developed and marketed at rapid and unprecedented rates, has made the development of fast, simple, and effective alternative test systems compulsory;
4. Although the normal rate of progression of any scientific discipline is determined by the progress within the scientific community, some areas have received more encouragement than others.

Specifically, public objections to and disapproval of animal testing have forced academic institutions, industrial concerns, and regulatory agencies to direct research initiatives toward the development of alternative methods of toxicity testing.

9.1.2 Organ System Cytotoxicity

Homogeneous cell populations isolated from the framework of an organ present unique opportunities to study the toxic effects on a single cell type. Among the organs that have proved to be of particular interest as target organs of toxicity are the lung, liver, skin, and kidney and the hematopoietic, nervous, and immune systems. Consequently, two systems are currently used to bridge the gap between animal and alternative testing methodologies:

1. *Primary cultures* represent freshly isolated cells from target organs; these cultured cells retain most of the metabolic capacity of the intact organ from which the cells are derived and are used in tests that more closely mimic the *in vivo* state of the organ; these cultures are valuable in assessing target organ toxicity.
2. *Continuous cell line cultures* are established from continuous propagation of primary cultures; consequently, the cells successively lose some of their specialized functions and selective metabolic capacity, a process referred to as *dedifferentiation*; the less specialized cells cease to be in a primary state but are useful as *in vitro* alternative models for acute systemic toxicity. Continuous cell lines are either mortal or immortal.

Often, classes of chemicals are tested so that their toxicity can be compared within the same system. This permits the establishment of toxicity data to be used as a reference for previously unknown toxicity of similar classes of chemicals. Hepatocyte cultures are essential to this testing principle because of the important role that the liver enjoys in xenobiotic metabolism—that is, metabolism of exogenous substances. Other cell types that are employed for assessing target organ toxicity include lung macrophages and epithelial cells and cardiac, renal, neuronal, and muscle cells.

9.1.3 Applications to Clinical Toxicology

Although primary cultures offer sensitive and specific avenues for detecting target organ toxicity, the system has several disadvantages:

1. Primary cultures require sacrificing animals to establish the cells;
2. Primary cultures are limited to understanding the mechanisms of only one of the many possible organ pathologies that could be determined using animals;
3. Primary systems established from human donors are not standardized unless efforts are made to match the age and sex of the donors; and
4. The response of primary cultures to chemical insult may not reflect the complex nature of chemical–target interaction as inherent in animal species.

Nevertheless, regulatory agencies in Europe and the United States have promulgated the use of select *in vitro* tests to be incorporated into routine industrial toxicity testing programs. Many of the systems currently in validation programs are *in vitro* models for local acute toxicology testing (Barile, 2013). The main goal of these procedures is to replace, reduce, or refine the skin and ocular (Draize) test.

As noted earlier, established continuous and finite cell lines are also employed to analyze a wide spectrum of mechanisms and effects and are generally used to detect acute systemic toxicity. In contrast to animal studies, these systems measure the concentration of a substance that damages components, alters structures, or influences biochemical pathways within specific cells. The range of injury is further specified by the length of exposure (incubation time) and the parameter used to detect an insult (indicator). Careful validation of test methods and repetitive interlaboratory confirmation allows the tests to be predictive of risks associated with toxic effects *in vivo* for those concentrations of the effective substances tested. All established cell lines, however, are prone to dedifferentiate in culture with time, thus limiting their usefulness for assessing effects other than general toxicity.

9.1.4 Relationship to Animal Experiments

The clinical toxicity of chemicals is known to involve various physiological targets interacting with a variety of complex toxicokinetic factors. Chemicals induce injury through an assortment of toxic mechanisms. As a result, *in vitro* testing requires several relevant systems, such as human hepatocytes and heart, kidney, lung, and nerve cells, as well as other cell lines of vital importance, to obtain information necessary for human risk assessment. These cell systems are also used to measure basal cytotoxicity—that is, the toxicity of a chemical that affects basic cellular functions and structures common to all human specialized cells. The basal cytotoxicity of a chemical may reflect the measurement of cell membrane integrity, mitochondrial activity, or protein synthesis, since these are examples of basic metabolic functions common to all cells. The level of activity varies from one organ to another, depending on its metabolic rate and contribution to homeostasis, but the processes do not differ qualitatively among organs. Recent multicenter studies have now shown that human cell lines predict basal cytotoxicity to a greater extent than animal cell lines. These *in vitro* tests are as accurate in screening for chemicals with toxic potential as LD_{50} methods performed in rodents. In addition, validation studies have proposed that these tests be used to supplement animal testing protocols when screening for toxic substances.

9.2 CORRELATION WITH HUMAN EXPOSURE

9.2.1 Risk Assessment

Together with animal testing, genotoxicity tests, toxicokinetic modeling studies, and microarray technology, *in vitro* cellular test batteries could be applied as biological markers of chemically induced risks. Such risks include biomonitoring for environmental toxicants and biohazards and for the development of specific antagonists. The batteries of tests can be validated for their ability to screen for or predict human toxic effects, as the measurements can establish a database for monitoring environmental and clinical biohazards and general toxicity.

9.2.2 Extrapolation to Human Toxicity

Standardized batteries of *in vitro* tests have been proposed for assessing acute local and systemic toxicity. This battery of selected protocols may serve as a primary screening tool before the implementation of routine animal testing. The correlation of *in vitro* with *in vivo* data functions as a basis for predicting human toxicity. As it becomes apparent that several batteries of tests reflect the acute toxicity of defined classifications of substances, the numbers of animals used in toxicity testing will be considerably reduced.

9.2.3 Predictive Toxicology

Many relatively common and widely available chemicals have never been tested in any system. *In vitro* systems offer the possibility for use as screening methods for carcinogenic, mutagenic, or cumulative toxicity. It is possible to reduce or replace the toxicity testing of *existing* chemicals with risk assessment based on the results from these batteries. In combination with clinical studies of dermal and ocular irritancy, and cumulative regressive and progressive longitudinal studies, *in vitro* test databases can contribute to reliable, clinically relevant human toxicity information.

REFERENCES

SUGGESTED READINGS

Barile, F.A., *Introduction to In Vitro Cytotoxicology: Mechanisms and Methods*, CRC Press, Florida, 1994.

Barile, F.A., *Principles of Toxicology Testing*, 2nd ed., CRC Press, New York, 2013.

Calabro, A.R., Konsoula, R., and Barile, F.A., Evaluation of *in vitro* cytotoxicity and paracellular permeability of intact monolayers with mouse embryonic stem cells. *Toxicol. In Vitro* 22, 1273, 2008.

Davidge, K.S. and Wilkinson, J.M., Training and good science are the foundation stones for animal replacement. *Altern. Lab Anim.* 45, 281, 2017.

Grindon, C., Combes, R., Cronin, M.T., Roberts, D.W., and Garrod, J.F., An integrated decision-tree testing strategy for repeat dose toxicity with respect to the requirements of the EU REACH legislation. *Altern. Lab. Anim. (ATLA)* 36, 139, 2008.

Gura, T., Toxicity testing moves from the legislature to the petri dish—and back. *Cell* 134, 557, 2008.

Janer, G., Verhoef, A., Gilsing, H.D., and Piersma, A.H., Use of the rat postimplantation embryo culture to assess the embryotoxic potency within a chemical category and to identify toxic metabolites. *Toxicol. In Vitro* 22, 1797, 2008.

Nowak, J.A. and Fuchs, E., Isolation and culture of epithelial stem cells. *Methods Mol. Biol.* 482, 215, 2009.

Salem, H. and Katz, S.A., *Alternative Toxicological Methods*, CRC Press, Florida, 2003.

REVIEW ARTICLES

Blaauboer, B.J., The integration of data on physicochemical properties, *in vitro*-derived toxicity data and physiologically based kinetic and dynamic as a modeling tool in hazard and risk assessment. A commentary. *Toxicol. Lett.* 138, 161, 2003.

Carere, A., Stammati, A., and Zucco, F., *In vitro* toxicology methods: impact on regulation from technical and scientific advancements. *Toxicol. Lett.* 127, 153, 2002.

Combes, R.D., The ECVAM workshops: a critical assessment of their impact on the development, validation and acceptance of alternative methods. *Altern. Lab. Anim. (ATLA)* 30, 151, 2002.

Costa, D.L., Alternative test methods in inhalation toxicology: challenges and opportunities. *Exp. Toxicol. Pathol.* 60, 105, 2008.

Donnelly, T.M., Free Web resources on alternatives. *Lab. Anim. (NY)* 33, 46, 2004.

Foth, H. and Hayes, A., Concept of REACH and impact on evaluation of chemicals. *Hum. Exp. Toxicol.* 27, 5, 2008.

Gerde, P., How do we compare dose to cells *in vitro* with dose to live animals and humans? Some experiences with inhaled substances. *Exp. Toxicol. Pathol.* 60, 181, 2008.

Hakkinen, P.J. and Green, D.K., Alternatives to animal testing: information resources via the Internet and World Wide Web. *Toxicology*, 173, 3, 2002.

Indans, I., The use and interpretation of *in vitro* data in regulatory toxicology: cosmetics, toiletries and household products. *Toxicol. Lett.* 127, 177, 2002.

Kojima, H., Trend on alternative to animal testing. *Int. J. Cosmet. Sci.* 29, 331, 2007.

Liebsch, M. and Spielmann, H., Currently available *in vitro* methods used in the regulatory toxicology. *Toxicol. Lett.* 127, 127, 2002.

Pazos, P., Fortaner, S., and Prieto, P., Long-term *in vitro* toxicity models: comparisons between a flow-cell bioreactor, a static-cell bioreactor and static cell cultures. *Altern. Lab. Anim. (ATLA)* 30, 515, 2002.

Putnam, K.P., Bombick, D.W., and Doolittle, D.J., Evaluation of eight *in vitro* assays for assessing the cytotoxicity of cigarette smoke condensate. *Toxicol. In Vitro* 16, 599, 2002.

Scholz, S.Sela, E., Blaha, L., Braunbeck, T., Galay-Burgos, M., García-Franco, M., Guinea, J., Klüver, N., Schirmer, K., Tanneberger, K., Tobor-Kapłon, M., Witters, H., Belanger, S., Benfenati, E., Creton, S., Cronin, M.T., Eggen, R.I., Embry, M., Ekman, D., Gourmelon, A., Halder, M., Hardy, B., Hartung, T., Hubesch, B., Jungmann, D., Lampi, M.A., Lee, L., Léonard, M., Küster, E., Lillicrap, A., Luckenbach, T., Murk, A.J., Navas, J.M., Peijnenburg, W., Repetto, G., Salinas, E., Schüürmann, G., Spielmann, H., Tollefsen, K.E., Walter-Rohde, S., Whale, G., Wheeler, J.R., and Winter, M.J., A European perspective on alternatives to animal testing for environmental hazard identification and risk assessment. *Regul. Toxicol. Pharmacol.* 67, 506, 2013.

Sewell, F., Doe, J., Gellatly, N., Ragan, I., and Burden, N., Steps towards the international regulatory acceptance of non-animal methodology in safety assessment. *Regul. Toxicol. Pharmacol.* 89, 50, 2017.

Snodin, D.J., An EU perspective on the use of *in vitro* methods in regulatory pharmaceutical toxicology. *Toxicol. Lett.* 127, 161, 2002.

Stacey, G. and MacDonald, C., Immortalisation of primary cells. *Cell Biol. Toxicol.* 17, 231, 2001.

Stokes, W.S., Kulpa-Eddy, J., Brown, K., Srinivas, G., and McFarland, R., Recent progress and future directions for reduction, refinement, and replacement of animal use in veterinary vaccine potency and safety testing: a report from the 2010 NICEATM-ICCVAM International Vaccine Workshop. *Dev Biol (Basel)* 134, 9, 2012.

Ukelis, U., Kramer, P.J., Olejniczak, K., and Mueller, S.O., Replacement of *in vivo* acute oral toxicity studies by *in vitro* cytotoxicity methods: opportunities, limits and regulatory status. *Regul. Toxicol. Pharmacol.* 51, 108, 2008.

Vedani, A., Dobler, M., Spreafico, M., Peristera, O., and Smiesko, M., VirtualToxLab—*in silico* prediction of the toxic potential of drugs and environmental chemicals: evaluation status and internet access protocol. *ALTEX* 24, 153, 2007.

10 Toxicokinetics

10.1 TOXICOKINETICS

10.1.1 Relationship to Pharmacokinetics

The study of drug disposition in physiological compartments is traditionally referred to as *pharmacokinetics*. Toxicokinetics includes the more appropriate study of exogenous compounds (xenobiotics), the adverse effects associated with their concentrations, their passage through the physiological compartments, and the chemicals' ultimate fate. Compartmental toxicokinetics involves administration or exposure to a chemical and rapid equilibration to the central compartment (plasma and tissues), followed by distribution to peripheral compartments. Consequently, the principles of pharmacokinetics and toxicokinetics historically have been used interchangeably.

10.1.2 One-Compartment Model

The one-compartment model is the simplest representation of a system that describes the *first-order* extravascular absorption of a xenobiotic. The absorption rate is constant (k_a) and allows entry into a central compartment (plasma and tissues). Constant *first-order elimination* (k_{el}) follows. The reaction is illustrated in Reaction 1. Few compounds follow one-compartment modeling—agents such as creatinine equilibrate rapidly, distribute uniformly between blood, plasma, and tissues, and are eliminated as quickly.

Reaction 1: Chemical $\xrightarrow{k_a}$ **One-Compartment** $\xrightarrow{k_{el}}$ **Excreted**

Drug or chemical elimination processes, in general, follow either zero-order or first-order kinetics. *First-order elimination*, or first-order kinetics, describes the rate of elimination of a drug in proportion to the drug concentration—that is, an increase in the concentration of the chemical results in an increase in the rate of elimination. This rate usually occurs at *low* drug concentrations. Alternatively, ***zero-order elimination*** refers to the removal of a fixed amount of drug independently of the concentration—that is, increasing the plasma concentration of the chemical does not result in a corresponding increase in elimination. Chemical reactions that trigger zero-order kinetics usually occur when biotransforming enzymes are saturated. Clinically, this situation is the most common cause of drug accumulation and toxicity. *Michaelis–Menten kinetics* occurs when a compound is metabolized according to both processes. For instance, at low concentrations, ethanol is metabolized by first-order kinetics (dose-dependently). As the blood alcohol concentration (BAC) approaches alcohol dehydrogenase saturation levels, ethanol kinetics switches to zero-order (dose-independent) elimination.

10.1.3 Two-Compartment Model

In a two-compartment model, the xenobiotic chemical requires a longer time for the concentration to equilibrate with tissues and plasma. Essentially, molecules enter the central compartment at a constant rate (k_a) and may also be eliminated through a constant *first-order elimination* (k_{el}) process. However, a parallel reaction (reaction 2) occurs, whereby the chemical enters and exits peripheral compartments through first-order rate constants for distribution (k_{d1} and k_{d2}, respectively). The distribution phases are variable but rapid relative to elimination. Most drugs and xenobiotics undergo two- or multi-compartment modeling behavior.

$$\text{Reaction 2: Chemical} \xrightarrow{k_a} \text{First-Compartment} \xrightarrow{k_{el}} \text{Excreted}$$

$$\text{First-Compartment} \underset{k_{d2}}{\overset{k_{d1}}{\rightleftharpoons}} \text{Second Compartment}$$

10.1.4 Application to Clinical Toxicology

Clinical applications of kinetics, and the phenomena of absorption, distribution, metabolism, and elimination (ADME), are useful in the monitoring of pharmacological and toxicological activity of drugs and chemicals. Assigning a compound to compartmental systems, particularly a two-compartment or the more complex multi-compartment model, allows prediction of adverse effects of the agents based on distribution and equilibration within the body. The ADME factors that influence the fate of a chemical are discussed in the following sections.

10.2 ABSORPTION

10.2.1 Ionic and Nonionic Principles

How does the ionic or nonionic species of a chemical relate to absorption? Figure 10.1 illustrates the fluid mosaic model of the cell membrane. The membrane is characterized as a flexible, semi-permeable barrier that maintains a homeostatic environment for normal cellular functions by preventing chemicals, ions, and water from traversing through easily.

The membrane consists of phospholipid polar heads, glycolipids, and integral proteins organized toward the periphery of the membrane, and the nonpolar tails and cholesterol directed inward. The rest of the membrane is interspersed with transmembrane channel proteins and other intrinsic proteins. The fluidity of the membrane is due to the presence of cholesterol and integral proteins. The phospholipid polar groups and nonpolar tails are derived from triglycerides. Figure 10.2 illustrates the structure of a triglyceride, detailing the positions of the polar environment contributed by the carboxyl moieties relative to the nonpolar saturated carbon chains.

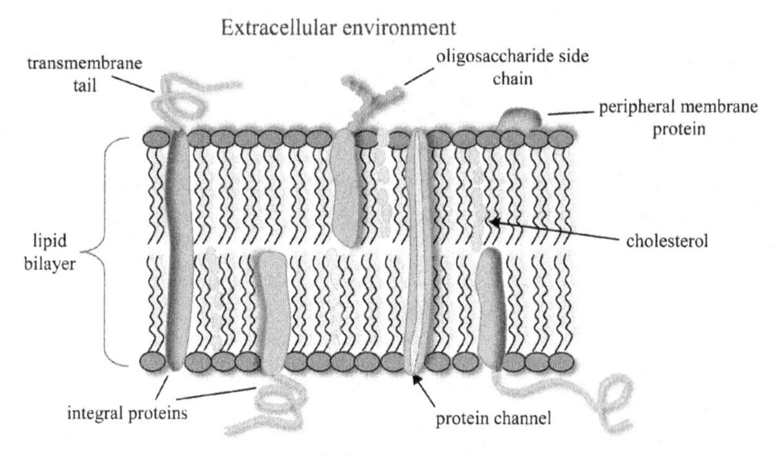

FIGURE 10.1 Fluid mosaic model of the cell membrane.

Because of the selectively permeable arrangement of the cell membrane, the ability of a chemical to bind to or penetrate intact membranes is determined by its polar or nonpolar characteristics in solution. In turn, the relative contribution of a chemical to its polarity is influenced by the acidic or basic nature of the chemical in solution, which is determined by the proportion of the chemical's ability to dissociate or associate. According to the Brønsted–Lowry theory of acids and bases, an acidic compound dissociates—that is, it donates a proton. Similarly, a basic molecule associates—that is, it accepts a proton. The dissociation constant for an acid can be summarized according to the following formula:

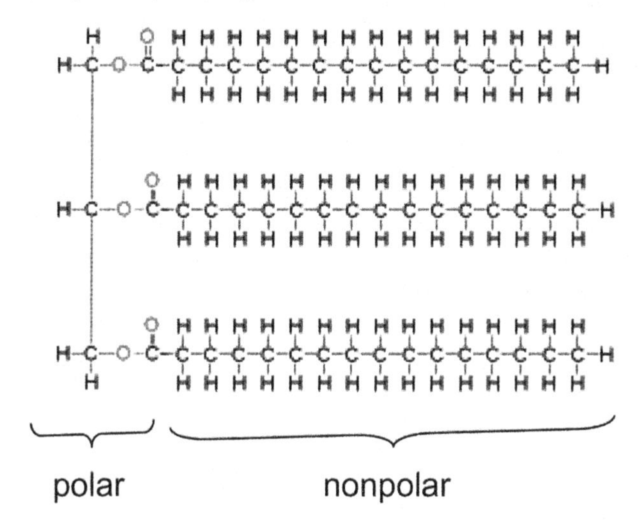

FIGURE 10.2 Structure of triglyceride showing the positions of polar and nonpolar contributions.

$$K_a = \frac{[H^+][A^-]}{[HA]}$$

where

K_a is the dissociation constant for an acid

$[H^+]$, $[A^-]$, and $[HA]$ represent the hydrogen ion, conjugate acid, and undissociated acid concentrations, respectively

The pK_a, derived from the negative logarithm of the acid dissociation constant, is calculated according to the following formula:

$$pK_a = -\log K_a$$

The pK_a of a drug, therefore, is defined as the pH at which 50% of the compound is ionized. Similarly, pK_b is the negative logarithm of the base dissociation constant (K_b). In general, acids and bases are classified according to their ability to dissociate relative to the pH of the solution. Independently of the pH, however, the lower the pK_a of an acid, the stronger the acid. Similarly, the lower the pK_a (or the higher the pK_b) of a base, the stronger the base. Most therapeutic drugs are weak acids, weak bases, or neutral or amphoteric compounds.

The associated and dissociated structures of a base and an acid are illustrated in Figure 10.3 with aniline and benzoic acid, which possess strong basic and acidic characteristics, respectively. The structure of benzoic acid shows the carboxyl anion after dissociation. Since its $pK_a = 4$, the $K_a = 1 \times 10^{-4}$, resulting in a higher dissociation constant and strong acidic properties. Similarly, aniline is a highly protonated species, with a pK_b of 10 and corresponding K_b of 1×10^{-10} ($pK_a + pK_b = 14$; thus, although it is a base, the pK_a of aniline can also be expressed as 4). The low dissociation constant of the base indicates that the H^+ ions are closely held to the nitrogen and contribute to its strongly basic nature.

10.2.2 Henderson–Hasselbalch Equation and Degree of Ionization

The relationship between the ionization of a weakly acidic drug, its pK_a, and the pH of the solution is given by the Henderson–Hasselbalch equation. The equation allows the prediction of nonionic and ionic states of a compound at a given pH. For acids and bases, the formulas are

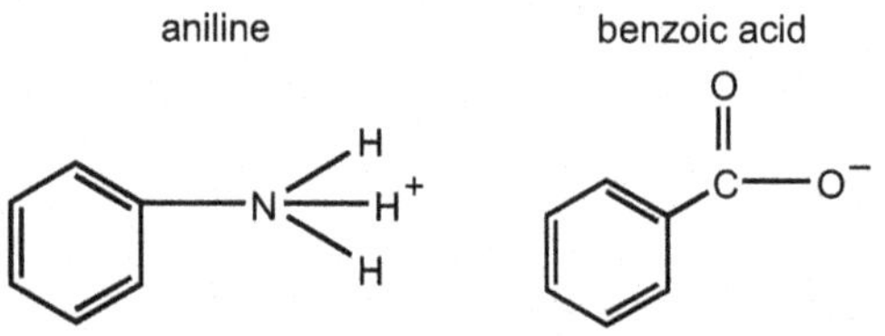

FIGURE 10.3 Structures of aniline and benzoic acid.

$$\text{acid}: \text{pH} = \text{pK}_a + \log([A^-]/[HA])$$

$$\text{base}: \text{pH} = \text{pK}_a + \log([HA]/[A^-])$$

The equations are derived from the logarithmic expression of the dissociation constant formula in the previous subsection. Small changes in pH near the pK_a of a weakly acidic or basic drug markedly affect its degree of ionization. This is more clearly shown with rearrangement of the Henderson–Hasselbalch equations such that

$$\text{for an acid}: \text{pK}_a - \text{pH} = \log([HA]/[A^-])$$
$$\text{for a base}: \text{pK}_a - \text{pH} = \log([A^-]/[HA])$$

For an acidic compound present in the stomach (e.g., $pK_a = 4$ in the stomach, where the average pH ~ 2), inserting the numbers into the Henderson–Hasselbalch equation results in a relative ratio of *nonionized to ionized* species of 100:1. This transforms the compound predominantly to the nonionic form within the acidic environment of the stomach, rendering it more lipophilic. Lipophilic compounds have a greater tendency for absorption within that compartment.*

In the proximal small intestine areas of the duodenum and jejunum, where the pH is approximately 8, the same compound will be predominantly in the ionized state. The relative ratio of nonionized to ionized species is reversed ($1{:}10^4$). Thus, within the weakly basic environment of the proximal intestine, a strongly acidic drug is less lipophilic, more ionized, and slower to be absorbed.

Conversely, for a strong basic compound with a $pK_a = 4$ ($pK_b = 10$) in the stomach, the Henderson–Hasselbalch equation predicts that the ratio of *ionized to nonionized* species equals 100:1. Thus, a basic compound is more ionized, less lipophilic, and slower to be absorbed in the stomach. In the basic environment of the proximal intestine (pH = 8), however, the ionized to nonionized species ratio is $1{:}10^4$, rendering it more lipophilic and imparting a greater propensity for absorption.

The knowledge of the pK_a of a therapeutic drug is useful in predicting its absorption in these compartments. In addition, as will be explained, this information is helpful in determining distribution and elimination functions. Tables 10.1 and 10.2 summarize the chemical properties and behavior of strong acidic and basic drugs, such as aspirin and amphetamine hydrochloride, respectively. Based on the extremely acidic environment of an empty stomach, such compounds are either completely nonionized and highly lipophilic or ionized and hydrophilic. In the basic environment of the small intestine, although some ionization is present, amphetamine absorption is favored over that of aspirin. Conversely, the behavior of weakly basic or acidic drugs, such as morphine sulfate and sodium phenobarbital, respectively, is predicted as determined by the pH of the environment (Tables 10.3 and 10.4).

It should be noted that some of the pharmacokinetic principles that govern drug absorption are not always significant factors in the determination of toxic effects. This is primarily due to the circumstances surrounding toxic chemical exposure.

* It should be noted that physiologically, the overall absorption rate of the stomach is limited and is secondary to its digestion and churning functions.

TABLE 10.1
Structure, Chemical Properties, and Behavior in the Stomach Environment (pH = 2) of Aspirin and Amphetamine Hydrochloride

Compound Properties	ASA	Amphetamine HCl
Structure	COOH, O—C(=O)—CH_3	H, CH_3; —C—C—$NH_3^+Cl^-$; H, H
pKa	~3	10
Nonionic:ionic ratio	10:1	$1:10^8$
Acidic/basic nature	Free acid, some lipophilic properties	Protonated, hydrophilic, very low lipophilicity
Absorption	Favorable	Not favorable

ASA: acetylsalicylic acid (aspirin); HCl: hydrochloride salt.

Consequently, some of the circumstances that influence drug absorption in a therapeutic setting will not be considered here. These factors include formulation and physical characteristics of the drug product; drug interactions; presence of food in the intestinal tract; gastric emptying time; and concurrent presence of gastrointestinal diseases. Individual categories of toxic substances, and special circumstances that alter their effects, will be considered in their respective chapters.

TABLE 10.2
Structure, Chemical Properties, and Behavior in Proximal Small Intestine (pH = 8–10) of Aspirin and Amphetamine Hydrochloride

Compound Properties	ASA	Amphetamine HCl
Structure	COO-, O—C(=O)—CH_3	H, CH_3; —C—C—$NH_3^+Cl^-$; H, H
pKa	~3	10
Nonionic:ionic ratio	$10^{-5}:1$ to $10^{-7}:1$	$10^{-2}:1$ to 1:1
Acidic/basic nature	Proton donor, very low lipophilicity	Good balance between ionized and lipophilic species
Absorption	Not favorable	Favorable

ASA: acetylsalicylic acid (aspirin); HCl: hydrochloride salt.

TABLE 10.3
Structure, Chemical Properties, and Behavior in Stomach Environment (pH = 2) of Morphine Sulfate and Sodium Phenobarbital

Compound Properties	Morphine SO_4 Pentahydrate	Na Phenobarbital
Structure	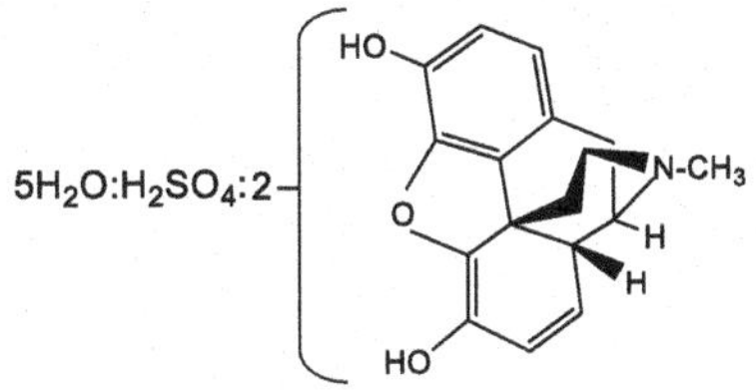 $5H_2O{:}H_2SO_4{:}2$–[morphine structure: HO, O, HO, N-CH$_3$, H, H]	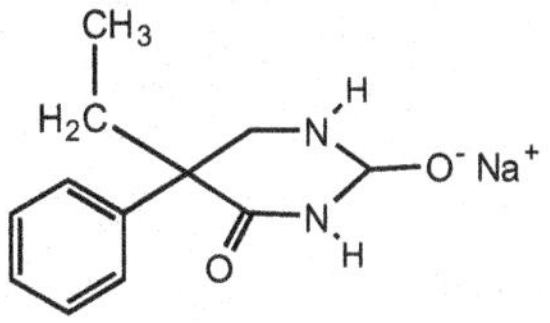 [phenobarbital structure: CH$_3$, H$_2$C, N, H, O$^-$ Na$^+$, N, H, O]
pKa	8.2–9.9	7.3
Nonionic:ionic ratio	$1{:}10^7$	$10^5{:}1$
Acidic/basic nature	Amphoteric nature (weak base), highly ionized, low lipophilicity	Weak acid, mostly nonionic, lipophilic tendencies
Absorption	Not favorable	Favorable

H_2SO_4: sulfate salt; Na: sodium salt.

TABLE 10.4
Structure, Chemical Properties, and Behavior in Proximal Small Intestine (pH = 8–10) of Morphine Sulfate and Sodium Phenobarbital

Compound Properties	Morphine SO_4 Pentahydrate	Na Phenobarbital
Structure	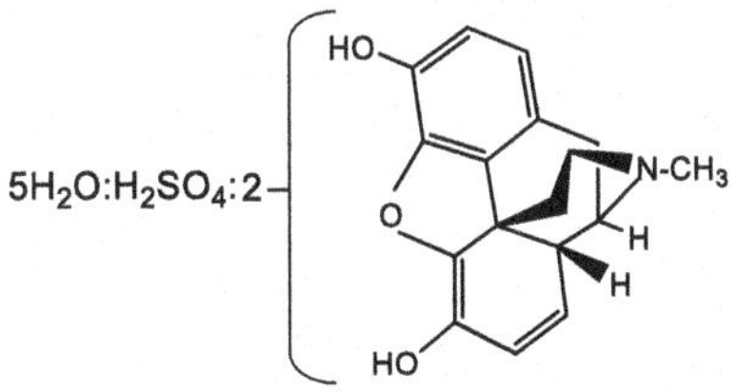 $5H_2O{:}H_2SO_4{:}2$–[morphine structure: HO, O, HO, N-CH$_3$, H, H]	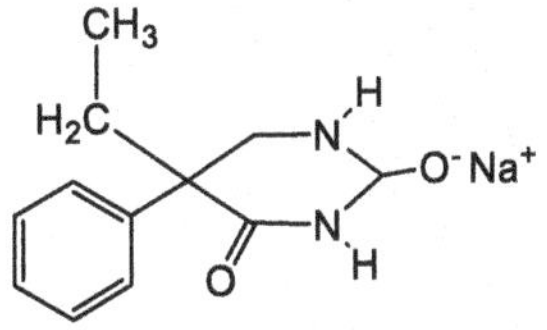 [phenobarbital structure: CH$_3$, H$_2$C, N, H, O$^-$ Na$^+$, N, H, O]
pKa	8.2–9.9 (mean = 9.0)	7.3
Nonionic:ionic ratio	$10{:}1$ to $10^{-1}{:}1$	$10^{-1}{:}1$ to $10^{-3}{:}1$
Acidic/basic nature	Amphoteric nature (acts as very weak acid), mostly protonated, lipophilic	Weak acid, good ionization, low lipophilicity
Absorption	Favorable	Some absorption over length of intestinal tract

H_2SO_4: sulfate salt; Na: sodium salt.

10.2.3 Route of Administration and Solubility

As noted earlier, the degree of ionization and the ultimate physiological fate of ionic and nonionic substances are eloquently described by the principles of Henderson–Hasselbalch. In addition, the relationship between the route of administration and exposure to drugs or toxins is extensively explained in Section 5.2. Thus, the influence of the route through which a chemical gains access to target systems is an initial step in predicting toxic effects. The bridge connecting the physiological route of administration and the degree of ionization is constructed from the model of solubility.

A solution is a chemically and physically homogeneous mixture of two or more substances in a liquid. Similarly, suspensions are pharmaceutical preparations of finely divided, insoluble solids dispersed in a liquid vehicle for oral, inhalation, or parenteral administration. The extent to which a solute dissolves in a liquid is referred to as *solubility*. Solubility is also described as the concentration of a compound in a solution that is in equilibrium with a solid phase at a specified temperature. For any defined solid chemical or drug, the degree of solubility is determined by the innate chemical characteristics of the solid and the temperature. Thus, solubility is constant for a solid in a particular solvent at a given temperature. In addition, different categories of solutions or suspensions are possible depending on the size of the dispersed particle.* Consequently, it is conceivable to formulate homogeneous solutions of solids, liquids, or gases (solute) in liquids (solvent). The concept can be further extended into the preparation of gases in gases and solids in solids. More detailed discussion of the chemical properties of true solutions, solute–solvent interaction, and dispersion systems can be obtained from the Suggested Readings, some of which present the reader with quantitative data on the behavior and properties of solutions). This section is concerned with basic definitions of solutes and solvents, the influence of temperature on solution stability, and the eventual movement of a dissolved substance toward absorption.

Pharmaceutically, formulations of solids in liquids are clinically most significant. The principles and the chemical foundation of these formulations define the field of pharmaceutical chemistry. The solubilities of solids are expressed in a variety of formats, usually in water at room temperature (25 °C). Table 10.5 provides the descriptive terms for solubility primarily as information for pharmaceutical workers who manufacture, prepare, or dispense drugs.† Solubility is a physical attribute of pharmaceuticals and has traditionally served as one of a series of standards for purity and as a critical determinant of physiological availability. The latter criterion is of value for toxicological purposes, particularly in determination of the amount of drug absorbed.

As implied throughout this section, the solubility of a solid in a liquid depends on the temperature of the mixture. In the process of solution, an equilibrium is

* The degree to which the solid disperses in the liquid, and the uniformity of the dispersion throughout the solvent, can be visualized, thus allowing classification of different categories of solutions—e.g., true solutions, colloidal solutions, and heterogeneous dispersions.

† More precisely, solubility is calculated and expressed as mass percent of solute (w_2) according to the following formula: $w_2 = m_2/(m_1 + m_2)$, where m_2 is the mass of water and m_2 is the mass of solute.

TABLE 10.5
Descriptive Terms for Solubility

Descriptive Term	Parts of Solvent for 1 Part of Solute
Very soluble	Less than 1
Freely soluble	From 1 to 10
Soluble	From 10 to 30
Sparingly soluble	From 30 to 100
Slightly soluble	From 100 to 1,000
Very slightly soluble	From 1,000 to10,000
Practically insoluble, or insoluble	More than 10,000

From *Remington: The Science and Practice of Pharmacy*, Chapter 16, 21st edition, University of the Sciences in Philadelphia, 2005.

established such that the *heat of solution* (ΔH) is a constant for a specific temperature.* In general, if heat is absorbed with increasing temperature, ΔH increases, and the solubility of the solute improves. Such is the case with most salts. Alternatively, if a solute in liquid releases heat during the process of solution (e.g., calcium hydroxide), ΔH decreases, and the solubility of the solute declines with increasing temperature. With respect to clinical toxicology and pharmacology, pharmaceutical dosage forms take advantage of the former, more common phenomenon. This guides the observation that at the higher physiologic temperature of 37 °C, most solid dosage forms readily enter solution, especially when they encounter the volume and capacity of the stomach fluids.

Thus, drug solubility is influenced by physical and chemical forces interacting with the physiological compartments. In practice, these same forces influence the pharmacological and toxicological effects of the solute—that is, the greater the solubility of a drug or chemical in the liquid compartments, the higher the proportion of absorption for that compartment and the greater the availability at the target site.

10.2.4 Absorption in Nasal and Respiratory Mucosa

The vestibule of the nasal cavity and nasopharynx is covered by olfactory epithelium. This mucous membrane lining consists of an extensive network of capillaries and pseudostratified columnar epithelium interspersed with goblet cells. Inspired air enters the nasal conchae and is warmed by circulating blood in the capillaries, while mucus secreted by the goblet cells moistens the air and traps airborne particles. Thus, several factors facilitate rapid dissolution and absorption of drugs administered through the nasal route: 1. the vast surface area provided by the capillary

* An equilibrium (solid ↔ liquid ↔ solution) is established from the overall process of formulation of solutions and is based on the Law of Mass Action for interactions between solids and liquids. For a more detailed description of heat of solution and temperature dependency, see the Suggested Readings.

circulation and epithelial cells in the mucous membranes; 2. the ability to trap dissolved and particulate substances; and 3. the phospholipid secretions of the nasal mucosa (resulting in a relatively neutral pH). Nonionized drugs are more rapidly absorbed in this compartment relative to their ionized counterparts.

The upper and lower respiratory tracts extend from the nasal cavity to the lungs. The *conducting portion*, responsible for ventilation, consists of the pharynx, larynx, trachea, bronchi, bronchioles, and terminal bronchioles. The *respiratory portion*, the main site for gas exchange, consists of the respiratory bronchioles, alveolar ducts, alveolar sacs, and alveoli. Most of the upper and lower tracts are covered by ciliated secretory epithelial cells. Aside from the nasal epithelium, however, significant absorption of airborne substances does not occur above the level of the respiratory bronchioles. As respiratory bronchioles penetrate deeply into the lungs, the epithelial lining changes from simple cuboidal to simple squamous, enhancing the surface absorptive capability of the mucosa. In addition, the thin 0.5 μm respiratory membrane is composed of 300 million alveoli and further supported by an extensive network of capillary endothelium. This anatomy allows rapid diffusion of chemicals delivered as gases, within aerosols, or carried by humidified air.

10.2.5 Transport of Molecules

The absorption of chemicals requires basic mechanisms of transporting the molecules across fluid barriers and cell membranes. These transport vehicles include *diffusion, active transport, facilitated diffusion, and filtration.*

The simplest mechanism for the transportation of molecules involves *diffusion*, defined as the transport of molecules across a semi-permeable membrane. The most common pathway of diffusion is *passive transport.* In passive transport, molecules are transported from an area of high to low solute concentration—that is, down the concentration gradient. It is not an energy-dependent process, and no electrical gradient is generated. Lipophilic molecules, small ions, and electrolytes generally gain access through membrane compartments by passive diffusion, since they are not repelled by the phospholipid bilayer of the cell membrane. Most passive diffusion processes, however, are not molecularly selective.

In contrast to passive transport, polar substances, such as amino acids, nucleic acids, and carbohydrates are moved by an *active transport* process. The mechanism shuttles ionic substances from areas of low solute concentration to high solute concentration—that is, against a concentration gradient. Active transport is an energy-dependent process and requires macromolecular carriers. The latter two requirements render this process susceptible to metabolic inhibition and consequently, a site for pharmacological or toxicological influence. In addition, active transport processes are saturable systems, implying that they have a limited number of carriers, making them competitive and selective.

Facilitated diffusion resembles active transport with the exception that substrate movement is not compulsory against a concentration gradient, and it is not energy dependent. For example, macromolecular carriers augment the transfer of glucose across gastrointestinal epithelium through to the capillary circulation.

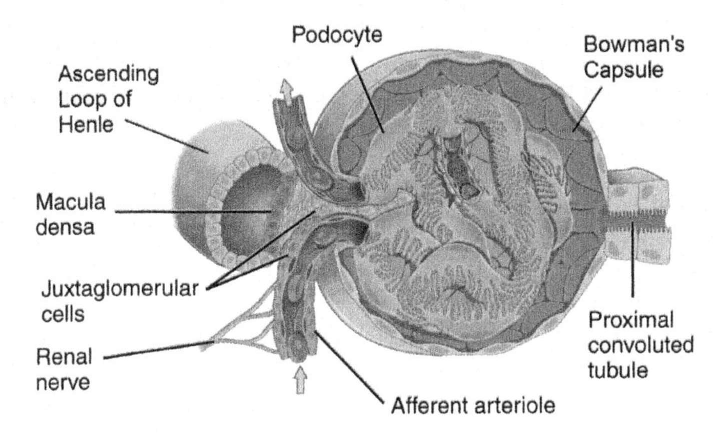

FIGURE 10.4 Diagram of glomerular apparatus of the nephron, illustrating the podocytes and interdigitating foot processes that comprise the filtering unit. (From: http://cnx.org/content/col11496/1.6/. Creative Commons Attribution 3.0 Un-ported license.)

Filtration allows the movement of solutes across epithelial membrane barriers along with water movement. The motive force here is hydrostatic pressure. In general, the molecular weight cutoff for solutes transported along with water is about 100–200 mw units. Under normal circumstances, several membranes throughout the body, with pores or fenestrations averaging 4 Å in diameter, are designed to allow entry for larger molecules. These special membranes permit coarse filtration of lipophilic or hydrophilic substances. Such filtration processes are described for the nephron and illustrated in Figure 10.4. The filtration slits, formed by the interdigitating podocytes in the glomerular apparatus and Bowman's capsule, prevent the passage of cells and solutes above 45 kD. Under normal circumstances, most smaller molecules pass through the filter only to be reabsorbed in the proximal and distal convoluted tubules and collecting tubules.

Many organs exhibit passive filtration as a fundamental process for the passage of water. The intestinal absorption of water involves a transport method whereby water and dissolved nutrients are coarsely filtered and absorbed through the intestinal lumen. At the tissue level, the inotropic action of the cardiac system is the driving force for capillary perfusion. For instance, the thin endothelial/epithelial membranes forming the walls of the pulmonary alveoli represent the only barrier to the diffusion of gases into the alveolar spaces.

10.3 DISTRIBUTION

10.3.1 Fluid Compartments

Anatomically, the body is not a homogeneous entity. Since approximately 70% of total body weight is composed of water, the physiological water compartments play significant roles in drug distribution and ultimate fate. Table 10.6 illustrates the composition and contribution of water-based compartments relative to body weight (the shaded areas in Table 10.6 represent the sum of the two boxes immediately above them). As

TABLE 10.6
Components and Relative Percentages of Water-Based Compartments of Total Body Weight

Compartment	Description	Relative % of Body Weight
Plasma water (PL)	Noncellular water component of blood	10%
Interstitial water (IS)	Fluids relating to the interstices (spaces) of tissue	15%
Extracellular water (EC)	PL + IS	= 25%
Intracellular water (IC)	Confined to cytosol (within cells)	45%
Total body water (TBW)	EC + IC	= 70%
Solids	Tissue binding capacity	30%
Total body weight	Solids + TBW	= 100%

Note: the shaded areas represent the sum of the two rows immediately above them.

such, the ability of a compound to gain access to these compartments is determined by a number of factors that influence the compound's subsequent fate. The distribution of the substance to the target organ (receptor site), as well as to the sites of metabolism and excretion, is reasonably predicted by such factors. These include 1. the *physicochemical properties* of the substance, such as lipid solubility and ionic/nonionic principles; 2. *binding to macromolecules*; and 3. the *physiological forces* of blood flow and capillary perfusion. The cooperation of these properties determines the relative bioavailability of a compound. These concepts are discussed in the following sections.

10.3.2 Ionic and Nonionic Principles

The same mechanisms that govern the absorption of molecules across semi-permeable membranes also determine the bioavailability of a compound. Thus, a compound with a propensity for a nonionic existence is more likely to be lipophilic. Consequently, it has greater ability to penetrate phospholipid membranes and more access for distribution to liver, kidney, muscle, adipose, and other tissues. Conversely, molecules that are highly ionic in plasma have diminished access to organs and tissues. As with absorption, pH and pK_a have similar influence over the ionization, or lack of it, at the tissue level. As an indicator of distribution to organs, the apparent volume of distribution (V_d) is used as a measure of the amount of chemical in the entire body (in milligrams per kilograms) relative to that in the plasma (milligrams per liter) over a specified period of time. Thus, V_d (L/kg) is useful in estimating the extent of distribution of a substance in the body. Table 10.7 shows the V_d for 50 chemicals whose toxicity is well established. Examination of these numbers indicates that a compound with a large V_d (e.g., chloroquine phosphate, 200 L/kg) is highly lipophilic and sequesters throughout tissue compartments. It necessarily demonstrates a low plasma concentration relative to dose. Alternatively, a compound with a low V_d (warfarin, 0.15 L/kg) is highly bound to circulating plasma proteins and thus, has a plasma concentration seven times that of the mean total body concentration.

TABLE 10.7
Human V_d of 50 Chemicals Tested in the Multicenter Evaluation for *In Vitro* Cytotoxicity (MEIC) Program

MEIC Chemical	V_d (µg/ml)	MEIC Chemical	V_d (µg/ml)
Paracetamol	1.0	Acetylsalicylate	0.18
Ferrous sulfate	1.0	Diazepam	1.65
Amitriptyline	8.0	Digoxin	6.25
Ethylene glycol	0.65	Methanol	0.6
Ethanol	0.53	Isopropanol	0.6
1,1,1-TCE	1.0	Phenol	1.0
NaCl	1.0	Na fluoride	0.6
Malathion	1.0	2,4-DCP	0.1
Xylene	1.0	Nicotine	1.0
K cyanide	1.0	Lithium SO_4	0.9
Theophylline	0.5	*d*-Propoxyphene HCl	19.0
Propranolol HCl	4.0	Phenobarbital	0.55
Paraquat	1.0	Arsenic O_3	0.2
Copper SO_4	1.0	Mercuric Cl_2	1.0
Thioridazine HCl	18.0	Thallium SO_4	1.0
Warfarin	0.15	Lindane	1.0
Chloroform	2.6	Carbon tetrachloride	1.0
Isoniazid	0.6	Dichloromethane	1.0
Barium nitrate	1.0	Hexachlorophene	1.0
Pentachlorophenol	0.35	Verapamil HCl	4.5
Chloroquine PO_4	200	Orphenadrine HCl	6.1
Quinidine SO_4	2.4	Fenytoin	0.65
Chloramphenicol	0.57	Na oxalate	1.0
Amphetamine SO_4	4.4	Caffeine	0.4
Atropine SO_4	2.3	K chloride	1.0

From Barile, F.A. et al., *Cell Biol. Toxicol.*, 10, 155, 1994. With permission.

Another criterion for estimating bioavailability is the calculation of the area under the plasma concentration versus time curve (AUC; shaded area in Figure 10.5). Dividing AUC after an oral dose by AUC after an intravenous dose and multiplying by 100 yields the percentage bioavailability. This value is used as an indicator for the fraction of chemical absorbed from the intestinal tract that reaches the systemic circulation after a specified period of time. As will be explained later, other factors, such as first-pass metabolism, rapid protein uptake, or rapid clearance, may alter the calculation. The ability to distribute to target tissues is then determined by the V_d.

10.3.3 Plasma Protein Binding

Polar amino acids projecting from the surface of circulating proteins induce nonspecific, reversible binding of many structurally diverse small molecules. Circulating

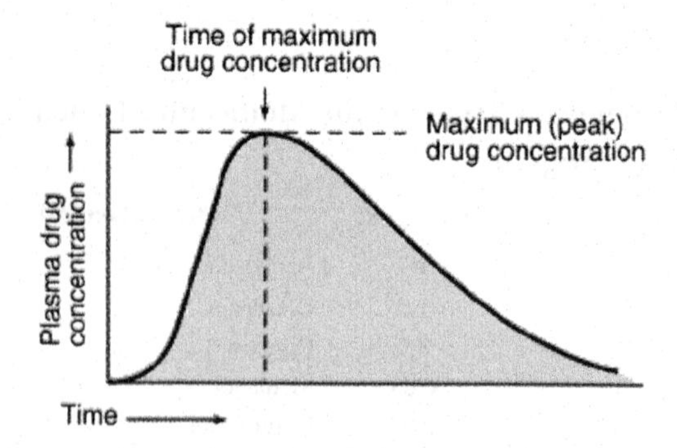

FIGURE 10.5 Plasma concentration versus time curve for drug X. The shaded region represents the area under the curve (AUC).

proteins transport chemicals through the circulation to target receptor sites. Such proteins include serum albumin; hemoglobin (binds iron); α1 globulins (such as transcortin; binds steroid hormones and thyroxin); α2 globulins (ceruloplasmin; binds various ions); β1 globulins (transferrin; binds iron); lipoproteins; and hemoglobin. The factors that decide the extent of protein binding are dictated by nonionic and ionic forces of the radicals. Most binding is reversible and displaceable, primarily because it involves noncovalent, hydrogen, ionic, or van der Waals forces. The binding renders the molecule nonabsorbable and unable to attach to the target receptor site.

Clinically, serum albumin can cause significant alteration of drug toxicokinetics and pharmacokinetics by virtue of its high affinity for acidic drugs, such as warfarin and penicillin. The binding of a substance to a circulating transport protein forms a drug–protein complex, such that its activity, metabolism, and rate of elimination are determined by the strength of this bond and diffusion into various compartments. The rate at which a drug–protein complex dissociates, therefore, determines its toxicological, pharmacological, and metabolic fate.

The plasma concentration of a chemical is driven to equilibrium by the rate of cardiac output and assuming steady-state conditions, reaches a plateau concentration. In addition, consider that only unbound drug can diffuse into tissues for access to receptor sites or for biotransformation, since a drug–protein complex cannot pass through semi-permeable cell membranes. Then, as the unbound transformed molecule is excreted, the dissociation of the drug–protein complex increases, shifting the equilibrium toward unbound chemical. This shift continues until the free drug concentration is equal to that in plasma and tissues (i.e., steady state). Thus, this complex acts as a depot, releasing enough active ingredient as determined by its chemical properties.

Chemicals that are extensively bound to circulating or tissue proteins pose a significant toxicological risk. The toxic effects of such chemicals extend well beyond acute exposures. Chronic, subclinical exposure to lead, for instance, has significant long-term neurological consequences by virtue of its affinity for depositing in bone. Therapeutically, steady-state concentrations of warfarin are easily disrupted when another acidic drug, such as acetylsalicylic acid (aspirin), displaces it from circulating albumin binding sites. Since warfarin is extensively protein bound (97%), a displacement of as little as 3% can double the amount of warfarin available for interaction with target organs, resulting in significant adverse hemodynamic events. Consequently, displacement of protein-bound chemicals results in increased unbound drug concentration in plasma and increased pharmacological and toxicological effects, followed by enhanced metabolism.*

10.3.4 Lipids

The distribution of chemicals is also determined, in part, by their lipid/water partition coefficient. Two chemicals with similar pK_a values, such as amphetamine (9.8) and chlorpromazine (9.3), have different lipid affinities in their nonionized forms. At basic pH (8–9), both chemicals are completely nonionized. However, at pH 10 or above, the partition coefficient of chlorpromazine in lipid medium ($K_d = 10^{6.3}$) is $10^{3.5}$ times greater than that of amphetamine ($K_d = 10^{1.8}$), allowing greater absorption of the former in the contents of the intestinal tract. Similarly, the polychlorinated biphenyls (PCBs) and the contact insecticide DDT have high lipid solubilities, dissolve in neutral fats, and thus, can readily accumulate in adipose tissue.

10.3.5 Liver and Kidney

The high capacity of the liver and kidney for binding and concentrating toxicants, coupled with their extensive circulation and metabolic activity, render these organs susceptible to chemical attack. The contribution of the metabolic roles played by the liver and kidney to chemical detoxification are further discussed in section 10.4.

10.3.6 Blood–Brain Barrier

Formed by the tight junctions of capillary endothelial cells and surrounded by basement membrane and an additional layer of neuroglial cells, the blood–brain barrier (BBB) poses a formidable hindrance to the passage of molecules. Although often misconstrued as an anatomical structure, it is primarily a physiologically functional entity, selectively allowing the passage of a few water-soluble molecules, ions, and glucose (by active transport). It is sufficiently impermeable to most compounds, including cells, proteins, and chemicals. Lipid-soluble chemicals, such as oxygen, carbon dioxide, alcohol, anesthetics, and the opioid analgesics, enjoy relatively easy access to the central nervous system via the intact BBB. When the BBB loses its barrier function, as in bacterial or viral encephalitis, the compromised membrane is then susceptible to unregulated entry.

10.3.7 Placenta

The placenta is fully developed by the third month of pregnancy and is formed by the embryonic chorion (syncytiotrophoblast) and the maternal endometrium

* It is important to note that the availability of the chemical at the receptor site is influenced more by the rate of diffusion (*Vd*) than by the rate of dissociation. This is especially true when the displacement of an acidic drug, with a small apparent *Vd*, displays clinically important consequences.

(decidua basalis). Although gases, vitamins, amino acids, small sugars, and ions, as well as waste products, diffuse across or are stored in the placental membrane, viruses and lipid-soluble chemicals can access the fetus. Thus, the same principles of ionization that govern the absorption of chemicals influence their distribution and passage through the placenta. Nonionized, lipid-soluble substances gain access more easily than ionized, water-soluble materials. In addition, passage through to the fetal membranes does not depend on the number of placental layers formed as the fetus matures. The biotransformation properties of the placenta are addressed in Section 10.4.

10.3.8 Other Factors Affecting Distribution

As noted earlier, the distribution of a drug or chemical is influenced by a variety of dynamic features. Regardless of concentration, when a drug meets the two minimum conditions for distribution—namely, relatively minor size of the molecule and high water solubility—then the main requirements for favorable absorption and distribution are satisfied. If molecular size and solubility were the only important criteria, however, then most drugs would traverse biomembranes without the prerequisite of carrier proteins or active transport. In addition, the presence of charges on drug molecules affects ionization, association, dissociation, and eventually, unequal distribution and passage through membranes. Consequently, other factors that shape the distribution of chemicals into physiologic compartments are listed and briefly described in Table 10.8. These include the pH partition principle, the electrochemical and Donnan distribution effect, binding and storage, and nonequilibrium and redistribution. These principles are not functionally independent of other pressures but contribute to determining distribution kinetics. Further discussion of these factors is found in the Suggested Readings.

10.4 BIOTRANSFORMATION (METABOLISM)

10.4.1 Principles of Detoxification

Biotransformation refers to the alteration of a chemical (xenobiotic) in biological systems.* Xenobiotic transformation plays a crucial role in maintaining homeostasis during chemical exposure. This is accomplished by converting lipid-soluble (nonpolar), nonexcretable xenobiotics to polar, water-soluble compounds accessible for elimination in the bile and urine. Thus, the major desirable outcomes of biotransformation include facilitation of renal excretion; conversion of toxic parent compounds to nontoxic metabolites (i.e., detoxification); conversion of nontoxic parent compounds to toxic metabolites (i.e., bioactivation); and conversion of nonreactive compounds to reactive metabolites (pharmacological bioactivation). The catalysts for xenobiotic transformation are incorporated into *Phase I* and *Phase II* enzyme reactions. Although occurring primarily in the liver, other organs, such as kidney,

* Exogenous chemicals are often referred to as *xenobiotics*, foreign substances accessing biological systems.

TABLE 10.8
Other Factors That Affect Distribution

Factor	Description	Consequence
pH partition principle	Difference in pH between two compartments influences nonionic diffusion of weakly acidic or basic drugs	Concentration of nonionized form *only* is equally distributed between two compartments, whereas concentration of ionized form depends on pH differential
Electrochemical potential	Positively charged outer membrane and negatively charged inner cell membrane affect migration of an ionized substance	Higher concentration of cations, for example, once inside the cell, remain inside and thus are moving along the *electrochemical gradient or potential*
Donnan distribution (equilibrium)	Presence of an impermeable charged macromolecule on one side of a cell membrane will alter permeability of other ionized particles	Molecules with the same charge as the impermeable macromolecule are committed to the opposite side of the membrane—due to mutual ionic repulsion or *Donnan equilibrium*
Binding and storage	Drugs bound to intracellular or circulating macromolecules cannot negotiate partitioning among physiologic compartments	Partitioning is determined by binding capacity and binding constant (K) as long as the concentration of the drug is not increased beyond K
Nonequilibrium and redistribution	Physiologically, true equilibrium among compartments does not exist; equilibrium between plasma and other compartments (e.g., fat, liver, muscle) does not occur at equal rates	This *nonequilibrium* exists because of *redistribution* to the rate-limiting compartment (e.g., fat) and is affected by biotransformation, elimination, and the practicality of dosing intervals

lung, and dermal tissue, have large capacities for these reactions. Table 10.9 outlines the major differences between Phase I and Phase II reactions, which will be discussed in detail in the following subsections.

10.4.2 Phase I Reactions

Phase I reactions involve the hydrolysis, reduction, and oxidation of chemicals to more hydrophilic, usually smaller, entities. The most important enzymes used in Phase I hydrolysis reactions include carboxylesterase, azo and nitro reductases, and quinone reductases. Carboxylesterases cleave carboxylic acid esters, amides, or thioesters. Azo and nitro reduction reactions incorporate two enzymes: azo (–N=N–) reductases, from intestinal flora, and liver cytochrome P-450 reductases. Quinone reduction reactions involve quinone reductases and NADPH cytochrome P-450 reductases. The former are involved in *two-electron* reduction reactions (detoxification) and the latter in *one-electron* reduction reactions (bioactivation).

Oxidative enzymes include alcohol dehydrogenases and cytochrome P-450 oxidases. For example, alcohol dehydrogenase (ADH) and aldehyde dehydrogenases convert alcohols to the corresponding aldehydes and acids, respectively.

TABLE 10.9
Major Features of Phase I and Phase II Biotransformation Reactions

Features	Phase I Reactions	Phase II Reactions
Chemical modification	Oxidation, dealkylation, deamination, N-oxidation, reduction, hydrolysis, dehalogenation	Conjugation reactions: glucuronidation, sulfation, acetylation, with glutathione
Enzymatically mediated	Primarily involves hepatic microsomal and mitochondrial enzymes; cytochrome-P450 enzyme family	Specific transfer enzymes; microsomal, cytosolic enzymes
Chemical modification of metabolite	Increased polarity, hydrophilic tendencies	Further increase in polarity and preparation for renal elimination
Pharmacologic or toxicologic activity of metabolite	Variable (increase or decrease)	Mostly inactivation; some activation or conversion to more toxic species
Primary biotransformation organ	Liver; also kidney and skin	Liver

Among Phase I biotransformation enzymes, the cytochrome P-450 (designated *CYP*) superfamily of enzymes is the most versatile and ubiquitous, accounting for 60% of drug and chemical biotransformation. The enzyme system is predominantly composed of mono-oxygenase, heme-containing proteins, associated mostly with the hepatic microsomal (liver homogenate fraction) of the smooth muscle endoplasmic reticulum, as well as with mitochondria. The enzymes are classified into families, sub-families, or subtypes of sub-families according to their amino acid sequences. They catalyze the incorporation of one atom of oxygen, or facilitate the removal of one atom of hydrogen, in the substrate. Other types of enzymes involved in the P-450 system are hydroxylases, dehydrogenases, esterases, N-dealkylation, and oxygen transferases.

10.4.3 Phase II Reactions

In general, Phase II reactions usually follow Phase I with glucuronidation, sulfation, acetylation, or methylation of the Phase I metabolites or conjugation of the molecules with amino acids. Phase II enzymes act on hydroxyl, sulfhydryl, amino, and carboxyl functional groups present on either the parent compound or the Phase I metabolite. Therefore, if a functional group is already present on the parent compound, Phase I reactions are circumvented.*

Phase II biotransformation enzymes include UDP-glucuronosyl transferase, sulfotransferases, acetyltransferases, methyltransferases, and glutathione S-transferases.

* For instance, because of the presence of hydroxyl groups, morphine undergoes direct Phase II conversion to morphine-3-glucuronide, whereas codeine must undergo transformation with both Phase I and II reactions.

Except for acetylation and methylation reactions, Phase II reactions, along with enzyme cofactors, interact with different functional groups to increase hydrophilic properties. Glucuronidation is the main pathway for Phase II reactions. It converts nonpolar, lipid-soluble compounds to polar, water-soluble metabolites. Sulfation reactions involve the transfer of sulfate groups from PAPS* to the xenobiotic, yielding sulfuric acid esters that are eliminated into the urine.

Glutathione (GSH) conjugation is responsible for the detoxification (or bioactivation) of xenobiotics. GSH conjugates are excreted intact in bile or converted to mercapturic acid in the kidney and excreted in the urine.

Table 10.10 lists some important drugs and chemicals involved primarily with Phase I and/or Phase II biotransformation reactions. Table 10.11 lists examples of molecular changes that occur with Phase I reactions. Lastly, Table 10.12 outlines some frequently cited drugs and chemicals that are predominantly metabolized via the cytochrome P-450 enzyme system.

TABLE 10.10
Predominant Biotransformation Pathways and Their Drug or Chemical Substrates

Biotransformation Pathway	Reactions	Drug or Chemical Substrates
Phase I Reactions	Oxidation	Amitriptyline; erythromycin; dextromethorphan; indomethacin; barbiturates; cyclosporine; chlorpheniramine; APAP; amphetamines
	Dealkylation, deamination, N-oxidation	Diazepam; tamoxifen; theophylline
	Reduction	Chloramphenicol
	Hydrolysis	ACE inhibitors; local anesthetics (e.g., cocaine, lidocaine, procaine)
	Dehalogenation	Halogenated general anesthetics; aromatic hydrocarbon gases (e.g., lacrimating agents)
Phase II Reactions	Glucuronidation	Sulfonamide antibiotics; APAP; morphine; THC; chloramphenicol
	Sulfation	Estrogens; methyldopa; APAP
	Acetylation	Sulfonamide antibiotics; procainamide; sulfonamides; isoniazid; clonazepam
	With glutathione	APAP; ROS/RNS

ACE: angiotensin-converting enzyme; APAP: acetyl para-amino phenol (acetaminophen); ROS: reactive oxygen species; RNS: reactive nitrogen species; THC: tetrahydrocannabinol.

* 3′-phosphoadenosine-5′-phosphosulfate.

TABLE 10.11
Examples of Chemical Changes That Occur with Phase I Reactions

Type of Reaction	Process	Substrate	Metabolite(s)
Oxidation	Aromatic hydroxylation	R—C_6H_5	R—C_6H_4—OH
	Aliphatic hydroxylation	R—CH_3	R—CH_2—OH
	Epoxidation	R—CH=CH—R'	R—CH_2—O—CH_2—R'
	N-hydroxylation	C_6H_5—NH_2	C_6H_5—NH—OH
	O-dealkylation	R—O—CH_3	R—OH + HC(H)=O
	N-dealkylation	R—NH—CH_3	R—NH_2 + HC(H)=O
	S-dealkylation	R—S—CH_3	R—SH + HC(H)=O
	Deamination	R—CH(NH_2)—CH_3	R—C(CH_3)=O + NH_3
	S-oxidation	R—S—R'	R—S(=O)—R'
	Dechlorination	CCl_4	$CHCl_3$
	Amine oxidation	R—CH_2—NH_2	R—CH=O + NH_3
	Dehydrogenation	R—CH_2—OH	R—C(H)=O + H^+
Reduction	Azo-	R—N=N—R'	R—NH_2 + 'R—NH_2
	Nitro-	R—NO_2	R—NH_2
	Carbonyl-	R—C(=O)—R'	R—CH(OH)—R'
Hydrolysis	Ester	R—C(=O)—O—R'	R—C(OH)=O + 'R—OH
	Amide	R—C(=O)—NH_2	R—C(OH)=O + NH_3

R: aromatic, alkyl, allyl, or other long-chain hydrocarbon.

10.5 ELIMINATION

10.5.1 URINARY

Urinary elimination (excretion)* represents the most important route of drug, chemical, and waste elimination in the body. Normal glomerular filtration rate (GFR, ~110–125 ml/min) is determined by the endothelial fenestrations (pores) formed by the visceral layer of the specialized endothelial cells of the glomerulus. The pores are coupled with the overlying glomerular basement membrane and the filtration slits of the pedicels. This "three-layer" membrane permits the passage of noncellular material with a molecular weight cutoff of ~45–60 K. The filtrate is then collected in Bowman's capsule. Reabsorption follows in the proximal convoluted tubules (PCT), loop of Henle, and distal convoluted tubules (DCT). The collecting ducts have limited reabsorption and secretion capabilities. Control of the transfer of molecules back to the capillary circulation or elimination in the final filtrate depends predominantly on reabsorption and secretion processes.

As with absorption through biological membranes and distribution in physiological compartments, the urinary elimination of toxicants is guided by ionic and nonionic principles. While Na^+ and glucose are actively transported or facilitated by symporters across the proximal tubules, other, larger polar compounds are prevented from reabsorption back to the capillary circulation. They are eventually eliminated in the urine. Nonionic, nonpolar compounds are transported by virtue of their ability to move across phospholipid membranes. Clinically, these principles are somewhat susceptible to manipulation for reducing chemical toxicity. For instance, it has long been proposed that overdose of barbiturates (pK_a between 7.2 and 7.5) could be effectively treated with sodium bicarbonate. The mechanism of this therapeutic intervention relies, in theory, on the alkalinization of the urine, rendering the acidic drug ionic, water soluble, and more amenable to elimination. In practice, however, the buffering capacity of the plasma, coupled with a large dose of the depressant, makes this a treatment modality of last resort.†

Other organic ions undergo active tubular secretion. Organic anion transporter proteins (*oat*, organic acid transporters) and organic cation transporter proteins (*oct*, organic base transporters) both actively secrete polar molecules from capillaries to the PCT lumen. Some classic examples include the active tubular secretion of para-aminohippuric acid (PAH) by *oat* proteins, as well as sulfonamide antibiotics and methotrexate, which compete for the same secretory pathways. Probenecid, a uricosuric agent used in the treatment of gout, competes for *oat* proteins as part of its therapeutic mechanism. While small doses of the drug actually depress the excretion of uric acid by interfering with tubular secretion but not reabsorption, larger doses

---

* *Elimination* refers to the general process of removal or clearance of a substance from a compartment, whereas *excretion* is a physiologic function referring to the discharge of a substance externally. The terms are often used interchangeably.

† Interestingly, abusers of cocaine or narcotic analgesics, both of which are weak bases, have taken advantage of the principles of elimination. To prolong intoxication with the drugs, the addictive behavior dictates ingesting large amounts of antacid preparations, thus alkalinizing the urine and increasing their reabsorption potential.

TABLE 10.12
Other Important Drug and Chemical Substrates Predominantly Metabolized through the Major Cytochrome P-450 Family Subdivisions of the Superfamily Enzyme System

CYP[a] Family and Subfamily	Drug or Chemical Substrates
CYP1A2	APAP, caffeine, theophylline, warfarin
CYP2C9	Diclofenac, ibuprofen, naproxen, phenytoin, tolbutamide, warfarin
CYP2C19	Diazepam, mephenytoin, omeprazole
CYP2D6	Codeine, dextromethorphan, fluoxetine, haloperidol, loratadine, metoprolol, paroxetine, risperidone, thioridazine, venlafaxine
CYP2E1	Ethanol, isoniazid
CYP3A4	Alprazolam, carbamazepine, cyclosporine, diltiazem, erythromycin, felodipine, fluconazole, lidocaine, lovastatin, nifedipine, quinidine, tacrolimus, terfenadine, verapamil

[a] CYP is the general designation of the cytochrome-P-450 superfamily enzyme system, followed by the number (family) and letter (subfamily) subdivisions.

APAP: acetyl para-amino phenol (acetaminophen); ROS: reactive oxygen species; RNS: reactive nitrogen species; THC: tetrahydrocannabinol.

depress the reabsorption of uric acid and lead to increased excretion and a fall in serum levels. Similarly, small doses of probenecid decrease the renal excretion of penicillins, which compete for the same *oat* proteins, necessitating an adjustment of the antibiotic dosage regimen.

10.5.1.1 First-Order Elimination

As noted previously with the general principles of toxicokinetics (Section 10.1), the elimination of chemical metabolites follows either first-order or zero-order elimination or both. *First-order* describes the rate of elimination of a drug in proportion to the drug concentration—that is, an increase in the concentration of the chemical results in an increase in the rate of removal. This rate usually occurs at low drug concentrations, or concentrations at or below the threshold for optimum enzyme metabolism, and accounts for the majority of toxicokinetic excretion. In addition, the half-life ($t_{½}$) of the chemical agent varies as a function of the dose administered. Figure 10.6 illustrates the graph of a chemical undergoing first-order elimination. Assuming that the effects of absorption, distribution, and metabolism are minimal on exposure, as with the administration of a rapid intravenous bolus dose, the concentration at time 0 is 100%. At this point, a constant fraction or percentage of the concentration of the substance in the plasma fraction (Cp) decreases per unit of time (e.g., hours). The elimination rate constant (k) is derived over a specified period of time as long as the initial plasma concentration remains below the metabolic threshold. Thus, over the time period, the rate of elimination is represented according to the following formula:

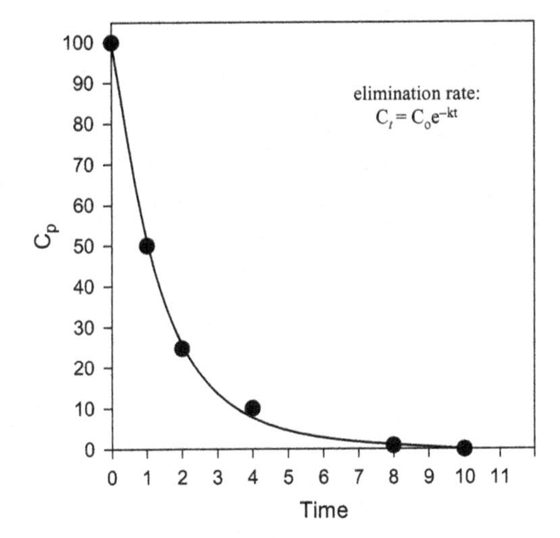

FIGURE 10.6 Graph of first-order elimination rate for a chemical or drug per unit time (see text for description and abbreviations).

$$C_t = C_0 e^{-kt}$$

where

- C_t is the plasma concentration of the drug at time t
- C_0 is the plasma concentration of the drug at time 0
- e^{-k} is the rate constant of elimination

Extrapolating the logarithmic plot of the first-order linear graph of elimination for the same chemical results in a straight line graph (Figure 10.7). The initial plasma concentration ($\log C_0$) and the plasma concentration at any given time point ($\log C_t$) are calculated according to the following formula:

$$\operatorname{Log} C_t = \log C_0 - (k/2.303)t$$

where

- $-k/2.303$ is the slope
- $\log C_0$ is the y-intercept

The estimated concentration of drug at any given time point or time range is based on the concentration at time 0 (which is theoretically instantaneous). In addition, the half-life ($t_{1/2}$)* of the substance is also calculated based on the slope. Thus, $t_{1/2}$ is constant, since the sum of all first-order kinetic processes is independent of the initial concentration of drug in the plasma and is inversely proportional to the elimination rate.

* Time required to reduce the amount of the drug in the plasma compartment by 50%.

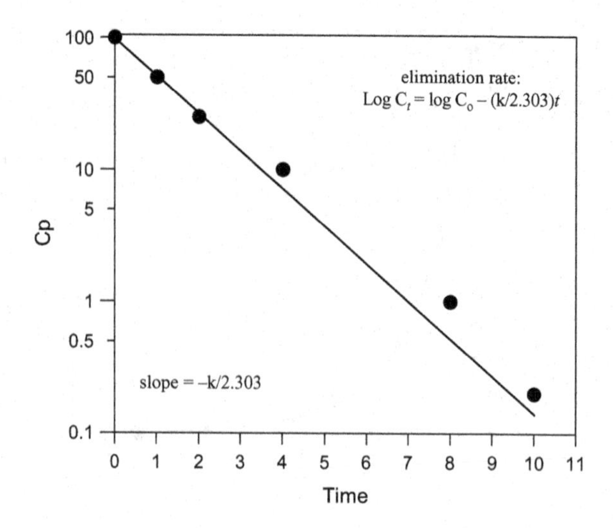

FIGURE 10.7 Semi-logarithmic graph of first-order elimination rate for a chemical or drug per unit time. Note that the ordinate is plotted on a logarithmic scale (see text for description and abbreviations).

10.5.1.2 Zero-Order Elimination

Alternatively, *zero-order elimination* refers to the removal of a fixed amount of drug independently of the concentration—that is, an increasing plasma concentration of the chemical does not result in a corresponding increase in elimination. As depicted in Figure 10.8, the disappearance or elimination of the drug is constant with respect to *time* and is expressed as

$$C_t = -k_0 t + C_0$$

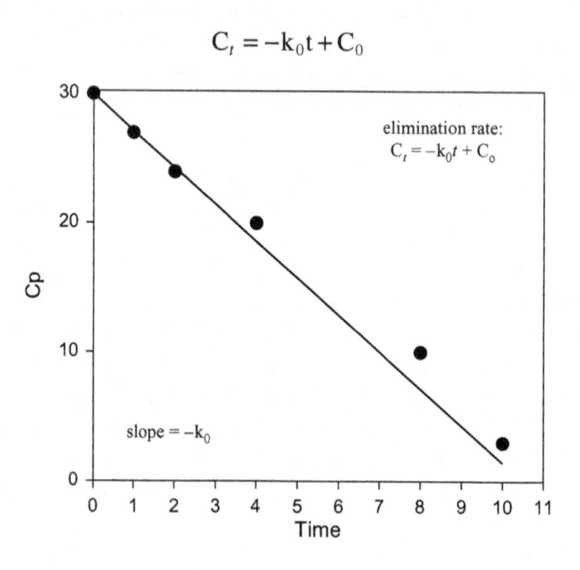

FIGURE 10.8 Graph of zero-order elimination rate for a chemical per unit time. Note that the ordinate scale is linear (see text for description and abbreviations).

where

$-k_0$ is the slope

C_0 is the y-intercept

Chemical reactions that trigger zero-order kinetics usually occur when biotransforming enzymes or transport systems are saturated. The rate process becomes constant and cannot be accelerated by increasing the concentration of the substrate. Clinically, zero-order elimination is the most common cause of drug accumulation and toxicity; it is typically encountered with administration of drugs through constant-rate intravenous infusions as well as prolonged-release (*PR*) or extended-release (*XR*) dosage forms.

10.5.2 Fecal

With the exception of cholestyramine, a cholesterol-binding drug, and the herbicide paraquat, the presence of nonabsorbable, nontransformed xenobiotics in the feces is rare. This is because many metabolites of compounds formed in the liver are excreted into the intestinal tract via the bile. Furthermore, organic anions and cations are biotransformed and actively transported into bile by carrier mechanisms similar to those in the renal tubules.

Orally administered drugs and compounds enter the hepatic portal system and liver sinusoids, only to be extensively cleared by the liver. This pharmacokinetic phenomenon is known as *hepatic first-pass elimination.* According to the principles of elimination, compounds may be classified based on their bile-to-plasma (B/P) ratio. Table 10.13 outlines the classification of representative chemicals based on this ratio. Substances with high B/P ratios are readily eliminated in bile. Drugs such as propranolol and phenobarbital are extensively removed by the liver, which explains why the oral bioavailability of these drugs is low despite complete absorption. Other drugs such as lidocaine, a local anesthetic, are extensively eliminated by first-pass elimination, such that oral administration is precluded. Compounds with low B/P values are not subject to this phenomenon and maintain relatively high plasma and other compartmental concentrations.

10.5.3 Pulmonary

Gases, aerosolized molecules, and liquids with low vapor pressure are eliminated by simple diffusion via exhalation. The rate of pulmonary elimination is inversely

TABLE 10.13
Bile-to-Plasma Ratio of Representative Xenobiotics

Classification	Bile-to-plasma ratio (B/P)	Representative Compounds
Class A	1/1 to 10/1	Na^+, K^+, glucose
Class B	10/1 to 1000/1	Arsenic, bile acids, lidocaine, phenobarbital, propranolol
Class C	less than 1/1	Albumin, zinc, iron, gold, chromium

proportional to the blood solubility of the compound. Thus, compounds with low blood solubility, such as ethylene dioxide, have high vapor pressure and are readily volatilized and eliminated as pulmonary effluents. Conversely, compounds with high blood solubility, such as ethanol, carbon tetrachloride, and chloroform, have low vapor pressures. Their elimination through the lungs is more a function of concentration than of solubility. Similarly, lipid-soluble general anesthetic agents have long half-lives in plasma because of their low vapor pressure and high plasma solubility.

10.5.4 Mammary Glands

The mammary glands are modified sudoriferous glands that produce a lipid-laden secretion (breast milk), which is excreted during lactation. The lobules of the glands consist of alveoli surrounded by one or two layers of epithelial and myoepithelial cells anchored to a well-developed basement membrane. Unlike the BBB or the placenta, the cells do not interpose a significant barrier to the transport of molecules. In addition, breast milk is a watery secretion with a significant lipid component. Consequently, both water-soluble and lipid-soluble substances are transferred through the breast milk. In particular, basic chemicals readily accumulate in the weakly acidic/alkaline environment of breast milk (pH 6.5–7.5). Accordingly, nursing infants are exposed to almost any substance present in the maternal circulation.

10.5.5 Secretions

The elimination of chemicals through saliva, tears, hair follicles, or sudoriferous/sebaceous (sweat) glands is a function of the plasma concentration. Elimination is also influenced by liver metabolism, by passage through the intestinal tract and pulmonary system, and by limited access to local glandular secretions. Thus, the secretion of compounds through mucous, ophthalmic, and dermal glands is not significant. In forensic toxicology, however, some chemicals may be detected in minute concentrations in hair follicles and nails when plasma levels seem to be nonexistent. Chronic arsenic and lead poisoning have been detected by measurement of the metals bound to hair follicles and hair shafts.

REFERENCES

SUGGESTED READINGS

Bhatia, P., Kolinski, M., Moaddel, R., Jozwiak, K., and Wainer, I.W., Determination and modeling of stereoselective interactions of ligands with drug transporters: a key dimension in the understanding of drug disposition. *Xenobiotica* 38, 656, 2008.

Burg, M.B., Ferraris, J.D., and Dmitrieva, N.I., Cellular response to hyperosmotic stresses. *Physiol. Rev.* 87, 1441, 2007.

Combes, R., Grindon, C., Cronin, M.T., Roberts, D.W., and Garrod, J.F., Integrated decision-tree testing strategies for acute systemic toxicity and toxicokinetics with respect to the requirements of the EU REACH legislation. *Altern. Lab. Anim. (ATLA)* 36, 91, 2008.

Galetin, A., Gertz, M., and Houston, J.B., Potential role of intestinal first-pass metabolism in the prediction of drug-drug interactions. *Expert Opin. Drug Metab. Toxicol.* 4, 909, 2008.

Kuentz, M., Drug absorption modeling as a tool to define the strategy in clinical formulation development. *AAPS J.* 10, 473, 2008.

Kim, B.H., Beskok, A., and Cagin, T., Molecular dynamics simulations of thermal resistance at the liquid-solid interface. *J. Chem. Phys.* 129, 174701, 2008.

Leucuta, S.E. and Vlase, L., Pharmacokinetics and metabolic drug interactions. *Curr. Clin. Pharmacol.* 1, 5, 2006.

Lu, H., Stereoselectivity in drug metabolism. *Expert Opin. Drug Metab. Toxicol.*, 3, 149, 2007.

Mann, H.J., Drug-associated disease: cytochrome P450 interactions. *Crit. Care Clin.* 22, 329, 2006.

Medinsky, M.A. and Valentine, J.L., Principles of toxicology, in *Casarett and Doull's Toxicology: The Basic Science of Poisons*, Klaassen, C.D. (Ed.), 7th ed., McGraw-Hill, New York, 2006, Chapter 7.

Sui, X., Sun, J., Wu, X., Li, H., Liu, J., and He, Z., Predicting the volume of distribution of drugs in humans. *Curr. Drug Metab.* 9, 574, 2008.

Sousa, M., Pozniak, A., and Boffito, M., Pharmacokinetics and pharmacodynamics of drug interactions involving rifampicin, rifabutin and antimalarial drugs. *J. Antimicrob. Chemother.* 62, 872, 2008.

Urso, R., Blardi, P., and Giorgi, G., A short introduction to pharmacokinetics. *Eur. Rev. Med. Pharmacol. Sci.* 6, 33, 2002.

Yamaoka, K. and Takakura, Y. ,Analysis methods and recent advances in nonlinear pharmacokinetics from in vitro through in loci to in vivo. *Drug Metab. Pharmacokinet.* 19, 397, 2004.

REVIEW ARTICLES

Barile, F.A., Dierickx, P.J., and Kristen, U., *In vitro* cytotoxicity testing for prediction of acute human toxicity. *Cell Biol. Toxicol.* 10, 155, 1994.

Beasley, G.M., Kahn, L., and Tyler, D.S., Current clinical and research approaches to optimizing regional chemotherapy: novel strategies generated through a better understanding of drug pharmacokinetics, drug resistance, and the development of clinically relevant animal models. *Surg. Oncol. Clin. North Am.* 17, 731, 2008.

Bodo, A., Bakos, E., Szeri, F., Váradi, A., and Sarkadi, B., The role of multidrug transporters in drug availability, metabolism and toxicity. *Toxicol. Lett.* 140, 133, 2003.

Buchanan, J.R., Burka, L.T., and Melnick, R.L., Purpose and guidelines for toxicokinetic studies within the National Toxicology Program. *Environ. Health Perspect.* 105, 468, 1997.

Edginton, A.N., Theil, F.P., Schmitt, W., and Willmann, S., Whole body physiologically-based pharmacokinetic models: their use in clinical drug development. *Expert Opin. Drug Metab. Toxicol.* 4, 1143, 2008.

El-Tahtawy, A.A., Tozer, T.N., Harrison, F., Lesko, L., and Williams, R., Evaluation of bioequivalence of highly variable drugs using clinical trial simulations. II: Comparison of single and multiple-dose trials using AUC and Cmax. *Pharm. Res.* 15, 98, 1998.

Fowler, S. and Zhang, H., *In vitro* evaluation of reversible and irreversible cytochrome P450 inhibition: current status on methodologies and their utility for predicting drug-drug interactions. *AAPS J.* 10, 410, 2008.

Krzyzanski, W. and Jusko, W.J., Mathematical formalism for the properties of four basic models of indirect pharmacodynamic responses. *J. Pharmacokinet. Biopharm.* 25, 107, 1997.

Langer, O. and Müller, M., Methods to assess tissue-specific distribution and metabolism of drugs. *Curr. Drug Metab.* 5, 463, 2004.

Li, H., Sun, J., Fan, X., Sui, X., Zhang, L., Wang, Y., and He, Z., Considerations and recent advances in QSAR models for cytochrome P450-mediated drug metabolism prediction. *J. Comput. Aided Mol. Des.* 22, 843, 2008.

Lin, J.H., Species similarities and differences in pharmacokinetics. *Drug Metab. Dispos.* 23, 1008, 1995.

Lipscomb, J.C. and Poet, T.S., *In vitro* measurements of metabolism for application in pharmacokinetic modeling. *Pharmacol. Ther.* 118, 82, 2008.

Lock, E.A. and Smith, L.L., The role of mode of action studies in extrapolating to human risks in toxicology. *Toxicol. Lett.* 140, 317, 2003.

Lohse, M.J., Maiellaro, I., and Calebiro, D., Kinetics and mechanism of G protein-coupled receptor activation. *Curr. Opin. Cell Biol.* 27, 87, 2014.

Mahmood, I. and Miller, R., Comparison of the Bayesian approach and a limited sampling model for the estimation of AUC and Cmax: a computer simulation analysis. *Int. J. Clin. Pharmacol. Ther.* 37, 439, 1999.

Matteucci, M.E., Miller, M.A., Williams, R.O., and Johnston, K.P., Highly supersaturated solutions of amorphous drugs approaching predictions from configurational thermodynamic properties. *J. Phys. Chem. B.*, 112, 16675, 2008.

Mor, A.L., Kaminski, T.W., Karbowska, M., and Pawlak, D., New insight into organic anion transporters from the perspective of potentially important interactions and drugs toxicity. *J. Physiol. Pharmacol.* 69, 2018.

Pelkonen, O., Kapitulnik, J., Gundert-Remy, U., Boobis, A.R., and Stockis, A., Local kinetics and dynamics of xenobiotics. *Crit. Rev. Toxicol.* 38, 697, 2008.

Perri, D., Ito, S., Rowsell, V., and Shear, N.H., The kidney—the body's playground for drugs: an overview of renal drug handling with selected clinical correlates. *Can. J. Clin. Pharmacol.* 10, 17, 2003.

Plusquellec, Y., Courbon, F., and Nogarede, S., Consequence of equal absorption, distribution and/or elimination rate constants. *Eur. J. Drug Metab. Pharmacokinet.* 24, 197, 1999.

Poulin, P. and Theil, F.P., Prediction of pharmacokinetics prior to in vivo studies. II. Generic physiologically based pharmacokinetic models of drug disposition. *J. Pharm. Sci.* 91, 1358, 2002.

Ramanathan, M., Pharmacokinetic variability and therapeutic drug monitoring actions at steady state. *Pharm. Res.* 17, 589, 2000.

Rogers, J.F., Nafziger, A.N., and Bertino, J.S. Jr., Pharmacogenetics affects dosing, efficacy, and toxicity of cytochrome P-450-metabolized drugs. *Am. J. Med.* 113, 746, 2002.

Sporty, J.L., Horálková, L., and Ehrhardt, C., *In vitro* cell culture models for the assessment of pulmonary drug disposition. *Expert Opin. Drug Metab. Toxicol.* 4, 333, 2008.

Warrington, J.S. and Shaw, L.M., Pharmacogenetic differences and drug-drug interactions in immunosuppressive therapy. *Expert Opin. Drug Metab. Toxicol.* 1, 487, 2005.

Zandvliet, A.S., Schellens, J.H., Beijnen, J.H., and Huitema, A.D., Population pharmacokinetics and pharmacodynamics for treatment optimization in clinical oncology. *Clin. Pharmacokinet.* 47, 487, 2008.

11 Chemical- and Drug-Receptor Interactions

Zacharoula Konsoula

11.1 TYPES OF CHEMICAL AND DRUG RECEPTORS

11.1.1 ION CHANNELS

Ion channels represent a diverse group of integral membrane proteins that regulate the passage of ions across cell membranes. They are ubiquitously expressed throughout the body and are essential for important physiological processes such as regulation of membrane potential, signal transduction, and cellular plasticity. Defective ion channel proteins are associated with cystic fibrosis, the long-QT syndrome, heritable hypertension (Liddle's syndrome), hereditary nephrolithiasis (Dent's disease), and a variety of hereditary myopathies, including generalized myotonia (Becker's disease) and myotonia congenita (Thomsen's disease). Table 11.1 illustrates drugs that target ion channels.

Ion channels are classified based on ion selectivity and gating. Selectivity refers to the capacity of some channels to distinguish between certain ions (e.g., sodium [Na^+], potassium [K^+], calcium [Ca^{+2}], or chloride [Cl^-]), allowing only selected ions to pass through the pore. Gating refers to the course of transformation between open and closed states. The opening and closing of most channels are determined by biological signals, such as alterations in membrane potential (voltage-gated channels), binding of intracellular or extracellular ligands (ligand-gated channels), change in temperature, or mechanical stress.

Voltage-gated ion channels are made up of either one α-subunit, which is an adjacent polypeptide that consists of four repeats (domains I–IV) (i.e., sodium [Nav] and calcium [Cav] channels), or four α-subunits, each with a single domain (i.e., potassium [Kv] channels). A single domain contains six α-helical transmembrane segments. The fourth transmembrane segment of each domain consists of positively charged arginines, which are mostly responsible for sensing changes in membrane potential. The four main groups of voltage-gated ion channels include

1. Sodium voltage-gated ion channels (Nav);
2. Potassium voltage-gated ion channels (Kv);
3. Calcium voltage-gated ion channels (Cav);
4. Voltage-gated proton ion channels.

1. *Sodium voltage-gated ion channels.* In excitable cells such as neurons and myocytes, sodium channels are associated with the rising phase of action potentials. The influx of Na^+ through Na^+ channels during the striking of

the action potential depolarizes the membrane, which also enhances the P_0 (the probability that a channel is open) and opens more Na^+ channels. The Na^+ influx is not maintained, because the P_0 spontaneously declines while the membrane is depolarized by inactivation. The decrease in Na^+ conductance promotes the repolarization phase by K^+ channels and effectively compresses the duration of the action potential.

2. *Potassium voltage-gated ion channels.* These gates represent an integral part of all cells and are responsible for the maintenance of resting potential. Cells employ potassium channels to modulate pacemaker potentials and mediate their overall excitability.
3. *Calcium voltage-gated ion channels.* These channels are electrophysiologically divided into L (long-lasting), T (transient), and N (for their involvement in norepinephrine) types. They are found in skeletal and heart muscles. There is also a fourth type (*P type*) found in Purkinje cells. They are all structurally homologous to voltage-gated Na^+ channels.
4. *Voltage-gated proton ion channels.* These channels are responsible for counterbalancing the elevation of positive charge caused by the efflux of electrons. They are induced by the formation of reactive oxygen species by NADPH oxidase and are localized in eosinophils and other phagocytic cells.

Ligand-gated ion channels, also known as *ionotropic receptors*, open in response to specific ligand binding to the extracellular domain of the receptor protein. Ligand binding induces a conformational change in the structure of the

TABLE 11.1
Ion Channels and Interacting Drugs

Channels	Drugs	Pharmacological Class
Sodium	Carbamazepine, Phenytoin, valproic acid	Anticonvulsants
	Bupivacaine, Lidocaine, Mepivacaine	Local anesthetics
		Class I antiarrhythmics
	Procainamide, Quinidine	IA
	Phenytoin, Tocainide	IB
	Flecainide, Propafenone	IC
	Amiloride	Diuretics
Potassium	Glipizide, Tolazamide	Antidiabetics
	Diazoxide, Minoxidil	Antihypertensives
Chloride	Amiodarone, *N*-acetylprocainamide, Sotalol	Class III antiarrhythmics
	Clonazepam, Phenobarbital	Anticonvulsants
Calcium	Clonazepam, Diazepam, Lorazepam	Hypnotics-Anxiolytics
	Amlodipine, Nifedipine, Verapamil	Antianginals
	Diltiazem, Felodipine, Isradipine	Antihypertensives
	Diltiazem, Verapamil	Class IV antiarrhythmics

channel protein that ultimately leads to the opening of the channel gate and the influx of ions across the plasma membrane. Examples of such channels include 1. the nicotine acetylcholine receptor (nAChR), 2. glutamate-gated receptors, 3. γ-aminobutyric acid ($GABA_A$) and glycine receptors, and 4. serotonin type 3 ($5\text{-}HT_3$) receptors.

1. *Nicotine acetylcholine receptor* (nAChR): It modulates neurotransmission at the neuromuscular junction and autonomic ganglia. The nAChR is activated by the neurotransmitter acetylcholine, the natural product nicotine, or synthetic compounds. It is a target against a variety of diseases, such as cognitive and attention deficits, Parkinson's disease, Alzheimer's disease, anxiety, and pain. The poison curare acts as an antagonist of the nAChR.
2. *Glutamate receptors*: Abnormal function of the glutamatergic system has been linked to the pathophysiology of many disorders, including amyotrophic lateral sclerosis (ALS), Huntington's chorea, epilepsy, Alzheimer's disease, schizophrenia, and anxiety disorders. There are three main types of glutamate receptors: i). AMPA (α-amino-3-hydroxy-5-methyl-4-isoxazole propionic acid) receptors, ii). kainate receptors, and iii). NMDA (N-methyl-D-aspartate) receptors.
   a. *AMPA receptors*: They moderate the desensitizing excitation at most synapses. Their induction opens the pore, allowing the inflow of Na^+ ions, and leads to the depolarization of the neuronal membrane.
   b. *Kainate (KA) receptors*: Like AMPA receptors, KA receptors modulate the influx of Na^+ ions and mediate excitatory neurotransmission.
   c. *NMDA (NR) receptors*: They exist primarily as tetrameric complexes and consist of two obligatory NR1 subunits and two NR2 subunits. The binding site for glutamate has been located in the NR2 subunit, and the site for the co-agonist glycine has been localized to the NR1 subunit. NRs are hindered under resting conditions by Mg^{2+} ions. However, once the membrane is depolarized, these receptors are stimulated by the combined binding of two molecules of glutamate and two molecules of glycine.
3. *$GABA_A$ and glycine receptors*: They are both families of chloride channels. They have complex gating characteristics and are made of five subunits, with each containing four transmembrane helices (TMs). GABA and glycine neurotransmitters function primarily on inhibitory neurons. GABA blocks the ability of neurons to fire action potentials. Alcohol and barbiturates (i.e., diazepam) operate on $GABA_A$ receptors as agonists (activators); they bind at sites where $GABA_A$ does not normally act and potentiate the effect of GABA when it binds at its own site. These types of drugs are used for epilepsy.
4. *Serotonin type 3 ($5\text{-}HT_3$) receptors*: These share structural features with $GABA_A$, glycine, and nACh receptors. The $5\text{-}HT_3$ receptor antagonists display anxiolytic and atypical antipsychotic properties.

11.1.2 G-Protein-Coupled Receptors

G-protein-coupled receptors (GPCRs) constitute the largest group of cell-surface proteins that are linked to signal transduction—they account for 2% of the proteins that are encoded by the human genome. They are related to a number of human diseases, since the GPCRs are the target of 50–60% of all current therapeutic agents.

The G-proteins are composed of three nonidentical subunits: αα, ββ, and γγ. The α subunit is the largest (33–55 kDa). When it is GDP-bound, the Gα subunit forms a complex with one Gβ (~35 kDa) and one Gγ (~15 kDa) subunit and functionally dissociates from the Gβγ complex when it binds GTP. Following GPCR activation, the Gβγ and GTP-bound Gα subunits of the G-protein stimulate a number of effector molecules, the nature of which depends on the coupling specificity of the receptor for the four families of mammalian G-protein α-subunits: Gαs, Gαi, Gαq, or Gα12/13. The G-protein subunits are depicted in Table 11.2.

Typically, receptors that are coupled to Gαs stimulate adenylyl cyclases and elevate the levels of cyclic AMP (cAMP), whereas those that trigger the Gαi family inhibit adenylyl cyclases, lower cAMP levels, and activate phospholipases and phosphodiesterases. Receptors that are coupled to the Gαq family bind to and activate phospholipase C-β (PLCβ), which hydrolyses phosphatidylinositol-4,5-bisphosphate to form inositol trisphosphate (IP3) and diacylglycerol (DAG). IP3 regulates the mobilization of intracellular calcium ions (Ca^{2+}) and the activation of Ca^{2+}/calmodulin-activated protein kinase II (CaMKII), whereas DAG activates protein kinase C (PKC). Both CaMKII and PKC can phosphorylate neighboring substrates. Figure 11.1 illustrates GPCR signaling.

The Gββ and Gγγ subunits function as a dimer to activate signaling molecules, such as phospholipases, ion channels, and lipid kinases. In addition to the regulation of these classical second-messenger-generating systems, Gβγ subunits and Gα subunits such as Gα12 and Gαq can modulate the actions of key intracellular signal-transducing molecules, including small GTP-binding proteins of the Ras and Rho

TABLE 11.2
G-Proteins and Subunits

Family	$G\alpha_s$	$G\alpha_i$	$G\alpha_q$	$G\alpha_{12/13}$	$G\beta\gamma$
Members	$G\alpha_s$	$G\alpha_{i1/i2/i3}$	$G\alpha_q$, $G\alpha_{14}$	$G\alpha_{12}$	β (1–5)
	$G\alpha_{sXL}$	$G\alpha_o$, $G\alpha_t$	$G\alpha_{11}$, $G\alpha_{15/16}$	$G\alpha_{13}$	γ (1–8)
	$G\alpha_{solf}$	$G\alpha_z$, $G\alpha_{gust}$			
Effectors	Adenylyl cyclases	Adenylyl cyclases	PLC_β	Rho GEFs	Ion channels
	Increased cAMP	Inhibition of cAMP	DAG	Rho	$PI3K_\gamma$
	PKA	Ion channels, phosphodiesterases, phospholipases	Ca^{2+} PKC		PLC_β, adenylyl cyclases

DAG: diacylglycerol; $G\alpha_i$: inhibit adenylyl cyclases; $G\alpha_s$: stimulate adenylyl cyclases; PKA: protein kinase A; PKC: protein kinase C; $PI3K_\gamma$: phosphoinositide-3 kinase γ, PLC_β: phospholipase C-β.

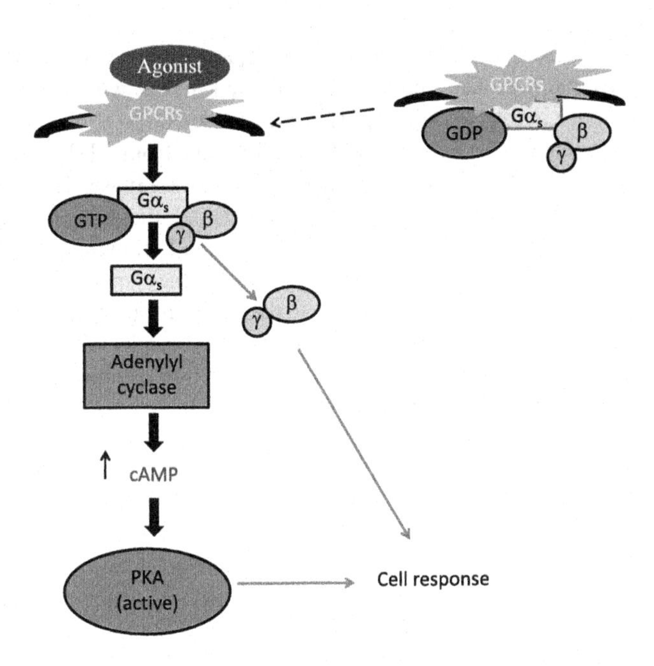

FIGURE 11.1 G-protein-coupled receptor (GPCR) signaling. Following agonist binding, GDP is discharged from the G-protein and substituted by GTP. This results in dissociation of the G-proteins complexes into αα subunits and ββγ dimers, which both stimulate several effectors. For example, Gαα triggers adenylyl cyclase, which leads to an increase in cyclic AMP (cAMP). This elevation in cAMP in turn turns on protein kinase A (PKA), a serine/threonine kinase that phosphorylates other kinases and transcriptional factors.

families and members of the mitogen-activated protein kinase (MAPK) family of serine-threonine kinases, including extracellular signal-regulated kinase (ERK), c-jun N-terminal kinase (JNK), and p38.

The termination or desensitization of GPCR signaling is highly regulated and can take place through several mechanisms. On ligand binding, GPCR kinases (GRKs) or second-messenger-regulated kinases (protein kinase C [PKC] and protein kinase A [PCA]) phosphorylate serine/threonine residues within the intracellular loops or C-terminal tail of GPCRs. PKC and PKA phosphorylation dissociates GPCRs from their respective G-proteins and functions as part of a negative-feedback mechanism. In some instances, this process can result in one receptor being desensitized following the activation of another. By contrast, GRKs phosphorylate only the activated or agonist-occupied form of the receptor, thereby leading to homologous desensitization. GRK-phosphorylated receptors then adhere to arrestins, which hinder the interaction with G-proteins and basically abrogate the receptor from the cell surface by endocytosis. Once internalized, the receptors can be recycled or degraded.

11.1.3 Kinases and Enzyme-Coupled Receptors

Protein phosphorylation mediates many basic cellular processes, including metabolism, growth, division, differentiation, motility, organelle trafficking, membrane transport, muscle contraction, immunity, learning, and memory. Protein kinases are divided into two major groups: the Ser/Thr-specific kinases, which constitute about 80% of the protein kinases, and the Tyr-specific kinases.

Protein kinases catalyze the transfer of the γ-phosphate from ATP to specific amino acids in proteins. First, ATP binds to the active site of the kinase. This is succeeded by binding of the substrate to the active site. Once bound, the γ-phosphate of ATP is transferred to a Ser, Thr, or Tyr residue of the substrate. Following phosphorylation, the substrate is discharged from the kinase. The order of the steps alternates for different kinases. For instance, some kinases bind to their protein substrates before binding ATP, and others produce ADP before releasing the protein substrate. Table 11.3 illustrates inhibitors of protein kinases that are undergoing human clinical trials.

The receptor tyrosine kinase consists of an extracellular ligand-binding domain, a transmembrane domain, and an intracellular tyrosine kinase domain. This class of receptor is associated with signaling by insulin, transforming growth factor-αα (TGFαα), epidermal growth factor (EGF), platelet-derived growth factor (PDGF), and many other trophic hormones. Ligand (hormone) binding to the receptor results in receptor dimerization, which triggers the tyrosine kinase activity in the intracellular domain. This causes autophosphorylation of the receptor and the induction of signal transduction cascades implicated in cell proliferation and survival. Figure 11.2 illustrates the mechanism of activation of the EGF receptor, a representative tyrosine kinase receptor.

TABLE 11.3
Inhibitors of Protein Kinases That Are Undergoing Human Clinical Trials

Kinase Targeted	Inhibitor	Disease	Status
	Tyrosine Kinases		
ABL	Gleevec	Cancer	FDA approved
EGFR	ZD1839 (Iressa)	Cancer	FDA approved
VEGFR	PTK787	Cancer	FDA approved
NGFR	CEP2583	Cancer	Phase II
	Serine/Threonine Kinases		
PKC	PKC412	Cancer	Phase I
MKK1	PD184352	Cancer	Phase I
mTOR	Rapamycin	Immunosuppression	FDA approved
	RADOO1	Cancer	Phase I
PKC_{β}	LY333531	Diabetic retinopathy	Phase III
MLK	CEP1347	Neurodegeneration	Phase I

ABL: Abelson tyrosine kinase;
EGFR, epidermal-growth-factor receptor;
FDA: U.S. Food and Drug Administration; MKK1: mitogen-activated protein kinase kinase 1; MLK: mixed-lineage protein kinase; mTOR: target of rapamycin (mammalian); NGFR: nerve-growth-factor receptor; PKC: protein kinase C; VEGFR: vascular-endothelial-growth-factor receptor.

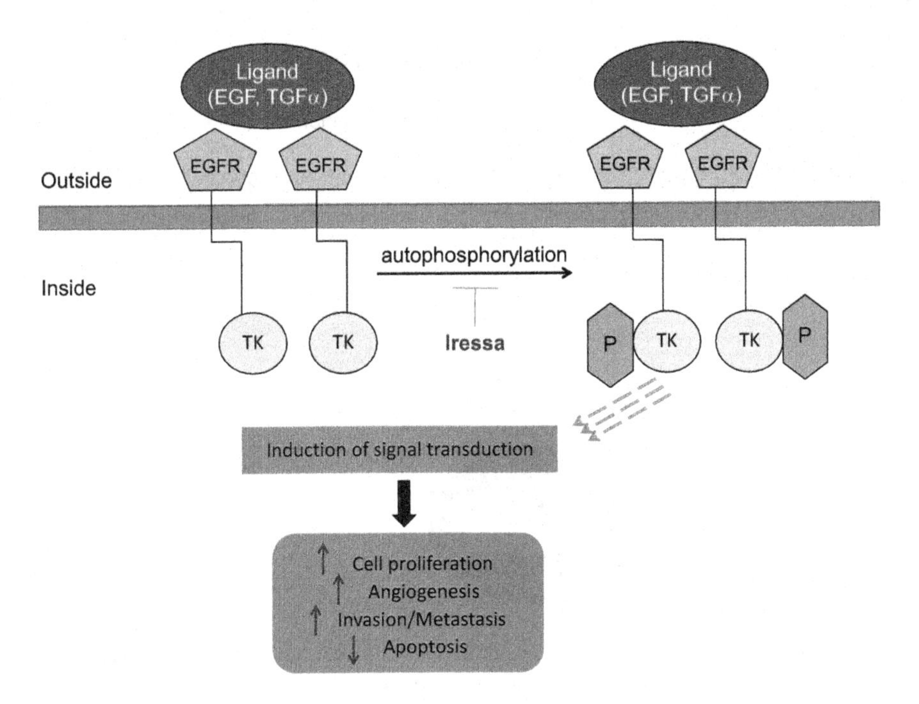

FIGURE 11.2 Signal transduction through the epidermal growth factor receptor (EGFR). Ligand binding leads to receptor dimerization. This results in receptor autophosphorylation, which is blocked by gefitinib. TGFαα: transforming growth factor; TK: tyrosine kinase domain; P: phosphorylation.

Unlike growth-factor receptors, many cytokine receptors do not possess intrinsic tyrosine kinase activity. Instead, ligand (i.e., growth hormone, erythropoeitin, or interferon [IFN]) binding results in the recruitment of receptor-associated tyrosine kinases, which are often members of the Janus kinases (JAKs). These kinases phosphorylate the cytoplasmic tails of the cytokine receptors on specific tyrosine residues to supply docking sites for the recruitment of monomeric STATs (signal transducers and activators of transcription). Once they have been recruited, STATs themselves become substrates for tyrosine phosphorylation. Phosphorylated STATs dimerize and translocate to the nucleus, where the dimers regulate the expression of target genes. Figure 11.3 illustrates the mechanism of signal transduction through the cytokine receptor. Protein drugs in these classes include erythropoietin (EPO), granulocyte colony-stimulating factor (G-CSF), IFN-αα, IFN-ββ, human growth hormone (HGH), and soluble tumor necrosis factor-αα(TNF-αα) receptor.

11.1.4 Intracellular Receptors

Nuclear receptors are transcription factors that are fundamental in embryonic development, maintenance of differentiated cellular phenotypes, metabolism, and cell death. Defective nuclear receptor signaling is implicated in cancer, infertility, obesity, and diabetes. Nuclear receptors can be subdivided into two groups according to

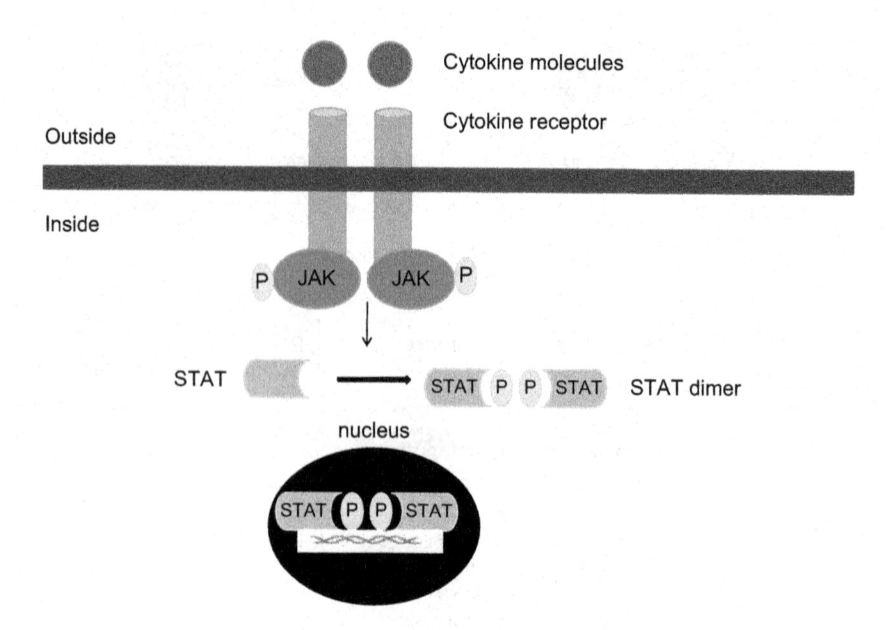

FIGURE 11.3 Signal transduction through the cytokine receptor. Binding of a ligand to the cytokine receptor results in the activation of the receptor-associated kinases, the Janus kinase (JAK). These tyrosine kinases sequentially phosphorylate the cytoplasmic tails of the receptor to accommodate the recruitment of monomeric STATs. Once they have been recruited, STATs themselves become substrates for tyrosine phosphorylation. Phosphorylated STATs dimerize and translocate to the nucleus and regulate the expression of target genes.

whether they create cytoplasmic complexes with heat shock protein (Hsp90) and are active as a monomer (steroidogenic factor-1 [SF1]), a homodimer (steroid receptors), or a heterodimer (thyroid hormone receptor, vitamin D receptor, and several orphan nuclear receptors). Translocation of steroid hormone receptors into the nucleus is illustrated in Figure 11.4.

It must be emphasized that the nuclear receptor superfamily consists of many ligand receptors (Table 11.4) as well as "orphan" receptors, for which no ligand has yet been identified.

All nuclear receptors operate as ligand-activated zinc-finger transcription factors with a modular structure that comprises six domains. Figure 11.5 presents a schematic illustration of the structural and functional organization of nuclear receptors.

The amino terminus (also known as the *A/B region*) comprises a trans-activation domain (AF-1), which is of variable length and sequence in the different family members and is perceived by co-activators and/or other transcription factors. The central DNA-binding domain (DBD) has two zinc-finger motifs that are typical of the entire family. The carboxy-terminal ligand-binding domain (LBD), whose overall architecture is well protected between the various family members, differs adequately to guarantee selective ligand recognition. The central DNA-binding domain is the most conserved, whereas the carboxy-terminal ligand binding domain is the most variable.

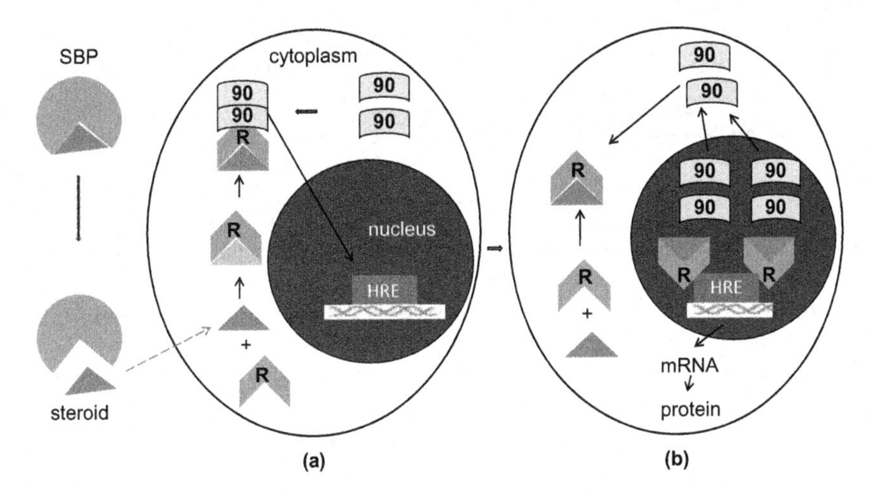

FIGURE 11.4 Translocation of steroid hormone receptors into the nucleus. (a) Disengaged from its steroid binding protein (SBP), the steroid ligand is transported into the cytoplasm by passive diffusion or active transport. When bound to the non-ligand-bound receptor (R), it causes a conformational change that allows the interaction with the Hsp90 dimer, which functions as its chaperone. The nuclear localization signal of the receptor allows the Hsp90–R complex to translocate into the nucleus. (b) Once in the nucleus, the ligand–receptor complex disconnects from the Hsp90 and dimerizes. This allows the receptor to connect with the hormone response element (HRE) in the target gene and induce transcription.

TABLE 11.4
Human Nuclear Receptors

Name	Abbreviation	Ligand
Thyroid hormone receptor	TRα	Thyroid hormone
	TRβ	Thyroid hormone
Retinoic acid receptor	RARα	Retinoic acid
	RARβ	Retinoic acid
	RARγ	Retinoic acid
Peroxisome proliferator-activated receptor	PPARα	Fatty acids, leukotrieneB4, fibrates
	PPARβ	Fatty acids
	PPARγ	Fatty acids, prostaglandin J2
Vitamin D receptor	VDR	1,25-dihydroxy vitamin D_3, lithocholic acid
Human nuclear factor 4	HNF4α	Orphan
	HNF4γ	Orphan
Estrogen receptor-related receptor	ERRα	Orphan
	ERRβ	4-OH tamoxifen
	ERRγ	4-OH tamoxifen
Glucocorticoid receptor	GR	Cortisol, dexamethasone
Mineralocorticoid receptor	MR	Aldosterone, spironolactone
Progesterone receptor	PR	Progesterone
Androgen receptor	AR	Testosterone

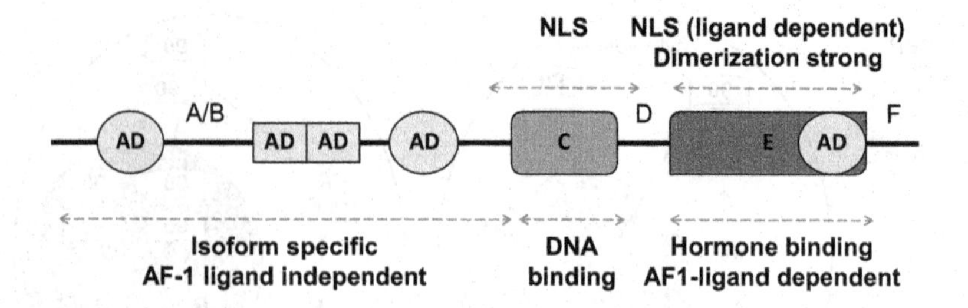

FIGURE 11.5 Schematic illustration of the structural and functional organization of nuclear receptors. The conserved regions C and E are presented as boxes, and a black bar indicates the divergent regions A/B, D, and F. Domain functions are depicted above and below the scheme. AD: activation domain; AF1:activation function 1; NLS: nuclear localization signal.

11.2 SIGNAL TRANSDUCTION

During the last 20 years, four major signaling pathways have been identified; most hormones and growth factors were induced by one of these. They are now considered classical signaling pathways. Figure 11.6 illustrates the four classical pathways. Many hormones function through the G-protein-coupled receptors (GPCRs) that couple to adenylyl cyclase, causing increased cAMP levels, or to phospholipase C, resulting in elevated diacylglycerol (DAG) and inositol-1,4,5-trisphosphate (Ins(1,4,5)P3). cAMP activates protein kinase A (PKA), leading to alterations in gene expression. DAG stimulates protein kinase C (PKC. Ins(1,4,5)P3 triggers calcium elevation, which in turn mediates calcium-binding proteins. Growth factors act via receptor tyrosine kinases, recruiting proteins, such as phospholipase Cγ (PLCγ), phosphoinositide-3 kinase (PI3K), and Ras, the latter of which induces mitogen-activated protein kinase (MAPK) and transcription factors, such as activator protein 1 (AP1). Nuclear hormone receptors (activated by steroids) bind to cytosolic receptors and then translocate to the nucleus, where they function as transcription factors.

The first pathway is linked to the stimulation of heterotrimeric G-proteins by receptors consisting of seven transmembrane domains (GPCRs), resulting in induction of adenylyl cyclase and alterations in the second messenger, cyclic AMP, followed by activation of PKA.

In the second classical signaling pathway, induction of phospholipase C by GPCRs triggers the formation of second messengers such as DAG and inositol-1,4,5-trisphosphate (Ins-1,4,5-P3). Subsequently, DAG activates PKC, and Ins-1,4,5-P3 induces calcium release from the endoplasmic reticulum, thus activating calcium-dependent protein kinases.

The third pathway is associated with signaling by growth factors (e.g., epidermal growth factor) and is followed by the induction of receptor tyrosine kinases and effector proteins. These effector proteins regulate the activation of the monomeric G-protein Ras, followed by the activation of MAPKs, which leads to the commencement of the cell cycle. Other effectors include phospholipase Cγ, which generates DAG and Ins-1,4,5-P3.

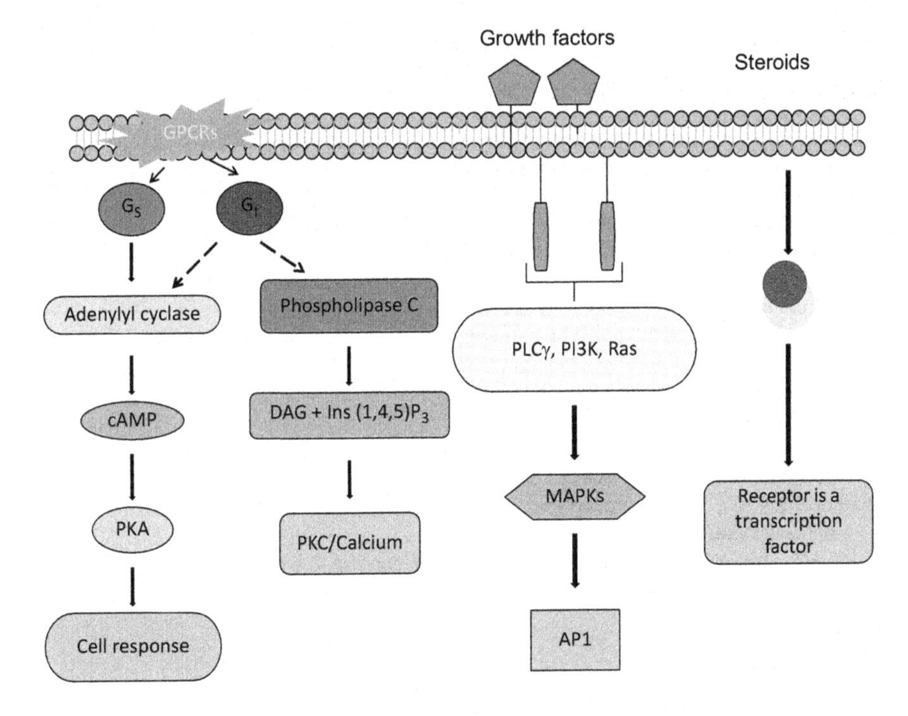

FIGURE 11.6 The four classical signaling pathways.

The fourth pathway is associated with the nuclear hormone receptor system. This conduit is stimulated by the binding of a hormone (e.g., a steroid) to a receptor complex that is a transcription factor—after translocation to the nucleus, binding of this complex to the promoter causes expression of the target gene.

Compounds targeting the classical pathways are illustrated in Table 11.5.

TABLE 11.5
Selected Compounds Targeting Classical Signaling Pathways

Compound	Target	Disease	Status
Cilomilast (Ariflo)	cAMP PDE	Asthma, COPD	Approved
Roflumilast	cAMP PDE	Asthma, COPD	Approved
Sildenafil (Viagra)	cGMP PDE	Erectile dysfunction	Approved
Gefitinib (Iressa)	EGFR (tyrosine kinase)	Non-small cell lung carcinoma	Approved
Erlotinib (Tarceva)	EGFR (tyrosine kinase)	Non-small cell lung carcinoma	Approved
Trastuzumab (Herceptin)	HER2/neu	HER2/neu-positive breast cancer	Approved
Imatinib (Gleevec)	BCR-ABL (tyrosine kinase)	CML	Approved

cAMP: cyclic AMP; cGMP: cyclic GMP; CML: chronic myelogenous leukemia; COPD: chronic obstructive pulmonary disease; EGFR: epidermal growth factor receptor; HER2/neu: human EGFR2; PDE: phosphodiesterase.

REFERENCES

SUGGESTED READINGS

Chanda, B., Asamoah, O.K., Blunck, R., Roux, B., and Bezanilla, F., Gating charge displacement in voltage-gated channels involves limited transmembrane movement. *Nature* 436, 852, 2005.

Cohen, P., Protein kinases—the major drug targets of the twenty-first century? *Nat. Rev. Drug Discov.* 1, 309, 2002.

Kliewer, A., Reinscheid, R.K., and Schulz, S., Emerging paradigms of G protein-coupled receptor dephosphorylation. *Trends Pharmacol. Sci.* 38, 621, 2017.

Long, S.B., Campbell, E.B., and MacKinnon, R., Crystal structure of a mammalian voltage-dependent Shaker family K^+ channel. *Science* 309, 897, 2005.

Molokanova, E. and Savchenko, A., Bright future of optical assays for ion channel drug discovery. *Drug Discov. Today* 13, 14, 2008.

O'Shea, J.J., Pesu, M., Borie, D.C., and Changelian, P.S., A new modality for immunosuppression: targeting the JAK/STAT pathway. *Nat. Rev. Drug Discov.* 3, 555, 2004.

Rochette-Egly, C., Nuclear receptors: integration of multiple signaling pathways through phosphorylation. *Cell Signal* 15, 355, 2003.

Yeagle, P.L. and Albert, A.D., A conformational trigger for activation of a G protein by a G protein-coupled receptor. *Biochemistry* 42, 1365, 2003.

REVIEW ARTICLES

Calebiro, D. and Godbole, A., Internalization of G-protein-coupled receptors: implication in receptor function, physiology and diseases. *Best Pract. Res. Clin. Endocrinol. Metab.* 32, 83, 2018.

Dorsam, R.T. and Gutkind, J.S., G-protein-coupled receptors and cancer. *Nat. Rev. Cancer* 7, 79, 2007.

Grigoriadis, D.E., Hoare, S.R.J., Lechner, S.M., Slee, D.H., and Williams, J.A., Drugability of extracellular targets: discovery of small molecule drugs targeting allosteric, functional, and subunit-selective sites on GPCRs and ion channels. *Neuropsychopharmacology* 34, 106, 2009.

Gronemeyer, H., Gustafsson, J.A., and Laudet, V., Principles for modulation of the nuclear receptor superfamily. *Nat. Rev. Drug Discov.* 3, 950, 2004.

Malbon, C.C., G-proteins in development. *Nat. Rev. Mol. Cell Biol.* 6, 689, 2005.

Neves, S.R., Ram, P.T., and Iyengar, R., G protein pathways. *Science* 296, 1636, 2002.

Pierce, K.L., Premont, R.T., and Lefkowitz, R.J., Seven-transmembrane receptors. *Nat. Rev. Mol. Cell Biol.* 3, 639, 2002.

Ubersax, J.A. and Ferrell Jr., J.E., Mechanisms of specificity in protein phosphorylation. *Nat. Rev. Mol. Cell Biol.* 8, 530, 2007.

12 Toxicogenomics

Anirudh J. Chintalapati, Zacharoula Konsoula, and Frank A. Barile

12.1 INTRODUCTION

Toxicogenomics is a relatively new field oriented toward the response of an entire genome to toxicants or environmental stressors. Toxicogenomics encompasses three major objectives: to elucidate the connection between environmental stress and disease susceptibility; to establish proper biomarkers of disease and exposure to toxic substances; and to clarify the molecular mechanisms of toxicity. Toxicogenomics incorporates traditional toxicological and histopathologic endpoint-evaluation with the data surges derived from transcriptomics, proteomics, metabolomics, and metabonomics. The terminology of these terms is described in Table 12.1.

12.2 HUMAN GENOMIC VARIATION

The analysis of human DNA genome disclosed a high degree of DNA sequence similarity among individuals from around the world: any two humans are expected to be about 99.9% alike in their DNA sequence. The differences in the genome are known as *genetic variations* and are triggered by different types of mechanisms, including 1. nucleotide substitutions; 2. insertion or deletion of nucleotides; 3. alterations in the number of tandem-repeat sequences; 4. variations in the copy number of genomic segments; and 5. the combinations of these changes. A subtype of these genetic variations is defined as *genetic polymorphisms* when they are recognized at a frequency of ≥1% in a certain population. The DNA polymorphisms are divided into five major classes:

1. RFLP (restriction fragment length polymorphism);
2. VNTR (variable number of tandem repeat or minisatellite);
3. STR (short tandem repeat or microsatellite);
4. SNP (single-nucleotide polymorphism); and
5. CNV (copy-number variation).

1. *RFLP* is formed by variations in the size of the DNA fragment digested by a restriction endonuclease due to the base substitutions at the site identified by the endonuclease. RFLP patterns allow differentiation between the parental alleles (whether the allele is of maternal or paternal origin) of the specific loci in the genome.
2. *VNTR* is a point in a genome where a short nucleotide sequence is aligned as a tandem repeat (two or more nucleotides are duplicated, and the repetitions are next to each other). VNTR are significantly polymorphic with

TABLE 12.1
Terminology

TRANSCRIPTOMICS: Techniques that assess the full complement of activated genes, mRNAs, or transcripts in a specific tissue at a specific time, usually through the use of cDNA or oligonucleotide microarrays.

PROTEOMICS: A group of techniques used to assess the structural and functional attributes of proteins through the use of two-dimensional gel electrophoresis or liquid chromatography mass spectrometry (LC-MS).

METABONOMICS: Techniques that ascertain alterations in the concentration of low–molecular weight metabolites present in a cell or organism at a particular time (the metabonome) by using nuclear magnetic resonance (NMR) or mass spectrometry coupled to liquid (LC-MS) or gas chromatography (GC-MS).

METABOLOMICS: The use of quantitative analytical methods for analyzing the whole metabolic composition of a cell or organism at a given time.

BIOMARKER: A pharmacological or physiological assessment used to detect an event in a test specimen.

high heterozygosity in given populations because of the wide range of copy number of tandem-repeat DNA sequences.

3. Unlike VNTR loci, which constitute a few thousand in the human genome, *STR* or microsatellite (10 and 60 nucleotides are repeated) loci are accessible at >100,000 regions, which comprise most of the genomic regions. For that reason, microsatellite markers have been successfully employed for linkage analysis or population genetics; they require a very small quantity of DNA.
4. *SNPs* are genetic variations that are most commonly present in the human genome. They have been confirmed to be found in one in every 500–1000 base pairs (bp) on average. SNPs have been accepted as a very valuable tool for population genetics, particularly for the analysis of genes prone to various diseases.
5. *CNV*, a mode of genomic structural variation, is a segment of DNA that is 1 kb or larger and exists at a variable copy number in comparison with a reference genome. Categories of CNVs encompass insertions, deletions, and duplications. The implication of CNVs in human phenotypes was first introduced as "genomic disorders," which are frequently activated by *de novo* structural changes, and so far, dozens of genomic disorders have been reported.

12.2.1 Genomic Variation in Target Molecules

Target identification aims to distinguish new targets, normally proteins, whose modulation might suppress or prevent disease progression. The advent of technologies that endeavor to associate changes in gene (genomics) and protein (proteomics) expression or genetic variation with human disease metamorphosed many aspects of

TABLE 12.2
Pharmacogenetics vs. Pharmacogenomics

Pharmacogenetics	Pharmacogenomics
Differential effects of a drug *in vivo*, in different patients, that are dependent on the presence of inherited gene variants	Differential effects of compounds—*in vivo* or *in vitro*—on gene expression among all expressed genes
Genetic approaches (single-nucleotide polymorphisms [SNPs] but also expression profiles and biochemical measures)	Gene expression profiling
Provides more patient-/disease-specific health care	A tool for compound selection/drug discovery
One drug, many genomes (that is, patients)	Many drugs, one genome (that is, the "normative" genome of the experimental system used)
Patient variability	Compound variability

TABLE 12.3
Classification of Pharmacogenetic Responses

Pharmacokinetics	Pharmacodynamics
Absorption	Causative drug action: related to molecular pathology
Metabolism	
Distribution	Palliative drug action: related to molecular physiology
Elimination	

medicine. Pharmacogenetics and pharmacogenomics are acknowledged as the new "disciplines" of biomedical science. Their comparisons are presented in Table 12.2.

Pharmacogenetics is the study of how genetic attributes result in disparity in individuals' responses to drugs. A reaction to a specific medicine may be altered by factors that vary in agreement with the individual's versions of a gene or genes (alleles). These determinants cover the amount and rate of drug absorption, the rate of drug metabolism and elimination, and the variability in the pharmacological site of drug action (Table 12.3).

Response variation has appeared during clinical trials of new drugs. For example, the anticancer drug trastuzumab (Herceptin) has been found to be effective only in women who carry multiple copies of the *ERBB2* gene (also known as HER2/*neu*). Therefore, pharmacogenetics allows the early determination of an assortment of potential responses to a new drug and genetic factors that can escalate the risk of detrimental effects and ameliorate patient outcome.

Pharmacogenomics is associated with the detection and determination of the genome (DNA) and its products (RNA and proteins) as they are linked to drug response. Pharmacogenomics involves the orderly evaluation of how chemical compounds alter the complete expression motif in defined tissues of interest. For

TABLE 12.4
Examples of Polymorphisms Reported to Mediate Drug Action

Gene Product	Drug	Drug Action Linked to Minor Allele
CYP2D6	Codeine	Decreased analgesia
CYP2D6	Propranolol	Augmented β-blockage
CYP3A5	Many	Function not yet determined
CYP2C9	Warfarin	Reduced anticoagulant effect
CYP2C9	Phenytoin	Increased toxicity
P-glycoprotein	Digoxin	Altered blood level
N-acetyl transferase	Isoniazid	Increased risk of hepatotoxicity

instance, gene expression profiling (employing a variety of microarray technologies) has enabled the establishment of an explicit subset of genes that may be represented distinctly in diseased and healthy tissues.

Although both pharmacogenetics and pharmacogenomics involve the assessment of drug effects using nucleic-acid technology, the approaches are unambiguously different: pharmacogenetics designates the analysis of differences between several individuals with attention to clinical response to a particular drug, whereas pharmacogenomics outlines the study of differences between several compounds with respect to gene expression response in a single genome (i.e., a cell line). Pharmacogenetics will aid in the clinical setting to identify the best medicine for an individual, whereas pharmacogenomics will help in pharmaceutical research and development to select the best candidate from a library of compounds that are undergoing evaluation.

12.2.2 Genomic Variation in Drug Metabolism

The activity of cytochrome P450s determines the duration of action and the amount of a drug that remains in the body. In humans, different forms of cytochrome P450 (i.e., *CYP1A2, CYP2C9, CYP2C19, CYP2D6, CYP2E1*, and *CYP3A4*) play a profound role in eliminating drugs. Polymorphism accounts for the majority of variation among *CYP2C9, CYP2C19*, and *CYP2D6* individuals. Some people have polymorphisms (i.e., deletions) that cause enzyme deficiencies—these individuals are recognized as "poor" or "slow" metabolizers and are at risk of concentration-associated toxicity. Others have polymorphisms (e.g., gene amplifications) that prompt enzyme activity; they are identified as "extensive" or "ultra-rapid" metabolizers and can be unresponsive to therapy. Table 12.4 illustrates examples of polymorphisms documented to mediate drug activities.

12.3 GENE STRUCTURE AND FUNCTION

A genome consists of deoxyribonucleic acid (DNA). Each genome encompasses all of the information required to assemble and support an organism. In the nucleus of each cell, the DNA molecule is wrapped into thread-like structures named

TABLE 12.5
Human Syndromes with Defective Genome Maintenance

Syndrome	Affected Mechanism	Type of genome Instability	Cancer Predisposition
BRCA1/BRCA2	HR	Chromosome aberrations	Breast/ovarian cancer
HNPCC	MMR	Point mutations	Colorectal cancer
Xeroderma pigmentosum	NER	Point mutation	UV-induced skin cancer
Ataxia telangiectasia (AT)	DSB response	Chromosome aberrations	Lymphoma
Nijmegen breakage syndrome	DSB response	Chromosome aberrations	Lymphoma

DSB: double-strand break; HNPCC: hereditary nonpolyposis colorectal cancer; HR: homologous repair; MMR: mismatch repair NER: nucleotide excision repair.

chromosomes. Each chromosome is comprised of DNA compactly folded around proteins called *histones* that support its structure.

A nucleic acid is made up of a chain of nucleotides connected together through an alternating series of sugar and phosphate residues from which nitrogenous bases project. The sugar residues in DNA and ribonucleic acid (RNA) are 2-deoxyribose and ribose, respectively. DNA comprises four types of base: two purines, adenine (A) and guanine (G), and two pyrimidines, cytosine (C) and thymine (T). The same bases exist in RNA with the exception that uracil (U) substitutes for thymine.

A gene is the fundamental unit of heredity. Genes, which are composed of DNA, provide directions to create molecules called proteins. In humans, genes alternate in size from a few hundred DNA bases to more than 2 million bases.

The borderlines of a protein-encoding gene are assigned as the positions at which transcription initiates and terminates. The coding domain is the center of the gene, which consists of the nucleotide sequence that is translated into the sequence of amino acids in the protein. The coding domain appears with the initiation codon (ATG) and completes with one of the termination codons (TAA, TAG, or TGA). On both sides of the coding domain are DNA sequences that are transcribed but are not translated. Both the coding domain and the untranslated domains are interspersed by introns. Genes are grouped into exons and introns. The exons are the components that are present in the mature transcript (messenger RNA [mRNA]), while the introns are abolished from the primary transcript by a procedure called *splicing* (Figure 12.1).

12.3.1 DNA Alterations and Genotoxic Effects

All cells can suffer DNA damage from endogenous or exogenous sources.

Endogenous damage involves (a) incorrect integration of nucleotides during DNA replication; (b) automatic or intended de-amination of cytosine, adenine, guanine, or 5-methylcytosine, modifying these bases to the miscoding uracil, hypoxanthine, xanthine, and thymine, respectively; (c) automatic hydrolysis of the N-glycosidic

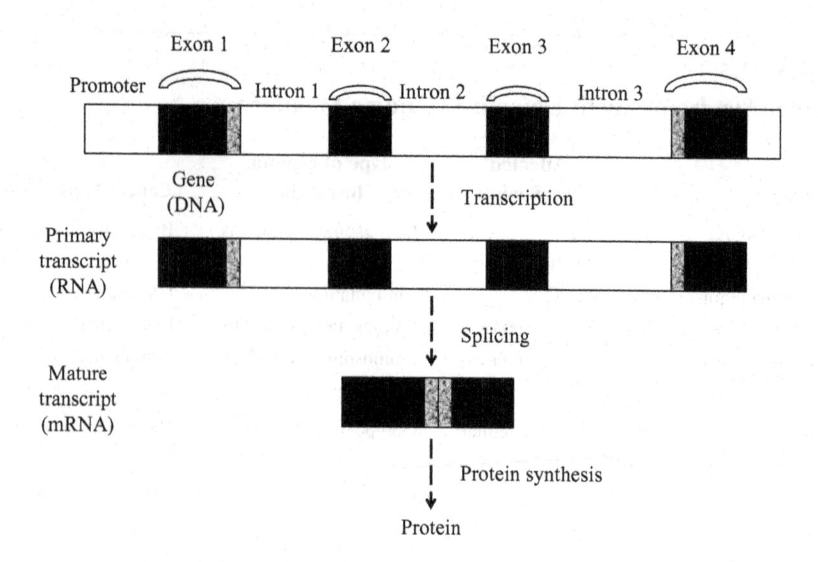

FIGURE 12.1 Structure of a gene.

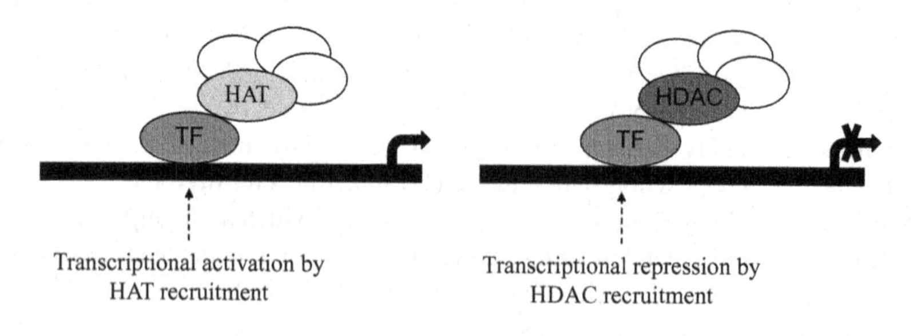

FIGURE 12.2 Endogenous and Exogenous DNA damage pathways.

bond, leading to depurination or depyrimidination; and (d) various chemical conversions in nuclotide bases, stimulated by reactive oxygen species (ROS) (superoxide anions, hydroxyl radicals, and hydrogen peroxide). Figure 12.2 illustrates DNA damage sources, their repair, and their consequences.

Environmental (exogenous) damage includes ionizing radiation (IR); ultraviolet radiation (UV); and various genotoxic chemicals that cause modifications in DNA structure, which can result in mutations that advance cancer risk.

Nucleotide base damage triggers a variety of cellular responses in eukaryotic cells. These cover (a) the induction of cell-cycle checkpoint pathways that prevent cell-cycle progression and provide time for DNA repair; (b) the transcriptional activation of an elevated number of genes; (c) the activation of programmed cell death (apoptosis); (d) the stimulation of biochemical pathways for circumventing the base damage; and (e) the stimulation of explicit DNA repair pathways. Summaries of endogenous and exogenous DNA damage pathways and DNA damage responses are depicted in Figures 12.2 and 12.3, respectively.

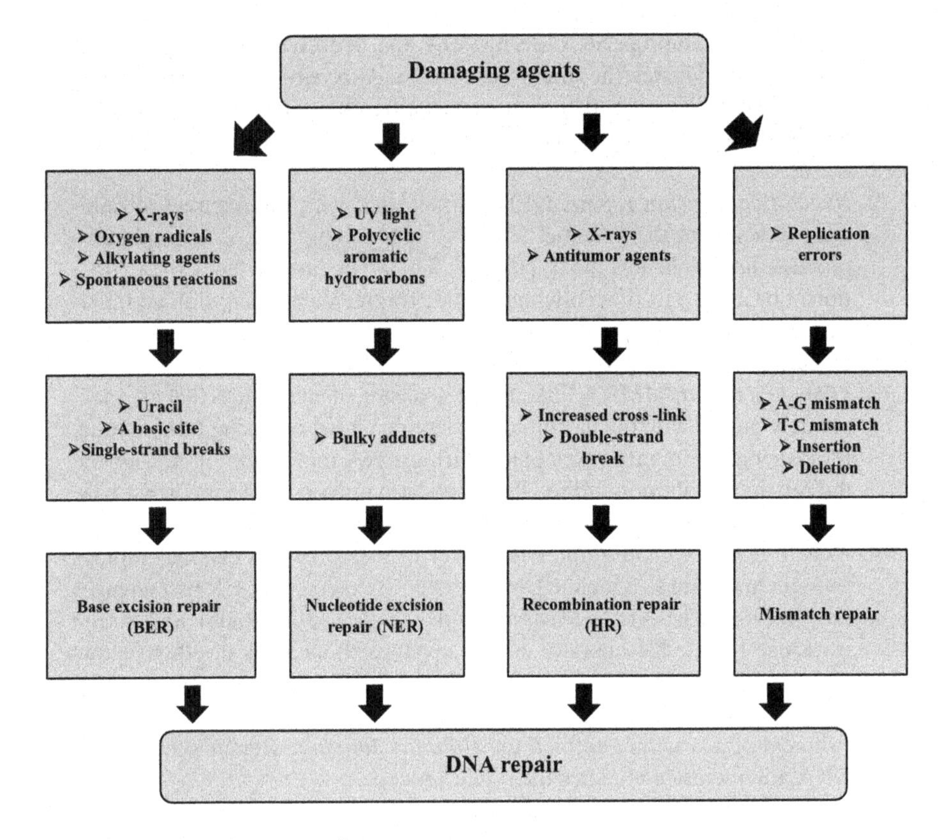

FIGURE 12.3 DNA Damage Responses.

12.3.2 DNA Repair Mechanisms

Eukaryotic cells can restore many types of DNA damage. The major DNA repair pathways are 1. base excision repair (BER); 2. nucleotide excision repair (NER); 3. mismatch repair (MMR); 4. homologous recombination (HR); and 5. nonhomologous end-joining (NHEJ) repair. The first three pathways lead to the excision of the damaged or mispaired bases, whereas HR and NHEJ enable the repair of double-strand breaks (DSBs). These, considered the most lethal of all DNA lesions, can be induced endogenously by cellular processes, including fracture during immunoglobulin gene rearrangement (V (D) J recombination) and meiotic recombination. DSBs are also formed by exposure to IR as well as radiomimetic drugs (e.g., Bleomycin). Unrepaired DSBs can elicit cell-cycle checkpoint arrests and induce cell death. Table 12.5 summarizes diseases associated with different DNA repair mechanisms.

1. *Base excision repair*: BER is induced by DNA repair–specific enzymes (DNA glycosylases), which distinguish a single or a subset of chemically altered or false bases. For instance, uracil DNA glycosylase precisely selects

uracil as an inappropriate base in DNA and hydrolyzes the N-glycosidic bond that conjugates the uracil base to the deoxyribose–phosphate backbone of DNA. Uracil is then removed as a free base, leaving a site of base loss in the DNA. This site is further repaired by an array of biochemical events.

2. *Nucleotide excision repair*: NER is associated with the removal of damaged bases from the genome. NER in mammalian cells is a complicated process that implicates many proteins. The procedure includes the extraction of bases as part of an oligonucleotide fragment. Humans with an inherited deficiency in NER suffer from xeroderma pigmentosum (XP) and have an elevated propensity to skin cancer induced by sunlight exposure.
3. *Mismatch repair*: MMR is linked to the excision of nucleotides that are inaccurately paired with the (accurate) nucleotide on the opposing DNA strand. Mispairing commonly takes place during DNA replication on account of the restricted adhesion of the DNA replication process. The incorrect base appears in the recently incorporated DNA strand.
4. *Homologous recombination*: HR is an error-free DSB process that employs a sister chromatid as template DNA to achieve exact repair. HR first engages an Mre11–Rad50–Nbs1 (MRN) complex, followed by Rad51 and strand invasion. The Rad51 operates cooperatively with an array of other factors (e.g., Rad52 and Rad54 proteins) to foster strand invasion and sequential recombination. Strand invasion and migration are connected with the establishment of a structure entitled the *Holliday Junction*. Repair synthesis by DNA polymerase completes the repair process.
5. *Nonhomologous end joining*: NHEJ is an error-prone DSB repair mechanism that involves the DNA-dependent protein kinase (DNA-PK) and the DNA ligase IV complex, which together expedite the interconnecting of broken inappropriate DNA ends. The DNA-PK complex encompasses the Ku70 and Ku80 proteins, which pinpoint the breach and control the use of the DNA-PK catalytic subunit. DNA-PK is a nuclear serine/threonine protein kinase that consists of a catalytic subunit (DNAPKcs), with the Ku subunits operating as the regulatory element.

12.3.3 Experimental Monitoring for Genetic Toxicity

The ability to delve into thousands of genes (i.e., possibly all of the genes in a cell type) conveys new insights into the responses of biological systems to chemicals or drugs. Microarray technology will be imperative to validate toxic substances, appraise whether toxic effects appear at low doses, and extrapolate responses from one species to another.

The cDNA microarray can be applied to examine gene expression changes as indications of chemical reactions. Following treatment with chemicals (i.e., polycyclic aromatic hydrocarbons, peroxisome proliferators, or estrogenic chemicals), a gene expression "trademark" on a cDNA microarray is presented, which corresponds to the cellular or tissue chain reactions to these chemicals. The response to different chemicals precipitates changes in the expression levels of many genes, which connote a general toxic response; however, a subpopulation of the expressed genes is designated as being exclusive to a distinct class of compounds, specifically at low doses. This application signals compounds as potential carcinogens/toxicants and unfolds their mode of action by elucidating the mediated signal transduction pathways. Table 12.6 summarizes the applications of microarrays in toxicology.

12.3.4 Clinical Monitoring for Genetic Toxicity

By accommodating pharmacogenetics early into clinical evaluation, it is possible to appraise response characteristics to deduce efficacy during large Phase II clinical trials (i.e., in advance of enrollment, it could be determined whether patients have the SNP profile related to efficacy). This approach will reduce inter-subject variation and the haphazard exposure of "unresponsive" patients to a new drug.

Pharmacogenetics can be implemented to develop medicines as well as to advance current drug surveillance strategies. For example, patients on medication would have a blood spot drawn and stored in a designated place. This approach to surveillance would translate to high volume and improve valuable medical data for thousands of patients. Figure 12.4 summarizes the fields and tools of epigenetic toxicology.

TABLE 12.6
Applications of Microarrays in Toxicology

Analyze toxicants on the basis of tissue-specific patterns of gene expression by establishing molecular signatures for chemical exposures
Assess mechanisms of effect of environmental agents through the analysis of gene expression networks
Use toxicant-induced gene expression as a biomarker to assess human exposure
Examine interaction of mixtures of chemicals
Study the effects of low-dose exposures versus high-dose exposures
Compare the effects of toxicants from one species to another

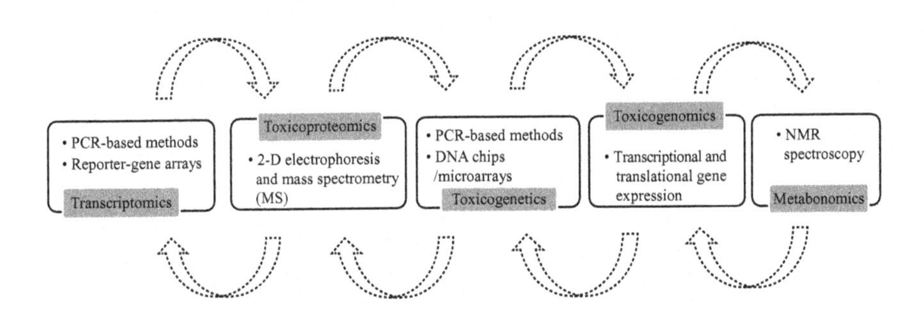

FIGURE 12.4 Tools of epigenetic toxicology.

12.4 EPIGENETIC TOXICOLOGY

Epigenetics is the study of heritable modifications resulting in altered gene function leading to variable phenotype while maintaining an intact genotype. These changes influence the expression of a gene or the properties of its product. Fundamental to regulation of gene expression is the chromatin structure. The essential repeating unit of chromatin is the nucleosome, which consists of 147 bp of DNA wrapped around a core of eight histone proteins, two copies of each, denominated H2A, H2B, H3, and H4. Chromatin exists in two conformational states: heterochromatin, which is conventionally "compact" and transcriptionally disengaged, and euchromatin, which is an "open" configuration suitable for gene transcription. Epigenetic modifications such as DNA methylation and posttranslational modification of histone core proteins govern the differential affinity of the DNA wrapped around them. Reduced affinity between the histone core and the DNA wrapped around it favors euchromatin conformation, resulting in exposure of DNA to the transcriptional machinery and enhancing gene expression, whereas increased affinity results in clustered heterochromatin formation and transcriptional repression. Noncoding RNAs are nontranslatable products of non-protein-coding segments of the mammalian genome, which regulate and influence several aspects of gene transcription and translation in addition to regulating developmental processes, such as embryogenesis and tissue morphogenesis. The ability of noncoding RNAs to modulate gene expression in an extra-genomic pathway makes them the predominant class of epigenetic modifiers.

12.4.1 Mechanisms of Epigenetic Toxicity

12.4.1.1 DNA Methylation

DNA methylation has long been regarded as an epigenetic silencing mechanism of cardinal significance in transcription, chromatin structure, genomic imprinting, and chromosome instability. Modification by DNA methylation arises by the covalent attachment of a methyl group to position 5 of the cytosine ring, forming 5-methylcytosine. DNA methylation induces transcription, in which the methyl group that extends out from the cytosine nucleotide substitutes for transcription factors that commonly attach to the DNA, or entices methyl-binding domains, which in turn are implicated in gene suppression and chromatin condensation. DNA methyl transferases (DNMTs) are the enzymes that catalyze this reaction by substituting the hydrogen with a methyl group on position 5 of the pyrimidine ring in the cytosine molecule. DNMT1, DNMT2, DNMT3a, and DNMT3b are the most commonly recognized mammalian DNA methyltransferases; hemimethylated DNA is the target of DNMT1 (maintenance methylation), and DNMT3a and 3b are responsible for *de novo* methylation. Biochemical analyses of DNMT2 enzymes that contain motifs similar to DNA methyltransferases demonstrated very weak, albeit no specific, DNA methyltransferase activity. However, recent studies confirm RNA methylation activity of DNMT2 at cytosine38 (C38) of tRNAAsp^{GTC} in mice, *Drosophila melanogaster*, and *Arabidopsis thaliana*. Furthermore, sequence alignment analysis of methyltransferase catalytic domains suggests that eukaryotic DNMT1,

DNMT2, and DNMT3 probably evolved from common ancestral DNMT2-like RNA methyltransferases.

12.4.1.2 Posttranslational Modifications of Histone Proteins

The basic amino-terminal tails of histones protrude out of the nucleosome and encounter numerous posttranslational modifications, including 1. acetylation, 2. methylation, 3. phosphorylation, 4. ubiquitination, and 5. sumoylation.

1. *Histone acetylation/deacetylation*: Histone acetylation is the process whereby an acetyl group is presented to lysine residues at the N terminus of the histone protein. The reversed procedure is histone deacetylation, that is, the detachment of an acetyl group. Histone acetylation is triggered by histone acetyltransferases (HATs), while histone deacetylation is stimulated by histone deacetylases (HDACs). HATs and HDACs are mobilized to the DNA by transcription factors (TF) that identify particular DNA sequences (Figure 12.5).

   Transcriptionally active chromatin is associated with acetylated histones, while inactive chromatin is linked to deacetylated histones. Acetylation is also implicated in replication and nucleosome adjustment, chromatin packing, and communications of nonhistone proteins with nucleosomes.
2. *Histone methylation/demethylation*: Histone methylation is involved in the remodeling of amino acids in a histone protein by the joining of one, two, or three methyl groups. Histone methylation takes place on lysine or arginine residues and causes modifications in chromatin architecture by condensing or loosening its structure. The methylation of histones is modulated by

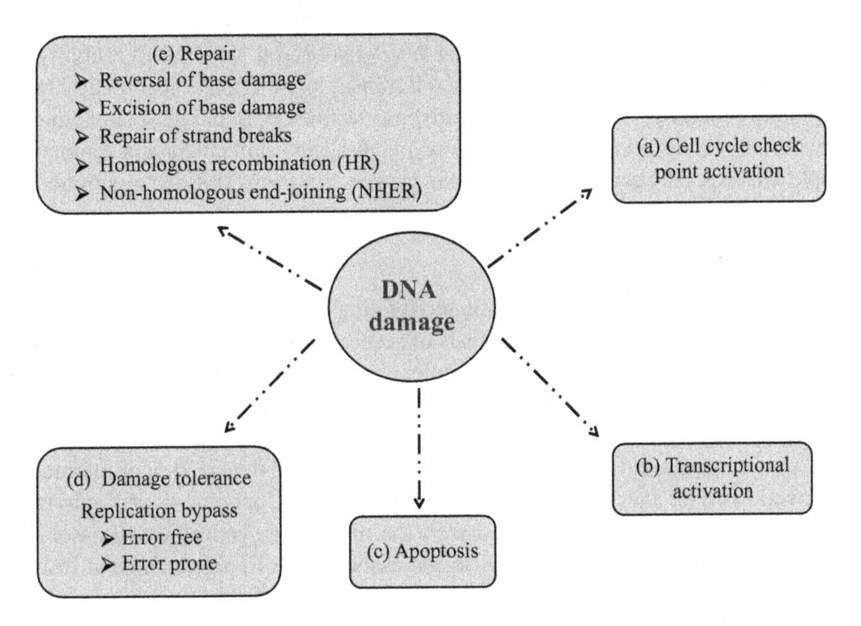

FIGURE 12.5 Transcriptional HAT/HDAC Recruitment.

histone methyltransferases (HMT), and the donor of the methyl group is *S*-adenosylmethionine (SAM). Histone methylation plays a significant role in configuring between the active (euchromatin) and silent states of chromatin (heterochromatin).
3. *Histone phosphorylation/dephosphorylation*: The addition and removal of phosphate groups are engaged in histone modification. The attachment of negatively charged phosphate groups to histone tails abolishes their basic charge and reduces their affinity for DNA. Phosphorylation is activated by histone kinases, whereas dephosphorylation is stimulated by histone phosphatases.
4. *Histone ubiquitination/deubiquitination*: Histone ubiquitination is a modification correlated with the attachment of ubiquitin (Ub) (a 76-amino-acid protein). Ubiquitin has been implicated in protein degradation, cell cycle regulation, endocytosis, and transcription.
5. *Histone sumoylation*: The covalent attachment of a small ubiquitin modifier, SUMO, to lysine residues leads to histone modifications. Histone sumoylation is mediated via a peptide bond between the carboxyl terminus of SUMO and an epsilon (ε) amino group of a lysine residue of the target protein.

12.4.1.3 Noncoding RNA

Genome-wide analysis in eukaryotes reveals a very small amount, 1 to 2%, of the transcribed genome to be functionally active, encoding proteins essential for regulating biological processes; the transcribed, non-protein-coding segment of DNA is referred to as *junk DNA*, and its nontranslatable product is labeled *noncoding RNA*. microRNAs (miRNAs), small nucleolar RNAs (snoRNAs), transcribed ultraconserved regions (T-UCRs), PIWI-interacting RNAs (piRNAs), and a diverse group of long noncoding RNAs (lncRNAs) have been discovered, which differentially target and modulate genomic transcription and translation. miRNAs specifically target and bind to functional mRNA, which leads to posttranscriptional gene silencing and suppression of translation. lncRNAs are postulated to recruit chromatin remodeling complexes to specific gene loci, resulting in progressive transcriptional manipulation.

12.4.2 Epigenetics and Disease

Selective preprogrammed or stochastic modulation of epigenetic information by the environment is considered to be the basis of several diseases, including cancer, type 1 diabetes mellitus, the metabolic syndrome, and autoimmune disorders such as systemic lupus erythematosus and multiple sclerosis. Diet is the most commonly studied environmental factor in relation to transgenerational epigenetic modifications and disease predisposition. For instance, folate deficiency and deprivation of methionine are linked to colon cancer in animals and humans. Also, genome-wide methylation changes in blood cells are correlated with the metabolic syndrome and associated symptoms. Despite current advances in the association of epigenetics and disease, cancer remains as a disease resulting from multiple quantifiable perturbations in the epigenome. Most tumors contain widespread losses and minor gains in DNA methylation throughout the genome. Interestingly, epigenetic changes precede cancer rather than occurring post tumor formation, notably exemplified in Beckwith–Wiedemann syndrome, caused by loss of imprinting of the gene encoding insulin-like growth factor 2 (IGF-2) due to an altered methylation pattern, causing Wilms tumors of the kidney. Cancer can also be a consequence of mutations in EG modulators, which include *writers* (DNA methyltransferases, histone methyltransferases, and HATs); *readers* (bromo-domain, chromo-domain, and tudor protein); *erasers* (HDACs and histone demethylases [HDMs]); and *chromatin remodelers* (SWItch/Sucrose NonFermentable [SWI/SNF]) complex, all of which result in abnormal processing patterns of the genetic code. Whole-genome RNA sequencing analysis led to the discovery of several noncoding RNAs that can serve as potential biomarkers in several disease states. For instance the miRNAs miR-33, -34a, and -22 are considered biologically significant in regulating lipid metabolism and are proposed to serve as valuable epigenetic biomarkers to monitor the metabolic syndrome.

REFERENCES

SUGGESTED READINGS

Bennett, R.L. and Licht, J.D., Targeting epigenetics in cancer. *Ann. Rev. Pharmacol. Toxicol.* 58, 187, 2018.

Clayton, T.A., Lindon, J.C., Cloarec, O., Antti, H., Charuel, C., Hanton, G., Provost, J.P., Le Net, J.L., Baker, D., Walley, R.J., Everett, J.R., and Nicholson, J.K., Pharmaco-metabonomic phenotyping and personalized drug treatment. *Nature* 440, 1073, 2006.

Collis, S.J., DeWeese, T.L., Jeggo, P.A., and Parker, A.R., The life and death of DNA-PK. *Oncogene* 24, 949, 2005.

Engle, L.J., Simpson, C.L., and Landers, J.E., Using high-throughput SNP technologies to study cancer. *Oncogene* 25, 1594, 2006.

Friedberg, E.C., A brief history of the DNA repair field. *Cell Res.* 18, 3, 2008.

Gadhia, S.R. and Barile, F.A., Epigenetic modeling and stem cells in toxicology testing, in *Handbook ofNanotoxicology, Nanomedicine and Stem Cell Use in Toxicology*Handbook of Nanotoxicology, Nanomedicine and Stem Cell Use in Toxicology. 2014, pp. 359–378. Edited by Saura C. Sahu, and Daniel A. Casciano. © 2014 John Wiley & Sons, Ltd.

Gadhia, S.R., Calabro, A.R., and Barile, F.A., Trace metals alter DNA repair and histone modification pathways concurrently in mouse embryonic stem cells. *Toxicol. Lett.* 212, 169, 2012.

Hellebrekers, D.M.E.I., Griffioen, A.W., and van Engeland, M., Dual targeting of epigenetic therapy in cancer. *Biochim. Biophys. Acta* 1775, 76, 2007.

Jirtle, R.L. and Skinner, M.K., Environmental epigenomics and disease susceptibility. *Nat. Rev. Genet.* 8, 253, 2007.

Phillips, E.R., McKinnon, P.J., DNA double-strand break repair and development. *Oncogene* 26, 7799, 2007.

Sinha, A., Singh, C., Parmar, D., and Singh, M.P., Proteomics in clinical interventions: achievements and limitations in biomarker development. *Life Sci.* 80, 1345, 2007.

REVIEW ARTICLES

Bártová, E., Krejčí, J., Harničarová, A., Galiová, G., and Kozubek, S., Histone modifications and nuclear architecture: a review. *J. Histochem. Cytochem.* 56, 711, 2008.

Branzei, D. and Foiani, M., Regulation of DNA repair throughout the cell cycle. *Nat. Rev. Mol. Cell Biol.* 9, 297, 2008.

Chintalapati, A.J. and Barile, F.A., Chapter 46: Epigenetic biomarkers in toxicology, in *Biomarkers in Toxicology*, Gupta, S. (Ed.), 2nd ed., Academic Press, U.K., U.S., Elsevier, 2019.

Feuk, L., Carson, A.R., and Scherer, S.W., Structural variation in the human genome. *Nat. Rev. Genet.* 7, 85, 2006.

Glasspool, R.M., Teodoridis, J.M., and Brown, R., Epigenetics as a mechanism driving polygenic clinical drug resistance. *Br. J. Cancer* 94, 1087, 2006.

Huang, T.T. and D'Andrea, A.D., Regulation of DNA repair by ubiquitylation. *Nat. Rev. Mol. Cell Biol.* 7, 323, 2006.

Konsoula, Z., Barile, F.A. Chapter 19. Epigenetic Modifications and Stem Cell Toxicology: Searching for the Missing Link,in *Handbook of Nanotoxicology, Nanomedicine and Stem Cell Use in Toxicology*. pp. 347–357. Edited by Saura C. Sahu and Daniel A. Casciano. © 2014 John Wiley & Sons, Ltd.

Leon, J., Pharmacogenomics: the promise of personalized medicine for CNS disorders. *Neuropsychopharmacol. Rev.* 34, 159, 2009.

Szyf, M., The dynamic epigenome and its implications in toxicology. *Toxicol. Sci.* 100, 7, 2007.

Waters, M.D. and Foster, J.M., Toxicogenomics and systems toxicology: aims and prospects. *Nat. Rev. Genet.* 5, 936, 2005.

Section II

Toxicity of Therapeutic Agents

13 Sedative/Hypnotics

Frank A. Barile and Anirudh J. Chintalapati

13.1 BARBITURATES

13.1.1 History and Classification

From the beginning of the twentieth century, bromides enjoyed a significant role in the United States as sedatives. With the advent of barbiturates in the 1940s, the use of bromides waned. Since then, barbiturates have been used traditionally for the alleviation of anxiety, pain, hypertension, epilepsy, and psychiatric disorders. Eventually, the therapeutic use of barbiturates would be replaced by other sedative-hypnotics (S/H) of comparable strength, namely, chloral hydrate and meprobamate. With time, benzodiazepines were marketed as therapeutic substitutes for the traditional S/H as more effective, less toxic, and less addictive anxiolytics (sedatives). Today, the most popular therapeutic agents marketed for anxiety (sedation), hypnosis (sleep), or anesthesia (deep sleep) include the benzodiazepines (BZ) as well as zolpidem and the nonnarcotic alternatives such as buspirone (Buspar®). With strict enforcement of federal and state controlled substances regulations, barbiturates have lost significant influence in both the therapeutic and illicit drug markets. Nevertheless, they are still the cause of 50,000 accidental and intentional poisonings per year, and their synergistic effects with ethanol are among the leading causes of hospital admissions.

13.1.2 Epidemiology

The current occurrence of barbiturate abuse and toxicity is relatively low; barbiturates have been replaced with therapeutic alternatives, such as BZs. In contrast to barbiturates, BZs possess a lower incidence of cardiorespiratory toxicity. However, recent reports suggest a gradual increase in the abuse of barbiturates among young adults. The drugs also are referred to by street names such as "purple hearts" and "golfballs" (phenobarbital), "nembies," "yellow jackets" and "Mexican yellows" (pentobarbital), and "red devils" (secobarbital) (U.S. NIDA, 2017).

13.1.3 Medicinal Chemistry

Barbituric acid was first synthesized in 1864 by the German researcher Adolf von Baeyer by condensing urea with diethyl malonate. As such, barbiturates are malonylurea derivatives (diureides; Figure 13.1). The electronegative carbonyl carbons confer an acidic nature to the molecule, thus classifying them as weak acids. Allyl, alkyl, and allocyclic side chains determine their pharmacological classification.

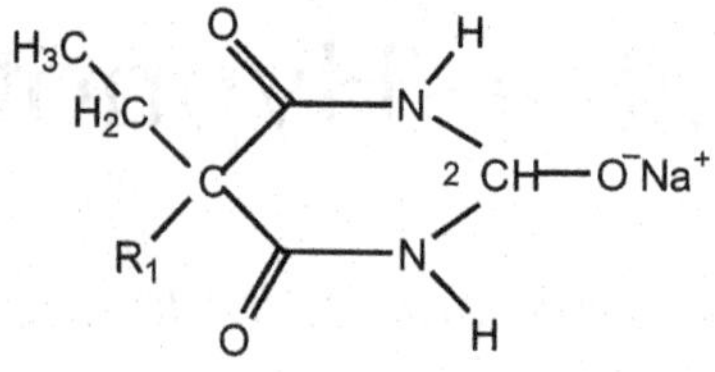

FIGURE 13.1 Structure of prototype barbiturate.

13.1.4 Pharmacology and Clinical Use

Pharmacologically and clinically, barbiturates are classified according to their duration of effect. The presence of longer and bulkier aliphatic and allocyclic side chains (Table 13.1) produces compounds that range from ultrashort, short, and intermediate to long acting. Their major action is the production of sedation, hypnosis, or anesthesia through central nervous system (CNS) depression. The effect, however, depends largely on the dose, the mental status of the patient or individual at the time of ingestion, the duration of action of the drug, the physical environment while under the influence, and the tolerance of the individual to this class of drugs. These factors determine the probability of a therapeutic or euphoric response. For instance, a 100 mg dose of secobarbital ingested in the home at bedtime to induce sleep would, most likely, only cause mild euphoria without significant sedation in a habitual user at a social gathering.

13.1.5 Toxicokinetics and Metabolism

In general, an increase in the number of carbons and bulkier side chains results in enhanced lipid solubility with a corresponding increase in toxicity. The attachment of a methylbutyl (1-mb) and replacement of the C2 with a sulfur group (as with thiopental) decreases its electronegativity, making it less acidic and more lipid soluble. Consequently, thiopental is an ultrashort-acting barbiturate used exclusively as a preoperative S/H.

As a class, the barbiturates are largely nonionic, lipid-soluble compounds, and their pK_a ranges between 7.2 and 7.9. The dissociation constants, therefore, do not account for differences in duration of action, especially with the long-acting compounds. Rapid movement into and out of the CNS appears to determine rapid onset and short duration. Conversely, the barbiturates with the slowest onset and longest duration of action contain the most polar side chains (ethyl and phenyl with phenobarbital structure; Table 13.1). Thus, the structure in Table 13.1 dictates that phenobarbital enters and leaves the CNS very slowly compared with the more lipophilic thiopental (with intermediate pK_a). In addition, the lipid barriers to drug-metabolizing enzymes lead to a slower metabolism for the more polar barbiturates, considering that phenobarbital is metabolized to the extent of 10% per day. Similarly, distribution in biological compartments, especially the CNS, is governed by lipid solubility.

The metabolism of oxybarbiturates occurs primarily in the liver, whereas thiobarbiturates are also metabolized to a limited extent in the kidney and brain.

Phase I reactions introduce polar groups at the C5 position of oxybarbiturates, transforming the radicals to alcohols, ketones, and carboxylic acids. These inactive metabolites are eliminated in urine as glucuronide conjugates. Thiobarbiturates undergo desulfuration to the corresponding oxybarbiturates, followed by opening of the barbituric acid ring. Side chain oxidation at the C5 position is the most important biotransformation reaction leading to drug detoxification.

13.1.6 Mechanism of Toxicity

CNS and cardiovascular (CV) depression accounts for the major toxic manifestations of barbiturate poisoning. The drugs bind to an allosteric site on the GABA-Cl^- ionophore complex (γ-aminobutyric acid), an inhibitory neurotransmitter, in presynaptic or postsynaptic neuronal terminals in the CNS. This complex formation prolongs the opening of the chloride channel. Ultimately, by binding to $GABA_A$ receptors, barbiturates diminish the action of facilitated neurons and enhance the action of inhibitory neurons. Barbiturates concomitantly stimulate the release of GABA at sensitive synapses. Thus, the chemicals have GABA-like effects by decreasing the activity of facilitated neurons and enhancing inhibitory GABA-ergic neurons.

Two major consequences account for the toxic manifestations: 1. barbiturates decrease postsynaptic depolarization by acetylcholine, with ensuing postsynaptic block, resulting in smooth, skeletal, and cardiac muscle depression; 2. at higher doses, barbiturates depress medullary respiratory centers, resulting in inhibition of all three respiratory drives. The *neurogenic drive*, important in maintaining respiratory rhythm during sleep, is initially inhibited. Interference with carotid and aortic chemoreceptors and pH homeostasis disrupts the *chemical drive*. Lastly, interruption of carotid and aortic baroreceptors results in a decreased *hypoxic drive* for respiration. Thus, with increasing depth of depression of the CNS, the dominant respiratory drive shifts to the chemical and hypoxic drives.

13.1.7 Signs and Symptoms of Acute Toxicity

As noted earlier, the signs and symptoms of barbiturate poisoning are related directly to CNS and CV depression. Reactions are dose dependent and vary from mild sedation to complete paralysis. Clinical signs and symptoms are more reliable indicators of clinical toxicity than plasma concentrations. This is especially true when CNS depression does not correlate with plasma concentrations, an indication that other CNS depressants may be involved.

At the highest doses, blockade of sympathetic ganglia triggers hypotension, bradycardia, and decreased inotropy, with consequent decreased cardiac output. In addition, inhibition of medullary vasomotor centers induces arteriolar and venous dilation, further complicating the cerebral hypoxia and cardiac depression. Respiratory acidosis results from accumulation of carbon dioxide, shifting pH balance to the formation of carbonic acid. The condition resembles alcoholic inebriation as the patient presents with hypoxic shock, rapid but shallow pulse, cold and sweaty skin (hypothermia), and either slow or rapid, shallow breathing. Responsiveness and depth of coma are evaluated according to the guidelines for the four stages of coma (see Chapter 3).

TABLE 13.1
Properties of Barbiturates

Compounds[a]	R_1	Classification	Sedative/Hypnotic Dose (Total mg Daily)	Toxic Concentration (mg/dL)	$t_{½}$ (h)	pKa
Barbital	ethyl	LA	100–200/300–500	6–8	57–120	7.8
Mephobarbital	phenyl	LA	32–100/—	1.5	10–70	7.9
Phenobarbital	phenyl	LA	30–90/100–200	4–6	24–140	7.2
Amobarbital	isopentyl	IA	30–150/100–200	1–3	8–42	7.8
Butabarbital	1-mp	IA	15–30/50–100	5	35–50	7.9
Aprobarbital[c]	propene	SA to IA	120/40–160	3–4	14–40	8
Pentobarbital	1-mb	SA	20–150/100	0.5–1.0	16–48	7.9
Secobarbital[b]	1-mb	SA	30–200/100	0.5–1.0	20–34	7.9
Methohexital[d]	1-m2-p	UA	500–2500/—	—	~0.1	7.8
Thiopental	1-mb, C2=S	UA	3–5 mg/kg	<0.5	<1	7.4

[a] Grouped according to classification.
[b] Ethyl group for secobarbital is replaced by an allyl side chain.
[c] Ethyl group for aprobarbital is replaced by isopropyl side chain.
[d] Ethyl group for methohexital is replaced by propene.
1-m2-p: 1-methyl 2-pentyne; 1-mb: 1-methylbutyl; 1-mp: 1-methylpropyl; IA: intermediate acting; LA: long acting; R_1: side group substitutions corresponding to Figure 11.1; SA: short acting; UA: ultrashort acting.

13.1.8 Emergency Guidelines

The following changes during assessment of vital signs are noted in barbiturate poisoning: decrease in blood pressure, decrease in ventilation, bradycardia, and drop in body temperature. Pupil size and reactivity are variable, although miosis (pupillary constriction) is consistent with long-acting barbiturate intoxication (e.g., phenobarbital). Although barbiturate serum levels are generally not useful in moderate to severe poisoning, the serum concentration may establish the diagnosis in a stable patient. It may also distinguish between short- and long-acting barbiturates, since the treatment approach for the former is more critical.

13.1.9 Clinical Management of Acute Overdose

Treatment of overdose is symptomatic and follows the general guidelines adapted from the Scandinavian method for symptomatic treatment for CNS depressants. This includes maintaining adequate ventilation, keeping the patient warm, and supporting vital functions. To this end, oxygen support, forced diuresis, and administration of volume expanders have been shown to maintain blood pressure and adequate kidney perfusion and prevent circulatory collapse. If less than 24 h has elapsed since ingestion, gastric lavage, induction of apomorphine emesis,* delivery of a saline cathartic, or administration of activated charcoal enhances elimination and decrease absorption. In particular, multidose activated charcoal (MDAC) increases the clearance and decreases the half-life of phenobarbital. Alkalization of urine to pH 7.5–8.0 increases the clearance of long-acting barbiturates, while short- and intermediate-acting compounds are not affected by changes in urine pH.

Should renal or cardiac failure, electrolyte abnormalities, or acid–base disturbances occur, hemodialysis is recommended. Although most cases of phenobarbital overdose respond well to cardiopulmonary supportive care, severe cases will also require hemodialysis or charcoal hemoperfusion. Neither of these procedures will remove significant amounts of short- or intermediate-acting barbiturates. Through ion exchange, hemodialysis is more effective in removing long-acting barbiturates than short-acting compounds, because there is less protein and lipid binding of the former. Recently, continuous venovenous hemodiafiltration was used successfully in a case of severe phenobarbital poisoning. The protocol effectively uses continuous renal replacement therapy in phenobarbital clearance associated with severe coma and hypotension.

If the renal and cardiac functions are stabilized, alkalization of urine and plasma with sodium bicarbonate promotes ionization of the acidic compounds. This maneuver hastens their excretion provided that reabsorption and protein and lipid binding are minimized. The incorporation of CNS stimulants or vasopressors into the

* Guidelines issued by the American Association of Poison Control Centers (AAPCC, 2017) suggest that the effectiveness of ipecac syrup as an emetic in preventing drug absorption is questionable. There are potentially significant contraindications and adverse effects associated with the use of ipecac syrup. The panel concluded that its routine administration following oral poisoning, in both inpatient and outpatient departments, is rarely the appropriate or desired method of gastric decontamination (see Chapter 3 for a detailed discussion on ipecac administration).

treatment protocol is not recommended, since mortality rates from pharmacological antagonism are as high as 40%.

Even with complete recovery, complications affect prognosis. These include the development of pulmonary edema and bronchopneumonia, infiltration with lung abscesses, and renal shutdown.

13.1.10 Tolerance and Withdrawal

Pharmacodynamic tolerance (functional or adaptive) contributes to the decreased effect of barbiturates, even after a single dose. With chronic administration of gradually increasing doses, pharmacodynamic tolerance continues to develop over a period of weeks to months, as the homeostatic feedback response to depression is activated in the presence of the compounds. Interestingly, no concomitant increase in LD_{50} is observed; rather, as tolerance increases, the therapeutic index decreases.

Pharmacokinetic tolerance (drug disposition) contributes less to the development of this phenomenon. Enhanced metabolic transformation and induction of hepatic microsomal enzymes ensue subsequent to pharmacodynamic tolerance, the former of which reaches a peak after a few days. In either situation, the effective dose of a barbiturate is increased six-fold, a rate which is primarily accounted for by the production of pharmacodynamic tolerance.

Sudden withdrawal of barbiturates from addicted individuals runs the risk of development of hallucinations, sleeplessness, vertigo, and convulsions, depending on the degree of dependency.

13.1.11 Methods of Detection

Several gas chromatographic methods are used for the analysis of barbiturates. With the advent of modern fused-silica capillary columns, the historical need for derivatization has decreased. The procedure is used to avoid adsorption and peak tailing on the chromatogram; however, some barbiturates display similar mass spectra, and diligent use of retention times is needed for additional resolution, necessitating some derivatization. Sample preparation requires acid followed by organic solvent extraction, or a direct extraction into ethanol; sample drying; and injection into a methyl- or phenylmethylsilicone chromatographic column.

13.2 BENZODIAZEPINES (BZ)

13.2.1 Epidemiology

The introduction of chlordiazepoxide in 1961 started the era of BZ. Although this class of S/H did not render the barbiturates obsolete, they have since enjoyed wide therapeutic utility as S/H, anxiolytics, anticonvulsants, preanesthetic sedatives, and muscle relaxants. In particular, their increased therapeutic index relative to barbiturates and their lack of anesthetic properties have promoted the substitution of BZs for barbiturates. Nonetheless, about 27,060 BZ exposures were reported to U.S. poison control centers in 2014, of which 370 resulted in major toxicity and 18 cases in death (Mowry et al., 2015). Concurrent misuse of S/H drugs and/or alcohol with BZs results in synergistic toxicity. Analysis of a West Virginia database showed that alprazolam contributed to 17% of drug-related deaths, 97.5% of which involved co-administration of CNS depressants (Shah et al., 2012).

13.2.2 Medicinal Chemistry

Figure 13.2 illustrates the 5-aryl-1,4-benzodiazepine structure. All the important BZs currently in use (Table 13.2) contain a 5-aryl or 5-cyclohexenyl moiety. The chemical nature of substitutions at positions 1 and 3 varies widely. Their half-life correlates well with their duration of action as well as the duration of toxic manifestations. The structure–activity relationship, however, has failed to correlate the pharmacological and toxicological profiles with the chemical structure.

13.2.3 Pharmacology and Mechanism of Toxicity

BZs bind to all three omega receptor subtypes in the areas of the limbic system, thalamus, and hypothalamus. BZs bind to an allosteric site of the α and/or β subunits of the $GABA_A$–Cl^- ionophore complex. This action increases the frequency of the opening of the chloride channels. Ultimately, the drugs enhance the affinity of GABA for $GABA_A$ receptors and potentiate the effects of GABA throughout the nervous system. The effects of GABA-mediated actions account for BZs' S/H, anticonvulsant, and skeletal muscle relaxation properties.

At high doses, BZs induce neuromuscular blockade and cause vasodilation and hypotension (after intravenous administration). The compounds do not significantly alter ventilation except in patients with respiratory complications, in the elderly population, and in the presence of alcohol or other S/H. There is also minimal effect on cardiovascular integrity.

13.2.4 Toxicokinetics

With the exception of clorazepate (Table 13.2), all the BZs are completely absorbed, primarily due to their high nonionic/ionic ratio. Lipophilicity, however, does not account for the variation in polarity and electronegativity conferred by the various substituents.

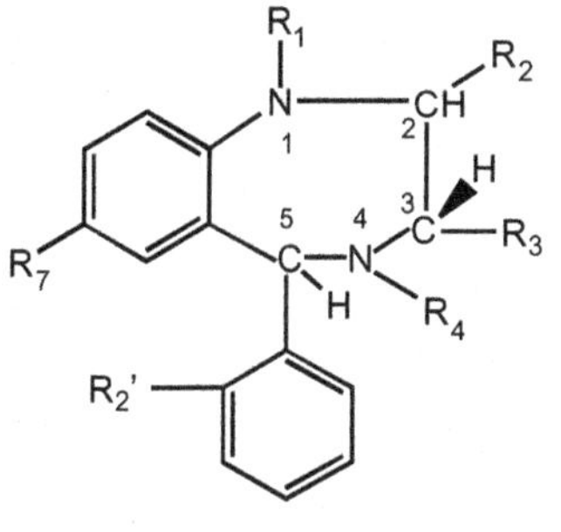

FIGURE 13.2 Structure of prototype benzodiazepine.

13.2.5 Signs and Symptoms of Acute Toxicity

Although signs and symptoms are generally nonspecific, apparent toxicity depends on the extent of intoxication. Serum toxic concentrations of BZs do not correlate well with signs and symptoms and are generally listed as "cut-off" points (Table 13.2).

Mild toxicity is characterized by ataxia, drowsiness, and motor incoordination. Psychologically, the patient displays different degrees of paranoia or erratic behavior and is easily aroused. In moderate toxicity, the patient is aroused by verbal stimulation, although they may enter coma stage one or two. Patients in severe toxicity are unresponsive except to deep pain stimulation, consistently with coma stage one or two. In general, respiratory depression and hypotension are rare.

13.2.6 Emergency Guidelines

BZ overdose is associated with relatively low morbidity and mortality. Although rare, fatalities are more likely to occur with short-acting benzodiazepines (e.g., triazolam, alprazolam, or temazepam). During assessment, vital sign changes occur with BZ poisoning, including drowsiness, slurred speech, confusion, and cognitive impairment. Respiratory depression and hypotension occur more often with parenteral administration. Pupil size and reactivity are not reliable. Screenings for BZ serum concentrations are generally not useful in overdose.

13.2.7 Clinical Management of Acute Overdose

Clinical management is symptomatic and may also incorporate the use of a specific antidote. Flumazenil (Romazicon®, Roche Pharmaceuticals, NJ, United States), 1,4-imidazobenzodiazepine, is a BZ antagonist. It competitively antagonizes the binding and allosteric effects of BZs. Flumazenil completely reverses sedative, anxiolytic, anticonvulsant, ataxic, anesthetic, comatose, and muscle relaxant effects. Intravenous administration of 0.2–1.0 mg has an acute onset of 1–3 minutes and a peak effect at 6–10 minutes. Routine use of the antagonist, however, must allow the determination of concomitant drug ingestion. The presence of stimulants such as cocaine, amphetamine, or tricyclic antidepressants precludes the routine use of the antidote, since the anticonvulsant effects of BZs may be neutralized by flumazenil.

13.2.8 Tolerance and Withdrawal

Cross-tolerance to barbiturates and alcohol has been noted with BZs, although the withdrawal syndrome is not as severe as with the former and is slower in onset (up to 2 weeks). Interestingly, depending on the dose, flumazenil administration will precipitate acute withdrawal in BZ-addicted individuals. As with barbiturates and alcohol, BZ initial withdrawal signs and symptoms are similar.

13.2.9 Methods of Detection

Although BZs have relatively long retention times, stable fused-silica capillary chromatographic columns with better stationary phases have improved separation.

TABLE 13.2
Properties of Benzodiazepines

Generic Name[a]	Proprietary Name	Classification	Sedative/Hypnotic Total Daily Dosage (mg)	Toxic Concentration	$t_{½}$ (h)
Alprazolam	Xanax	Anxiolytic	0.25–1.5/0.5–1.0	0.4 µg/dL	12–19
Bromazepam	Lectopam	Anxiolytic	6–9/–	30–40 µg/dl	10–22
Camazepam	Paxor	Anxiolytic–S/H	0.5–2/10–20	0.2 mg/dL	20–24
Chlordiazepoxide	Librium	Anxiolytic	5–100/25–50	3.5–10 mg/L	7–28
Clorazepate	Tranxene	Anxiolytic–S/H	3.75–22.5/7.5–15	–	30–60
Diazepam	Valium	Anxiolytic–S/H, SKR, anticonvulsant	2–40/5–10	0.5–2.0 mg/dL	20–90
Flurazepam[b]	Dalmane	Hypnotic	–/15–30	0.25 mg/dL	24–100
Halazepam	Paxipam	Anxiolytic-S/H	20–40/–	–	30–40
Lorazepam	Ativan	Anxiolytic–S/H, anticonvulsant	0.5–3.0/2–4	0.3 µg/dL	10–20
Midazolam	Versed	Anxiolytic–S/H, amnestic	1–5/–	0.1–0.15 mg/dL	1.5–2.5
Oxazepam	Serax	Anxiolytic–S/H	10–120/10–30	3–5 mg/L	5–10
Prazepam	Centrax	Anxiolytic, anticonvulsant	20–60/-	–	63–70
Temazepam[b]	Restoril	S/H	—/7.5–30	1.0 mg/L	9–12
Triazolam[b]	Halcion	Anxiolytic-S/H	0.25–1.5/0.125–0.5	7 µg/kg (toxic dose)	2–3

[a] Grouped by generic name.

[b] Used to treat insomnia.

S/H: sedative/hypnotic; SKR: skeletal muscle relaxant.

Compounds such as chlordiazepoxide and clorazepate are inclined to form derivatives during preparation that interfere with proper identification.* Nevertheless, OV-1 and SE-30 phases are used for packed columns with corresponding retention indices.

13.3 MISCELLANEOUS SEDATIVE/HYPNOTICS

13.3.1 Chloral Hydrate

Although it has no analgesic effect, and more effective and less toxic S/H drugs are available, chloral hydrate is still used therapeutically, primarily as a hypnotic.† Currently available in oral liquid dosage forms only, the compound is lipid soluble and is a metabolite of chloral betaine and triclofos (see structure in Figure 13.3). The long-acting metabolite of chloral hydrate, trichloroethanol, is responsible for most of its toxicity, low therapeutic index, and undetectable presence in plasma. Signs and symptoms of toxicity include CNS depression, ataxia, gastrointestinal irritation, cardiovascular instability, and proteinuria. In addition, chloral hydrate significantly impairs myocardial contractility by sensitizing the myocardium to catecholamines. An increased risk of sudden death with chloral hydrate intoxication is a result of the development of arrhythmias. Emergency guidelines and assessment of vital signs include pear-like breath odor, hypotension, and dysrhythmias. Gastrointestinal irritation and bleeding are also a common occurrence.

Abdominal X-rays may be useful in diagnosis, since the compound is radio-opaque.

Beta-blockers, such as propranolol, are recommended for ameliorating chloral hydrate–induced cardiac arrhythmias. Recently, chloral hydrate has been implicated as a carcinogen, although this finding remains controversial.

13.3.2 Meprobamate (Miltown®, Equanil®)

A propanediol carbamate derivative, meprobamate is still marketed as an alternative S/H to barbiturates. It enjoyed some popularity until the introduction of benzodiazepines, yet its toxicity is similar to that of barbiturates, including the production of ataxia and coma. Emergency guidelines and assessment of vital signs with meprobamate intoxication include hypotension, seizures, cardiac arrhythmias, and pulmonary edema. Gastrointestinal concretion formation with high doses of the drug

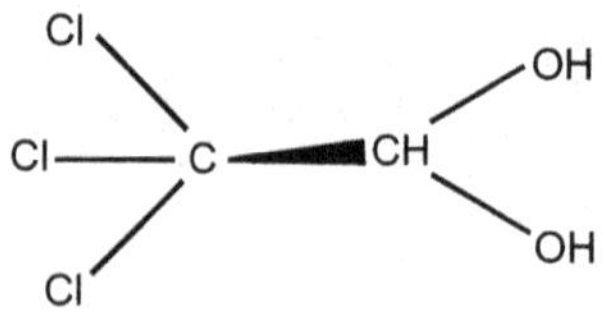

FIGURE 13.3 Structure of chloral hydrate.

* Consequently, HPLC may be a valuable alternative to GC methods of analysis.

† The mixture of chloral hydrate and ethanol is called a *Mickey Finn* or knock-out drops. The drink was popularized as such in classic early twentieth-century films for its dramatic and instantaneous hypnotic effect.

may promote continuous absorption, thus complicating treatment and stabilization of vital signs. Chronic use of meprobamate is associated with severe hematopoietic disturbances, manifested primarily as aplastic anemia and thrombocytopenia.

13.3.3 Zolpidem Tartrate (Ambien®)

Zolpidem is a non-benzodiazepine hypnotic of the imidazopyridine class. Pharmacologically, it interacts with the GABA–BZ receptor complex and shares some pharmacological properties of BZs. In contrast to the latter, zolpidem binds preferentially to the ω_1 receptor, which is found primarily in the midbrain, brain stem, and thalamic regions. Selective binding to ω_1 receptors explains the relative absence of muscle relaxation, anticonvulsant activity, and induction of deep sleep. The drug is relatively safe, as large ingestions produce mostly drowsiness, while coma and respiratory depression are rare. As with BZs, flumazenil reverses some effects of zolpidem overdose, but its use as an antidote is not routinely recommended.

13.3.4 Buspirone (Buspar®)

Buspirone is an antianxiety agent that is not chemically or pharmacologically related to the BZs, barbiturates, or other S/H drugs. It does not exert anticonvulsant or muscle relaxant effects, nor does it interact with GABA receptors. Instead, it has high affinity for serotonin ($5\text{-}HT_{1A}$) and dopamine (D_2) receptors. It is indicated for the management of anxiety disorders characterized by motor tension, autonomic hyperactivity, apprehensive expectation, and vigilance and scanning (hyperattentiveness). Overdoses of up to 150 times the average anxiolytic dose (approximately 3 g) have not been associated with lasting untoward effects. In addition, additive effects with ethanol are minimal. It has not been associated with the production of dependence or withdrawal, and unlike the barbiturates and BZs, buspirone is not a federally controlled substance.

13.3.5 Flunitrazepam (Rohypnol®)

Although flunitrazepam is used internationally in the treatment of insomnia as well as a preanesthetic benzodiazepine, it is a U.S. federal schedule I controlled substance with no legitimate therapeutic or prescribing uses. The tablets are tasteless, odorless, and relatively inexpensive, dissolve rapidly in alcohol, and are easily administered by intranasal or oral routes. Flunitrazepam has been used as a "date-rape" drug because of its intoxicating and amnestic effects.* The agent causes euphoria, hallucinations, and disinhibiting effects and has also been used to enhance heroin and cocaine euphoria. It is highly lipid soluble with a quick onset (20–30 min) and duration of up to 12 h. It affects $GABA_A$ receptors with a potency of up to 10 times that of diazepam. Consequently, drowsiness, disorientation, and dizziness ("DDD"), slurred

* According to the U.S. Department of Justice, "drug-facilitated rape," or acquaintance rape, is defined as the sexual assault of a woman who is incapacitated or unconscious while under the influence of a mind-altering drug and/or alcohol. The victim of a drug-facilitated sexual assault may exhibit signs of confusion, memory loss, drowsiness, impaired judgment, slurred speech, and uninhibited behavior. Suspicion of acquaintance rape is based on clinical history and emergency department presentation.

speech, and nausea can progress rapidly to amnesia, psychomotor impairment, respiratory depression, and coma. Treatment is supportive with the maintenance of patent airway and respiration of primary importance (airway, breathing, and circulation [ABCs]). Flumazenil is used as an antidote for the respiratory depression except in the presence of stimulants.

13.3.6 Gamma-Hydroxybutyrate (GHB)*

Since its synthesis in 1960, GHB has been promoted and abused for sleep disorders (narcolepsy), as an anesthetic agent, as treatment for alcohol and controlled substance withdrawal, and currently, as an inducer of growth hormone (bodybuilding) and aphrodisiac. The tasteless, odorless, liquid or gel form makes it convenient for use as a "date-rape" drug.

Pharmacologically, GHB interacts with GABA-B and GHB receptors, as well as opioid receptors, resulting in a predominant CNS depression. It has a quick onset (15–30 min) and relatively short duration (3 h). Most users of GHB report that it induces a pleasant state of relaxation and tranquility. Frequent effects are a tranquil mental state, mild euphoria, emotional warmth, well-being, and a tendency to verbalize. Ingestion of GHB in an alcoholic drink by an unsuspecting victim, however, causes hallucinations and amnesia, as well as symptoms not unlike flunitrazepam. Alcohol potentiates the depressant and adverse reactions, including dizziness, nausea, vomiting, respiratory collapse, seizures, coma, and possibly death.

Mild to moderate GHB intoxication eventually outweighs the euphoric effects, since the chemical produces a sudden onset of aggressive behavior followed by drowsiness, dizziness, rapid reawakening, and amnesia within 15 minutes of ingestion. With higher doses, marked agitation on physical stimulation, apnea, hypoxia, and vomiting, followed by dose-dependent respiratory depression, bradycardia, miosis, involuntary muscle contractions, decreased cardiac output, coma, and amnesia, are common.

As with flunitrazepam, treatment of GHB toxicity is supportive, with respiratory maintenance of primary importance (ABCs). Resolution of signs and symptoms occurs within 2–8 hours.

Withdrawal symptoms are significant. The addicted individual experiences profuse sweating, anxiety attacks, and increased blood pressure a few hours after the last dose. The reactions may subside after several days, depending on the depth of addiction, or may be followed by a second phase of hallucinations and altered mental state.

GHB is detected in most biological fluids using HPLC following derivatization with a coumarin analogue coupled with florescence detection.

13.3.7 Ethchlorvynol (Placidyl®), Methaqualone (Quaalude®), Glutethimide (Doriden®), Methyprylon (Noludar®)

Historically, these compounds were associated with significant toxic potential and abuse, and their popularity in the illicit drug market soared in the 1960s. The compounds, however, are no longer available therapeutically and have not resurfaced as

* Also known as gamma-hydroxybutyric acid or GHBA.

significant drugs of abuse. The manifestations of their toxicology are similar to those of barbiturate toxicity. Individually, they possess distinct diagnostic features, which are thoroughly discussed in Section 13.5.2 (Review Articles).

13.4 METHODS OF DETECTION AND LABORATORY TESTS FOR S/H

Urine screening is useful for detecting many illicit drugs, except flunitrazepam, which requires specific analysis. Because of the short half-life and low concentrations of most S/H drugs, most commercially available toxicology screens are unable to detect flunitrazepam or GHB used in acquaintance rape. The most definitive detection of flunitrazepam is gas chromatography-mass spectrometry (GC-MS), although an LC method has been recently described (Maublanc et al., 2015). Many community districts are developing programs to analyze for date-rape drugs as part of sexual assault kits.

More recently, gas and liquid chromatography (GC and LC, respectively) methods have supplanted the immunoassays and are more reliable for the identification, screening, and confirmation of barbiturates. These include GC with flame ionization detection after derivatization (GC/FID), GC with nitrogen-phosphorus detection (GC/NPD), GC-MS, and liquid chromatography with mass spectroscopy (LC-MS). The immunoassays, enzyme-multiplied immunoassay technique (EMIT) and radioimmunoassay (RIA), are designed to identify unmetabolized secobarbital in the urine. Both assays will detect other commonly encountered barbiturates as well, depending on the concentration of drug present in the sample. With GC, LC, and immunoassays, positive tests for phenobarbital have been noted in chronic users up to several weeks after discontinuation of the drug. Thin layer chromatography (TLC) is less sensitive than most chromatographic techniques but can be used as a screening method to demonstrate the presence of barbiturates within a specimen for up to 30 h.

BZs are generally detected as a structural class. A broad spectrum of analytical methods, including GC, HPLC, TLC, GLC/MS, RIA, and EMIT, has been reported for the analysis of the BZs. Many BZs have common metabolites; therefore, it is not always possible to determine the specific drug taken through the use of urine testing. The EMIT screening procedure rapidly recognizes oxazepam, a common and specific metabolite of many benzodiazepines. Other screening procedures, such as TLC and HPLC, are used to detect benzodiazepines in urine alone or in combination with other major drugs of abuse. RIA methods are sensitive enough for the determination of diazepam directly in microsamples of blood, plasma, and saliva for up to 16 h following oral administration of a single 5 mg dose. Confirmation of positive results from tests performed with immunoassay or TLC may be difficult. More specific GC and GC-MS procedures may not be able to confirm the metabolites screened with TLC.

REFERENCES

Gummin, D.D., Mowry, J.B., Spyker, D.A., Brooks, D.E., Osterthaler, M.P.H., and Banner, W. 2017 Annual Report of the American Association of Poison Control Centers' National Poison Data System (NPDS): 35th Annual Report. *Clin Toxicol (Phila).* 56, 1, 2018.

Mowry, J.B., Spyker, D.A., Brooks, D.E., McMillan, N., and Schauben, J.L., 2014 Annual Report of the American Association of Poison Control Centers' National Poison Data System (NPDS): 32nd Annual Report. *Clin. Toxicol. (Phila)* 53, 962, 2015.

Shah, N.A., Abate, M.A., Smith, M.J., Kaplan, J.A., Kraner, J.C., and Clay, D.J., Characteristics of alprazolam-related deaths compiled by a centralized state medical examiner. *Am. J. Addict.* 21, S27, 2012.

SUGGESTED READINGS

Abanades, S., Farré, M., Barral, D., Torrens, M., Closas, N., Langohr, K., Pastor, A., and de la Torre, R., Relative abuse liability of gamma-hydroxybutyric acid, flunitrazepam, and ethanol in club drug users. *J. Clin. Psychopharmacol.* 27, 625, 2007.

Beynon, C.M., McVeigh, C., McVeigh, J., Leavey, C., and Bellis, M.A., The involvement of drugs and alcohol in drug-facilitated sexual assault: a systematic review of the evidence. *Trauma Violence Abuse* 9, 178, 2008.

Bay, T., Eghorn, L.F., Klein, A.B., and Wellendorph, P., GHB receptor targets in the CNS: focus on high-affinity binding sites. *Biochem. Pharmacol.* 87, 220, 2014.

Brettell, T.A., Chapter 16. Forensic science applications of gas chromatography, in *Modern Practice of Gas Chromatography*, 4th ed., Grob, R.L. and Barry, E.F. (Eds.), John Wiley & Sons, New York,2004.

Busardò, F.P. and Jones, A.W. GHB pharmacology and toxicology: acute intoxication, concentrations in blood and urine in forensic cases and treatment of the withdrawal syndrome. *Curr. Neuropharmacol.* 13, 47, 2015.

Coomber, R., Moyle, L., Belackova, V., Decorte, T., Hakkarainen, P., Hathaway, A., Laidler, K.J., Lenton, S., Murphy, S., Scott, J., Stefünkova, M., van de Ven, K., Vlaemynck, M., Werse, B. The burgeoning recognition and accommodation of the social supply of drugs in international criminal justice systems: an eleven-nation comparative overview. *Int. J. Drug Policy* 58, 93, 2018.

Dias, A.S., Castro, A.L., Melo, P., Tarelho, S., Domingues, P., and Franco, J.M., A fast method for GHB-GLUC quantitation in whole blood by GC-MS/MS (TQD) for forensic purposes. *J. Pharm. Biomed. Anal.* 150, 107, 2018.

Dziadosz, M., LC-MS small-molecule drug analysis in human serum: could adducts have good prospects for therapeutic applications? *Bioanalysis* 10, 371, 2018.

Egger, S.S., Bachmann, A., Hubmann, N., Schlienger, R.G., and Krahenbuhl, S., Prevalence of potentially inappropriate medication use in elderly patients: comparison between general medical and geriatric wards. *Drugs Aging* 23, 823, 2006.

Heyerdahl, F., Hovda, K.E., Bjornaas, M.A., Brørs, O., Ekeberg, O., and Jacobsen, D., Clinical assessment compared to laboratory screening in acutely poisoned patients. *Hum. Exp. Toxicol.* 27, 73, 2008.

Isbister, G.K., O'Regan, L., Sibbritt, D., and Whyte, I.M., Alprazolam is relatively more toxic than other benzodiazepines in overdose. *Br. J. Clin. Pharmacol.* 58, 88, 2004.

Krenzelok, E.P. Ipecac syrup-induced emesis … no evidence of benefit. *Clin. Toxicol. (Phila)* 43, 11, 2005.

Kriikku, P.Hurme, H., Wilhelm, L., Rintatalo, J., Hurme, J., Kramer, J., Ojanperä, I. Sedative-hypnotics are widely abused by drivers apprehended for driving under the influence of drugs. *Ther. Drug Monit.* 37, 339, 2015.

Manoguerra, A.S. and Cobaugh, D.J., Guidelines for the Management of Poisoning Consensus Panel, Guideline on the use of ipecac syrup in the out-of-hospital management of ingested poisons. *Clin. Toxicol. (Phila)* 43, 1, 2005.

Qureshi, A.M., Mumtaz, S., Rauf, A., Ashraf, M., Nasar, R., and Chohan, Z.H., New barbiturates and thiobarbiturates as potential enzyme inhibitors. *J. Enzyme Inhib. Med. Chem.* 30, 119, 2015.

Sigmarsdottir, B.D., Gudmundsdottir, T.K., Zoëga, S., and Gunnarsson, P.S., Drugs that can cause respiratory depression with concomitant use of opioids. *Scand. J. Pain* 16, 186, 2017.

Zacharis, C.K., Raikos, N., Giouvalakis, N., Tsoukali-Papadopoulou, H., and Theodoridis, G.A., A new method for the HPLC determination of gamma-hydroxybutyric acid (GHB) following derivatization with a coumarin analogue and fluorescence detection: application in the analysis of biological fluids. *Talanta* 75, 356, 2008.

REVIEW ARTICLES

Alagarsamy, V., Chitra, K., Saravanan, G., Solomon, V.R., Sulthana, M.T., and Narendhar, B., An overview of quinazolines: pharmacological significance and recent developments. *Eur. J. Med. Chem.* 151, 628, 2018.

Bourcier, E., Korb-Savoldelli, V., Hejblum, G., Fernandez, C., and Hindlet, P., A systematic review of regulatory and educational interventions to reduce the burden associated with the prescriptions of sedative-hypnotics in adults treated for sleep disorders. *PLoS One* 13, e0191211, 2018.

Brohan, J. and Goudra, B.G., The role of GABA receptor agonists in anesthesia and sedation. *CNS Drugs* 31, 845, 2017.

Castelli, M.P., Multi-faceted aspects of gamma-hydroxybutyric acid: a neurotransmitter, therapeutic agent and drug of abuse. *Mini Rev. Med. Chem.* 8, 1188, 2008.

Druid, H., Holmgren, P., and Ahlner, J., Flunitrazepam: an evaluation of use, abuse and toxicity. *Forensic Sci. Int.* 122, 136, 2001.

Elko, C.J., Burgess, J.L., and Robertson W.O., Zolpidem-associated hallucinations and serotonin reuptake inhibition: a possible interaction, *J. Toxicol. Clin. Toxicol.* 36, 195, 1998.

ElSohly, M.A. and Salamone, S.J., Prevalence of drugs used in cases of alleged sexual assault. *J. Anal. Toxicol.* 23, 141, 1999.

Fluyau, D., Revadigar, N., and Manobianco, B.E., Challenges of the pharmacological management of benzodiazepine withdrawal, dependence, and discontinuation. *Ther. Adv. Psychopharmacol.* 8, 147, 2018.

Freese, T.E., Miotto, K., and Reback, C.J., The effects and consequences of selected club drugs. *J. Subst. Abuse Treat.* 23, 151, 2002.

Hon, K.L., Ho, J.K., Hung, E.C., Cheung, K.L., and Ng, P.C., Poisoning necessitating pediatric ICU admissions: size of pupils does matter. *J. Natl. Med. Assoc.* 100, 952, 2008.

Ilomäki, J. Paljärvi, T., Korhonen, M.J., Enlund, H., Alderman, C.P., Kauhanen, J., and Bell, J.S., Prevalence of concomitant use of alcohol and sedative-hypnotic drugs in middle and older aged persons: a systematic review. *Ann. Pharmacother.* 47, 257, 2013.

Lee, S.J. and Levounis, P., Gamma hydroxybutyrate: an ethnographic study of recreational use and abuse. *J. Psychoactive Drugs* 40, 245, 2008.

Mandrioli, R., Mercolini, L., and Raggi, M.A., Benzodiazepine metabolism: an analytical perspective. *Curr Drug. Metab.* 9, 827, 2008.

Marchi, I., Schappler, J., Veuthey, J.L., and Rudaz, S., Development and validation of a liquid chromatography-atmospheric pressure photoionization-mass spectrometry method for the quantification of alprazalam, flunitrazepam, and their main metabolites in haemolysed blood. *J. Chromatogr. B Analyt. Technol. Biomed. Life Sci.* 877, 2275, 2009.

Maublanc, J., Dulaurent, S., Morichon, J., Lachâtre, G., and Gaulier, J.M., Identification and quantification of 35 psychotropic drugs and metabolites in hair by LC-MS/MS: application in forensic toxicology. *Int J Legal Med.* 129, 259, 2015.

Miller, N.S. and Gold, M.S., Benzodiazepines: reconsidered, *Adv. Alcohol Subst. Abuse* 8, 67, 1990.

Mo, Y., Thomas, M.C. and Karras, G.E. Jr., Barbiturates for the treatment of alcohol withdrawal syndrome: a systematic review of clinical trials. *J. Crit. Care* 32, 101, 2016.

Nakamura, M., Ohmori, T., Itoh, Y., Terashita, M., and Hirano, K., Simultaneous determination of benzodiazepines and their metabolites in human serum by liquid chromatography-tandem mass spectrometry using a high-resolution octadecyl silica column compatible with aqueous compounds. *Biomed. Chromatogr.* 23, 357, 2009.

O'Connell, T., Kaye, L., and Plosay, J.J. 3rd, Gamma-hydroxybutyrate (GHB): a newer drug of abuse. *Am. Fam. Physician* 62, 2478, 2000.

Santos, C. and Olmedo, R.E. Sedative-hypnotic drug withdrawal syndrome: recognition and treatment. *Emerg. Med. Pract.* 19, 1, 2017.

Smith, K.M., Larive, L.L., and Romanelli, F., Club drugs: methylenedioxymethamphetamine, flunitrazepam, ketamine hydrochloride, and gamma-hydroxybutyrate. *Am. J. Health Syst. Pharm.* 59, 1067, 2002.

Tanaka, E., Toxicological interactions between alcohol and benzodiazepines. *J. Toxicol. Clin. Toxicol.* 40, 69, 2002.

Thornton, S.L., Oller, L., and Coons, D.M., 2016 Annual Report of the University of Kansas Health System Poison Control Center. *Kans. J. Med.* 11, 1, 2018.

Traub, S.J., Nelson, L.S., and Hoffman, R.S., Physostigmine as a treatment for gamma-hydroxybutyrate toxicity. *J. Toxicol. Clin. Toxicol.* 40, 781, 2002.

U.S. NIDA, DrugFacts: Prescription CNS depressants. 2017. From: https://www.drugabuse.gov/publications/drugfacts/prescription-cns-depressants.

14 Opioids and Related Agents

14.1 OPIOIDS

14.1.1 U.S. Public Health and Historical Use

There has been no greater disruption of modern civilizations than the insidious havoc brought on them by the addictive potential of opioid compounds and their derivatives. From the introduction of opium into China in the seventeenth century, resulting in the undermining of its organized system, to the modern-day pharmaceutical production of synthetic narcotic analgesics, these compounds have infiltrated urban and rural societies alike. Fast forwarding to current trends reveals that narcotic addiction permeates all socioeconomic classes, from economically underserved communities to affluent neighborhoods, the U.S. armed forces, and daily occupational settings.

Opioid analgesics are readily and easily available. These compounds are not necessarily obtained only through illicit drug dealing (street drugs); their supply is abundant due to fraudulent and illegitimate prescriptions as well as in the course of over-prescribing practices. Health-care professionals are also particularly vulnerable to the addictive potential of narcotics, principally due to easy accessibility.

The variety of opioid derivatives encountered in the twentieth and twenty-first centuries reflects the cyclical appearance and disappearance of individual compounds within the category, mostly because of popularity among users and availability. In the 1980s, opioid addicts inadvertently ingested what they thought was a designer derivative of meperidine (4'-methyl-α-pyrrolidinopropiophenone [MPPP]). Instead, inadvertent contamination of the synthetic protocol with 1-methyl-4-phenyl-1,2,3, 6-tetrahydropyridine (MPTP) led to the development of an idiopathic Parkinson-like state in individuals exposed to the narcotic. In the same decade, a new, more potent form of heroin from Mexico (black tar heroin) made its appearance, resulting in an increase in acute overdose fatalities. Simultaneously during this period, heroin usage began to wane, only to be replaced with the more versatile forms of cocaine. Despite the counter-effects of narcotic law enforcement efforts to prevent or discourage their nontherapeutic use, opioid abuse is still a major public health problem.

Initial narcotic ingestion is often an unpleasant experience. Patients usually complain of nausea, dizziness, and muscular weakness. With continued use, individuals build tolerance to the unpleasant adverse reactions in preference to the euphoric effects. Opioid compounds are ingested orally in tablet or capsule form, the most common method of administration (considering both therapeutic and illicit drug use). As tolerance develops, ingesting the same amount of drug does not produce euphoria as initially experienced, necessitating either higher doses or an alternate, more immediate method of administration. This includes rendering the tablets to a powder, or using a reformulated powder form, for nasal insufflation (*snorting*), subcutaneous (S.C.) injection (*skin popping*), or intravenous (I.V.) injection (*mainlining*).

The northeast and southwest regions of the United States have shown significant spikes in opioid use in the last 5 years. In fact, heroin use predominates in urban areas, as well as in suburban and rural communities, near large cities in the northeast, southwest, and middle states. Law enforcement departments have reported increasing amounts of heroin seized by officials as well as increasing numbers of overdose deaths near the U.S. cities of Chicago and St. Louis.

Drug overdose deaths involving prescription opioid pain relievers have increased dramatically since 1999. The U.S. Centers for Disease Control and Prevention (U.S. CDC) reported 63,632 drug overdose deaths in the United States in 2016 representing a 21.4% increase from 2015; two-thirds of these deaths involved an opioid. From 2015 to 2016, drug overdose deaths increased in all drug categories examined; the largest increase occurred among deaths involving synthetic opioids other than methadone, which includes illicitly manufactured fentanyl (Scholl et al., MMWR 2019). Concerted U.S. federal and state efforts have been made to curb this epidemic. In 2011, the White House released an interagency strategy for "Responding to America's Prescription Drug Crisis." Enacting this strategy, federal agencies have worked with states to educate providers, pharmacists, patients, parents, and youth about the dangers of prescription drug abuse. Specifically, guidelines focused on the need for the proper prescribing, dispensing, use, and disposal of opioid mediations; implementing effective prescription drug monitoring programs; facilitating proper medication disposal through prescription take-back initiatives; supporting aggressive enforcement to address "physician shopping" and "pill mills"; and supporting the development of abuse-resistant formulations for opioid pain relievers.

Consequently, some regulatory and public health changes in various regions of the United States have resulted in decreasing availability of prescription opioid drugs. A decline in O.D. deaths is noted in U.S. states with the most aggressive policies (Johnson et al., 2018). In addition, in an effort to combat the intertwined problems of prescription opioid misuse and heroin abuse, the U.S. Secretary of Health and Human Services (DHHS) announced the "Secretary's Opioid Initiative" (HHS takes strong steps, March 2015). The platform aims to reduce addiction and mortality related to opioid drug abuse by

- reforming opioid prescribing practices;
- expanding access to the overdose-reversal drug, naloxone;
- expanding access to medication-assisted treatment for opioid use disorder.

The relationship between prescription opioid abuse and increases in heroin use in the United States is under scrutiny. These substances are classified as part of the same opioid drug category and overlap in important ways. Accordingly, current available research demonstrates that prescription opioid use is a risk factor for heroin use; globally, heroin consumption is rare among prescription drug users; prescription opioids and heroin have similar effects yet different risk factors; a small subset of individuals who misuse prescription opioids progress to heroin use; increased drug availability is associated with increased use and overdose; heroin use is driven by its low cost, high potency, and high availability; and emphasis is needed on both prevention and treatment.

According to public health and government databases, heroin use is also on the rise in many urban areas among young adults aged 18–25. Individuals in this age group seeking treatment for heroin abuse increased from 11% of total hospital and emergency department (ED) admissions in 2008 to 26% in the first half of 2012. Specifically, although heroin use in the general population is rather low, the number of younger individuals experimenting with the narcotic has steadily risen to 669,000 since 2007 . This may be due in part to a shift from abuse of prescription pain relievers to heroin as a readily available, cheaper alternative. For example, roughly 3%, 5%, and 5% of 8th-, 10th-, and 12th-graders, respectively (the last constituting high school seniors), reported some "past-year" heroin use in 2015 (NIDA, Monitoring the Future Study, National Survey on Drug Use and Health [NSDUH], 2018). Moreover, there has been an alarming jump from adolescent groups (ages 12–17) to adults (ages 18–25), from 1% to 8%, respectively, for those using heroin in the past year.

The statistics have been interpreted with confounding conclusions. For instance, the number for at least one-time use of heroin among adults 26 or older in a lifetime is as high as 20% (NIDA, Monitoring the Future Study, National Survey on Drug Use and Health [NSDUH], 2018). The trend appears to be driven largely by young adults, among whom the 18–25-year age group, as noted earlier, has demonstrated the greatest increase. Overall, an estimated 65 million people (20% of those aged 12 and older) have used opioid prescription drugs for nonmedical reasons at least once in their lifetimes . However, while these numbers are unacceptably high, especially with 156,000 people starting heroin use in 2012 (double the number from 90,000 in 2006), heroin use is reported to be steadily declining among teens, suggesting that past-year heroin use among U.S. adolescent 8th-, 10th-, and 12th-graders is at its lowest historical levels (NIDA, MTF, 2015).

Among the more particularly alarming U.S. statistics are the drug-related ED visits, which show that heroin was involved in 8% (47,604) of all cases, while an additional 4% of all hospital admissions related to drug use (24,623 cases) involved unspecified opiates (DAWN, NIDA, 2019). From 2007 to 2014, overdose deaths related to heroin increased to 10,574, which represents more than a fivefold increase of the heroin death rate from 2002 to 2014 (U.S. Centers for Disease Control and Prevention [USCDC], Opioid Overdose Data, 2019).

14.1.2 Neonatal Abstinence Syndrome (NAS)

Neonatal abstinence syndrome (NAS), or neonatal withdrawal, is an array of signs and neurobehavioral symptoms experienced by human newborns that occur after abrupt discontinuation of gestational exposure to opioids (narcotics) during maternity. The U.S. Centers for Disease Control and Prevention (USCDC, 2015) report a five-fold increase in the proportion of babies born with NAS from 2000 to 2012 (an estimated 21,732 infants born with NAS). The number of children born with NAS per 1000 hospital births was 1.2 in 2000, 1.5 in 2003, 1.96 in 2006, 3.4 in 2009, and 5.8 in 2012 (Figure XX, NIDA, 2018). Moreover, newborns with NAS were more likely than other babies to also have low birthweight and respiratory complications.

In 2012, newborns with NAS remained in the hospital for an average of 16.9 days (compared with 2.1 days for other newborns; Figure 14.1, NIDA, 2018), and the New York City area and vicinity has suffered a disproportionately higher rate than most other large counties in the United States.

Figure 14.1

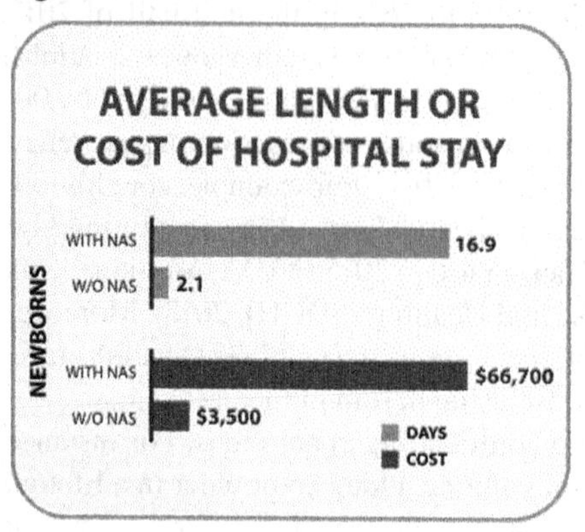

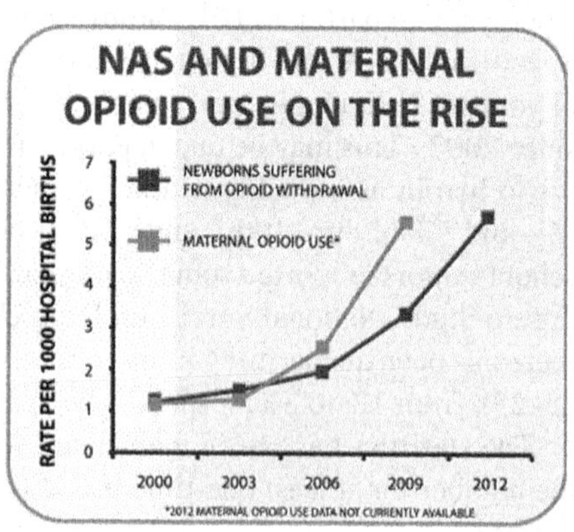

Reprinted from www.drugabuse.gov/related-topics/trends-statistics/infographics/dramatic-increases-in-maternal-opioid-use-neonatal-abstinence-syndrome, Advancing Addiction Science, National Institute on Drug Abuse (NIDA), National Institutes of Health (NIH); U.S. Department of Health and Human Services (USDOHHS), 2016.

Other substances pertinent to addiction science and clinical toxicology also produce neurobehavioral dysregulation in the neonatal period consistent with the withdrawal syndrome. These substances include sedative/hypnotics (benzodiazepines and barbiturates), sympathomimetics (amphetamines), hallucinogens, and cocaine. Some of these can also potentiate the infant's expression of opioid-induced NAS.

14.1.3 Classification

By definition, opiates, as a class, exert their pharmacological effects at opioid receptors, whereas opioids are alkaloid extracts of the opium poppy.* The opioids are traditionally classified according to their source, as summarized in Table 14.1. Opium, the parent crude form of the naturally occurring compounds, is derived from the milky exudates of the unripe capsule of *Papaver somniferum* L. (opium poppy). The plant is cultivated in the Mediterranean and Middle East regions, India, and China. About two dozen alkaloids, of which morphine comprises about 10%, are formed primarily in various cells of the poppy plant and excreted into the lactiferous ducts. Depending on diurnal variations, the isolated latex undergoes alkaloid biosynthesis and metabolic destruction, which contributes to the variability in alkaloid composition of crude opium samples.† The narcotic, antispasmodic, sedative, hypnotic, and

* The terms *opiate* and *opioid* are often used interchangeably and will not be otherwise distinguished in this context.

† In non-Western medicine, opium refers to the dried capsule of the poppy plant from which the latex has been extracted.

TABLE 14.1
Categories, Structure (of Morphine), and Proprietary Names of Opiate Analgesics, Derivatives, and Narcotic Antagonists Currently Available

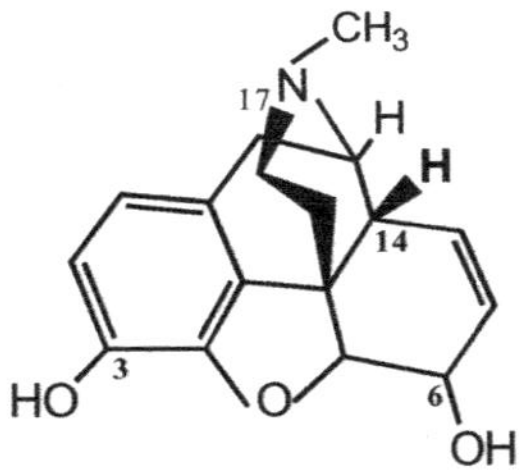

Category	Compound	Proprietary Name	3[a]	6[a]	14[a]	N17[a]	Other[e]
Naturally occurring	Morphine	Various	OH	OH	H	CH_3	——
	Codeine (methylmorphine)	Various	O-CH_3	OH	H	CH_3	——
Semi-synthetic	Buprenorphine	Buprenex	OH	O-CH_3	OH	CH_3-cp[c]	Single bond
	Diacetylmorphine	Heroin[b]	OCO-CH_3	OCO-CH_3	H	CH_3	——
	Oxycodone	Percodan[d] Percocet[d] Oxycontin	O-CH_3	=O	OH	CH_3	Single bond
	Oxymorphone	Numorphan	OH	=O	OH	CH_3	Single bond
	Hydrocodone	Vicodin, Lorcet, Hycodan[d]	O-CH_3	=O	H	CH_3	Single bond
	Hydromorphone	Dilaudid	OH	=O	H	CH_3	Single bond

(Continued)

TABLE 14.1 (CONTINUED)
Categories, Structure (of Morphine), and Proprietary Names of Opiate Analgesics, Derivatives, and Narcotic Antagonists Currently Available

Category	Compound	Proprietary Name	3[a]	6[a]	14[a]	N17[a]	Other[e]
Synthetic	Fentanyl	Sublimaze, Durgesic, Ionsys	Table 14.2				——
	Levorphanol	Levo-dromoran	OH	H	H	CH_3	Single bond, no O
	Methadone	Various	Table 14.2				——
			Tramadol				
Antagonist	Naloxone	Narcan	OH	=O	OH	$CH_2CH=CH_2$	Single bond
	Naltrexone	Trexan	OH	=O	OH	CH_2-cp[c]	Single bond
	Nalmefene	Revex	OH	$=CH_2$	OH	CH_2-cp[c]	Single bond

[a] Numbers 3, 6, 14, and 17 refer to the positions in the phenanthrene nucleus; only morphine and diacetylmorphine have the C7-C8 double bond.
[b] Street name.
[c] cp: cyclopropane.
[d] Component of a formulation.
[e] Other = single bond between C7–C8, no O between C4–C5.

analgesic properties of the extract have been recognized for centuries. Interestingly, the numerous and very small seeds of the plant do not contain opium.

Few pharmacological and toxicological differences exist between the classes. Some toxicokinetic properties, however, distinguish the compounds, especially among the many narcotic derivatives (Section 14.1.7).

14.1.4 Medicinal Chemistry

Table 14.1 illustrates the structure of morphine and the side chains of the derivatives. The opioids are composed of six-membered saturated heterocyclic rings forming the phenanthrene nucleus (in bold) to which is attached a piperidine ring. The structure represents the prototype for all opioids except methadone and meperidine and their derivatives (Table 14.2). Although the more important opiate alkaloids exhibit a phenanthrene nucleus, the majority of the derivatives have the isoquinoline ring structure. Esterification of the phenolic functions, such as in the formation of diacetylmorphine, results in a compound with increased lipid solubility and increased potency and toxicity.

14.1.5 Mechanism of Toxicity

The mechanism of opiate toxicity is an extension of its pharmacology and is directly related to interaction with stereospecific and saturable binding sites or receptors in central nervous system (CNS) tissues, including the cerebrum, limbic system, basal ganglia, and brain stem. These receptors are classified according to the empirical observations noted for the variety of opioid effects. The opioid receptors are biologically active sites of several endogenous ligands, including the two pentapeptides, methionine-enkephalin and leucine-enkephalin. Several larger polypeptides that bind to opioid receptors, such as β-endorphin, are the most potent of the endogenous opioid-like substances.* In addition, three receptor classes have been identified:

1. Compounds that selectively bind to the *mu-receptor* (μ) exhibit morphine-like analgesia, euphoria, respiratory depression, miosis, partial gastrointestinal (GI) inhibition, and sedative effects;
2. Narcotic antagonists such as pentazocine, nalorphine, and levorphanol appear to bind to the *kappa-receptor* (κ), although analgesia, sedation, delusion, hallucinations (psychotomimesis), GI inhibition, and miotic effects still persist;
3. Pentazocine and nalorphine are also described as having affinity for the *delta-receptors* (δ), although this binding is primarily associated with dysphoria and mood changes (inhibition of dopamine release). The roles of epsilon and zeta receptors have yet to be delineated in humans. The sigma receptor (σ), for which pentazocine has been purported to have affinity, was once understood to represent an opioid receptor.

* Collectively, the term *endorphin* refers to the three families of endogenous opioid peptides: the enkephalins, the dynorphins, and the β-endorphins.

TABLE 14.2
Categories, Structural Features, and Proprietary Names of Meperidine and Methadone Congeners Currently Available

Meperidine

Methadone

Category[a]	Compound	Proprietary Name	R_1	R_2
Phenylpiperidine analgesics	Meperidine	Demerol	CH_3	$COOCH_2CH_3$
	Fentanyl	Sublimaze Duragesic	CH_2CH_2-phenyl	N-(phenyl)-$COCH_2CH_3$[b]
	Diphenoxylate	Lomotil[c]	CH_2CH_2C-(phenyl)$_2$-CN	$COOCH_2CH_3$
Methadone congeners	Methadone	Dolophine	Substituted diphenyl pseudopiperidine derivative	
	Pentazocine	Talwin-NX[c]	Benzazocin derivative	
	Propoxyphene	Darvon, Darvocet-N[c]	Substituted diphenyl pseudopiperidine derivative	

[a] All synthetic compounds.
[b] Sole *para*-substitution on piperidine ring.
[c] Component of a formulation.

14.1.6 Brain Chemistry

For much of the past century, research on understanding drug abuse labored in the shadows of powerful myths and misconceptions about the nature of addiction. When addictive behavior was first identified in the 1930s, addicted individuals were thought to be morally flawed and lacking in willpower. Those views shaped society's responses to drug abuse, treating it as a moral failing rather than a health problem, which misled regulators into emphasizing punishment rather than prevention and treatment. Current views and responses to addiction and other substance use disorders have changed dramatically. It is now widely accepted that addiction is a condition that affects sufferers both physiologically and behaviorally. Many biological and environmental factors suggest that genetic variations contribute to the development and progression of the disease. Consequently, this knowledge is used to develop effective prevention and treatment approaches that reduce the toll drug abuse takes on individuals, families, and communities. Interestingly, the National Institute on Drug Abuse (NIDA) has posted general information on opioids and their interactions among brain regions, particularly the amygdala central nucleus, which regulates and allocates reward and the emotional responses accordingly (NIDA, Monitoring the Future Study, National Survey on Drug Use and Health [NSDUH], 2018).

14.1.7 Toxicokinetics

Morphine is rapidly absorbed from an oral dose and from intramuscular (I.M.) and S.C. injections. Peak plasma levels occur at 14–60 and 14 min, respectively. Morphine is metabolized extensively, with only 2–14% excreted as the parent molecule, while 60–80% is excreted in the urine as the conjugated glucuronide. Heroin is rapidly biotransformed first to monoacetylmorphine and then to morphine. Both heroin and monoacetylmorphine disappear rapidly from the blood ($t_{½}$ = 3 and 5–10 min, respectively). Thus, morphine levels rise slowly, persist longer, and decline slowly. Codeine is extensively metabolized, primarily to the 6-glucuronide conjugate. About 10–14% of a dose is demethylated to form morphine and norcodeine conjugates. Therefore, codeine, norcodeine, and morphine in free and conjugated form appear in the urine after codeine ingestion.

14.1.8 Mechanism of Toxicity

Clinical signs and symptoms correlate with the highest concentrations of binding sites in CNS and other tissues. In particular, the limbic system (frontal and temporal cortex, amygdala, and hippocampus), thalamus, corpus striatum, hypothalamus, midbrain, and spinal cord have the highest concentrations. Analgesia appears to affect spinal ascending and descending tracts, extending up to the medullary raphe nuclei (midbrain). Effects on mood, movement, and behavior correlate with interaction with receptors in the globus pallidus (basal ganglia) and locus ceruleus, while mental confusion and euphoria (or dysphoria) are related to altered neuronal activity in the limbic system. Hypothalamic effects are responsible for hypothermia.

Miosis (pinpoint pupils) is thought to occur due to μ-receptor stimulation at the Edinger–Westphal nucleus of the oculomotor nerve.

14.1.9 Signs and Symptoms of Clinical Toxicity

The clinical presentation of the opioid toxidrome (triad) is characterized by CNS depression (including coma), miosis, and respiratory depression. Miosis is generally an encouraging sign, since it suggests that the patient is still responsive. Respiratory depression is a result of depressed brain stem and medullary respiratory centers, responsible for maintenance of normal rhythm. μ-Receptor agonists depress respiration in a dose-dependent manner and can lead to respiratory arrest within minutes. Fifty percent of acute opioid overdose is accompanied by a frothy, noncardiogenic pulmonary edema, responsible for the majority of fatalities. The condition involves loss of consciousness and hypoventilation, probably resulting from hypoxic, stress-induced pulmonary capillary fluid leakage. Peripheral effects include bradycardia, hypotension, and decreased GI motility. Urine output also diminishes as a consequence of increased antidiuretic hormone (ADH) secretion.

14.1.10 Clinical Management of Acute Overdose

Maintenance of vital functions, including respiratory and cardiovascular integrity, is of paramount importance in the clinical management of acute opioid toxicity. Gastric lavage and induction of emesis are effective if treatment is instituted soon after ingestion. It is possible to reverse the respiratory depression with opioid antagonists. Naloxone (Narcan®) is a pure opioid antagonist available as an injectable only. A 2 mg bolus repeated every 5 minutes, followed by 0.4 mg every 2–3 minutes as needed (up to 24 mg total), dramatically reverses the CNS and respiratory depression (in this capacity, naloxone is also indicated in the diagnosis of suspected acute opioid overdose).* Depending on the extent of narcotic overdose, a continuous infusion of naloxone may be required, especially in the presence of opioids with longer half-lives, such as propoxyphene or methadone. As respiration improves, naloxone, which has a half-life of 60–90 minutes, may be discontinued and resumed as necessary. If there is no response after 10 mg of naloxone, concomitant ingestion with other depressants is likely. It should be noted that naloxone is of little benefit in reversing noncardiogenic pulmonary edema.

Naltrexone (Revia®) is also a pure opioid antagonist available as an oral tablet dosage form only. A 50 mg dose of naltrexone blocks the pharmacological effects of opioids by competitive binding at opioid receptors. It is also indicated in the treatment of alcohol dependence. Naltrexone has been noted to induce hepatocellular injury when given in excess.

Nalmefene (Revex®), available in 100 μg/ml and 1 mg/ml ampoules, is indicated for the complete or partial reversal of natural or synthetic opioid effects. It is a 6-methylene analog of naltrexone. Nalmefene has been associated with cardiac

* In the presence of respiratory or CNS depression, a test dose of 0.2 to 0.5 mg I.V., I.M., or through an endotracheal tube may be necessary to avoid the risk of violent withdrawal symptoms and agitation.

instability, although this reaction appears to be the result of abrupt reversal of opioid toxicity.

Several drugs have agonist activity at some receptors (κ) and antagonist activity at other (μ) receptors. Nalbuphine (Nubain®) is a potent analgesic with narcotic agonist and antagonist actions. Other mixed agonist–antagonist compounds are designated as partial agonists, such as butorphanol (Stadol®), buprenorphine (Buprenex®), and pentazocine (Talwin® and various tablet combinations). These compounds are potent analgesics and weakly antagonize the effects of opioids at the μ-receptor while maintaining some agonist properties at the κ- and δ-receptors.

Drug enforcement personnel and customs officials respond to different conditions of opioid overdose, especially those involving *body packers* and *body stuffers*. The drug carriers, who differ only in the apparent manner of sealing and concealing illicit drug packets, encounter problems when the packets leak or burst. The overall clinical response to the situation requires rapid detection with body cavity searches and abdominal radiographs. Decontamination with activated charcoal, gastric lavage, high-dose continuous infusion with naloxone, and attention to the ABCs of emergency management of toxicity in anticipation of a developing opioid syndrome are also warranted.

14.1.11 Tolerance and Withdrawal

The Department of Mental Health and Substance Dependence at the World Health Organization (WHO), in collaboration with the U.S. National Institute on Drug Abuse (NIDA), defines several terms important in understanding drug abuse and the phenomena of tolerance and withdrawal. *Addiction* involves compulsive psychoactive drug use with an overwhelming involvement in the securing and using of such drugs. As described later, the withdrawal syndrome occurs as a result of sudden or abrupt discontinuation of the substance. *Compulsive drug use* involves the psychological need to procure and use drugs, often referred to as *craving*. In this case, the uncontrollable drive to obtain the drugs is necessary to maintain an optimum state of well-being. *Habituation* refers to psychological dependence. *Physical (physiological) dependence* involves the need for repeated administration to prevent withdrawal (abstinence) syndrome. In fact, with repeated chronic dosing, the seizure threshold for opiate narcotics is elevated, threatening the precipitation of seizure on withdrawal (rebound effect). Cross-dependence occurs with all opioids, regardless of category.

The more complex phenomenon of *tolerance* requires the satisfaction of several criteria. With repeated administration, addicted individuals require greater amounts of drug to achieve the desired effect. Conversely, the euphoric effect is markedly diminished with continued use of the same amount of drug. Since various pharmacological effects on different organ systems are not uniformly distributed, tolerance is not evenly demonstrated. While a diminished effect continues with progressive tolerance, the increasing doses threaten the induction of respiratory depression. Increased metabolism and adjustment to the sedative, analgesic, and euphoric effects are proposed as possible mechanisms for the development of tolerance—that is, the physiological drive to achieve homeostasis.

Depending on the drug and the intensity and severity of the addiction, the *withdrawal syndrome* is precipitated hours after the last narcotic dose, with peak intensity occurring at about 72 hours (Table 14.3). The intensity of the syndrome is greatest with heroin, followed by morphine and methadone. Heroin withdrawal is characterized by acute, sudden symptoms of greater vigor, while methadone withdrawal is distributed over 7 to 10 days and of lower intensity. The development of muscle spasms has come to define the syndrome, commonly known as "kicking the habit." Although the syndrome is rarely fatal, administration of an opioid at any time and at relatively low doses during withdrawal alleviates the condition.

A more subtle form of withdrawal symptoms accompanies mild dependence on opioids. Casual, habitual, and slightly controlled use of prescription narcotics often follows the "medical addict."* The cyclic behavior is characterized by excessive use of the drugs followed by a period of abstinence and is often accompanied by subtle signs and symptoms of withdrawal. The withdrawal experience in this case is often indistinguishable and unrecognizable by both the patient and the health-care professional and may be mistaken for other low-grade indicators. For example, within 24 hours after the last opioid dose, patients may experience anxiety, irritability, abdominal cramps, diaphoresis, insomnia, and joint pain. Under any other circumstances, these symptoms would largely proceed uneventfully. Nevertheless, these are warning signs of the development of tolerance and withdrawal and should be deemed, understood, and accepted as the commencement of a chronic disease. Interestingly,

TABLE 14.3
Characterization of the Opioid Withdrawal Syndrome

Stage	Time after Last Dose	Signs and Symptoms
Anticipatory	3–4 hours	Withdrawal fear, craving, compulsive drug-seeking behavior
Early withdrawal	8–12 hours	Lacrimation, sweating, listless behavior, anxiety, restlessness, stomach cramps
	12–16 hours	Restless sleep, nausea, vomiting, mydriasis, anorexia, tremors; cold clammy skin, fever, chills; compulsive drug-seeking behavior
	48–72 hours	Peak intensity: tachycardia, hypertension, hypothermia, piloerection (goose-flesh appearance of skin, "cold turkey"), muscle spasms; continued nausea, vomiting, dehydration, compulsive drug-seeking behavior; risk of cardiovascular collapse
Protracted abstinence	6 months	Stimulus-driven cravings, anorexia, fatigue, bradycardia, hypotension

* This term is used to describe a patient in medically necessary treatment for a disorder, such as a pain-related condition, who has become addicted to the prescribed drug.

by avoiding the risk of overdose and major behavioral changes, opioid addicts have survived for decades with this pattern of behavior.

14.1.12 Clinical Management of Addiction

NIDA publishes *The Principles of Drug Addiction Treatment—A Research Based Guide*. The Guide outlines the social and clinical approach associated with drug addiction treatment in the United States. Outpatient drug-free treatment, long- and short-term residential treatment, scientifically based counseling, and psychotherapeutic and community-based programs are discussed as approaches to drug addiction treatment. Among these modalities, the risks and benefits of traditional medical detoxification associated with the use of methadone and narcotic antagonists are presented.

Although methadone treatment programs have shown some success in opioid detoxification, significant progress in therapeutic intervention has resulted more recently using buprenorphine.* Buprenorphine is the first opioid drug approved under DATA 2000† for the treatment of opioid dependence in an *office-based setting*. The drug, a partial opioid agonist, is dispensed for outpatient use and administered sublingually, sometimes in combination with naloxone, the full opioid antagonist. The latter drug discourages the patient from dissolving the tablet for parenteral injection of the formulation. The antagonist blocks μ-receptors, thus precipitating the withdrawal syndrome. Buprenorphine reduces illicit opioid use and helps patients remain in treatment programs by increasing the availability of treatment options, by suppressing and/or ameliorating symptoms of opioid withdrawal, and by decreasing cravings for opioids. Opioid addiction treatment in a primary care setting also allows increased clinical attention to other health conditions among opioid-dependent individuals.

14.2 SPECIFIC OPIOID DERIVATIVES

14.2.1 Codeine

Codeine (methylmorphine) is available in combination with other ingredients as an analgesic (Tylenol with Codeine®) and as an antitussive in prescription cough, cold, antihistaminic, and expectorant formulas. The usual dosage form contains 14–60 mg/tablet or 10 mg/5 ml liquid. About 140 mg of codeine is equivalent to 10 mg of morphine. The compound produces the same triad of signs and symptoms with high doses, although tolerance and toxicity are less severe. Interestingly, in the 1950s and early 1960s, codeine cough and cold preparations (such as Cheracol Syrup®) could be purchased without a prescription, and quantities were only monitored with a signature.

---

* Buprenorphine (Subutex®) is also used in combination with naloxone (Suboxone®).

† Title XXXV, Section 3502 of the U.S. Children's Health Act of 2000 permits physicians who meet certain qualifications to treat patients with opioid addiction with Schedule III, IV, and V narcotic medications that have been specifically approved by the U.S. Food and Drug Administration for that indication. Such medications may be prescribed and dispensed by "waived" physicians in treatment settings other than the traditional opioid treatment program.

14.2.2 Diphenoxylate

A synthetic opiate chemically related to meperidine, diphenoxylate is combined with atropine (Lomotil®) for the treatment of diarrhea. The toxicity of this combination, therefore, is primarily due to the presence of the anticholinergic. Children are especially sensitive to the effects of atropine, including the production of tachycardia, flushing, hallucinations, and urinary retention. The narcotic toxicity demonstrates as miosis, respiratory depression, and in severe cases, coma.

14.2.3 Fentanyl

In the 1990s, fentanyl enjoyed increasing popularity as the narcotic of choice among illicit drug users, principally because of its enhanced potency (*China white*). At 200 times and 7000 times greater potency than morphine, respectively, α-methylfentanyl and 3-methylfentanyl also display greater potential for toxicity. The median lethal dose is about 145 μg for the former and 5 μg for the latter.* Therapeutically, fentanyl is marketed in the form of medicated patches (Duragesic Transdermal System®) for the management of chronic pain. Depending on the size of the patch and the amount of fentanyl delivered (10–40 cm^2 containing 2.5–20 mg total per patch), the transdermal system can release up to 200 μg per hour.

A solid formulation of fentanyl citrate (Actiq®) is available on a stick that dissolves slowly in the mouth for transmucosal absorption. This formulation of the drug is intended for opiate-tolerant individuals and is effective in treating breakthrough pain in cancer patients. Carfentanil (Wildnil®) is an analogue of fentanyl with an analgesic potency 10,000 times that of morphine and is used in veterinary practice to immobilize certain large animals.

14.2.4 Meperidine

The first synthetic opioid (1939), meperidine is equianalgesic with morphine. In the liver, the compound is hydrolyzed to meperidinic acid and normeperidine by carboxyesterases and by N-demethylation and microsomal enzymes, respectively. Both the metabolites are active, although they possess half of the analgesic effects and twice the neurotoxic activity. Consequently, chronic oral ingestion of meperidine tablets is associated with CNS stimulation, resulting in tremors, muscle twitching, nystagmus, and convulsions. Neurotoxicity correlates directly with opioid plasma concentrations and requires several days before onset. Benzodiazepines are recommended for treatment of CNS excitation. The use of naloxone is cautioned with chronic meperidine use, since the antagonist may decrease the seizure threshold (increasing the potential for convulsions).

---

* In October 2002, Russian commandos pumped an aerosol derivative of fentanyl into a Moscow theater to end a hostage crisis. All but two of 140 deaths occurred as a result of the effects of the opioid (Miller and Broad, NY Times 2002).

14.2.5 Pentazocine

Pentazocine is a benzomorphan derivative of morphine with three to four times its analgesic potency and the same addictive potential. It is presumed to exert its agonistic actions at the κ- and δ-receptors and may precipitate withdrawal symptoms in patients taking narcotic analgesics regularly. I.V. injection of oral preparations of pentazocine and tripelenamine, an H_1-blocking antihistamine, was a common form of drug abuse.* The tablets were crushed, dissolved in tap water, heated over a flame, and injected. The combination purportedly produced an effect similar to heroin at much lower cost. Because the method of sterilization was less than optimal, and the solution contained undissolved pieces of tablet binders and fillers, addicted individuals often developed skin decubiti, abscesses, and cellulitis. Continued injection resulted in serious pulmonary artery occlusion, pulmonary hypertension, and neurologic complications. As a consequence, oral pentazocine tablets were replaced with Talwin-NX® (pentazocine plus naloxone) to decrease this practice. Similarly to the combination with buprenorphine, the inhibitory action of naloxone on pentazocine's analgesic effect is experienced only when the tablets are crushed and injected, since naloxone is not absorbed orally.†

14.2.6 Propoxyphene

A methadone analog, propoxyphene is implicated in cardiotoxicity. The parent compound and its metabolite, norpropoxyphene, cause dose-dependent widening of the QRS complex similarly to tricyclic antidepressants. This quinidine-like effect results from the inhibition of cardiac fast sodium channels, causing tachydysrhythmias. In addition, propoxyphene is frequently used as the napsylate salt in combination with acetaminophen (Darvocet-N®). The unique salt form stimulates hepatic mixed function oxidase (MFO) enzymes, increasing the presence of toxic metabolites of acetaminophen. Consequently, in chronic repeated administration, it often masks acetaminophen toxicity.

14.2.7 Hydrocodone/Oxycodone

Hydrocodone and oxycodone are powerful μ-receptor agonists with addictive and analgesic potential equivalent to those of morphine and heroin, respectively.

Hydrocodone is used as an analgesic in oral dosage forms (Vicodin®, Lorcet®, Lortabs®, and Tylox®) for mild to moderate pain associated with minor surgical procedures, chronic joint and muscle pain, and inflammatory conditions. It is also used as an antitussive (in Hycodan®). Consequently, its addictive potential is significant when administered chronically.

---

* The combination of the crushed tablets was known as *Ts and Blues*: *T* for Talwin® and *Blues* for the blue color of the antihistamine tablet.

† By itself in high doses, pentazocine increases plasma epinephrine concentrations, risking the development of hypertension and increased heart rate.

Oxycodone, in combination with aspirin or acetaminophen (Percodan® and Percocet®, respectively, 2.5 mg per tablet), has enjoyed popularity as an effective analgesic for the relief of moderate to severe pain of chronic inflammation and surgery. It is particularly useful in the alleviation of chronic pain of many cancers. In 1985, MS Contin® was introduced as a delayed-release morphine tablet, with the advantage of decreasing the frequency of dosing in patients with chronic pain. This formulation was especially convenient for elderly individuals. By 1994, morphine consumption in the United States had risen by 75%. Based on this success, Oxycontin® was introduced in 1995 as a delayed-release oral dosage form of the more powerful oxycodone. Revenues from Oxycontin® rose from $55 million in 1996 to $1.14 billion in 2000, at which time it became the number one opioid analgesic, with 6.5 million prescriptions in 2000. By 1995, the first cases of Oxycontin® abuse ("oxys") appeared in rural Missouri and spread throughout the rust belt states of Pennsylvania, Ohio, West Virginia, Virginia, and Appalachian Kentucky. Increasing unemployment rates in these states and the large numbers of chronically ill and disabled elderly people who were unable to relocate, coupled with the remoteness of the regions, created an environment conducive to illicit drug distribution (the drug became known as *hillbilly heroin*). Economically poor, the elderly would readily sell their Oxycontin® medication to young teens offering money, producing a captive market of nontraditional drug abusers. Unlike heroin, "oxys" are regarded as legal compounds, more easily available, and with less ambiguity associated with "copping dope" on the street. The allure of the substance was not in the potency of the tablet form but in the large quantities of active ingredient immediately accessible when a 10 to 40 mg delayed release tablet is crushed and either "snorted" or injected.

By 1998, Oxycontin® abuse had spread to suburban and urban metropolitan areas. Since 2000, several hundred fatalities due to injected Oxycontin® overdose have been reported. Its relative purity and abundance in crushed form created an immensely desirable compound.*

14.2.8 Tramadol

Tramadol is a centrally acting synthetic analog of codeine with low affinity for the μ-receptor. It is used for moderate to severe pain control. In 2014, the U.S. Drug Enforcement Administration (DEA) announced that tramadol would be placed into schedule IV of the Controlled Substances Act (CSA)(DEA, CSA 21 CFR Part 1308, 2014 (https://www.deadiversion.usdoj.gov/fed_regs/rules/2014/fr0702.htm). Many of its effects appear to be through modulation of central monoamine pathways by inhibiting the reuptake of 5-hydroxytryptamine and norepinephrine. In overdose, the effects are similar to those of other opioids, with convulsions predominating in susceptible individuals.

* Interestingly, since then, Purdue Pharma, the maker of Oxycontin®, and three of its executives have pleaded guilty in a U.S. federal court to criminal charges of "misbranding" the product—the defendants had misled doctors and patients when claiming that the drug was less likely to be abused than traditional narcotics (Meier, *New York Times*, 2007).

14.2.9 Clonidine

Clonidine (Catapres®) primarily stimulates central postsynaptic α_2-receptors that inhibit neuronal activity and decrease sympathetic overtone. Clonidine shares some pharmacological properties (μ-receptors) and clinical features with the opioids. Symptoms of overdose with clonidine occur within 60 to 90 min after ingestion, producing bradycardia, hypotension, arrhythmias, CNS depression, decreased respiration, and miosis. Although the mechanism is poorly understood, it is believed to involve antagonism of the μ-receptors. Patients who demonstrate opioid-like toxicity with clonidine respond to naloxone administration, particularly the reversal of hypoventilation and CNS depression.

14.3 METHODS OF DETECTION

Opioids are detected using a radioactive or enzyme-linked immunoassay technique (enzyme multiplied immunoassay technique [EMIT]; Kinetic Interaction of Microparticles in Solution [KIMS]).* The principle of the assays is the reaction of morphine in an aliquot of the urine sample with its corresponding antibody. Significant cross-reactivity occurs with opioid derivatives (as well as with components of poppy seeds) because of the reaction of the antibody with the common phenanthrene structure. Radioimmunoassays (RIAs) are also very sensitive and can detect opioids at levels of 0.5–10 ng/ml. RIA, however, requires tritiated (radioactive) ligands as indicators. Other immunoassays for specific opioid derivatives, such as fentanyl, methadone, and meperidine, are also available. These drawbacks limit the usefulness of these screening methods, thus requiring confirmation with gas chromatography (GC) and identification with gas chromatography/mass spectrometry (GC-MS) protocols.

Both EMIT and the Abuscreen® RIA detect codeine and morphine in free and conjugated forms but do not distinguish between them. Based on the toxicokinetics of the opioids noted earlier, distinguishing morphine from heroin or codeine is difficult but clinically and forensically important. The presence of morphine alone or its conjugate can indicate either clinical morphine use or illicit morphine or heroin use (within the previous 1 to 2 days). The distinction is possible when the test is employed 2 to 4 days after the last dose. Other narcotics identified by the immunoassays for morphine include dihydrocodeine, dihydromorphine, and hydromorphone. Current U.S. Department of Defense (DoD) and U.S. Department of Health and Human Services (HHS) procedures use an initial opiate (codeine/morphine) immunoassay (IA) screen followed by GC-MS confirmation of the 6-acetylmorphine (6-AM) metabolite. Thus, confirmation of positive results and distinction between the compound derivatives are accomplished with principally with GC-MS. TLC, high-performance liquid chromatography (HPLC), and gas-liquid chromatography (GLC) are alternative methods. Acid or enzyme hydrolysis of the urine sample is necessary, however, when the latter testing techniques are used, since approximately 90% of codeine and morphine are found in urine in the conjugated glucuronide form.

* These methods have supplanted the traditional GC-MS and thin layer chromatography (TLC) methods used for urine and blood screening.

REFERENCES

SUGGESTED READINGS

Arthur, J. and Hui, D., Safe opioid use: management of opioid-related adverse effects and aberrant behaviors. *Hematol. Oncol. Clin. North Am.* 32, 387, 2018. doi:10.1016/j.hoc.2018.01.003.

Bronstein, A.C., Spyker, D.A., Cantilena, L.R., Jr., Green, J.L., Rumack, B.H., and Heard, S.E., 2007 Annual report of the American Association of Poison Control Centers' National Poison Data System (NPDS): 25th Annual Report., *Clin. Toxicol.* 46, 927, 2008.

Bronstein, A.C. Spyker, D.A., Cantilena, L.R. Jr., Rumack, B.H., and Dart, R.C., 2011 Annual report of the American Association of Poison Control Centers' National Poison Data System (NPDS): 29th Annual Report. *Clin. Toxicol. (Phila)* 50, 911, 2012.

Darcq, E. and Kieffer, B.L., Opioid receptors: drivers to addiction? *Nat. Rev. Neurosci.* 2018.

Gummin, D.D., Mowry, J.B., Spyker, D.A., Brooks, D.E., Fraser, M.O., and Banner, W., 2016 Annual Report of the American Association of Poison Control Centers' National Poison Data System (NPDS): 34th Annual Report. *Clin. Toxicol. (Phila)* 55, 1072, 2017.

Manchikanti, L. and Singh A., Therapeutic opioids: a ten-year perspective on the complexities and complications of the escalating use, abuse, and nonmedical use of opioids. *Pain Physician* 11, S63, 2008.

Maurer, H.H., Sauer, C., and Theobald, D.S., Toxicokinetics of drugs of abuse: current knowledge of the isoenzymes involved in the human metabolism of tetrahydrocannabinol, cocaine, heroin, morphine, and codeine. *Ther. Drug Monit.* 28, 447, 2006.

Meier, B., Narcotic maker guilty of deceit over marketing, *New York Times—Business*, May 11, 2007.

Miller, J., and Broad, W.J., *Hostage Drama in Moscow: The Toxic Agent: U.S. Suspects Opiate in Gas in Russia Raid*, New York Times Foreign Desk, October 29, 2002.

Modarai, F., Mack, K., Hicks, P., Benoit, S., Park, S., Jones, C., Proescholdbell, S., Ising, A., and Paulozzi, L., Relationship of opioid prescription sales and overdoses, North Carolina. *Drug Alcohol Depend.* 132, 81, 2013.

Moss, J. and Rosow, C.E., Development of peripheral opioid antagonists' new insights into opioid effects. *Mayo Clin. Proc.* 83, 1116, 2008.

Mowry, J.B., Spyker, D.A., Brooks, D.E., Zimmerman, A., and Schauben, J.L., 2015 Annual report of the American Association of Poison Control Centers' National Poison Data System (NPDS): 33rd Annual Report. *Clin. Toxicol. (Phila)* 54, 924, 2016.

Mowry, J.B., Spyker, D.A., Brooks, D.E., McMillan, N., and Schauben, J.L., 2014 Annual Report of the American Association of Poison Control Centers' National Poison Data System (NPDS): 32nd Annual Report. *Clin. Toxicol. (Phila)* 53, 962, 2015.

Mowry, J.B., Spyker, D.A., Cantilena, L.R. Jr., McMillan, N., and Ford, M., 2013 Annual Report of the American Association of Poison Control Centers' National Poison Data System (NPDS): 31st Annual Report. *Clin. Toxicol. (Phila)* 52, 1032, 2014.

Mowry, J.BV., Spyker, D.A., Cantilena, L.R. Jr., Bailey, J.E., and Ford, M., 2012 Annual Report of the American Association of Poison Control Centers' National Poison Data System (NPDS): 30th Annual Report. *Clin Toxicol (Phila)* 51, 949, 2013.

Muñoa, I., Urizar, I., Casis, L., Irazusta, J., and Subirán, N., The epigenetic regulation of the opioid system: new individualized prompt prevention and treatment strategies. *J. Cell Biochem.* 116, 2419, 2015.

Scholl, L., Seth, P., Kariisa, M., Wilson, N., and Baldwin, G., Drug and Opioid-Involved Overdose Deaths — United States, 2013–2017; Morbidity and Mortality Weekly Report (MMWR), *CDC MMWR Weekly*, 67, 1419, 2019.

Trescot, A.M., Datta, S., Lee, M., and Hansen, H., Opioid pharmacology. *Pain Physician* 11, S133, 2008.
Vowles, K.E., McEntee, M.L., Julnes, P.S., Frohe, T., Ney, J.P., van der Goes, D.N. Rates of opioid misuse, abuse, and addiction in chronic pain: a systematic review and data synthesis. *Pain* 156, 569, 2015.
Wax, P.M., Becker, C.E., and Curry, S.C., Unexpected "gas" casualties in Moscow: a medical toxicology perspective. *Ann. Emerg. Med.* 41, 700, 2003.

REVIEW ARTICLES

Budd, K., Buprenorphine and the transdermal system: the ideal match in pain management. *Int. J. Clin. Pract. Suppl.* 133, 9, 2003.
Choi, Y.S. and Billings, J.A., Opioid antagonists: a review of their role in palliative care, focusing on use in opioid-related constipation. *J. Pain Symptom Manage.* 24, 71, 2002.
Cohen, G., The "poor man's heroin". An Ohio surgeon helps feed a growing addiction to OxyContin. *U.S. News World Rep.* 130, 27, 2001.
Drug Abuse Warning Network (DAWN), Substance Abuse and Mental Health Services Administration (SAMHSA), National Institute on Drug Abuse, DHHS, U.S. Public Health Service. www.drugabuse.gov, last viewed March 2019.
Green, H., James, R.A., Gilbert, J.D., Harpas, P., and Byard, R.W., Methadone maintenance programs—a two-edged sword? *Am. J. Forensic Med. Pathol.* 21, 359, 2000.
Haile, C.N., Kosten, T.A., and Kosten, T.R., Pharmacogenetic treatments for drug addiction: alcohol and opiates. *Am. J. Drug Alcohol Abuse* 34, 355, 2008.
Hancock, C.M. and Burrow, M.A., OxyContin use and abuse. *Clin. J. Oncol. Nurs.* 6, 109, 2002.
Hassamal, S., Miotto, K., Dale, W., and Danovitch, I., Tramadol: understanding the risk of serotonin syndrome and seizures. *Am. J. Med.* 131, 11382, 2018.
Hino, Y., Ojanpera, I., Rasanen, I., and Vuori, E., Performance of immunoassays in screening for opioids, cannabinoids and amphetamines in post-mortem blood. *Forensic Sci. Int.* 131, 148, 2003.
Holler, J.M., Bosy, T.Z., Klette, K.L., Wiegand, R., Jemionek, J., and Jacobs, A. Comparison of the Microgenics CEDIA heroin metabolite (6-AM) and the Roche Abuscreen ONLINE opiate immunoassays for the detection of heroin use in forensic urine samples. *J. Anal. Toxicol.* 28, 489, 2004.
Iordanova, M.D., Deroche, M.L., Esber, G.R., and Schoenbaum, G., Neural correlates of two different types of extinction learning in the amygdala central nucleus. *Nat. Commun.* 7, 12330, 2016.
Johnson, K., Jones, C., Compton, W., Baldwin, G., Fan, J., Mermin, J., and Bennett, J.,Federal response to the opioid crisis. *Curr. HIV/AIDS Rep.* 15, 293, 2018.
Jones, H.E., Practical considerations for the clinical use of buprenorphine. *Sci. Pract. Perspect.* 2, 4, 2004.
Joshi, G.P., Morphine-6-glucuronide, an active morphine metabolite for the potential treatment of post-operative pain. *Curr. Opin. Investig. Drugs* 9, 786, 2008.
Juurlink, D.N. and Dhalla, I.A., Dependence and addiction during chronic opioid therapy. *J. Med. Toxicol.* 8, 393, 2012.
Kemp, P., Sneed, G., Kupiec, T., and Spiehler, V., Validation of a microtiter plate ELISA for screening of postmortem blood for opioids and benzodiazepines. *J. Anal. Toxicol.* 26, 504, 2002.
Latta, K.S., Ginsberg, B., and Barkin, R.L., Meperidine: a critical review. *Am. J. Ther.* 9, 53, 2002.

Lehmann, K.A., Opioids: overview on action, interaction and toxicity. *Support Care Cancer* 5, 439, 1997.

Lu, L. and Wang X., Drug addiction in China. *Ann. N. Y. Acad. Sci.* 1141, 304–317, 2008. Review.

Rook, E.J., Huitema, A.D., van den Brink, W., van Ree, J.M., and Beijnen, J.H., Pharmacokinetics and pharmacokinetic variability of heroin and its metabolites: review of the literature, *Curr. Clin. Pharmacol.* 1, 109, 2006.

Schranz, A.J., Barrett, J., Hurt, C.B., Malvestutto, C., and Miller, W.C., Challenges facing a rural opioid epidemic: treatment and prevention of HIV and hepatitis C. *Curr. HIV/AIDS Rep.* 15, 245, 2018.

Suffett, W.L., OxyContin abuse. *J. Ky. Med. Assoc.* 99, 72, 2001.

Theisen, K., Jacobs, B., Macleod, L., and Davies, B. The United States opioid epidemic: a review of the surgeon's contribution and health policy initiatives. *BJU Int.* 122, 754, 2018.

U.S. Centers for Disease Control and Prevention (USCDC), Opioid Overdose Data. From: https://www.cdc.gov/drugoverdose/ last viewed March 2019.

U.S. National Institute on Drug Abuse (NIDA), National Institutes of Health (NIH), National Survey of Drug Use and Health, Monitoring the Future 2018 Survey Results. From: https://www.drugabuse.gov/related-topics/trends-statistics/infographics/monitoring-future-2018-survey-results, last viewed March 2019.

U.S. National Institute on Drug Abuse (NIDA), National Institutes of Health (NIH), Research Monograph 73, *Urine Testing for Drugs of Abuse*, Hawks, R.L. and Chiang, C.N. (Eds.), Department of Health and Human Services (DHHS), Public Health Service, 1986.

U.S. National Institute on Drug Abuse (NIDA), National Institutes of Health (NIH), *Topics in Brief: Science of Addiction*, Department of Health and Human Services (DHHS), Public Health Service, 2007.

U.S. National Institute on Drug Abuse (NIDA), National Institutes of Health (NIH), *Topics in Brief: Buprenorphine: Treatment for Opiate Addiction Right in the Doctor's Office*, Department of Health and Human Services (DHHS), Public Health Service, 2006.

Vadivelu, N. and Hines, R.L., Management of chronic pain in the elderly: focus on transdermal buprenorphine. *Clin. Interv. Aging* 3, 421, 2008.

Wang, H.E., Street drug toxicity resulting from opioids combined with anticholinergics. *Prehosp. Emerg. Care* 6, 351, 2002.

Wright, C. 4th, Schnoll, S., and Bernstein, D., Risk evaluation and mitigation strategies for drugs with abuse liability: public interest, special interest, conflicts of interest, and the industry perspective. *Ann. N.Y. Acad. Sci.* 1141, 284, 2008.

15 Sympathomimetics

15.1 AMPHETAMINES AND AMPHETAMINE-LIKE AGENTS

15.1.1 Incidence

The sporadic use of amphetamines reflects the competition and availability of other drugs of abuse, the desires of the particular socioeconomic group (generation) interested in the substance, and the efforts of law enforcement to eliminate the accessibility of any particular substance of abuse. The amphetamines made their appearance for use in weight reduction and control of obesity. Their stimulant/euphoric effects improved their popularity, which eventually gave rise to common street names such as *ups, bennies, dexies, and speed,* among others. Today, related stimulants such as methamphetamine and methylphenidate are available as prescription drugs due to their frequent use in the treatment of attention-deficit hyperactivity disorder (ADHD). The stimulants are also approved for the treatment of short-term depression. Their "off-label" use is frequently exploited for reducing fatigue among truck drivers and college students.

The worldwide prevalence of amphetamine use is estimated to be 0.3% to 1.1% as per the United Nations Office of Drugs and Crime (UN World Drug Report, 2017). It is estimated that 3.5% of 8th-graders, 5.6% of 10th-graders, and 5.9% of 12th-graders used amphetamines during the past year (U.S. NIDA, 2018). Men have a higher prevalence of amphetamine misuse compared with women. The overall prevalence of methamphetamine use in the United States in individuals aged 12 or older was 4.7% in 2013 as per the National Survey of Drug Use and Health (NSDUH, NIDA, 2018). According to UNODC World Drug Report 2017, global seizures of amphetamine in 2015 rose 8% from the previous year, with increasing seizures reported in parts of Central America, southeastern Europe, and southwestern Asia (UNODC, World Drug Report, 2017).

In particular, 3700 overdose deaths were reported in the United States from methamphetamine (MA) abuse in 2014, and 4900 in 2015, more than double the number in 2010 (NSDUH, NIDA, 2018). Approximately 1.2 million persons used MA in 2012, a rise in drug use from previous years. In 2011, MA ingestion accounted for 103,000 emergency department (ED) visits, the fourth most popular drug for emergency treatment. However, overall use has declined from 8.1% in 2005 to 5.6% in 2011. Law enforcement reports indicate a decrease in clandestine laboratory incidents involving MA production in the United States from 15,220 in 2010 to 9338 for 2014. Notwithstanding this curtailment, MA misuse and abuse continue to strangle public health resources.

15.1.2 Classification

The homeostatic regulation of central and peripheral autonomic function is mediated principally through the actions of the sympathetic nervous system (SNS) and moderated through the opposing effects of the parasympathetic system (PNS). Stimulation of the sympathetic nervous system normally occurs in response to physical activity, psychological stress, generalized allergic reactions, and situations that require a heightened response. As a result, the physiological reaction of the SNS favors functions that support vigorous physical activity and response to stressors; thus, the SNS is labeled the "fight-or-flight" response. Increases or enhancements of cardiac contraction, vasomotor tone, blood pressure, bronchial airway tone, carbohydrate and fatty acid metabolism, psychomotor activity, mood, and behavior are physiological responses that occur following SNS stimulation. The variety of agents listed in Table 15.1 pharmacologically mimic or alter sympathetic activity. These substances were originally classified as catecholamines due to their structural resemblance to OH-substituted *o*-dihydroxybenzene (catechol). Although most are clinically useful, the toxicity of these compounds results from development of the adverse reactions from abuse or overuse and is manifested as an exaggeration of the pharmacological profiles.

The actions of the sympathomimetic amines are classified into six broad types: 1. peripheral excitatory smooth muscle action, including effects on peripheral blood vessels; 2. peripheral inhibitory smooth muscle, including effects on the gastrointestinal (GI) tract and bronchial tree, and in skeletal muscle blood vessels; 3. cardiac excitatory action, resulting in increased heart rate (chronotropy) and force of contraction (inotropy); 4. metabolic actions, resulting in stimulation of hepatic and muscle glycogenolysis and lipolysis; 5. endocrine actions, such as modulation of the secretion of insulin, renin, and pituitary hormones; and, 6. central nervous system (CNS) stimulation of respiration, enhancement of psychomotor activity, and appetite suppression. Thus, the pharmacological properties of amphetamines and amphetamine-like compounds mimic the actions of the prototype agent, epinephrine.

15.1.3 Medicinal Chemistry

Table 15.1 illustrates the β-phenethylamine parent structure of the sympathomimetic amines, consisting of a benzene ring and an ethylamine side chain. Substitutions on the aromatic ring and the side chain (R-groups) yield a variety of compounds whose pharmacological and toxicological profiles and relative potencies are distinguishable from those of epinephrine.

15.1.4 Pharmacology and Clinical Use

Figure 15.1 illustrates sympathetic neurons and their receptors in the SNS. Most sympathetic postganglionic neurons are adrenergic, active in synthesizing, storing, and releasing norepinephrine. On release, norepinephrine diffuses and binds to adrenergic receptors, alpha and beta (α_1, α_2, β_1, β_2 and β_3) on the postsynaptic membrane, resulting in chemical activation of the effector organs mentioned earlier. In general, stimulation of α_1 and β_1 receptors results in excitation, while activation of α_2 and β_2 receptors causes inhibition. The α_1 receptors predominate at the effector sites of peripheral vascular smooth muscle and glandular cells. Excitation of α_1 receptors results in vasoconstriction (pressor effects), pupillary dilation (mydriasis), and closing of sphincters. The α_2 receptors are present in some effector sites of visceral

TABLE 15.1
Categories, Structures, and Proprietary Names of Currently Available Naturally Occurring Sympathomimetics, Amphetamines and Amphetamine-like Stimulants

$$\text{C}_6\text{H}_5\text{—CH(R}_1\text{)—CH(R}_2\text{)—NH(R}_3\text{)}$$

Category	Compound	Proprietary Name	R_1[b]	R_2	R_3
Catecholamines (naturally occurring)[a]	Epinephrine	Various	OH	H	CH_3
	Norepinephrine	None	OH	H	H
	Dopamine	Various	H	H	H
Amphetamines and derivatives (non-catecholamines)	Amphetamine	Various	H	CH_3	H
	Dextroamphetamine	Dexedrine, Adderal[c]	d-isomer of d,l-amphetamine		
	Ephedrine[d], Pseudoephedrine[e]	Various	OH	CH_3	CH_3
	Methamphetamine	Desoxyn	H	CH_3	CH_3
	Diethylpropion	Tenuate	=O	CH_3	$(C_2H_5)_2$
	Phenylpropanolamine[d]	PPA	OH	CH_3	H
	Phentermine	Fastin[d]	H	$(CH_3)_2$	H
	Phentermine HCl	Adipex-P			
	Pemoline	Cylert	2-amino-5-phenyl-2-oxazolin-4-one		
	Phendimetrazine	Prelu-2, Bontril, Plegine[d], Preludin[d]	3,4-dimethyl-2-phenylmorpholine		
	Methylphenidate	Ritalin	Methyl α-phenyl-2-piperidine acetate		

[a] Catecholamines are normally found throughout the body and are classified here as naturally occurring sympathomimetics.
[b] R-substitutions refer to the positions on the ethylamine side chain.
[c] Amphetamine/dextroamphetamine combination.
[d] No longer marketed.
[e] Restricted nonprescription sales regulated by state and federal laws.

and vascular smooth muscle. Activation of α_2 receptors results in vasodilation and decreased insulin secretion. β_1 receptors predominate in cardiac muscle fibers and juxtaglomerular cells of the kidney. Stimulation of β_1 receptors precipitates positive cardiac inotropy and chronotropy as well as increased renin secretion (stimulating the renin–angiotensin pathway for elevating blood pressure). β_2 receptors are present primarily in respiratory smooth muscle, cardiac and skeletal muscle, and hepatic and adipose blood vessels. The net effect of binding of amphetamines to β_2 receptors is

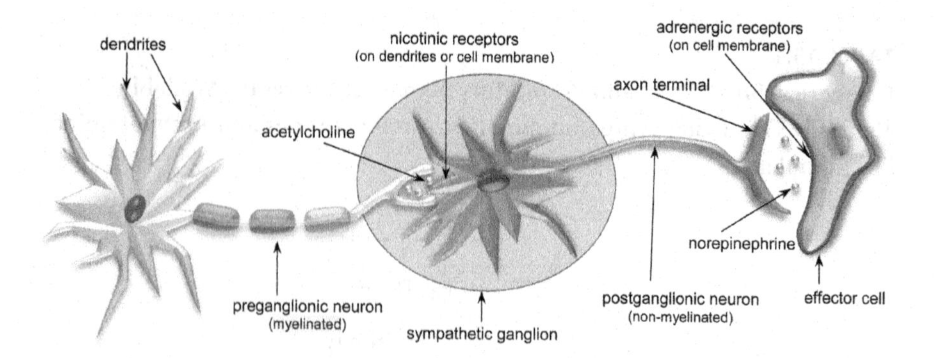

FIGURE 15.1 Sympathetic neurons and their receptors in the SNS.

relaxation of bronchial airways, vasodilation of coronary and skeletal muscle vasculature, and glycogenolysis. Stimulation of β_3 receptors, found primarily in adipose tissue, induces a thermogenic response (heat production). This action, in combination with the central anorexic effect, explains their use for weight reduction.

Pharmacologically, amphetamines and their derivatives are indirect agonists that mimic the actions of epinephrine and norepinephrine (Table 15.1). The agents either stimulate the release of, or block the reuptake of, naturally occurring sympathomimetics. The only approved indications for use of amphetamines and derivatives are for the treatment of attention-deficit hyperactivity syndrome (ADHD) and narcolepsy and the short-term management of obesity.

15.1.5 Toxicokinetics

The catecholamine sympathomimetics, epinephrine, norepinephrine, and dopamine, are rapidly conjugated and oxidized in the gastrointestinal lumen and liver, rendering them ineffective when taken orally. A variety of parenteral, solubilized, and aerosolized preparations, however, are available for inhalation, intravenous (I.V.), intramuscular, and intranasal (I.N.) administration.

Unlike the naturally occurring sympathomimetics, the amphetamine compounds are weak bases that are well absorbed orally ($pK_a \sim 10$), predominantly throughout the length of the basic environment of the small intestine.

15.1.6 Effects and Mechanism of Toxicity

Amphetamine and derivatives have powerful CNS stimulant actions (analeptic) in addition to sympathetic stimulation. Thus, the toxicity is an extension of the pharmacological properties. Table 15.2 summarizes the desirable and undesirable effects of the sympathomimetics as well as the treatment associated with the toxicity. The therapeutic uses follow the indications outlined previously, while the euphoric effects are the major reasons for psychological and physiological dependence. The psychic properties vary according to the mental state of the individual and the dose. An oral dose of 5 mg of dextroamphetamine sulfate results in a desirable feeling of alertness,

TABLE 15.2
Characterization of the Therapeutic, Euphoric, and Toxic Syndromes Associated with Acute Amphetamine Use or Overdose

Effects	Signs and Symptoms	Clinical Management of Acute Toxicity
Euphoric	Wakefulness, alertness, mood elevation, increased initiative and self-confidence, elation, improved motor and speech activities	Euphoric effects are overcome with S/H (benzodiazepines) or diminish with chronic use
Therapeutic	Anorexia, bronchodilation, improved cerebral circulation, improved attention span (in children with ADHD), alertness and wakefulness (for narcolepsy)	—
Neurologic	Tremors, hyperactive reflexes, seizures, convulsions, coma	Withdrawal of agent, S/H (diazepam), phenobarbital
Cardiovascular and circulatory	Headache, chilliness, palpitations; increased oxygen consumption, blood pressure, peripheral vasoconstriction; anginal pain, tachypnea, tachycardia, circulatory collapse	Withdrawal of agent or Na nitroprusside or Ca^{2+} channel blockers; digoxin, diuretics; β-blockers not recommended
Gastrointestinal	Dry mouth, metallic taste, anorexia, nausea, vomiting, diarrhea, abdominal cramps	Withdrawal of agent, S/H, or symptomatic
Renal and electrolyte disturbances	Increased blood sugar, excessive sweating, hyperkalemia, hypernatremia; renal failure	Withdrawal of agent, S/H, diuretic or symptomatic
CNS	Restlessness, dizziness, tremor, irritability, insomnia, anxiety, hyperpyrexia, mydriasis	Withdrawal of agent, S/H, or symptomatic
Toxic psychosis syndrome	Paranoid syndrome characterized by hallucinations, schizophrenic episodes, delirium, panic state, suicidal or homicidal tendencies, parasitosis	Self-limiting in absence of agent or phenothiazine (chlorpromazine, haloperidol), barbiturate

Parasitosis refers to the irrational feeling or hallucination of worms or insects crawling on the skin.
S/H: sedative/hypnotic.

wakefulness, mood elevation, and improved self-confidence, which induces a sense of well-being and euphoria. Prolonged use or overdose is invariably followed by depression and fatigue (the *crash*). Treatment modalities include clinical management of signs and symptoms, especially regulating cardiovascular effects, and therapeutic intervention for the *toxic psychosis syndrome*. Although acidification of urine enhances the renal elimination of amphetamines, it may worsen renal failure by exacerbating the effects of profound hyperthermia.

In 1999, a clinical investigation sought to determine the effect of the combination of phentermine and fenfluramine ("phen-fen") on the prevalence of valvular heart disease in obese subjects enrolled in a prospective, strict weight-loss research protocol. The "phen-fen" therapy was identified and associated with a low, but clinically significant, prevalence of aortic and mitral regurgitation, which may reflect age-related degenerative changes. This toxic finding of a frequently used stimulant forced the removal of fenfluramine (Pondimin®) from the market.

15.1.7 Chronic Methamphetamine Use

Chronic MA abuse significantly changes brain function. Noninvasive human brain imaging studies have reported alterations in dopamine activity in CNS neurons associated with reduced motor performance and impaired verbal learning. More recently, chronic users have also shown severe structural and functional changes in brain tissue associated with emotion and memory. Structural and neuropharmacological changes in brain physiology account for the emotional and cognitive problems observed in chronic users.

15.1.8 Tolerance and Withdrawal

As with many potentially addictive agents, chronic use of amphetamines leads to psychological dependence, characterized by compulsive drug seeking and craving not unlike other drugs of abuse. In the short term, it is not unusual for amphetamine abusers to occasionally augment their euphoria with an injection every 3 to 4 h continuously for 7 to 10 days. In addition, simultaneous administration of an opioid with amphetamine is a common practice among addicts to minimize the psychological withdrawal from the stimulant. To maximize the euphoric actions, habitual users will often ingest large quantities of antacids. The rationale for this behavior is to alkalinize the urine—that is, to maintain the basic compound in the nonionic state, force renal tubular reabsorption, and prolong the blood levels. In fact, alkalinization of urine prolongs the half-life from 7 to 8 h up to 35 h. Tolerance to the anorexic effect, as well as to the improvement of mood, is frequent but not invariable. Many patients, however, have been treated for narcolepsy for years without requiring major adjustments in dosage.

Induction of the withdrawal syndrome depends on the extent of habituation but varies between several hours to several days after discontinuation of the drug. It is characterized by prolonged sleep, lassitude, drug-craving behavior, and hyperphagia. A deep depression follows interruption of prolonged or chronic use. No clear physical symptoms of withdrawal are associated with amphetamines.

15.1.9 Clinical Management of Amphetamine Addiction

Currently, the most effective therapy incorporates a comprehensive behavioral treatment approach that combines behavioral therapy, family education, individual counseling and support, drug testing, and encouragement for nondrug-related activities. Contingency management interventions, which provide tangible incentives in exchange for engaging in treatment and maintaining abstinence, have also been shown to be effective. Unlike the use of methadone or buprenorphine in opioid addiction, there are no treatment programs for MA addiction that include therapeutic drug intervention.

15.1.10 Methods of Detection

To distinguish between the phenethylamine derivatives, including amphetamine, MA, phentermine, ephedrine, and other structurally related compounds, detection requires gas chromatography (GC) analysis with high resolution, selective liquid phases, and good sensitivity. Consequently, amphetamine salts are first converted to the freebase by extraction into an organic solvent, followed by derivatization to Schiff bases, amides, or halogenated products. The derivatives are then chromatographed on methylsilicone capillary columns and visualized with an electron capture detector.

Other methods used to measure phenethylamines depend on the sensitivity required and the sample available. Routine qualitative urine drug screening tests, such as thin layer chromatography (TLC) with fluorescent dye staining, are used for qualitative analysis. Quantitative analysis with TLC, however, is limited due to a lack of sensitivity, whereas radioimmunoassay (RIA) techniques have sensitivities down to 2.5 ng. High-performance liquid chromatography (HPLC) analysis, with a fluorescence detector, is less common for routine screening but can detect levels of amphetamines in biological fluids down to 100 fmol. Using the latter techniques, unchanged amphetamine has been identified in the urine up to 29 h after a single 5 mg oral dose.

RIA assays, such as the Roche Abuscreen®, identify only amphetamine, while Syva's EMIT (enzyme-multiplied immunoassay technique) is able to distinguish amphetamine from MA. A more specific Abuscreen amphetamine assay is also available and is sometimes used as a second screen. The new Abbott TDx Drug Detection System® is reported to sense both MA and amphetamine with little or no cross-reactivity to ephedrine or other alpha-adrenergic sympathetic agonists.

15.2 COCAINE

15.2.1 Incidence and Occurrence

The sporadic use of cocaine in the last 50 years is a reflection of its cyclic popularity as well as availability, cost, and competition with other illicit drugs. According to the National Survey on Drug Use and Health (NSDUH), the prevalence of misuse of cocaine has been relatively stable since 2009. Furthermore, data from the 2011 Drug Abuse Warning Network (DAWN) reported that cocaine was involved in 505,224 of the nearly 1.3 million visits to EDs for drug misuse or abuse. In 2014, there were an estimated 1.5 million current (past-month) cocaine users aged 12 or older (0.6% of the population); adults aged 18 to 25 years have a higher rate of current cocaine use than other age groups, with 1.4% of young adults reporting past-month cocaine use; 1.1% of 12th-graders and only 0.8% of 10th- and 0.5% of 8th-graders reported using cocaine in the past month (U.S. NIDA, 2018).

The drug is derived from the dried leaves of the plant *Erythroxylon coca* (Figure 15.2), which appears to grow best at higher elevations (1500–6000 ft.) in the South American countries of Peru, Bolivia, and Colombia. Historically, the ancient Incas of Peru chewed on the leaves of the plant to decrease fatigue and reduce appetite during long journeys, although it was still considered an inferior substitute for

FIGURE 15.2 Cocaine plant on a farm outside Cuzco, a city in southeastern Peru. (From personal archives.)

tea. Drug analysis of ancient Egyptian ruins detected the presence of cocaine and nicotine. Also, cocaine was sold medicinally as a brain tonic in the early 1900s (as an ingredient of Coca-Cola®). Today, the illicit form of the compound is transported to the United States principally in the same manner as described for opioids (Chapter 14)—that is, by *body packing*.* Before it arrives on the illicit market, the crystalline form is diluted (*cut*) with a pharmaceutically compatible diluent, such as lactose, mannitol, sucrose, or procaine HCl. The compounded substance is injected I.V. as a 10–20% solution, administered by intranasal insufflation (*snorted* as 50–75% crystalline powder), or burned and inhaled as the freebase (*crack* cocaine, 20% powder).†

15.2.2 Medicinal Chemistry

Coca leaves contain the alkaloids of ecgonine, tropine, and hygrine, of which only the derivatives of ecgonine are of commercial importance. Cocaine is an ester of benzoic acid and methylecgonine, the latter of which is related to the amino alcohol group found in atropine. The chemical structures of cocaine and related local anesthetics are illustrated in Figure 15.3.

15.2.3 Pharmacology and Clinical Use

Although not generally regarded as a sympathomimetic, cocaine has stimulant as well as anesthetic properties. Cocaine displays reversible CNS and peripheral actions. It

* Cocaine drug carriers, or *body packers*, seal and conceal illicit drug packets in body compartments either by swallowing or by insertion into body orifices. As with the opioids, the packets run the risk of leaking or bursting, making the overall clinical management of these individuals of immediate attention.

† What is believed to be the last (and perhaps a forgery) of the Sherlock Holmes novels, *The Seven Percent Solution*, describes the famous detective struggling with the addictive power of cocaine.

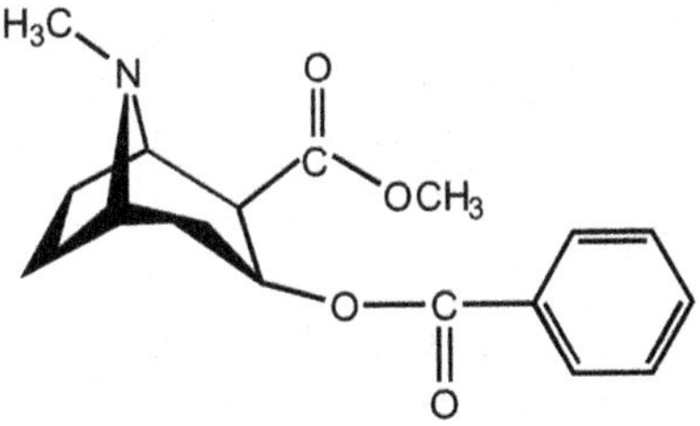

Cocaine

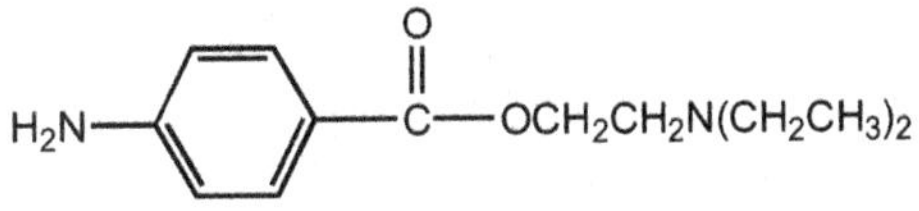

Procaine

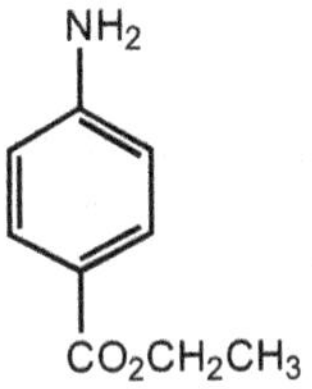

Benzocaine

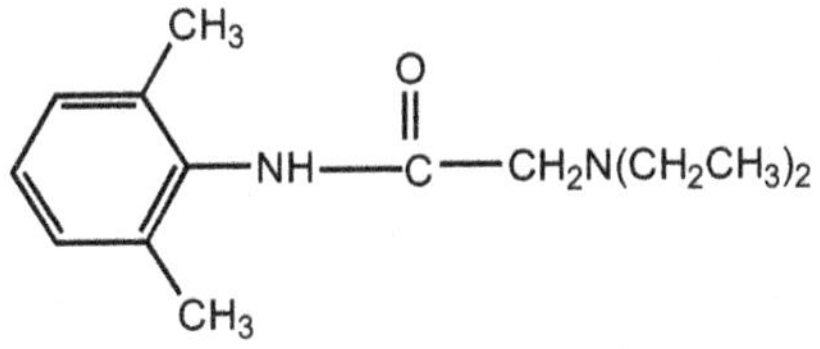

Lidocaine

FIGURE 15.3 Structure of cocaine and related local anesthetics.

blocks nerve conduction through its local anesthetic properties, primarily by inhibiting neuronal sodium permeability. It possesses local vasoconstrictor actions secondary to inhibition of local norepinephrine reuptake at adrenergic neurons. Unlike with other local anesthetics, the production of euphoria is due, in part, to inhibition of dopamine reuptake in central synapses. Because of its high toxicity and potential for abuse, its use is limited as a topical anesthetic/vasoconstrictor (1–10% solution) for surgical procedures involving the oral and nasal mucosal cavities.

15.2.4 Toxicokinetics

Cocaine is rapidly absorbed after I.V. or I.N. administration or by inhalation. Distribution depends largely on access to the systemic circulation, which is determined by protein binding and its vasoconstrictor effects, both of which limit systemic distribution. Thus, the distribution of the compound after local application is diminished by its vasoconstrictor activity. This effect, however, is obviated when exposure through inhalation or I.V. routes circumvents the local effects, resulting in significant potential for systemic toxicity.

15.2.5 Signs and Symptoms of Acute Toxicity

The central, peripheral, and parasympathetic nervous systems, as well as cardiovascular and smooth muscle, display initial signs of stimulation followed by depression. This characteristic is presumably due to selective depression of inhibitory neurons as well as inhibition of general neuronal activity in these areas. Initially, the stimulation mimics sympathetic activation exhibited by excitement, apprehension, nausea, vomiting, and tremors. Ensuing spasms and seizures are a result of high serum concentrations and overwhelming neuronal depression.

Its cardiovascular toxicity is evidenced by increased blood pressure and heart rate, resulting in an increase in myocardial work and oxygen demand. Thus, the drug has the potential for precipitating ventricular dysfunction and arrhythmias. Simultaneously, it induces coronary artery vasospasm as a result of sympathetic stimulation. This dual mechanism precipitates myocardial ischemia or infarction (*heart attack*). Ischemic and hemorrhagic strokes present as headaches or changes in mental status, often mistaken as psychiatric instability.*

Smoking *crack cocaine* hastens episodes of asthma or chronic obstructive pulmonary disease (COPD). Mouth and pharyngeal pain, drooling, and hoarseness accompany the severe upper airway burn injury resulting from inhaling the heat from pipe smoking. In moderate toxicity, cyanosis, dyspnea, and rapid, irregular respirations ensue. Cardiogenic or noncardiogenic pulmonary edema and respiratory failure are progressive sequelae.

Other important toxic features of cocaine toxicity include the production of acute dyskinesia (*crack dancing*). This syndrome, characterized by episodes of choreoathetoid movements of the extremities, lip-smacking, and repetitive eye blinking, occurs soon after cocaine use and lasts for several days. Acute renal failure, increased risk of spontaneous abortions, elevation of maternal hypertension during pregnancy, and profound hyperthermia are significant complications of cocaine toxicity. Effects on GI smooth muscle are not dramatic.

15.2.6 Clinical Management of Acute Overdose

The maintenance of airway, breathing, and circulation (ABCs) is the priority in managing patients with cocaine toxicity. Cardiovascular, neurologic, and psychiatric complications are effectively controlled with benzodiazepine administration, particularly diazepam or lorazepam. I.V. phenobarbital is used to control cocaine-induced seizures if the benzodiazepines are inadequate. Phenytoin is not useful in

* The consumption of cocaine and alcohol together induces the hepatic formation of cocaethylene. This combination intensifies cocaine's euphoric effects but is associated with a greater risk of sudden death than with cocaine alone.

cocaine-induced seizures, and since succinylcholine risks hyperkalemia and hyperthermia, a nondepolarizing neuromuscular blocker, such as pancuronium bromide, should be used in the event of intractable seizures.

15.2.7 Tolerance and Withdrawal

Addiction to cocaine is essentially indistinguishable from amphetamine habituation. Because of its rapid metabolism, larger doses of cocaine are required to maintain the euphoric effects in a chronic user. The development of tolerance, however, is inconsistent, and users will succumb to the toxic effects at low doses despite the continuous use. There is generally no cross-tolerance between cocaine and sympathetic stimulants.

15.2.8 Clinical Management of Cocaine Addiction

As with amphetamine and MA abuse, the most effective treatments for decreasing cocaine use and preventing relapse combine cognitive-behavioral interventions with social supports, family education, individual counseling, drug testing, and encouragement for nondrug-related activities. Treatment is individualized for the patient's needs to optimize outcomes.

Unlike opioid treatment programs, currently, there are no medications that incorporate therapeutic drug intervention for treating cocaine addiction.

15.2.9 Methods of Detection

Cocaine is extensively metabolized, primarily by liver and plasma esterases, and only 1% of a dose is excreted in the urine unchanged. Approximately 70% of a dose can be recovered in the urine over a period of 3 days. About 25–40% of cocaine is metabolized to benzoylecgonine, the major metabolite found in the urine. About 18–22% is excreted as ecgonine methyl ester and 2–3% as ecgonine. Immunochemical techniques such as EMIT and RIA are designed to detect benzoylecgonine. Unchanged cocaine is sometimes discovered by chromatographic methods for up to 24 hours after a given dose, while benzoylecgonine can generally be spotted by immunoassays for 24–48 hours. The presence of cocaine and its metabolites in urine is confirmed by liquid chromatography/mass spectrometry (LC-MS), HPLC, and gas chromatography/mass spectrometry (GC-MS). Benzoylecgonine is generally detected in urine up to 2 days after cocaine use using gas-liquid chromatography (GLC) or LC-MS, the latter of which is the most specific of the chromatographic techniques.

15.3 XANTHINE DERIVATIVES

15.3.1 Source and Medicinal Chemistry

Figure 15.4 illustrates the structures of the xanthine derivatives, caffeine, theophylline, and theobromine. The compounds contain the purine nucleus and are naturally occurring xanthine derivatives. Caffeine is the most active component of coffee and

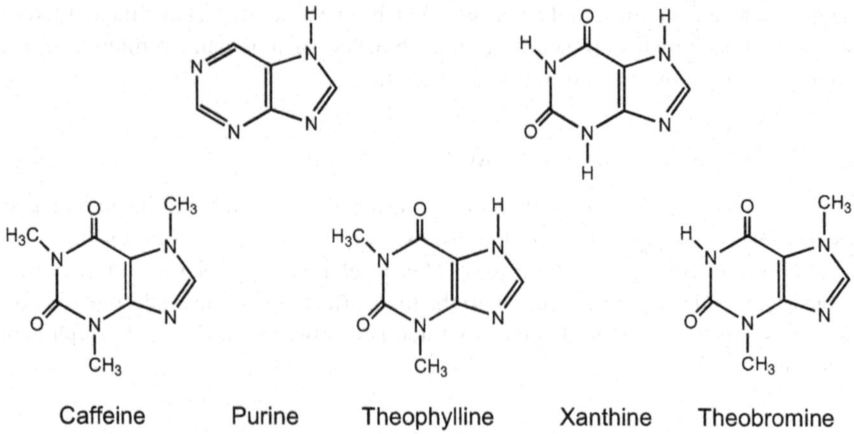

Caffeine Purine Theophylline Xanthine Theobromine

FIGURE 15.4 Structure of purine and xanthine derivatives.

the coffee beans (seeds) of *Coffea arabica*, a small evergreen shrub abundant in the tropical areas of South America, Central America, and the Middle East. It is also found in *Cola accuminata* and *Cola nitida*, tropical nuts of South America and Africa. Both caffeine and theophylline are distributed throughout the tea leaves of *Thea sinensis*, a shrub native to Japan, China, Southeast Asia, and Indonesia. The seeds of *Theobroma cacao*, the 12 meter cocoa plant, are indigenous to the South American, Asian, and African tropics. Table 15.3 organizes the xanthine derivatives with their major botanical sources and pharmacological actions.

15.3.2 Occurrence

Because of its ubiquitous occurrence, caffeine is distributed throughout coffee, tea, chocolate, and cola beverages, with coffee beans and tea leaves containing equivalent amounts of the stimulant (up to 2%). Brewed coffee boasts the highest amounts of caffeine (up to 100 mg); instant and decaffeinated coffees contain less (up to 75 and 5 mg, respectively); while a cola beverage possesses a significant dose of stimulant (up to 55 mg). A dose of 200 mg of theobromine is usually consumed in a cup of hot cocoa.

15.3.3 Pharmacology and Clinical Use

The history of the use of the xanthine derivatives for their medicinal and stimulant effects dates back throughout ancient civilizations. The xanthine derivatives, particularly caffeine, exert their pharmacological actions by increasing calcium permeability in the sarcoplasmic reticulum, inhibiting phosphodiesterase-promoting accumulation of cyclic AMP, and acting as competitive, nonselective antagonists of adenosine A_1 and A_{2A} receptors. Theophylline, in addition, inhibits the action of extracellular adenosine (bronchodilation effect), stimulates endogenous catecholamines (central stimulant effect), and directly promotes mobilization of intracellular calcium and β-adrenergic agonist activity (airway smooth muscle relaxation).

TABLE 15.3
Xanthine Derivatives and Their Major Properties and Actions

Chemical	Botanical Source	Predominant Pharmacological Actions	Therapeutic Uses
Caffeine	*Coffea arabica* (coffee beans, seeds, of the small evergreen shrub); *Cola accuminata* and *Cola nitida* (tropical nuts); *Thea sinensis* (leaves of the tea shrub); *Theobroma cacao* (seeds of the cocoa plant)	CNS stimulant; cardiac and skeletal muscle stimulant; smooth muscle relaxation; coronary vasodilator; increases BMR	Migraine headaches; weight loss; wakefulness; improve mental alertness; relieve drug-induced respiratory depression; short-term treatment of apnea in premature infants; emergency cardiac stimulant; diuretic
Theophylline	*Thea sinensis* (leaves of the tea shrub)	Bronchodilator; pulmonary smooth muscle relaxation; increases BMR	Treatment of bronchial asthma and other respiratory-related disorders
Aminophylline (Theophylline + ethylenediamine)	As for theophylline	Bronchodilator; pulmonary smooth muscle relaxation	COPD, emphysema, and other respiratory-related disorders
Theobromine	*Theobroma cacao* (seeds of the cocoa plant)	Similar to caffeine	Diuretic; smooth muscle relaxation; myocardial stimulant; vasodilator
Pentoxifylline	Dimethyl xanthine derivative	Vasodilator; anti-inflammatory	Chronic occlusive arterial diseases; intermittent claudication

BMR: basal metabolic rate; CNS: central nervous system; COPD: chronic obstructive pulmonary disease.

The predominant CNS stimulant properties of caffeine and the respiratory relaxation produced by theophylline are well documented. Caffeine stimulates cerebral activity, skeletal and cardiac muscle contraction, and general basal metabolic rate, while theophylline has less central stimulation but significant bronchial smooth muscle relaxation properties. Caffeine and theophylline enhance cardiac muscle contraction, induce coronary vasodilation, and promote diuresis. Caffeine is available in combination with ergotamine, belladonna alkaloids, or pentobarbital for the treatment of migraine headaches (Wigraine® tablets, Cafergot® suppositories), for its synergistic action with ephedrine for weight loss, and as an aid to wakefulness and restoring mental alertness. In combination with sodium benzoate (injectable), caffeine is used in the treatment of drug-induced respiratory depression, and as caffeine citrate (injectable) for the short-term treatment of apnea in premature infants. Theophylline (Elixophyllin®, Theolair®, Theo-Dur®, and various others) is employed in the treatment of bronchial asthma and other respiratory-related disorders. The calcium and sodium salicylate salts of theobromine are available for use as mild diuretics, vasodilators, smooth muscle relaxants, and cardiac stimulants.

15.3.4 Toxicokinetics

Xanthine derivatives are well absorbed orally, reaching peak distribution within 2 hours. The compounds are metabolized by the liver to methylxanthine and methyluric acid derivatives. Metabolic variability among different age groups, smokers, and individuals with pathologic complications is probably due to variable levels of cytochrome P450 and N-acetyltransferase systems. The drugs are eliminated by the kidney with a half-life of 3 to 15 hours in nonsmokers (4 to 5 hours in adult smokers).

15.3.5 Signs and Symptoms and Clinical Management of Caffeine Toxicity

The estimated LD_{50} of caffeine in humans is 5 to 10 g. Although fatalities are unlikely approaching this dose, individuals who ingest up to 10 mg/kg are at risk of developing dysrhythmias or strychnine-like seizures. More likely adverse effects of excessive caffeine intake are attributable to CNS stimulation, including insomnia, restlessness, sensory disturbances, and delirium. Increased skeletal muscle tension, tachycardia, premature ventricular contractions (PVCs), diarrhea, development of peptic ulcers, and GI bleeding complete the detrimental properties of acute and chronic ingestion. In addition, caffeine can neutralize the desired therapeutic effects of diuretics, antihypertensive agents, β-blockers, and sedative/hypnotics.

Myocardial tachyarrhythmias and development of seizures should be monitored in patients after an acute ingestion of 1 g or more of caffeine. A short-acting β-adrenergic blocker, such as esmolol (Brevibloc® injectable), is useful in the management of the former, while the seizures are controlled by a short-acting benzodiazepine, such as midazolam (Versed® injection).

15.3.6 Signs and Symptoms and Clinical Management of Theophylline Toxicity

Theophylline shares similar properties with caffeine, although its toxicity is more acute and more common; chronic toxicity, however, is unlikely. Therapeutic blood levels are strictly regulated at 10 to 20 μg/ml but vary depending on interactions with other drugs, inaccurate dosing, or accidental or intentional overdose ingestion. Seizures occur at between 25 and 40 μg/ml. As with caffeine toxicity, rapid I.V. administration of theophylline is associated with headache, hypotension, dizziness, restlessness, agitation, and arrhythmias. To relieve the gastric irritation and nausea related with oral tablets, most theophylline preparations are available as controlled-release dosage forms.

15.3.7 Tolerance and Withdrawal

Caffeine withdrawal is associated with chronic use and is demonstrated abruptly within 12 to 24 hours after the last dose. Initial symptoms, including headache, anxiety, fatigue, and craving behavior, last for about 1 week. There is demonstrated tolerance to the diuretic action and the insomnia produced with theophylline, but no tolerance develops to the CNS stimulation or bronchodilation.

15.4 OTHER SPECIFIC SYMPATHOMIMETIC AGENTS

15.4.1 Strychnine

Originally labeled as an analeptic agent (stimulant), strychnine was introduced into Europe in about the fifteenth century, mainly for poisoning animals. Its use in medicine began in about 1540, when it was found to possess CNS stimulant properties to combat depression and sleeplessness and improve respiration. It was also popular as a GI bitter tonic (appetite stimulant) and to increase muscle tone. Unfortunately, strychnine exhibits a narrow therapeutic index and consequently does not enjoy any favorable medicinal value. However, it is still used as a rodenticide.

Strychnine is an indole alkaloid* obtained from the dried seeds of the plant *Strychnos nux-vomica*, which grows to approximately 12 meters in Southeast Asia and northern Australia. It is extremely toxic as a central stimulant. Fatal poisoning in adults, especially from the ingestion of tonic tablets, results from ingesting 30 to 100 mg and in children from accidental ingestion of rodenticide (up to 15 mg is a lethal dose). Since the alkaloid is predominantly a lipophilic glycoside, it is well absorbed orally, and it is metabolized via the hepatic microsomal enzyme system.

Strychnine inhibits the postsynaptic receptor for glycine, an inhibitory neurotransmitter, allowing the development of spinal convulsions of the tonic type (characterized as extensor thrusts). Strychnine exerts its toxic abilities by blocking postsynaptic conduction in the inhibitory spinal Renshaw motor neurons, where it

* Brucine, also an indole alkaloid in *Nux vomica*, is structurally classified as dimethoxystrychnine and is used commercially as an alcohol denaturant.

interferes with ascending and descending motor tracts, resulting in convulsions of spinal cord origin.

Signs and symptoms of toxicity are similar to those seen in experimental animals. Early stages are characterized by a grimacing stiffness of the neck and face, followed by increased reflex excitability that is precipitated by sensory stimulation. Tonic convulsions (traditionally demonstrated as arched back or *opisthotonus*) are followed by coordinated extensor thrusts. Convulsions occur as full contractions of all voluntary muscles, including thoracic, abdominal, and diaphragmatic, which ultimately suppress respiration. Convulsive episodes are continuous or intermittent, with depression and sleep interspersed, depending on the depth of toxicity, and the patient is generally conscious and in pain. After several full convulsions, medullary paralysis due to hypoxia is the cause of death.

Treatment must be instituted swiftly and is aimed at preventing convulsions, maintaining ventilation, and the administration of an anticonvulsant/skeletal muscle relaxant, such as diazepam. Successful treatment with diazepam (10 mg I.V. repeated as needed) depends on the following criteria: 1. the dose of poison is below the lethal dose; 2. intervention is started soon after ingestion; and 3. convulsions have not ensued. Gastric lavage is only indicated when the compound is suspected to be in the stomach contents and if convulsions have subsided.

15.4.2 Nicotine

Nicotine is a pyridine alkaloid (1-methyl-2(3-pyridyl) pyrrolidine) obtained from the cured and dried leaves of the tobacco plant, *Nicotiana* sp., indigenous to the southern United States and tropical South America. It is a tall annual herb introduced into Europe from the American colonies, who obtained it from the Indians.* Well before the isolation of nicotine in 1828, the finely powdered tobacco leaves were dusted on vegetable crops as an insecticide. Tobacco leaves contain from 0.6 to 9% of the alkaloid along with other nicotinic derivatives. Although the demand for nicotine has declined recently, nicotine is the active ingredient of commercial anti-smoking products and some contact insecticides (see Chapter 26).

Nicotine stimulates nicotinic receptors of all sympathetic and parasympathetic ganglia, neuromuscular junction innervating skeletal muscle, and CNS pathways (Figure 15.1). In general, nicotine toxicity is a complex syndrome whose cardiovascular toxicity varies widely between bradycardia and tachycardia. Signs and symptoms associated with nicotine overdose occur 1 to 2 h after exposure and include initial generalized stimulation followed by predominance of parasympathetic overtone (salivation, lacrimation, urination, defecation, and vomiting). Continued exposure progresses to muscular weakness, tremors, bradycardia, hypotension, and dyspnea, which if untreated, advances to convulsions and respiratory paralysis. The compound is best absorbed through inhalation, although oral absorption of nicotine-containing insecticides demonstrates severe toxicity. Mild toxicity is also observed in children who accidentally ingest therapeutic nicotine patches (Nicoderm-CQ®), displaying symptoms of acute nicotinic syndrome (hypertension, hyperreflexia, hyperpyrexia, tremors, vomiting, and diarrhea).

* *Tobacco* refers to the American Indian name for the "pipe" or tube used for smoking.

Treatment of poisoning is primarily symptomatic and involves washing of the agent from the skin with water, induction of emesis if intervention starts soon after ingestion, and maintenance of vital signs.

Nicotine is a highly addictive drug. Psychological and physical tolerance develops with chronic smoking and use of tobacco products. Numerous products and programs are available to aid in smoking cessation. Abrupt withdrawal of tobacco use is associated with irritability, aggression, and depression. Changes in electroencephalographic recordings and autonomic functions are clearly demonstrated. Without some therapeutic intervention, a large percentage of attempts at stopping tobacco use result in failure.

15.4.3 Ephedrine

Ephedrine, an alkaloidal amine derived from the plant *Ephedra sinica*, has been used in Chinese medicine for more than 5000 years (see Chapter 21). The plant is predominantly native to southeastern China as well as coastal regions of India and Pakistan. Although its effects are less pronounced, ephedrine is a long-acting orally active adrenergic sympathetic stimulant whose actions mimic epinephrine. Its principal effects are CNS stimulation, hypertension, and mydriasis. It has sustained bronchial muscle relaxation, mydriatic, and decongestant properties, making it suitable for therapeutic prescription and nonprescription use. Precautions in the use of ephedrine are similar to those outlined for the amphetamines.

Recently, continuous ephedrine use in nonprescription formulations has been associated with arrhythmias, strokes, heart attacks, and deaths, particularly when used by patients with preexisting cardiac disorders or in combination with other dietary stimulants. The U.S. Food and Drug Administration (FDA) in 1997 proposed mandating warning labels and a sharp reduction of the dose of ephedrine in herbal supplements, citing reports of 17 deaths and 800 illnesses linked to products that promise to help people lose weight, build muscle and heighten sexual awareness.* In 2003, the FDA proposed plans that would sharply reduce the amount of ephedrine in dietary supplements and restrict the marketing of ephedrine-containing substances for weight loss and bodybuilding. The regulations would also require warning labels advising users of the dangers of high doses of ephedrine. In 2004, the FDA created a ban on ephedrine alkaloids that are marketed for reasons other than asthma, colds, or allergies or in other traditional Chinese herbal products. In April 2005, the U.S. District Court for the District of Utah ruled that the FDA did not have proper evidence that low doses of ephedrine alkaloids are actually unsafe. In August 2006, however, the U.S. Court of Appeals for the Tenth Circuit in Denver reversed the 2006 ruling and upheld the FDA's final rule declaring all dietary supplements containing ephedrine alkaloids adulterated and therefore illegal for marketing in the United States. Today, the legal sale of ephedrine is available but monitored and limited by state jurisdictions.

* Until reports of adverse reactions to ephedrine began mounting, it had been the most widely used herbal ingredient in the dietary supplement industry. It was also sold as an alternative to the illicit drug MA. Florida and New York banned ephedrine supplements after pills with such names as Herbal Ecstacy and Ultimate Xphoria promised a "natural high." The bans followed the death of a 20-year-old student after taking Ultimate Xphoria in 1996.

15.4.4 Phenylpropranolamine

Phenylpropranolamine (PPA) shares the pharmacological properties of ephedrine but causes less CNS stimulation. It is an α_1-agonist that can produce significant hypertension with a resultant reflex bradycardia.* Until 2001, PPA was found as an ingredient of numerous proprietary prescription and nonprescription formulations in combination with antihistamines, used for treatment of nasal and sinus congestion. Since then, the compound has been voluntarily removed from most oral dosage forms because of its association with an increase in cerebral ischemia and strokes. The withdrawal of cold remedies and appetite suppressants from market shelves, including some that had been used for decades, came after federal regulators announced that they would take steps to ban all use of the active ingredient in the products.

15.4.5 Pseudoephedrine

Pseudoephedrine (PE) shares the same pharmacological properties as ephedrine. It binds to central and peripheral α_1-receptors, resulting in significant CNS stimulation and vasoconstriction, respectively, particularly affecting arterioles of the mucous membranes, skin, kidneys, and abdomen. Until 2005, PE was found as an ingredient of numerous proprietary prescription and nonprescription formulations in combination with antihistamines and analgesics. The drug is indicated as a decongestant for the treatment of nasal, sinus, and Eustachian tube congestion as well as for vasomotor rhinitis and other inflammatory upper respiratory disorders.

As with ephedrine, its usefulness has been obviated by its propensity for abuse. In particular, PE has been abused for its CNS stimulant properties and its potential for dependence. In addition, its structural similarity to the amphetamines has made PE desirable as a chemical precursor in the illicit manufacture of MA and methcathinone.† As a result of the increasing regulatory restrictions on the sale and distribution of PE, many pharmaceutical firms have voluntarily reformulated products containing PE. Most have substituted PE with phenylephrine as the decongestant. As with ephedrine, state laws regulate over-the-counter (OTC) sales of PE-containing products.

REFERENCES

SUGGESTED READINGS

Ashare, R.L. and Schmidt, H.D., Optimizing treatments for nicotine dependence by increasing cognitive performance during withdrawal. *Expert Opin. Drug Discov.* 9, 579, 2014.

Burger, A.J., Sherman, H.B., Charlamb, M.J., Kim, J., Asinas, L.A., Flickner, S.R., and Blackburn, G.L., Low prevalence of valvular heart disease in 226 phentermine-fenfluramine protocol subjects prospectively followed for up to 30 months, *J. Am. Coll. Cardiol.* 34, 1153, 1999.

---

* This effect is also seen with phenylephrine (Neosynephrine® and various other products), an α-agonist used as a decongestant, which raises blood pressure but slows heart rate.

† α-methylamino propiophenone, a phenethylamine derivative used for its CNS stimulant properties, is currently a Drug Enforcement Administration (DEA) Schedule I controlled substance in the United States.

Grady, D., Search for Cause of Diet Pill's Risk Yields New Warning, *Science Desk, New York Times*, Jan. 1999.
Grasing, K.,Mathur, D., Newton, T.F., and Desouza, C., Individual predictors of the subjective effects of intravenous cocaine. *Psychiatry Res.* 208, 245, 2013.
Jalil, A.M. and Ismail, A., Polyphenols in cocoa and cocoa products: is there a link between antioxidant properties and health? *Molecules* 13, 2190, 2008.
Kay-Lambkin, F.J., Technology and innovation in the psychosocial treatment of methamphetamine use, risk and dependence. *Drug Alcohol Rev.* 27, 318, 2008.
London, E.D. et al., Mood disturbances and regional cerebral metabolic abnormalities in recently abstinent methamphetamine abusers. *Arch. Gen. Psychiatry* 61, 73, 2004.
Rimsza, M.E. and Newberry, S., Unexpected infant deaths associated with use of cough and cold medications. *Pediatrics* 122, e318, 2008.
Schmidt, B., Methylxanthine therapy for apnea of prematurity: evaluation of treatment benefits and risks at age 5 years in the international Caffeine for Apnea of Prematurity (CAP) trial. *Biol. Neonate* 88, 208, 2005.
Thompson, P.M. et al., Structural abnormalities in the brains of human subjects who use methamphetamine. *J. Neurosci.* 24, 6028, 2004.
U.S. National Institute on Drug Abuse (NIDA), National Institutes of Health (NIH), Monitoring the Future 2018 Survey Results. From: https://www.drugabuse.gov/related-topics/trends-statistics/infographics/monitoring-future-2018-survey-results, last viewed March 2019.
Wang, G.J. et al., Partial recovery of brain metabolism in methamphetamine abusers after protracted abstinence. *Am. J. Psychiatry* 161, 242, 2004.
Won, S., Hong, R.A., Shohet, R.V., Seto, T.B., and Parikh, N.I.,Methamphetamine-associated cardiomyopathy. *Clin. Cardiol.* 36, 737, 2013.
Wood, D.M., Dargan, P.I., and Hoffman, R.S., Management of cocaine-induced cardiac arrhythmias due to cardiac ion channel dysfunction. *Clin. Toxicol. (Phila)* 47, 14, 2009.

REVIEW ARTICLES

Albertson, T.E., Derlet, R.W., and Van Hoozen, B.E., Methamphetamine and the expanding complications of amphetamines, *West J. Med.* 170, 214, 1999.
Benowitz, N.L. and Gourlay, S.G., Cardiovascular toxicity of nicotine: implications for nicotine replacement therapy, *J. Am. Coll. Cardiol.* 29, 1422, 1997.
Berman, S., O'Neill, J., Fears, S., Bartzokis, G., and London, E.D., Abuse of amphetamines and structural abnormalities in the brain. *Ann. N.Y. Acad. Sci.* 1141, 195, 2008.
Chang, L., Ernst, T., Speck, O., and Grob, C.S., Additive effects of HIV and chronic methamphetamine use on brain metabolite abnormalities. *Am. J. Psychiatry* 162, 361, 2005.
Ciccarone, D., Stimulant abuse: pharmacology, cocaine, methamphetamine, treatment, attempts at pharmacotherapy. *Prim. Care* 38, 41, 2011.
Courtney, K.E. and Ray, L.A., Methamphetamine: an update on epidemiology, pharmacology, clinical phenomenology, and treatment literature, *Drug Alcohol Depend.* 143, 11, 2014.
Cunningham, J.K., Callaghan, R.C., and Liu, L.M. US federal cocaine essential ("precursor") chemical regulation impacts on US cocaine availability: an intervention time–series analysis with temporal replication. *Addiction* 110, 805, 2015.
Davidson, C. Gow, A.J., Lee, T.H., and Ellinwood, E.H., Methamphetamine neurotoxicity: necrotic and apoptotic mechanisms and relevance to human abuse and treatment. *Brain Res. Brain Res. Rev.* 36, 1, 2001.
Docherty, J.R., Pharmacology of stimulants prohibited by the World Anti-Doping Agency (WADA). *Br. J. Pharmacol.* 154, 606, 2008.

Eccles, R., Substitution of phenylephrine for pseudoephedrine as a nasal decongestant. An illogical way to control methamphetamine abuse. *Br. J. Clin. Pharmacol.* 63, 10, 2007.

Elkashef, A. et al., Pharmacotherapy of methamphetamine addiction: an update. *Subst. Abuse* 29, 31, 2008.

Garrido, E. et al., Chromogenic and fluorogenic probes for the detection of illicit drugs. *ChemistryOpen* 7, 401, 2018.

Geiger, J.D., Adverse events associated with supplements containing ephedra alkaloids. *Clin. J. Sport Med.* 12, 263, 2002.

Glasser, A.M.Collins, L., Pearson, J.L., Abudayyeh, H., Niaura, R.S., Abrams, D.B., and Villanti, A.C., Overview of electronic nicotine delivery systems: a systematic review. *Am. J. Prev. Med.* 52, e33, 2017.

Goldstein, R.A., DesLauriers, C., and Burda, A.M., Cocaine: history, social implications, and toxicity—a review. *Dis. Mon.* 55, 6, 2009.

Gudelsky, G.A. and Yamamoto, B.K., Neuropharmacology and neurotoxicity of 3,4-methylenedioxymethamphetamine. *Methods Mol. Med.* 79, 55, 2003.

Hanania, N.A. and Donohue, J.F., Pharmacologic interventions in chronic obstructive pulmonary disease: bronchodilators. *Proc. Am. Thorac. Soc.* 4, 526, 2007.

Heal, D.J. Smith, S.L., Gosden, J., and Nutt, D.J., Amphetamine, past and present—a pharmacological and clinical perspective. *J. Psychopharmacol.* 27, 479, 2013.

Hser, Y.I., Huang, D., Brecht, M.L., Li, L., and Evans, E., Contrasting trajectories of heroin, cocaine, and methamphetamine use. *J. Addict. Dis.* 27, 13, 2008.

Jackson, K.J. et al., New mechanisms and perspectives in nicotine withdrawal. *Neuropharmacology* 96, 223, 2015.

Jones, G., Caffeine and other sympathomimetic stimulants: modes of action and effects on sports performance. *Essays Biochem.* 44, 109, 2008.

Keller, R.W. Jr. and Snyder-Keller, A., Prenatal cocaine exposure. *Ann. N. Y. Acad. Sci.* 909, 217, 2000.

Kohut, S.J., Interactions between nicotine and drugs of abuse: a review of preclinical findings. *Am. J. Drug Alcohol Abuse.* 43, 155, 2017.

Koren, G., Hutson, J., and Gareri, J., Novel methods for the detection of drug and alcohol exposure during pregnancy: implications for maternal and child health. *Clin. Pharmacol. Ther.* 83, 631, 2008.

Knuepfer, M.M., Cardiovascular disorders associated with cocaine use: myths and truths. *Pharmacol. Ther.*, 97, 181, 2003.

Kutlu, M.G., Parikh, V., and Gould, T.J., Nicotine addiction and psychiatric disorders. *Int. Rev. Neurobiol.* 124, 171, 2015.

Lawyer, G. Bjerkan, P.S., Hammarberg, A., Jayaram-Lindström, N., Franck, J., and Agartz, I., Amphetamine dependence and co-morbid alcohol abuse: associations to brain cortical thickness. *BMC Pharmacol.* 10, 5, 2010.

Lea, T. Kolstee, J., Lambert, S., Ness, R., Hannan, S., and Holt, M., Methamphetamine treatment outcomes among gay men attending a LGBTI-specific treatment service in Sydney, Australia. *PLoS One* 12(2), e0172560, 2017.

Lekskulchai, V. and Mokkhavesa, C., Evaluation of Roche Abuscreen ONLINE amphetamine immunoassay for screening of new amphetamine analogues. *J. Anal. Toxicol.* 25, 471, 2001.

Levisky, J.A. Karch, S.B., Bowerman, D.L., Jenkins, W.W., Johnson, D.G., and Davies, D., False-positive RIA for methamphetamine following ingestion of an Ephedra-derived herbal product. *J. Anal. Toxicol.* 27, 123, 2003.

Maxwell, J.C. and Rutkowski, B.A., The prevalence of methamphetamine and amphetamine abuse in North America: a review of the indicators, 1992–2007. *Drug Alcohol Rev.* 27, 229, 2008.

Mersfelder, T.L., Phenylpropanolamine and stroke: the study, the FDA ruling, the implications. *Cleve. Clin. J. Med.* 68, 208, 2001.

McCabe, S.E. Veliz, P., Wilens, T.E., and Schulenberg, J.E., Adolescents' prescription stimulant use and adult functional outcomes: a national prospective study. *J. Am. Acad. Child Adolesc. Psychiatry* 56, 226, 2017.

Mladěnka, P. et al., Comprehensive review of cardiovascular toxicity of drugs and related agents. *Med Res. Rev.* 38, 1332, 2018.

Nawrot, P. Jordan, S., Eastwood, J., Rotstein, J., Hugenholtz, A., and Feeley, M., Effects of caffeine on human health. *Food Addit. Contam.* 20, 1, 2003.

Philippe, G., Angenot, L., Tits, M., and Frédérich, M., About the toxicity of some Strychnos species and their alkaloids. *Toxicon* 44, 405, 2004.

Romanelli, F. and Smith, K.M., Clinical effects and management of methamphetamine abuse. *Pharmacotherapy* 26, 1148, 2006.

Rose, M.E. and Grant, J.E., Pharmacotherapy for methamphetamine dependence: a review of the pathophysiology of methamphetamine addiction and the theoretical basis and efficacy of pharmacotherapeutic interventions. *Ann. Clin. Psychiatry* 20, 145, 2008.

Rawson, R.A. Marinelli-Casey, P., Anglin, M.D., Dickow, A., Frazier, Y., Gallagher, C., Galloway, G.P., Herrell, J., Huber, A., McCann, M.J., Obert, J., Pennell, S., Reiber, C., Vandersloot, D., and Zweben, J.,A multi-site comparison of psychosocial approaches for the treatment of methamphetamine dependence. *Addiction* 99, 708, 2004.

Rusyniak, D.E., Neurologic manifestations of chronic methamphetamine abuse. *Neurol. Clin.* 29, 641, 2011.

Shekelle, P.G. Hardy, M.L., Morton, S.C., Maglione, M., Mojica, W.A., Suttorp, M.J., Rhodes, S.L., Jungvig, L., and Gagné, J.,Efficacy and safety of ephedra and ephedrine for weight loss and athletic performance: a meta-analysis. *JAMA* 289, 1537, 2003.

Simmler, L.D., Wandeler, R., and Liechti, M.E., Bupropion, methylphenidate, and 3,4-methylenedioxypyrovalerone antagonize methamphetamine-induced efflux of dopamine according to their potencies as dopamine uptake inhibitors: implications for the treatment of methamphetamine dependence. *BMC Res. Notes* 6, 220, 2013.

Spear, L.P. Silveri, M.M., Casale, M., Katovic, N.M., Campbell, J.O., and Douglas, L.A.,Cocaine and development: a retrospective perspective. *Neurotoxicol. Teratol.* 24, 321, 2002.

Stedeford, T., Cardozo-Pelaez, F., Vultaggio, B., Muro-Cacho, C., Luzardo, G.E., and Harbison, R.D., Alpha1-adrenergic receptors and their significance to chemical-induced nephrotoxicity—a brief review. *Res. Commun. Mol. Pathol. Pharmacol.* 110, 59, 2001.

Sweeney, C.T., Sembower, M.A., Ertischek, M.D., Shiffman, S., and Schnoll, S.H., Nonmedical use of prescription ADHD stimulants and preexisting patterns of drug abuse. *J. Addict. Dis.* 32, 1, 2013.

U.S. National Institute on Drug Abuse (NIDA), National Institutes of Health (NIH), U.S., National Survey of Drug Use and Health (NSDUH), Monitoring the Future Study, 2018. From: www.monitoringthefuture.org, last viewed March 2019.

United Nations World Drug Report, UN Office of Drugs and Crime, United Nations 2017.

Vanscheeuwijck, P.M. et al., Evaluation of the potential effects of ingredients added to cigarettes. Part 4: subchronic inhalation toxicity. *Food Chem. Toxicol.* 40, 113, 2002.

Wang, Z. Zhao, J., Xing, J., He, Y., and Guo, D., Analysis of strychnine and brucine in postmortem specimens by RP-HPLC: a case report of fatal intoxication. *J. Anal. Toxicol.* 28, 141, 2004.

Wolfe, S.M., Medicine. Ephedra–scientific evidence versus money/politics. *Science* 300, 437, 2003.

Zanardi, A., Leo, G., Biagini, G., and Zoli, M.,Nicotine and neurodegeneration in ageing. *Toxicol. Lett.* 127, 207, 2002.

16 Hallucinogenic Agents

16.1 HISTORY AND DESCRIPTION

The hallucinogens are a descriptive pharmacological class of synthetic and naturally occurring derivatives of the ergot alkaloids. These include the tryptamines, amphetamines, and related sympathomimetics. The feature that distinguishes these agents from the parent compounds is the production of hallucinations,* which are depicted as changes in perception, thought, and mood and the occurrence of dreamlike feelings. An individual who ingests hallucinogenic substances describes the effects as having a *spectator ego*, whereby a person experiences a bond with nature and society, a sense of overwhelming revelation and truth, and a vivid awareness of his surroundings. The use of hallucinogens was popularized in the 1960s by artists, musicians, writers, poets, scientists, and other celebrities (such as Timothy Leary) who aimed to enlighten the world regarding the benefits of the substances. The compounds were purported to improve creative abilities, foster higher levels of thinking and consciousness, and enhance perception. The hallucinogens were thus labeled as *mind-expanding* drugs. In addition, they were claimed to have beneficial effects in the treatment of psychiatric disorders, cancer, and alcoholism—claims that have since been proved invalid.

16.2 ERGOT ALKALOIDS

16.2.1 Incidence and Occurrence

Ergot is derived from the dried sclerotium (resting body of fungus) of *Claviceps purpurea* (rye plant). Although preparations of the crude drug are seldom found in pharmaceutical formulas, the alkaloid derivatives are important medicinal agents resulting from parasitic and saprophytic alteration of the plant. Ergot produces a large number of alkaloids, the most important of which are the ergotamines, the ergonovines, and ergotoxine.

The availability and illicit use of ergotamines have waned since the 1990s in comparison with the major hallucinogens, including the phenethylamines and the phenylisopropylamines described later in this chapter. Consequently, information regarding the incidence and occurrence of illicit use of ergotamines is scarce.

---

* The term *hallucinogen* refers to a substance that produces visual images or auditory perceptions that may, in fact, not be real. Some have suggested that drugs that cause such distorted perceptions should be labeled *illusogens*. However, since this change has not received universal acceptance, this chapter will use the traditional terminology.

16.2.2 Medicinal Chemistry

Figure 16.1 illustrates the parent structure of *d*-lysergic acid and the principal toxicologically important derivatives, ergotamine and lysergic acid diethylamide (LSD). The ergotamines, natural or semi-synthetic, are all derivatives of *d*-lysergic acid.

16.2.3 Pharmacology and Clinical Use

Ergotamine, ergonovine, and ergotoxine compounds are readily soluble in water, although their absorption rate is erratic. The ergotamine derivatives have a variety of seemingly unrelated pharmacological properties. In general, ergotamine has partial agonist or antagonist activity against trypaminergic, dopaminergic, and α-adrenergic receptors. It constricts central and peripheral blood vessels,

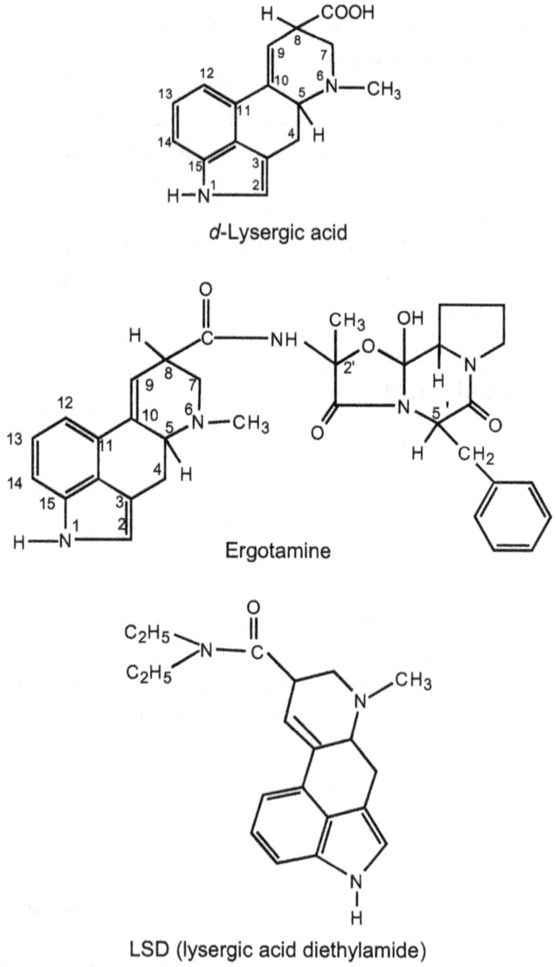

d-Lysergic acid

Ergotamine

LSD (lysergic acid diethylamide)

FIGURE 16.1 Structures of *d*-lysergic acid, ergotamine, and LSD.

depresses central vasomotor centers, and induces uterine contractions. For instance, methysergide (Sansert®) has no intrinsic vasoconstrictor properties in its actions against migraine and vascular headaches. Instead, it appears to block the effect of serotonin (Figure 16.2), a "headache substance" and a central neurochemical mediator, that may be involved in the production of vascular headaches. Methysergide displaces serotonin from pressor cranial arterial wall receptors during migraine headaches. Other ergot derivatives, such as methylergonovine (Methergine®), act directly on uterine smooth muscle and induce sustained contractions. Unlike the natural alkaloids, the semi-synthetic dihydrogenated ergot alkaloids (ergoloid mesylates, Hydergine®) increase brain metabolism presumably by increasing cerebral blood flow via α-adrenergic blockade, although this mechanism is contested.

Clinically, the agents are used for their oxytocin-like effect in the routine management of postpartum uterine contractions and bleeding (methylergonovine), for the prevention of vascular headaches, such as migraine or cluster headaches (dihydroergotamine, Migranal®; methysergide, Sansert®), and for the amelioration of symptoms associated with age-related mental capacity decline, as in progressive dementia and Alzheimer's dementia (dihydrogenated ergot alkaloids, Hydergine®).

16.2.4 Signs and Symptoms of Acute Toxicity

Adverse reactions occur in patients with acute ingestion of less than 5 mg of ergotamine derivatives. The most important signs and symptoms include nausea, vomiting, neuromuscular numbness and weakness, confusion, depression, drowsiness, and convulsions. Lightheadedness, disassociation, and hallucinatory experiences reflect central nervous system (CNS) toxicity.

16.2.5 Clinical Management of Acute Overdose

Treatment of acute overdose consists of withdrawal of the drug followed by symptomatic measures, including, but not limited to, the ABCs of emergency treatment.

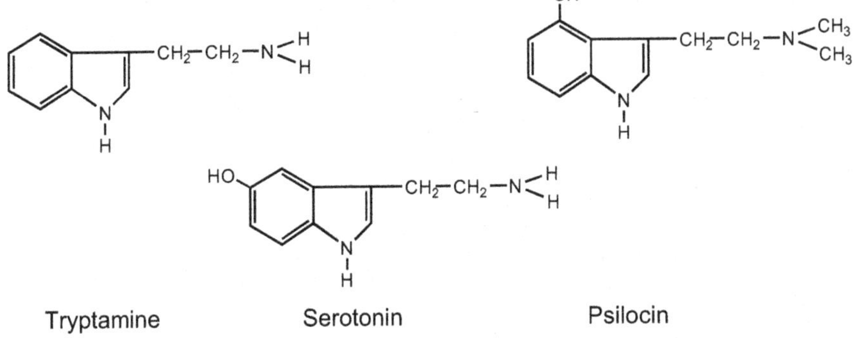

Tryptamine Serotonin Psilocin

FIGURE 16.2 Structures of tryptamine, serotonin (5-hydroxytryptamine), and psilocin (4-hydroxydimethyltryptamine).

16.3 LYSERGIC ACID DIETHYLAMIDE (LSD)

16.3.1 Incidence and Occurrence

Lysergic acid diethylamide (LSD) does not occur naturally but is a semi-synthetic preparation of *d*-lysergic acid (Figure 16.1). Discovered by A. Hoffman in 1943 during the course of experiments directed toward the synthesis of stimulants (analeptics), LSD produces opposing actions of powerful central stimulation with slight depression. Although it has no legitimate medical use today, the colorless, odorless, and tasteless compound is the most powerful psychotropic agent known, whose experimental value has proved it to be a model for psychosis.* The substance is available "on the street" in liquid, powder, and microdot dosage† forms, with 20 to 25 μg as an effective oral dose.

Large decreases in the use of LSD have been experienced since the mid-1990s when frequency peaked among youth. In 2016, 0.7% of Americans aged 12 and older had abused LSD at least once in the prior year, while 9.6% aged 12 or older reported they had used the substance at least once in their lifetime. Overall, 15.4% of Americans aged 12 or older have used a hallucinogen at least once in their lifetime (NIDA, National Survey of Drug Use and Health [NSDUH], Monitoring the Future, 2018).

16.3.2 Mechanism of Toxicity

Within minutes of ingestion, an effective dose causes hallucinogenic sensations that last between 6 and 24 h. LSD produces significant pyramidal and extrapyramidal effects as a result of interaction with central 5-hydroxytryptamine (5-HT; serotonin) receptors. Specifically, the compound mimics 5-HT at 5-HT_{1A} autoreceptors on raphe nucleus cell bodies and acts as a partial or full agonist at 5-HT_{2A} and 5-HT_{2C} receptors. Selective binding to serotonin receptors accounts for the hallucinogenic and behavioral alterations associated with the substance.

16.3.3 Hallucinogenic Effects

The desirable effects, or at least, the reason for intentional nontherapeutic use of the substance, vary between euphoria and dysphoria. Swift mood swings, even after a single dose within the same time period, are easily produced. The LSD *trip* is characterized predominantly by distortion in the realization of time and alteration of sensory perception, especially visual, auditory, tactile, olfactory, and gustatory. The individual experiences an intensified and altered perception of color. Objects are surrounded by halos and may leave a trail of individual afterimages when moving across a visual field. Other distortions of reality include *synesthesias*, whereby an

* The seeds of *Ololiuqui* (*Rivea corymbosa*, commonly known as bindweeds or morning glories) were used as an ancient Aztec hallucinogen, sporadically incorporated in Mexican religious ceremonies. The active ingredients of the seeds have been identified as ergot alkaloids, 0.05% of which include the amide and methylcarbinolamide derivatives of *d*-lysergic acid.

† Because of its relative potency, microgram quantities of the liquid are applied onto blotter (chromatographic or absorbent) paper and air dried. In this state, the substance is easily camouflaged and ingested by tearing and soaking the paper orally.

auditory stimulus is perceived "visually." Normal taste, smell, and touch sensations are often strange, distorted, grotesque, or uninterpretable.

The individual becomes depressed, anxious, and paranoid, and may experience *ego fragmentation* or feelings of depersonalization. The latter reaction may be pleasant, as the user experiences feelings of togetherness ("oneness") with the surroundings. Alternatively, the experience may be extremely uncomfortable, as the user feels a sense of loss of conscious control of thoughts and emotions and being "out of touch" with the surroundings. In extreme cases, frightening perceptions and psychological distortions may lead to unanticipated behavior in an otherwise normal person. Such *bad trips* may be accompanied by delirium-induced self-injury or suicide.

16.3.4 Toxicokinetics

Following oral ingestion of LSD, the effects are perceived within a few minutes and usually last for about 12 h. Although the cause is not clearly understood, the recurrence of *flashbacks* has been reported long after the detectable levels have disappeared. LSD is rapidly metabolized, and concentrations of the parent drug in the urine of a user do not exceed 1 to 2 ng/ml.

16.3.5 Signs and Symptoms of Acute Toxicity

Common adverse reactions occur within minutes after ingestion. Sympathetic stimulation results in mydriasis, hyperthermia, piloerection, tachycardia, hyperglycemia, and hypertension. Diaphoresis, anorexia, sleeplessness, dry mouth, and tremors also result from toxicologic, unregulated sympathetic activity. Although LSD has a high therapeutic index, first-time users may experience panic reactions and loss of psychological control of the immediate environment.

Flashbacks occur in 1 of 20 users. They are characterized as recurrence of the LSD experience in the *absence* of ingestion. The full range of perceptual and psychological distortions recurs, including somatic and emotional hallucinations. The lack of control over the experience often leads to the development of personality changes, prolonged psychosis, neurosis, and depression. The variety, intensity, and pattern of flashbacks are unpredictable and do not rely on dose or frequency of use. The flashbacks may be triggered in unsuspecting LSD users by the administration of selective serotonin reuptake inhibitors (fluoxetine, paroxetine).*

16.3.6 Clinical Management of Acute Overdose

Reduction of anxiety, and the reassurance that a *bad trip* is due to the effects of the drug and is not permanent, is important in managing LSD toxicity. Benzodiazepines are beneficial for sedation. At one time, phenothiazines were frequently used to ameliorate the psychological manifestations and perceptual distortions. The antipsychotic drugs, however, are not recommended because of the lowering of seizure thresholds. Management of symptoms of sympathetic stimulation, especially tachycardia and hyperglycemia, is also warranted.

---

* Prozac® and Paxil®, respectively.

16.3.7 Tolerance and Withdrawal

Chronic use of LSD is not associated with the physical and psychological tolerance or withdrawal seen with opioids or sedative-hypnotics (S/H), although cross-tolerance and sensitivity to mescaline and psilocybin have been reported (see Section 16.4). A return to normal functioning is common following a drug-free period.

16.3.8 Methods of Detection

Most published assays for LSD are intended for identification of the drug in illicit preparations. They do not offer the sensitivity and specificity required for detection in urine specimens. Radioimmunoassays (RIAs) for LSD appear to offer an effective means for detecting very recent drug use. However, confirmation of the presence of LSD or its metabolites and isomers requires specific analysis. High-performance liquid chromatography (HPLC) combined with fluorescence detection, and capillary column gas chromatography with electron ionization mass spectrometry (GC-MS), can measure urinary concentrations as low as 0.5 ng/ml. However, neither of these assays is useful for detection of LSD in urine for more than about 12 h after ingestion.

LSD and its isomers, iso-LSD and LAMPA (the methyl-propyl analog), have identical molecular weights and similar mass spectra, thus complicating its chromatography. Since the resolution of LSD and LAMPA is difficult, separation is best achieved with GC* and confirmed with mass spectrometry.

16.4 TRYPTAMINE DERIVATIVES

16.4.1 Incidence and Occurrence

Figure 16.2 illustrates the structure of tryptamine, serotonin (5-hydroxytryptamine), and psilocin (4-hydroxydimethyltryptamine), all of which are pharmacologically and toxicologically similar.

Based on reports that relate to the use of hallucinogens other than LSD (including peyote, psilocin, and others), 7.7% of high-school seniors in the United States had used other hallucinogens at least once in their lifetime, and 4.8% reported use in the last year (NIDA, National Survey of Drug Use and Health [NSDUH], Monitoring the Future 2018).

16.4.2 Mechanism of Toxicity

The mushroom *Psilocybe mexicana* (*food of the gods*) has been used for centuries by Mexican and Central American Indians in religious ceremonies.† The mushroom

---

* A short, nonpolar fused-silica capillary column with hydrogen as the eluent increases resolution and sensitivity.

† Common street names include *liberty caps*, *blue legs*, and *golden tops*.

can also be found on cattle manure and in forest areas throughout the world. In 1958, A. Hoffman identified the active ingredient of the mushroom as psilocybin (4-phosphoryloxy-dimethyltryptamine). The compounds are thermostable and cannot be inactivated by high temperature or freezing. As with LSD, the mechanism of toxicity is related to the stimulation of sympathetic and serotonergic activity, although with lower potency.

16.4.3 Signs and Symptoms of Acute Toxicity

The desirable pharmacological effects of psilocybin, and its more active metabolite psilocin, mimic those of LSD. A 20 mg dose of psilocin is equivalent to 100 μg of LSD. The effects, which last for about 3 h, begin with anxiety and nausea and then proceed to dream-like trances (with the inability to discern them from reality) and hallucinations. *Trailer* images, not unlike LSD, induce changes in color and shape perception.

Few reports of acute intoxication are available. Symptoms present within 30 to 60 min after mushroom ingestion and include agitation, hyperthermia, and possible hypotension and convulsions. Anticholinergic symptoms, such as mydriasis, blurred vision, and dizziness accompany visual disturbances. As with LSD, panic reactions and psychosis occur with larger doses.

Persistent effects from chronic ingestion, such as flashbacks, psychiatric illness, and impaired memory, have also been reported.

16.4.4 Clinical Management of Acute Overdose

As with LSD, benzodiazepines and cholinergic agents are useful in managing the undesirable effects of psilocybin mushroom poisoning. However, there are some special considerations surrounding the treatment of mushroom poisoning. For instance, depending on bioavailability, amount ingested, and individual susceptibility, not all persons ingesting the same meal will be symptomatic. Also, it is important to recognize that the anticholinergic effects are likely to mimic those from exposure to insecticides and may be delayed for up to 6 h.

16.4.5 Tolerance and Withdrawal

As with LSD, chronic ingestion of psilocybin-containing mushrooms is usually not associated with physical and psychological tolerance or withdrawal symptoms, although cross-tolerance and sensitivity to other hallucinogens have been reported.

16.4.6 Methods of Detection

Psilocybin requires derivatization before chromatographic analysis, and many extractions cleave the phosphoryl group from psilocybin, converting it to psilocin. HPLC is an alternative technique for separation of this compound from other impurities.

16.5 PHENETHYLAMINE DERIVATIVES

16.5.1 Incidence and Occurrence

The synthetic hallucinogenic amphetamine derivatives are widely used among high-school and college students at *rave** parties, alcoholic bars, and events requiring endurance and alertness, such as marathon dances. Data on abuse are as reported for "other hallucinogens," including psilocin, as noted earlier.

16.5.2 Medicinal Chemistry

The β-phenethylamine derivatives are synthetic analogs of amphetamine and are illustrated in Table 16.1. Since the compounds have amphetamine-like structures, their toxicity combines the properties of stimulants with those of the hallucinogens. Drugs, such as mescaline, with a long history of abuse, and comparatively newer drugs, such as DOM (*STP*) and MDMA (*ecstasy*), are discussed as prototypes for this class.

16.5.3 Mescaline

Mescaline is regarded as the first of a series of alkaloidal amine hallucinogens. The compound is derived from the dried tops of *Lophophora williamsii* (Peyote or mescal buttons) from the Peyote cactus.† The small, spineless plant is indigenous to northern Mexico and the southern United States, and its use is associated with Indian religious ceremonies. Its chief effect is the production of euphoria and hallucinations with concomitant decrease in fatigue and hunger. The mescal buttons, obtained from the crown of the cactus, contain mescaline, the most active of the peyote constituents.‡ Depending on the amount, ingestion, chewing, or boiling of the buttons releases 0.3 to 0.5 grams of mescaline and produces mydriasis, accompanied by unusual and bizarre color perception. Flashing lights and vivid configurations characterize the initial visions, followed by dimming of colors and sleep induction. Sensory alteration is 1000-fold less potent than with LSD, but the production of tremors, hypertension, increased heart rate, and deep tendon reflexes are important and similar to those of the sympathomimetic amines.

Mescaline is rapidly absorbed; plasma concentrations peak at about 2 h and last for about 12 h. Other than the risk of hypertensive crisis, acute toxicity with mescaline is rare. Long-term residual psychological and cognitive effects among Native Americans and users during religious ceremonies are relatively uncommon.

---

* Weekend-long dance parties.

† Mescaline is also easily produced synthetically with ephedrine, pseudoephedrine, or amphetamine as starting ingredients.

‡ Anhalanine is another active peyote ingredient.

TABLE 16.1
Structure of the β-Phenethylamine Derivatives

Compound	Chemical Name	Common Names	Substitutions				
			2	3	4	5	R_1
Amphetamine	β-phenyliso-propylamine	Speed, ups	H	H	H	H	H
Methamphetamine	β-phenyliso-propylmethylamine	Meth, crank, ice, glass	H	H	H	H	CH_3
Mescaline[a]	3,4,5-trimethoxy-β-phenethylamine	Mesc, buttons	H	OCH_3	OCH_3	OCH_3	H
DOM	2,5-dimethoxy-4-methylamphetamine	STP (serenity, tranquility, peace)	OCH_3	H	CH_3	OCH_3	H
MDA	3,4-methylene-dioxyamphetamine	harmony, love drug[b]	H	-O-CH_2-O-		H	H
MDMA	3,4-methylenedioxy-methamphetamine	XTC, ecstasy, Adam	H	-O-CH_2-O-		H	CH_3
MDEA	3,4-methylenedioxy-ethamphetamine	Eve	H	-O-CH_2-O-		H	CH_2CH_3

Numbered substitutions refer to carbon positions on the phenyl ring.

[a] CH_3 = H for mescaline.

[b] MDMA, also known as *the love drug*.

16.5.4 DOM (2,5-DIMETHOXY-4-METHYLAMPHETAMINE, STP) AND MDA (3,4-METHYLENE-DIOXYAMPHETAMINE)

DOM and MDA (the parent compound) are up to 100 times more potent than mescaline. Both have been associated with fatalities in the United States, particularly as a result of neurotoxic events. However, their effects are milder than those of MDMA, which are discussed in the following subsection.

16.5.5 MDMA (3,4-METHYLENEDIOXYMETHAMPHETAMINE, *ECSTASY*)

Originally introduced as an anorexiant and popularized in the 1980s, *ecstasy* is still promoted as a "good high, low risk" drug of abuse. MDMA use as part of a multi-drug experience, which includes marijuana, cocaine, methamphetamine, ketamine, and other substances, is also gaining popularity. In 2006, 2.1 million Americans aged 12 and older had ingested MDMA at least once in the prior year, which includes 3.5% of 10th-graders and 4.5% of 12th-graders (National Survey on Drug Use and Health, NSDUH, 2006; NIDA, 2006). In 2006, 528,000 persons in the United States aged 12 or older had used MDMA in the previous month, while 860,000 Americans had used the drug for the first time. All current statistics indicate an increase in use since 2005.

More recently, the Substance Abuse and Mental Health Services Administration (SAMHSA, 2015) reported that approximately 897,000 people aged 12 or older were current users of methamphetamine, representing 0.3% of the population aged 12 or older (NSDUH, 2015). About 13,000 (0.1%) adolescents aged 12 to 17 were current MA users in 2015. About 128,000 young adults aged 18 to 25 in 2015 used MA in the past month, corresponding to 0.4% of young adults. In the same year, an estimated 757,000 adults aged 26 or older used MA, which represents 0.4% of this age group.

The illicit substance is available in liquid, powder, capsule, and tablet forms, predominantly for oral or inhalation administration. An average dose of 100 mg produces initial symptoms similar to those seen with amphetamine use; namely, mydriasis and hyperventilation. Unlike LSD, individuals appear to have more control over the hallucinogenic experience, with feelings of great affection and desire to be with others. The individual displays an introspective behavior, complementing or overwhelming the euphoric effect (the *ecstatic* feeling). Even with small doses, however, toxicity soon follows. Muscle tension, especially of the neck and jaw, chills, sweating, rapid eye movement, and a trance-like state develop. Amnesia, delirium, and erratic behavior accompany the hypertensive crisis. In high doses, the drug induces a hyperthermic crisis capable of initiating liver, kidney, and cardiovascular collapse. The compound interferes with its own metabolism, thus potentiating an overdose with repeated administration within short periods. It is reasonable to understand, therefore, why the agent is responsible for several hundred deaths per year, especially among young adults.

Unlike other hallucinogens, long-term use of MDMA has been associated with potential for addiction.* Treatment of acute *ecstasy* toxicity is symptomatic and is

* Symptoms of which are associated with continuous use despite knowledge of physical or psychological harm, withdrawal effects, and tolerance.

managed as with the intoxication associated with amphetamine and/or LSD crisis. The most effective therapy for long-term abuse and addiction is similar to that for other drugs of abuse—that is, cognitive behavioral psychological interventions with the objective of modifying addictive behavior patterns and psychosocial activity.

16.5.6 Methods of Detection

As with amphetamines, qualitative drug screening will detect phenethylamine derivatives in urine for up to 72 h. Enzyme-linked immunoassays (ELISA) or RIA methods are sensitive to 1000 ng/ml of urine sample. Newer synthetic derivatives, however, may elude detection. In addition, false positives are likely in the event of concomitant ingestion of any of the sympathomimetic amines, including popular nonprescription decongestant drugs such as pseudoephedrine as well as restricted agents such as ephedrine and phenylpropranolamine. Common adulterants that mask a positive test for amphetamine derivatives (producing a false negative result) include bleach, baking soda, and detergents.

16.6 PHENCYCLIDINE (1-PHENYLCYCLOHEXYL PIPERIDINE; PCP)

16.6.1 Incidence and Occurrence

Originally marketed in the E.U. as an intravenous (I.V.) anesthetic (Sernyl®), the drug was soon identified with postanesthetic confusion and delirium. After a brief shift as an animal tranquilizer, its legitimate therapeutic use was banned, but its street popularity blossomed. Since the compound can be easily manufactured and packaged as tablets, liquid, capsules, or powder, PCP is usually sprinkled over marijuana and inhaled through smoking.*

In 2013, 33,000 persons reported current use of PCP. Statistics from 2016 indicate that 2.4% of Americans aged 12 and older reported PCP use in their lifetime. As with MDMA, abuse potential for PCP appears to be increasing from the previous years (National Survey on Drug Use and Health, 2016; NIDA, 2013).

16.6.2 Toxicokinetics

PCP is well absorbed following all routes of administration. Its structure is illustrated in Figure 16.3. Maximum plasma PCP concentrations are observed 5–15 min after smoking,† It is a lipophilic weak base ($pK_a = 8.6$), and its volume of distribution is high ($V_d = 6.0$ l/kg). Thus, the compound is widely distributed and produces unreliable plasma concentrations. Its long half-life (several hours to days) contributes to its prolonged toxic effect (4 to 6 hours), especially with chronic use.

PCP undergoes oxidation and conjugation in the body, primarily to hydroxylated metabolites. About 10% of PCP is excreted unchanged in the urine. Both parent and

---

* Common street names vary by region and often allude to its use as a veterinary anesthetic, including *Sherman*, *angel dust*, *killer weed*, *snorts*, *PeaCe Pill*, *rocket fuel*, and *embalming fluid* (named for its combination with marijuana when smoked through a formaldehyde-filled water pipe).

† The powder form of PCP is sprinkled on marijuana and then smoked.

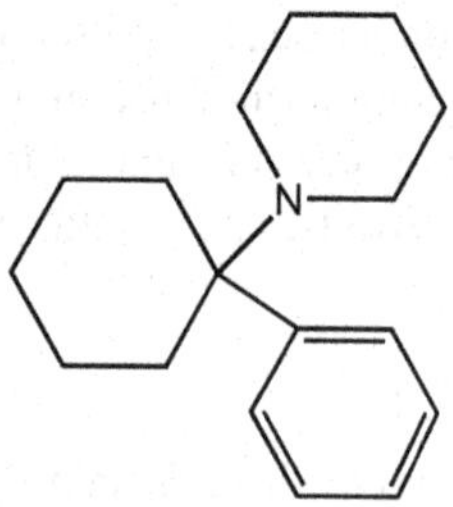

phencyclidine (PCP: 1-phenylcyclohexyl piperidine

FIGURE 16.3 Structures of phencyclidine (PCP: 1-phenylcyclohexyl piperidine)

conjugated metabolites may be detectable in urine for several days to several weeks. Some PCP is secreted in the saliva.

16.6.3 Signs and Symptoms of Acute Toxicity

Toxic effects are labeled as *dissociative anesthesia*, in which the patient experiences acute subjective feelings of depersonalization, detachment from the environment, and distortion of perceptions of sight and sound. Symptoms are variable and do not appear to correlate with PCP plasma concentrations. Initially, the patient exhibits excitation, paranoia, and dysphoria and may display aggressive tendencies. Disorientation, mood changes, catatonia,* and disorganized thoughts are displayed. Ataxia, impaired speech, myoclonus,† and choreoathetoid movements‡ are possible depending on the level of exposure. Hypertension and nystagmus§ are hallmark findings of PCP toxicity. With continued toxicity or exposure to higher doses, patients develop coma of brief duration, as well as hypertension, hyperthermia, and risk of rhabdomyolysis.¶ Agitation and depression are common after coma is resolved.

PCP-induced aggression and psychosis are frequently encountered and are characterized by aggressive and violent behavior of several days' duration. Short- and long-term psychological consequences resulting from continuous use involve memory loss, disordered thinking, and mood disturbances, the latter of which are responsible for about 50% of emergency department (E.D.) visits. Recovery is complete, although respiratory arrest, seizures, memory loss, and hallucinations are implicated in life-threatening (often fatal) self-inflicted behavior and may persist up to a year following discontinuation.

16.6.4 Clinical Management of Acute Overdose

In order to minimize aggression, patients are isolated from sensory stimuli. Supportive, symptomatic care is essential. In addition, I.V. benzodiazepines (diazepam) for

---

* A state of motor disturbance characterized by immobility with extreme muscular rigidity.

† Muscular spasms.

‡ Characterized by jerky, "tic-like" twitching (choreiform) and slow, writhing (athetoid) movements.

§ Involuntary, rhythmically oscillating movements of the eyes.

¶ Disintegration of muscle associated with excretion of myoglobin in urine.

seizure, along with haloperidol for agitation, are effective. Severe hypertension is treated aggressively with non-specific β-receptor antagonists (β-blockers), such as propranolol. Urinary acidification is useless, because only a small percentage is excreted unchanged in the urine, and increases the risk of myoglobinuric renal failure. Psychiatric reassurance, diuretics, and hemodialysis are ineffective, especially when concomitant multidrug and alcohol use is suspected. Activated charcoal can be administered if ingestion is recent. Recently, life-threatening PCP overdose has been managed with passive immunization approaches through the development of anti-PCP antibodies. Lastly, as with the other hallucinogens, cognitive behavioral psychiatric intervention is necessary for the treatment of PCP drug addiction.

16.6.5 Tolerance and Withdrawal

Psychological tolerance is noted with chronic abuse of PCP and is responsible for gradual development of symptoms of drug craving, paranoia, hallucinations, anxiety, and severe depression.

16.6.6 Methods of Detection

Because of its wide bioavailability and volume of distribution, PCP concentrations vary and are an unreliable monitor for clinical effects. In fact, urine samples can remain positive for up to 4 weeks after ingestion in chronic users. ELISA and RIA are the methods of choice for urine drug toxicology screening. Immunochemical methods are relatively specific for PCP, its metabolites, and some of the closely related analogs. For forensic analysis, many of the structurally similar analogs and homologs are impure by-products that are mixed in street samples during the synthesis of PCP. Separation of the moieties is facilitated with medium or nonpolar phases during chromatographic analysis. Methods for confirmation include gas-liquid chromatography with nitrogen-phosphorus detection (GLC/NPD), or gas- or liquid-chromatography-mass spectrometry (GC-MS, LC-MS). Other drugs, such as thioridazine, dextromethorphan, and chlorpromazine, may show false-positive reactions in immunochemical assays for PCP.

16.7 MARIJUANA

16.7.1 Incidence and Occurrence

Marijuana (Mexican cannabis, *pot*) is obtained from the dried flowering tops, seeds, and stems of the hemp plant variety of *Cannabis sativa* (Indian hemp). The plant is an annual herb indigenous to central and western Asia and is cultivated in India and other tropical and temperate regions for the fiber (manufacturing of rope) and hempseed. The oil of the hemp plant is expressed and used to make paints and soaps.*

* Cannabis, the ancient Greek name for hemp, was used in China and India (*charas*) and spread to Persia (*hashish*). It was introduced into European and American botanical formularies around the sixteenth century.

The amount of resin found in the pistil flowering tops markedly decreases as the plants are grown in more temperate regions, making the American crop of marijuana considerably less potent than the South American variety (also known as *sinsemilla*, without seeds).

The importation of crude marijuana cigarettes (*reefer, joint*) began in the 1950s and spread quickly through U.S. schools. This prompted federal and state law enforcement to start a campaign to eliminate and discourage their sale. The possession of marijuana was prohibited, and large areas of naturally growing American hemp were destroyed.* The effort resulted in discontinuance of the medicinal use of cannabis in the United States. Although a proprietary form of the active ingredient is available in capsule form as a prescription drug (dronabinol; see later), today, grassroots efforts have successfully prompted many U.S. state legislatures to legalize the use and possession of marijuana for medical as well as commercial purposes. Periodical attempts to overturn the federal ban against smoking of medical marijuana have not to date come to fruition. The current prevalence of marijuana use worldwide is outlined in Table 16.2, and general laws governing its use and possession in the United States are noted in Table 16.3.

Today, marijuana is the most commonly used and misused controlled substance in the United States and the number one cash crop, with earnings estimated at $32 billion yearly. In 2015, an estimated 11 million young adults aged 18 to 25 and 183 million persons of all age groups were past-year consumers of cannabis. Statistics at the same time point report that nearly 22% of high-school seniors claim to be current

TABLE 16.2
Current Prevalence of Cannabis Use Worldwide

Region	Estimated Prevalence of Worldwide Cannabis Use (%)
West/Central Africa	12.4
North America	11.6
Oceania	10.7
Americas	8.4
Africa	7.5
Chile	7.5
South America	5.9
Western/Central Europe	5.7
Europe and E.U.	4.3
Global	3.9
Colombia	3.3
Rest of Americas	2.5
Eastern/Southern Europe	2.3
Asia	1.9

* The herbicide paraquat was routinely sprayed on marijuana fields in northern Mexico and in the southern United States, precipitating a syndrome of pulmonary fibrosis (see Chapter 27).

TABLE 16.3
Current Status of Marijuana Use in United States

State	Personal Possession Limit
Alabama [N]	None
Alaska [M/R]	4 oz or less
Arizona [M]	None
Arkansas [M]	None
California [M/R]	Up to 1 oz
Colorado [M/R]	Up to 1 oz
Connecticut [M]	None
Delaware [M]	None
District of Colombia[M/R]	None
Florida [M]	None
Georgia [N]	None
Hawaii [M]	None
Idaho [N]	None
Illinois [M]	None
Indiana [N]	None
Iowa [N]	None
Kansas [N]	None
Kentucky [N]	None
Louisiana [M]	None
Maine [M/R/C]	Up to 2.5 oz
Maryland [M]	None
Massachusetts [M/R]	Up to 1 oz
Michigan [M]	None
Minnesota [M]	None
Mississippi [N]	None
Missouri [N]	None
Montana [M]	None
Nebraska [N]	None
Nevada [M/R]	Up to 1 oz
New Hampshire [M]	None
New Jersey [M]	None
New Mexico [M]	None
New York [M]	None
North Carolina [N]	None
North Dakota [M]	None
Ohio [M]	None
Oklahoma [N]	None
Oregon [M/R]	Up to 1 oz
Pennsylvania [M]	None
Rhode Island [N]	None
South Carolina [N]	None
South Dakota [N]	None

(*Continued*)

TABLE 16.3 (CONTINUED)
Current Status of Marijuana Use in United States

Tennessee [N]	None
Texas [N]	None
Utah [N]	None
Virginia [N]	None
Vermont [M]	None
Washington [M/R]	Up to 1 oz
West Virginia [M]	None
Wisconsin [N]	None
Wyoming [N]	None

[N] No broad laws legalizing marijuana.
[M] Marijuana possession limited to medical use.
[R] Marijuana possession allowed for medical as well as recreational use.
None: Possession of any amount of marijuana is subject to penalty, fines and/or imprisonment.

marijuana users, and at least 6% of these students smoke marijuana every day. From an international perspective, as per United Nations Office on Drugs and Crime estimates for 2015, 2.7–4.9% (corresponding to 125–227 million people, respectively) of the world's population aged 15–64 years were current consumers of cannabis (UN World Drug Report, 2017).

Although marijuana use decreased by more than 20% in 8th-, 10th-, and 12th-graders between 2000 and 2007, its presence in this age group still remains at unacceptably high levels, with more than 40% of high-school seniors reporting smoking the drug at least once in their lifetimes (National Survey on Drug Use and Health, NSDUH 2015; NIDA, 2018).

16.7.2 Medicinal Chemistry

Cannabis yields between 0.5% and 20% of a resin containing the major active euphoric principle, delta-9-*trans*-tetrahydrocannabinol (δ^9-THC), along with other cannabinoid constituents and derivatives (Figure 16.4).

16.7.3 Receptor Pharmacology and Toxicology

Smoking marijuana or oral ingestion of cannabinoids has complex central sympathomimetic effects, mediated in part by the presence of neural cannabinoid receptors. The receptors are part of the G-protein coupled receptor superfamily. Ligand binding to the external domain of the protein transmembrane peptide triggers an intracellular signal transduction pathway, expediting the opening of

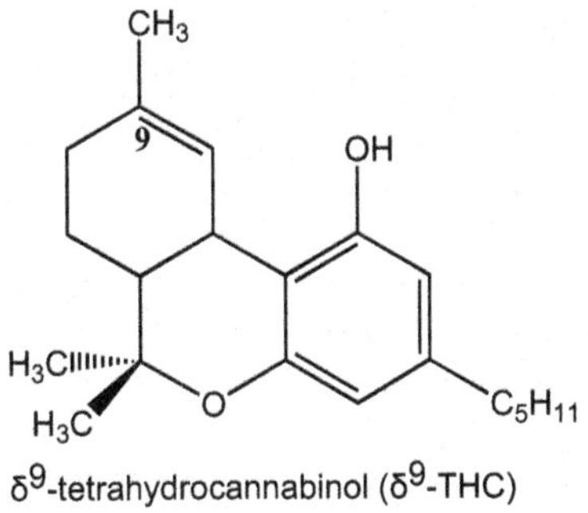

δ^9-tetrahydrocannabinol (δ^9-THC)

FIGURE 16.4 Structure of δ^9-tetrathydrocannabinaol (δ9-THC).

potassium and calcium channels. The reaction causes a depolarization-induced inhibition resulting in a reduction in GABA-mediated neurotransmission. Naturally occurring endocannabinoids released from depolarized neurons bind to CB_1 receptors in presynaptic neurons, resulting in a further reduction in GABA release.

Two subtypes of cannabinoid receptors are known: CB_1 and CB_2.

1. CB_1 is expressed mainly in brain, particularly in the olfactory bulb, cortical regions (neocortex, hippocampus, and amygdala), basal ganglia, thalamic and hypothalamic nuclei, cerebellar cortex, and brainstem nuclei. CB_1 receptors are also located in the lungs, liver, and kidneys. Their presence in these tissues accounts for the variety of physiologic effects noted with marijuana consumption.
2. CB_2 receptors are primarily expressed on immune cells, predominantly T-cells, macrophages, and B-cells, and in hematopoietic cells. They are also expressed in keratinocytes, on peripheral nerve terminals, and in mouse preimplantation embryos and have been implicated in mediating nociception.* In brain, they are mainly presented by microglial cells, the role of which is not pharmacologically clear.

16.7.4 Clinical Use and Effects

Dronabinol, a synthetic form of δ^9-THC, is available in 5 and 10 mg capsules for the treatment of nausea and vomiting associated with cancer chemotherapy. It is also indicated as an appetite stimulant for anorexia associated with weight loss in AIDS patients. Other off-label uses include reduction of intraocular pressure (IOP) in the treatment of glaucoma and control of seizures in epilepsy, as well as a variety of unsubstantiated clinical benefits. Nontherapeutic effects of dronabinol are identical to those of marijuana and other centrally active cannabinoids, albeit they are dose dependent.

* That is, the perception of pain.

Mood changes accompany the acute and routine habit of marijuana consumption, ranging from euphoria, depression, paranoia, and anxiety* to detachment. The most prominent effects are relaxation and sedation. Behavioral effects appear as loss of goal-oriented drive and short-term memory and a vague sense of time (*temporal disintegration*). The individual is prone to spontaneous laughter, hallucinations, and delusions, although the latter is usually seen with higher doses.

Several contrasting autonomic yet predominantly sympathetic-mediated symptoms involve dry mouth (sympathetic), stimulation of appetite (parasympathetic), muscular incoordination (skeletal cholinergic), decrease of testosterone levels (sympathetic), urinary retention (sympathetic), increase in heart rate (sympathetic or anticholinergic), and conjunctival injection (cholinergic). The last effect contributes to decreased IOP.

16.7.5 Toxicokinetics

An average cigarette (*joint*) holds about 500 to 1000 mg of the herb, containing 1% to 2% δ^9-THC. Only 5% to 10% of the available active ingredient is absorbed through inhalation of marijuana smoke, primarily because of loss to pyrolysis. Oral administration is almost completely absorbed (90%), yet inhalation results in more rapid accumulation (1 to 2 min vs. 0.5 to 1 h onset for oral), with a 2 to 4 h duration. Its high lipid solubility contributes to a large volume of distribution (V_d ~ 10 l/kg) and long half-life ($t_{1/2}$ ~ 7 days). This is accounted for by its leakage from lipid stores and detection in plasma and urine up to 8 weeks after the last dose in chronic users.

Dronabinol and δ9-THC undergo extensive *first-pass hepatic metabolism* yielding active and inactive principal metabolites, 11-hydroxy-δ9-THC and 8,11-dihydroxy-δ9-THC, respectively. The 9-carboxy metabolite is found principally in urine.

16.7.6 Acute Toxicity and Clinical Management

Toxic effects are an extension of the pharmacological and clinical effects. Although serious toxicity is uncommon, psychosis and dangers related to poor judgment are complications. Usually, the complications result from ingestion of high doses, excessive use, poor or contaminated street quality of the substance, or being taken in combination with other S/H or hallucinogens. For instance, when marijuana is sprinkled with PCP, it produces a substance known as *superweed*, a valuable street commodity. Smoking of marijuana impairs motor skills, making driving a motor vehicle hazardous. Pneumomediastinum, characterized by over-distention and rupture of pulmonary visceral and parietal pleura, is secondary to deep inhalation of marijuana smoke. In addition, the risk of heart attack more than quadruples in the first hour after smoking marijuana.

In general, psychiatric reassurance and supportive care are adequate treatment modalities for acute toxicity. Psychosis is transient and manageable with

* The aggression associated with PCP, however, is absent.

benzodiazepines. Interestingly, the latest treatment data indicate that marijuana was responsible for about 16% (more than 290,000) of all admissions to treatment facilities in the United States (NIDA, 2015).

16.7.7 Tolerance, Withdrawal, and Chronic Effects

Psychological tolerance and addiction, characterized by compulsive drug-seeking behavior and abuse despite its known harmful effects on social functioning, has been documented with long-term marijuana abuse. Some physical tolerance is also noted—withdrawal from long-term use results in irritability, sleeplessness, decreased appetite, weight loss, and anxiety. Withdrawal symptoms begin within about 1 day following abstinence, peak at 2–3 days, and subside within 1 or 2 weeks following drug cessation. Some pharmacological tolerance to the cardiovascular effects is also observable. In addition, increase in activation of the stress–response system and changes in activity of dopaminergic neurons have been noted with cannabinoid withdrawal in chronically exposed animals.

Long-term psychological effects have suggested a sixfold increase in the incidence of schizophrenia. There is anecdotal evidence for a higher incidence of mouth, throat, and lung cancer in young adults. There appears to be a decrease in sperm motility and number, and abnormal sperm morphology is associated with chronic marijuana smoking. Deleterious genetic effects, however, are debatable.

16.7.8 Clinical Management of Chronic Addiction

Behavioral interventions, including cognitive behavioral therapy and motivational incentives, have shown efficacy in treating marijuana dependence. As with most of the hallucinogenic agents, no pharmacological therapeutic interventions are currently available for marijuana addiction.

16.7.9 Methods of Detection

Cannabinoids are difficult to detect particularly because of their high lipid solubility and low concentrations in urine and plasma. Lipid solubility increases the difficulty of separating cannabinoids from the biological matrix for analysis. Qualitative drug screening, such as the labor-intensive thin-layer chromatographic (TLC) methods, will detect THC metabolites down to 50 ng/ml for several weeks post ingestion. The immunoassays (enzyme multiplied immunoassay [EMIT], RIA, and fluoroimmunoassay [FIA]), the preferred initial screening assays, detect the major metabolite of THC in urine (9-carboxy-THC). Some cross-reactivity occurs with many of the other glucuronide-conjugated metabolites. Confirmation of positive screening tests requires the use of chromatographic techniques (GLC, HPLC, TLC, GC-MS) that are capable of separating and detecting the major metabolites. GC-MS is the most reliable confirmatory method, especially when used with electron impact (EI) and chemical ionization (CI) detector modes. Common adulterants that mask a positive test for marijuana metabolites include detergents, salt, use of diuretics, and vinegar.

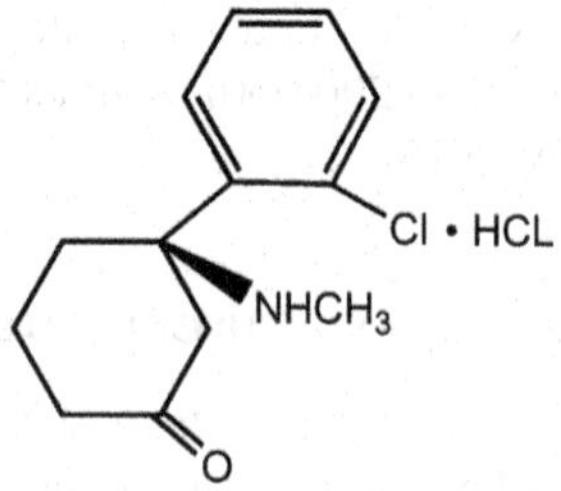

Cl • HCL
$NHCH_3$
O

ketamine (2-(o-chlorophenyl)-2-methylamino-cyclohexanone HCl)

FIGURE 16.5 Structure of ketamine (2-(o-chlorophenyl)-2-methylamino-cyclohexanone HCl).

16.8 MISCELLANEOUS HALLUCINOGENIC AGENTS

16.8.1 KETAMINE (SPECIAL K, VITAMIN K)

Ketamine (2-(o-chlorophenyl)-2-methylamino-cyclohexanone HCl) is a rapid-acting general nonbarbiturate anesthetic used alone or as a supplemental anesthetic in diagnostic and surgical procedures in emergency medicine and veterinary medicine (Figure 16.5). It produces a *dissociative anesthesia* similar to that of PCP. Ketamine selectively interrupts association pathways of the brain before producing somatesthetic sensory blockade. This activity is mediated through interaction with the NMDA receptor (a type of glutamate receptor), resulting in inhibition of neuronal transmission and anesthesia.

Ketamine has unfolded as a common recreational drug since 1965, especially for young adults. As a drug of abuse, it is administered orally, intravenously, or by inhalation. The powder or liquid formulations are tasteless, odorless, and colorless. These features render the drug suitable for use as a *club drug* since its administration is masked in an alcoholic beverage. When used as an acquaintance "date-rape" drug, the dissociative effect of ketamine makes it difficult for the victim to recall an account of recent events.

Drug users claim that the hallucinogenic effects are superior to those of LSD and PCP. Low doses require 15 to 20 min onset and produce profound anesthesia with normal laryngeal–pharyngeal reflexes and respiratory stimulation. The effects of higher doses vary from the induction of pleasant dream-like states with vivid imagery, hallucinations, and loss of sensitivity to music to delirium accompanied by confusion, nausea, excitement, and irrational behavior. There are reports that *binging** on ketamine, similar to that described for amphetamine or cocaine users, has resulted in impaired motor function, hypertension, and respiratory collapse. Although the effect of ketamine lasts for 20 to 45 min, the duration of psychological manifestations is ordinarily a few hours, and recurrence is seen up to 24 h postoperatively. Respiratory depression, seizures, arrhythmias, and cardiac arrest are toxic sequelae.

---

* Repeated, excessive administration of the drug within relatively short periods of time.

(Interestingly, a similar clinically useful anesthetic agent that has acquired significant popularity is Propofol [Diprivan®]. Its misuse and toxicity are commonly encountered among health professionals, who have access to considerable amounts. Acute administration of Propofol infusion causes *propofol infusion syndrome* [PRIS], characterized by severe metabolic acidosis and refractory bradycardia. The E.U. Medicines Agency registered 394 cases of PRIS between December 2001 and March 2015, of which 137 [35%] were fatal.)

General treatment of acute ketamine toxicity involves the ABCs of supportive care with particular attention to cardiac and respiratory functions. Prevention or treatment of hypertension, with β-receptor antagonists such as propranolol, is effective. Maintenance of a patent airway is of paramount importance. The patient should be in quiet recovery with minimal stimulation. For use as a general anesthetic in human surgical procedures, the incidence of psychological manifestations may be reduced by using lower doses of ketamine with I.V. diazepam.

The laboratory detection of ketamine follows an initial screening using basic drug extraction methods followed by gas or liquid chromatography with flame-ionization, nitrogen-specific, or MS confirmation.

16.8.2 Gamma-Hydroxybutyrate (GHB; Xyrem®)

Originally developed and used in the 1970s as an anesthetic agent and for sleep disorders, GHB is currently listed as an "orphan drug" for the treatment of narcolepsy and auxiliary symptoms of cataplexy, sleep paralysis, and hypnagogic hallucinations. It is a CNS depressant that was approved by the Food and Drug Administration (FDA) in 2002 exclusively for use in the treatment of narcolepsy. As a metabolite of gamma-aminobutyric acid (GABA), it is a naturally occurring inhibitory neurotransmitter found in low concentrations in areas of the brain with a high density of GABA-ergic neurons. GHB metabolites, such as gamma-butyrolactone (GBL) and 1,4-butanediol, have also been encountered as substances of abuse.

U.S. statistics indicate that use GHB has been increasing since 1992 (DAWN, 2010). The highest activity with GHB in the United States is reported in Florida, Texas, California, and Georgia. Recent data indicated that 6000 persons aged 12–17, 138,000 aged 18–25, and 131,000 aged 26 years and older had abused GHB in their lifetime (National Survey on Drug Use and Health, NSDUH 2013). The drug is obtainable illegally as a colorless, odorless liquid, gel, or crystalline powder, rendering it undetectable when combined with alcohol or other drugs. Although its use was popularized as an acquaintance date-rape drug, it is frequently presented for its anabolic effects as an aid in fat reduction and muscle building. Some nicknames include *G*, *liquid ecstasy*, *liquid X*, *somatox*, *grievous bodily harm*, and *invigorate*. Penalties for illegal possession are similar to those for DEA Schedule I controlled substances.*

* Interestingly, gamma-amino butyrolactone (GBL), the metabolite of GHB, is available from chemical suppliers and when combined with sodium hydroxide, forms the active parent compound. Similarly, butyrolactone, an industrial chemical, can also be converted to GHB with similar hallucinogenic effects.

TABLE 16.4
Characterization of GHB Effects Corresponding to Dosage

Dose (mg/kg)	Signs and Symptoms
Less than 10	Drowsiness, dizziness, disorientation
10–20	Vomiting, rapid onset of coma, amnesia
20–30	Induction of REM sleep, which cycles with non-REM sleep
30–50	Respiratory depression, bradycardia, clonic muscle contractions, anesthesia, depressed cardiac output

GHB is an agonist for GABA-B receptors* causing CNS depression. The depression is characterized by drowsiness, dizziness, and disorientation within 15 to 30 min. Apnea and hypoxia follow larger doses. Marked agitation on stimulation is a distinguishing feature of GHB ingestion. Table 16.4 lists the features of GHB toxicity corresponding to dosage (other depressant effects of GHB are discussed in Chapter 13).

As with ketamine, treatment of acute GHB toxicity involves maintenance of respiratory integrity and application of the ABCs in supportive care. Atropine is indicated for bradycardia, while bedside suction and supplemental oxygen are necessary for maintaining a patent airway.

Repeated use of GHB may lead to withdrawal effects, including insomnia, anxiety, tremors, and sweating. Severe withdrawal reactions have been reported among patients presenting with GHB overdose, especially in combination with other S/Hs or alcohol.

There are no GHB detection tests for use in emergency rooms, and as many clinicians are unfamiliar with the drug, many GHB incidents probably go undetected. Forensic routine screening for GHB is performed with urine samples only, since the endogenous neurotransmitter may interfere with serum detection. Positive urine tests are confirmed with follow-up quantitative blood analysis. Postmortem screening for GHB is also not routinely performed. Some extraction protocols include an acid hydrolysis for the conversion of the parent compound to gamma-butyrolactone, which is then detected and confirmed with GC-MS.

16.8.3 NBOMe

NBOMe represents a class of emerging new psychoactive substances that have gained prominence as illicit drugs of abuse. NBOMes are the N-2-methoxy-benzyl substituted 2C class of hallucinogens. They are marketed online as "research chemicals" under various names: *N-bomb*, *Smiles*, *Solaris*, and *Cimbi*. This category, along with the constant but slight alteration of the parent chemical structures, makes it difficult for law enforcement to monitor and regulate this class of chemicals. NBOMes

* GABA, gamma-amino butyric acid, is a naturally occurring inhibitory neurotransmitter and is discussed at length in Chapter 13.

are selective agonists of 5-HT2A receptors displaying considerably high potency. Unsuspecting users present with severe clinical intoxication, overdose, and traumatic fatalities. Several important reviews have reported new data, case reports, and toxicology of the NBOMes as well as the metabolites, suggesting that these have the potential to cause future public health problems.

REFERENCES

SUGGESTED READINGS

Brettell, T.A., Chapter 16. Forensic science applications of gas chromatography, in *Modern Practice of Gas Chromatography*, Grob, R.L. and Barry, E.F. (Eds.), 4th ed., 2004.

Chamberlain, J.C., *The Analysis of Drugs in Biological Fluids*, 2nd ed., CRC Press, Boca Raton, FL, 1995.

Cunningham, N., Hallucinogenic plants of abuse. *Emerg. Med. Australas.* 20, 167, 2008.

Dryburgh, L.M., Bolan, N.S., Grof, C.P.L., Galettis, P., Schneider, J., Lucas, C.J., and Martin, J.H., Cannabis contaminants: sources, distribution, human toxicity and pharmacologic effects. *Br. J. Clin. Pharmacol.* 84, 2468,2018.

Halpern, J.H., Sherwood, A.R., Hudson, J.I., Yurgelun-Todd, D., and Pope, H.G. Jr., Psychological and cognitive effects of long-term peyote use among Native Americans, *Biol. Psychiatry* 58, 624, 2005.

Jansen, K.L. and Darracot-Cankovic, R., The nonmedical use of ketamine, part two: a review of problem use and dependence. *J. Psychoactive Drugs* 33, 151, 2001.

Kosten, T. and Owens, S.M., Immunotherapy for the treatment of drug abuse. *Pharmacol. Ther.* 108, 75, 2005.

Lucas, C.J., Galettis, P., and Schneider, J., The pharmacokinetics and the pharmacodynamics of cannabinoids. *Br. J. Clin. Pharmacol* 84, 2477,2018.

Maeng, S. and Zarate, C.A. Jr., The role of glutamate in mood disorders: results from the ketamine in major depression study and the presumed cellular mechanism underlying its antidepressant effects. *Curr. Psychiatry Rep.* 9, 467, 2007.

Maxwell, J.C. and Spence, R.T., Profiles of club drug users in treatment. *Subst. Use Misuse* 40, 1409, 2005.

Nichols, D.E., Hallucinogens. *Pharmacol. Ther.* 101, 131, 2004.

NIDA, Hallucinogens, 2017. www.drugabuse.gov/drugs-abuse/hallucinogens, last viewed March 2019.

Snyder, S.H., Faillace, L., and Hollister, L., 2,5-dimethoxy-4-methyl-amphetamine (STP): a new hallucinogenic drug. *Science* 158, 669, 1967.

Yohn, N.L., Bartolomei, M.S., and Blendy, J.A., Multigenerational and transgenerational inheritance of drug exposure: the effects of alcohol, opiates, cocaine, marijuana, and nicotine. *Prog. Biophys. Mol. Biol.* 118, 21, 2015.

REVIEW ARTICLES

Araújo, A.M., Carvalho, F., Bastos Mde, L., Guedes de Pinho, P., and Carvalho, M., The hallucinogenic world of tryptamines: an updated review. *Arch. Toxicol.* 89, 1151, 2015.

Bigal, M.E. and Tepper, S.J., Ergotamine and dihydroergotamine: a review. *Curr. Pain. Headache Rep.* 7, 55, 2003.

Brennan, R. and Van Hout, M.C., Gamma-hydroxybutyrate (GHB): a scoping review of pharmacology, toxicology, motives for use, and user groups. *J. Psychoactive Drugs* 46, 243, 2014.

Brunt, T.M., van Amsterdam, J.G., and van den Brink, W., GHB, GBL and 1,4-BD addiction. *Curr. Pharm. Des.* 20, 4076, 2014.

Burgess, C., O'Donohoe, A., and Gill, M., Agony and ecstasy: a review of MDMA effects and toxicity. *Eur. Psychiatry* 15, 287, 2000.

Busardò, F.P. and Jones, A.W., GHB pharmacology and toxicology: acute intoxication, concentrations in blood and urine in forensic cases and treatment of the withdrawal syndrome. *Curr. Neuropharmacol.* 13, 47, 2015.

Deleu, D., Northway, M.G., and Hanssens, Y., Clinical pharmacokinetic and pharmacodynamic properties of drugs used in the treatment of Parkinson's disease. *Clin. Pharmacokinet.* 41, 261, 2002.

Dinis-Oliveira, R.J., Metabolism of psilocybin and psilocin: clinical and forensic toxicological relevance. *Drug Metab. Rev.* 49, 84, 2017. doi:10.1080/03602532.2016.1278228.

Fantegrossi, W.E., Murnane, K.S., and Reissig, C.J., The behavioral pharmacology of hallucinogens. *Biochem. Pharmacol.* 75, 17, 2008.

Freese, T.E., Miotto, K., and Reback, C.J., The effects and consequences of selected club drugs. *J. Subst. Abuse. Treat.* 23, 151, 2002.

Giorgetti, R., Tagliabracci, A., Schifano, F., Zaami, S., Marinelli, E., and Busardò, F.P., When "chems" meet sex: a rising phenomenon called "chemsex". *Curr. Neuropharmacol.* 15, 762, 2017.

Goss, J., Designer drugs. Assess and manage patients intoxicated with ecstasy, GHB or rohypnol—the three most commonly abused designer drugs. *J. Emerg. Med. Serv.* 26, 84, 2001.

Gudelsky, G.A. and Yamamoto, B.K., Neuropharmacology and neurotoxicity of 3,4-methylenedioxymethamphetamine. *Methods Mol. Med.* 79, 55, 2003.

Halpern, J.H. and Pope, H.G. Jr., Do hallucinogens cause residual neuropsychological toxicity? *Drug Alcohol Depend.* 53, 247, 1999.

Halpin, L.E., Collins, S.A., and Yamamoto, B.K., Neurotoxicity of methamphetamine and 3,4-methylenedioxymethamphetamine. *Life Sci.* 97, 37, 2014.

Iffland, K. and Grotenhermen, F., An update on safety and side effects of cannabidiol: a review of clinical data and relevant animal studies. *Cannabis Cannabinoid Res.* 2, 139, 2017.

Kyriakou, C., Marinelli, E., Frati, P., Santurro, A., Afxentiou, M., Zaami, S., and Busardo, F.P., NBOMe: new potent hallucinogens—pharmacology, analytical methods, toxicities, fatalities: a review. *Eur. Rev. Med. Pharmacol. Sci.* 19, 3270, 2015.

Logan, B.K., Mohr, A.L.A., Friscia, M., Krotulski, A.J., Papsun, D.M., Kacinko, S.L., Ropero-Miller, J.D., and Huestis, M.A., Reports of adverse events associated with use of novel psychoactive substances, 2013–2016: a review. *J. Anal. Toxicol.* 41, 573, 2017.

Miotto, K., Darakjian, J., Basch, J., Murray, S., Zogg, J., and Rawson, R., Gamma-hydroxybutyric acid: patterns of use, effects and withdrawal. *Am. J. Addict.* 10, 232, 2001.

Morland, J., Toxicity of drug abuse—amphetamine designer drugs (ecstasy): mental effects and consequences of single dose use. *Toxicol. Lett.* 112, 147, 2000.

Musselman, M.E. and Hampton, J.P., "Not for human consumption": a review of emerging designer drugs. *Pharmacotherapy* 34, 745, 2014.

O'Connell, T., Kaye, L., and Plosay, J.J. 3rd., Gamma-hydroxybutyrate (GHB): a newer drug of abuse. *Am. Fam. Physician* 62, 2478, 2000.

Papaseit, E., Torrens, M., Pérez-Mañá, C., Muga, R., and Farré, M., Key interindividual determinants in MDMA pharmacodynamics. *Expert Opin. Drug Metab. Toxicol.* 14, 183, 2018.

Parmar, J.R., Forrest, B.D., and Freeman, R.A., Medical marijuana patient counseling points for health care professionals based on trends in the medical uses, efficacy, and adverse effects of cannabis-based pharmaceutical drugs. *Res. Social Adm. Pharm.* 12, 638, 2016.

Pentney, A.R., An exploration of the history and controversies surrounding MDMA and MDA. *J. Psychoactive Drugs* 33, 213, 2001.

Pintori, N., Loi, B., and Mereu, M., Synthetic cannabinoids: the hidden side of Spice drugs. *Behav. Pharmacol.* 28, 409, 2017.

Rech, R.H. and Commissaris, R.L., Neurotransmitter basis of the behavioral effects of hallucinogens. *Neurosci. Biobehav. Rev.* 6, 521, 1982.

Rhee, D.J., Complementary and alternative medicine for glaucoma. *Surv. Ophthalmol.* 46, 43, 2001.

Schmid, R.L., Sandler, A.N., and Katz, J., Use and efficacy of low-dose ketamine in the management of acute postoperative pain: a review of current techniques and outcomes. *Pain* 82, 111, 1999.

Schwartz, R.H., Marijuana: a decade and a half later, still a crude drug with underappreciated toxicity. *Pediatrics* 109, 284, 2002.

Smith, K.M., Drugs used in acquaintance rape. *J. Am. Pharm. Assoc. (Wash.)* 39, 519, 1999.

Smith, K.M., Larive, L.L., and Romanelli, F., Club drugs: methylenedioxymethamphetamine, flunitrazepam, ketamine hydrochloride, and gamma-hydroxybutyrate. *Am. J. Health Syst. Pharm.* 59, 1067, 2002.

Teter, C.J. and Guthrie, S.K., A comprehensive review of MDMA and GHB: two common club drugs. *Pharmacotherapy* 21, 1486, 2001.

Tfelt-Hansen, P., Ergotamine, dihydroergotamine: current uses and problems. *Curr. Med. Res. Opin.* 17, S30, 2001.

U.S. National Institute on Drug Abuse (NIDA), National Institutes of Health (NIH), Department of Health and Human Services (DHHS), Public Health Service, *Monitoring the Future Study*, 2018. From: www.monitoringthefuture.org, last viewed March 2019.

U.S. National Survey on Drug Use and Health, 2006. www.samhsa.gov, last viewed March 2019.

Voth, E.A. and Schwartz, R.H., Medicinal applications of delta-9-tetrahydrocannabinol and marijuana. *Ann. Intern. Med.* 126, 791, 1997.

Zavaleta, E.G., Fernandez, B.B., Grove, M.K., and Kaye, M.D., St. Anthony's fire (ergotamine induced leg ischemia)–a case report and review of the literature. *Angiology* 52, 349, 2001.

17 Anticholinergic and Neuroleptic Drugs

17.1 INTRODUCTION TO DRUGS POSSESSING ANTICHOLINERGIC EFFECTS

A variety of chemicals, drugs, and herbal derivatives possess anticholinergic properties defined by their ability to block the neurotransmitter acetylcholine (ACh). This effect is a result of a direct interference with either of two types of cholinergic receptors—peripheral muscarinic or nicotinic receptors. Anticholinergic effects are also a consequence of adverse drug reactions (ADRs), as seen with the tricyclic antidepressants (TCA) and phenothiazine antidepressants. In addition, many anticholinergic compounds exert their action by occupying central cholinergic receptors, thus producing alterations in the central nervous system (CNS).

Autonomic neurons and their receptors govern sympathetic (SNS) and parasympathetic (PNS) activity throughout the body (see Figure 16.1 for an illustration of adrenergic neurons). *Nicotinic receptors* are present in the plasma membrane of dendrites and cell bodies of both SNS and PNS postganglionic neurons, at the neuromuscular junction, and in the spinal cord. Activation of these receptors triggers postsynaptic neuronal excitation and skeletal muscle contraction. *Muscarinic receptors* are present on the cell membranes of smooth and cardiac muscle, and sweat glands. Activation of these receptors by ACh delivers stimulation or inhibition, depending on the effector organ. Under normal circumstances, the effect of ACh is temporary and rapid due to inactivation by the enzyme acetylcholinesterase. Anticholinergic agents, therefore, have been traditionally referred to as *antimuscarinic agents* or *cholinergic blockers.*

17.2 ANTIHISTAMINE, GASTROINTESTINAL, AND ANTIPARKINSON DRUGS

17.2.1 Incidence

Table 17.1 organizes representative, currently available antihistamine, gastrointestinal (GI), and antiparkinson drugs that possess anticholinergic properties, along with their most important clinical adverse effects. Among these, the first-generation antihistamines (the H_1-antagonists) are commonly found in over-the-counter (OTC) cough and cold preparations and represent 5.5% of the most frequently ingested substances in children 6 years old and younger. Although seemingly innocuous, they frequently induce sedation and may adversely affect a child's learning ability. First-generation antihistamine-induced sedation has been described to occur in more than 50% of patients receiving therapeutic dosages. Serious adverse events are unusual

TABLE 17.1
Categories, Representative Compounds, and Proprietary Names of Currently Available Antihistamine, Antiparkinson, and Gastrointestinal Drugs Possessing Anticholinergic Properties

Category	Therapeutic Classification	Compound	Proprietary Name	Predominant Anticholinergic Effects
Antihistamines	H_1-antagonists	Brompheniramine	Dimetane	Dry mouth, mydriasis, drowsiness, dyspnea, facial flushing, sinus tachycardia
		Diphenhydramine	Benadryl	
		Dimenhydrinate	Dramamine	
		Chlorpheniramine	Chlortrimeton	
		Promethazine	Phenergan	
		Meclizine	Bonine, Antivert	
		Pyrilamine	In combination with nasal decongestants	
Antiparkinson agents	Anticholinergic[a]	Benztropine	Cogentin	Dry mouth, mydriasis, blurred vision, dyspnea, sinus tachycardia, toxic psychosis, coma, seizures, ataxia, EPS
		Trihexyphenydyl	Artane	
		Procyclidine	Kemadrin	
		Biperiden	Akineton	
Gastrointestinal anticholinergic agents	Antispasmodic	Atropine	Various	Dry mouth, mydriasis, drowsiness, dyspnea, excitement, agitation, constipation, blurred vision, sinus tachycardia
		Homatropine	Various	
		Belladonna alkaloids	Donnatal[b]	
		Clidinium bromide	Quarzan; Librax (with chlor-diazepoxide)	
		l-Hyoscyamine	Levsin, Levsinex	
		Scopolamine	Scopace	
		Glycopyrrolate	Robinul	
		Dicyclomine	Bentyl	
		Propantheline	Pro-Banthine	

Drugs such as tripelenamine have been removed from the therapeutic market.
[a] As distinguished from antidopaminergic agents.
[b] In combination with atropine, scopolamine, L-hyoscyamine, and phenobarbital.

following overdoses of these agents, although life-threatening adverse events have been described.

In contrast, the anticholinergic properties of antiparkinson agents are therapeutically desirable. They are indicated for the adjunctive treatment of all forms of Parkinsonism and are used to ameliorate the extrapyramidal symptoms (EPS) associated with traditional neuroleptic drugs.

Like the antiparkinson agents, the anticholinergic properties of the GI agents are also therapeutically advantageous. These substances, listed in Table 17.1, were originally developed as *spasmolytics* for the management of excessive intestinal activity associated with peptic ulcers, diarrhea, and related disorders.

17.2.2 Medicinal Chemistry, Pharmacology, and Clinical Use

Administered as tablets, capsules, and liquids and in tinctures, belladonna alkaloids are derived from the perennial herb *Belladonna* (deadly nightshade)* cultivated throughout the world. The belladonna compounds are semi-synthetic derivatives from the plant's naturally occurring tropane alkaloids. Today, their usefulness as GI agents has been largely replaced by histamine-2 (H_2) blockers, such as famotidine, and proton pump inhibitors, such as omeprazole. Closely related to the belladonna derivatives are the solanaceous alkaloids, L-hyoscyamine, scopolamine, and atropine, obtained from the extremely poisonous plant species of *Hyoscyamus Linné* (*Atropa belladonna*).†

17.2.3 Signs and Symptoms of Acute Toxicity

In general, anticholinergic compounds exert their effects by blocking central and peripheral muscarinic and cholinergic receptors. Table 17.1 displays the most common ADRs associated with antihistamine, antiparkinson, and GI drugs. Dry mouth and mydriasis are the most common ADRs of all anticholinergic agents, as well as headache, dysuria, and dyspnea (difficulty breathing due to tightness of the chest). Increased vascular permeability and capillary perfusion are mediated by unregulated SNS stimulation from the anticholinergic activity of these compounds and account for the facial and upper body flushing. Sinus tachycardia is the most sensitive sign of toxicity and may exacerbate other conduction abnormalities. Sedation is an effect common mostly to antihistamines and is mediated primarily through blocking of central H_1 receptors.

---

* In ancient Greece and Rome, the juice of the berry placed in the eyes caused mydriasis (dilation) of the pupils, thus producing a striking appearance; hence the name *Belladonna* (from the Italian, beautiful lady). The poisonous nature of the plant has been known since ancient times.

† Stramonium or Jimson weed (Jamestown weed, *Datura stramonium*) is indigenous to ancient Arabic lands but cultivated today throughout the world. The purple North American variety is noxious and frequently responsible for poisoning in children. It also has hallucinogenic properties, and is single-handedly accountable for overcoming British troops who ingested the stramonium-laden anticholinergic seeds during Bacon's rebellion in colonial America, 1676.

17.2.4 Clinical Management of Acute Overdose

Acute anticholinergic toxicity necessitates careful supportive treatment. Gastric lavage may be useful within 1 h of ingestion. With conscious patients, activated charcoal is useful to decrease drug absorption. Agitation, seizures, hyperthermia, and hypertension are treated conventionally in mild to moderate toxicity. For instance, benzodiazepines (e.g., diazepam) or barbiturates are effective in controlling seizures, while agitation is promptly treated with benzodiazepines only. Lidocaine is useful for controlling dysrhythmias. It should be noted, however, that antiarrhythmic drugs of class *IA*, such as quinidine, procainamide, and disopyramide, should be avoided.

Pharmacological cholinergic intervention is warranted when signs and symptoms result from moderate to severe anticholinergic toxicity. Although controversial, intravenous (I.V.) physostigmine (Antilirium®), a reversible cholinesterase inhibitor, is warranted only if conventional therapy fails to control seizures, agitation, and unstable dysrhythmia, coma with respiratory depression, malignant hypertension, or hypotension. Through its action at both central and peripheral cholinergic receptors, physostigmine reverses anticholinergic activity and ameliorates the coma, delirium, and seizures that accompany severe toxicity.* As discussed later, however, physostigmine is not recommended in reversing the anticholinergic actions of TCA and phenothiazine derivatives due to potential induction of fatal asystoles. Physostigmine administration is contraindicated in patients with cardiovascular or peripheral vascular disease, bronchospasm, intestinal obstruction, heart block, or bladder obstruction.

17.2.5 Methods of Detection

Thin-layer chromatography (TLC) is suitable for qualitative toxicological screening of antihistamines. In general, antihistamines are not usually part of a forensic toxicology screening protocol, and their structural class is sufficiently different from other drugs of abuse that they do not require extensive separation and identification. For TLC, clinical identification of antihistamines can be visualized on the whole biological sample in the same chromatographic system. Biological samples are treated according to the following general protocol:

1. Different pH solvents are used to extract the chemical substance from the sample to facilitate isolation of the corresponding acidic, basic, or neutral derivative;
2. The extracted sample is applied as a spot on the plate (coated with either silica gel or alumina fixed phase);
3. The relative migration of the sample components is influenced by the mobile phase; for antihistamines, acetic acid:butanol:butyl ether (10:40:40 proportional parts), acetic acid:ethanol:water (30:50:20), and *n*-butanol:methanol (40:60) are the most common mobile phases;

* Initial dose is 0.5 to 2.0 mg I.V. over 5 minutes.

4. The separated components of interest on the TLC plate are visualized and characterized according to the formation of characteristic colors. Ninhydrin is a typical spray reagent used for primary amines (violet-pink color) and secondary amines (yellow).

17.3 TRICYCLIC ANTIDEPRESSANTS (TCA)

17.3.1 Incidence

The term *neuroleptic* has come to replace older terminology that described the clinical effects of this major class belonging to the psychotherapeutic agents. These include the classes of compounds known as the major and minor tranquilizers and antipsychotic, antimanic, antipanic, and antidepressant drugs. Although the TCA and phenothiazine derivatives possess significant anticholinergic activity, they may still be identified with the antidepressant and antipsychotic categories, respectively. Newer atypical neuroleptic agents lack the parkinson-like effects.

The phenothiazine derivatives, chlorpromazine and promethazine, were the first neuroleptics synthesized in the 1950s and are the pharmacological prototypes for all psychoactive compounds. Because of their life-threatening toxicity and potential for inducing intentional suicide, especially in patients at increased risk for self-inflicted harm, treatment regimens are generally limited to 1 or 2 week supplies.

17.3.2 Medicinal Chemistry

The neuroleptic agents are grouped by chemical classification, yet their toxicities differ based on preferential receptor activity. Table 17.2 summarizes the categories, representative compounds, and proprietary names of representative currently available neuroleptic agents, including the TCA, phenothiazine, and newer antidepressant drugs, possessing anticholinergic properties.

17.3.3 Pharmacology and Clinical Use

Unlike the phenothiazines and related derivatives, TCA neuroleptic agents have predominantly antiserotonergic and anticholinergic activity. These central actions account for their utility as antipsychotics, particularly in producing sedation (for agitation) and in reducing hallucinations and delirium. In addition, anticholinergic and antihistaminic effects confer desirable antiemetic/antinausea properties (prochlorperazine) and antihistaminic properties (promethazine) on some agents, respectively.

17.3.4 Toxicokinetics

TCA drugs are rapidly absorbed with quick onset. They possess high protein-binding and high volume of distribution (V_d) properties, providing for a prolonged duration of action. Their therapeutic effects, however, require 5 to 7 days before benefits are observed because of necessary depletion of neurotransmitter storage.

17.3.5 Mechanism of Toxicity

The toxic manifestations of this class of drugs are explained as toxicologic extensions of their pharmacology. The effects are typically associated with the anticholinergic (antimuscarinic) blockade, their characteristic quinidine-like myocardial depressant action, and their ability to block the vagus nerve. In addition to interference with cardiac function, neuroleptic-induced hypotension occurs as a result of peripheral α-receptor (adrenergic) blockade associated with the TCAs.

17.3.6 Signs and Symptoms of Acute Toxicity

Table 17.2 outlines the most significant toxicologic effects of TCA neuroleptic drug ingestion. CNS depression, seizures, and cardiac arrhythmias are generally observed with acute overdose, while anticholinergic and some extrapyramidal symptoms (EPS) are common ADRs. Table 17.3 describes EPS associated with TCA and phenothiazine drugs.

Decreased cardiac output and circulatory collapse are potentially life-threatening and usually occur after ingestion of more than 10 mg/kg in adults. Serious toxicity may be seen at lower doses, especially in children, the elderly, and patients with preexisting cardiac conditions. Serious toxicity manifests within 6 hours of the overdose. The most common signs and symptoms appear as hypotension, respiratory depression, cardiac conduction delays, dysrhythmias, urinary retention, seizures, and coma.

Although therapeutic and toxic doses of TCA compounds vary with the individual agents, death from neuroleptic overdose is rare and is usually a consequence of multiple drug ingestion. Some adverse reactions, however, are life threatening, require emergency intervention, and are difficult to treat. Neuroleptic-induced arrhythmias are a consequence of the quinidine-like myocardial depressant action of the compounds. Patients with plasma concentrations approaching 100 μg/dl are in jeopardy of developing decreased atrioventricular (AV) conduction, vagal blockade, widening of the QRS interval, and prolongation of the QT interval.

CNS depression, agitation, delirium, confusion, and disorientation are frequent consequences of neuroleptic administration. In addition, abnormal tendon reflexes, hypothermia or hyperthermia, and myoclonus (spastic skeletal muscle contraction) contribute to central dystonias. Loss of short-term memory, seizures, and respiratory depression are complications.

Anticholinergic blockade results in typical mydriasis with concomitant blurred vision, vasodilation (aggravating the hypotension), urinary retention, constipation, and dyspnea (resulting from reduced bronchial secretions). In general, tertiary amines have greater antimuscarinic potency than secondary amines, tetracyclics, or triazopyridines, of which amitriptyline is the most potent (Table 17.2).

17.3.7 Clinical Management of Acute Overdose

TCA overdose requires that all patients receive activated charcoal (1 g/kg) orally or per nasogastric (NG) tube. Cardiovascular and respiratory functions are monitored

TABLE 17.2
Categories, Representative Compounds, and Proprietary Names of Currently Available Neuroleptic Drugs Possessing Anticholinergic Properties

Category	Neuroleptic Classification	Compound	Proprietary Name	Predominant Pharmacologic and Anticholinergic Effects
Tricyclic antidepressants	Secondary amines	Amoxapine	Asendin	Greater α-adrenergic blockade; low hypotensive, antimuscarinic, antihistaminic effects; mydriasis, ECG abnormalities; CNS depression, arrhythmias, hypothermia/hyperthermia, low EPS effects
		Desipramine	Norpramine	
		Nortriptyline	Pamelor	
		Protriptyline	Vivactyl	
	Tertiary amines	Amitriptyline	Elavil	Greater serotonergic blockade; high antimuscarinic effects, mydriasis, ECG abnormalities; CNS depression, cardiac arrhythmias, hypothermia/hyperthermia, low EPS effects
		Imipramine	Tofranil	
		Doxepin	Sinequan	
		Trimipramine	Surmontil	
	Tetracyclic	Maprotiline	Various	Low α-adrenergic blockade activity
	Triazopyridine	Trazodone	Desyrel	
	Aminoketone	Bupropion	Wellbutrin	Four-fold increased risk of seizure (0.4%)
Phenothiazine antipsychotics	Aliphatic	Chlorpromazine	Thorazine	Significant EPS, sedative, CNS, and hypotensive effects
	Piperidine	Thioridazine	Mellaril	Significant sedative and CNS effects, low EPS, moderate hypotensive effects
	Piperazine	Perphenazine	Trilafon	Significant EPS and CNS effects, low sedative, low hypotensive effects
		Fluphenazine	Prolixin	
		Trifluoperazine	Stelazine	
		Promethazine[a]	Phenergan	
		Perchlorperazine[b]	Compazine	

(*Continued*)

TABLE 17.2 (CONTINUED)
Categories, Representative Compounds, and Proprietary Names of Currently Available Neuroleptic Drugs Possessing Anticholinergic Properties

Category	Neuroleptic Classification	Compound	Proprietary Name	Predominant Pharmacologic and Anticholinergic Effects
Phenylbutyl-piperidines	Butyrophenone	Haloperidol	Haldol	Significant EPS and sedative, low hypotensive effects
	Diphenylbutyl-piperidine	Pimozide	Orap	
Thioxanthines	Thioxanthines[c]	Thiothixene	Navane	Moderate EPS, sedative, and hypotensive effects
Newer atypical neuroleptics	Dibenzodiazepine	Clozapine	Clozaril	Sedation, respiratory depression, or coma with O.D.; low EPS events; low frequency of arrhythmias, hypotension, seizures, NMS; *significant* blood dyscrasias (agranulocytosis)
		Olanzapine	Zyprexa	Lower incidence of blood dyscrasias but similar EPS and adverse effects to clozapine
	Benzisoxazole	Risperidone	Risperdal	Lower incidence of blood dyscrasias but similar EPS and adverse effects to clozapine
	Dibenzothiazepine	Quetiapine	Seroquel	

EPS = extrapyramidal symptoms.
[a] Promethazine is used predominantly as a cough suppressant.
[b] Prochlorperazine is used as an antiemetic/antinausea drug.
[c] Chlorprothixene (Taractan) is no longer marketed.

for 12 h up to 6 days because of the prolonged reactions. Treatment with I.V. fluids and vasopressors, such as norepinephrine or phenylephrine, is necessary for reversing neuroleptic-induced hypotension. Quinidine-like effects, especially ventricular dysrhythmias, are managed with lidocaine, while widening QRS complex requires sodium bicarbonate. Class *IA* antiarrhythmics, such as quinidine, procainamide, and disopyramide, should be avoided, as these compounds may aggravate AV conduction.

Ventilatory support is required in patients experiencing significant respiratory depression. Anticonvulsants, such as diazepam and phenobarbital, are beneficial for seizure management. Unlike in the treatment of anticholinergic overdose, physostigmine is not recommended for TCA toxicity due to the potential induction of fatal asystoles. Hemodialysis and hemoperfusion are also not useful because of the high V_d and high protein binding demonstrated with TCAs.

17.3.8 Tolerance and Withdrawal

Tolerance to the anticholinergic effects, such as dry mouth and tachycardia, tends to develop with continued use of TCAs. Occasionally, patients will show physical or psychic dependence on the antidepressant effects. It is important to note, however, that decades of evidence suggest that prolonged use of TCA therapy is not associated with tolerance to their desirable effects.

17.3.9 Methods of Detection

Plasma or urine samples are mixed with sodium bicarbonate (~33% in distilled water) and separated on preconditioned Sep-Pak® C17 chromatography cartridges.* The mixture is eluted with water:chloroform-2-propanol (9:1), and the eluate consists of an organic lower phase and an aqueous upper phase. The upper phase is discarded, the organic phase is evaporated to dryness, and the residue is dissolved in methanol. Aliquots of the methanol extract are injected into a gas chromatograph equipped with a silica SPB-1capillary column and a standard thermoionic ionization detector. Retention times vary from 8.9 minutes for amitryptyline to 32.3 minutes for chlorimipramine.

17.4 PHENOTHIAZINE, PHENYLBUTYLPIPERIDINE, AND THIOXANTHINE ANTIPSYCHOTICS

17.4.1 Classification and Indications

Formerly known as the major tranquilizers, the antipsychotic neuroleptics and related agents are grouped into several classes (Table 17.2). The drugs are indicated for the management of manifestations of psychotic disorders and as antiemetics, cough suppressants, and antivertigo agents. They are also effective in the treatment of phencyclidine (PCP)-induced psychosis, migraine headaches, and acute agitation in the elderly, although these are unlabeled indications. As with the TCAs, overdose is rarely fatal.

* The reusable cartridge is filled with solid nonpolar resin that separates components of mixtures.

17.4.2 Pharmacology and Clinical Use

Antipsychotic agents exert their actions primarily by antagonizing dopamine receptors. Varying degrees of selective dopamine blockade are seen in the cerebral limbic system and basal ganglia and along cortical dopamine tracts. Physiologically, these central pathways are associated with skeletal movement (nigrostriatal tracts), hallucinations and delusions (mesolimbic), psychosis (mesocortical), and prolactin release (tuberoinfundibular).

17.4.3 Toxicokinetics

Different colors and shapes of antipsychotics resemble small chocolate candies, which contributes to accidental ingestion by children. In addition, adult psychiatric patients intentionally or inadvertently combine antipsychotic medications with other S/H, alcohol, or antidepressants, thus precipitating additive or synergistic effects.

17.4.4 Mechanism and Signs and Symptoms of Acute Toxicity

The affinities of antipsychotic agents for nondopaminergic sites, such as cholinergic, α_1-adrenergic, and histaminic receptors, explain the varied ADRs produced.

As summarized in Table 17.2, phenothiazine, phenylbutylpiperidine, and thioxanthine derivatives display moderate to significant CNS reactions, including sedation, muscle relaxation, and lowering of the seizure threshold. The last sensitizes the individual to convulsions. In particular, antipsychotic drugs depress the reticular activating system (RAS) responsible for stimulating wakefulness (consciousness) and eliciting arousal reflexes. Unlike the barbiturates, these agents induce comatose-like states of consciousness that are more readily seen in children. Antimuscarinic and antihistaminic effects, such as mydriasis, dry mouth, tachycardia, and decreased GI activity, further complicate their toxic profiles. Their influence on hypothalamic nuclei is responsible for vasodilation, orthostatic hypotension, and hypothermia or hyperthermia. In addition, as with the TCAs, α-adrenergic blockade contributes to hypotension. Quinidine-like cardiovascular AV block precipitates potentially fatal ventricular arrhythmias and cardiac arrest.

EPS signs and symptoms develop when an imbalance between antidopaminergic (greater) and anticholinergic activity is created. In general, higher-potency phenothiazines are more likely to produce EPS than lower-potency TCA antidepressants or the newer neuroleptics. Table 17.3 summarizes EPS reactions, signs and symptoms, and treatment management associated with antipsychotic agents. Prolonged antipsychotic therapy, idiosyncratic reactions, or toxic overdose can precipitate any or most of the syndromes. In general, the more overwhelming the antidopaminergic properties, the greater the severity of EPS reactions.

Acute dystonic reactions appear in 95% of patients, predominantly young males, within 4 days of initiation of therapy or as dosage increases. In contrast, *akathisia* affects mostly elderly patients in early treatment (the first 60 days) and subsides with lower dosage. *Parkinsonism* develops within 10 weeks of therapy

TABLE 17.3
Characterization of the Extrapyramidal Symptoms (EPS) Associated with Neuroleptic Agents

Syndrome	Signs and Symptoms	Clinical Management
Acute dystonic reactions	Oculogyric crisis (upward gaze paralysis); muscular spasms of neck (torticollis), back (opisthotonos), and tortipelvis (abdominal wall)	Anticholinergics or benzodiazepines; antihistamine (diphenhydramine); dosage reduction or discontinuation
Akathisia	Motor restlessness and discomfort, inability to sit still	Anticholinergics or benzodiazepines; dosage reduction or discontinuation
Parkinsonism (akinesia)	Bradykinesia, shuffling gait, resting tremor ("pill rolling movements"), "masked face," perioral tremors ("rabbit syndrome")	Anticholinergics or benzodiazepines; antihistamine (diphenhydramine); dosage reduction or discontinuation
Tardive dyskinesia (older males) and tardive dystonia (younger males)	Choreoathetoid movements: involuntary, repetitive, spasmodic movements of face, tongue, lips (chorea); slow, writhing, involuntary movements of fingers and hands (athetoid); may occur after years of neuroleptic therapy and are irreversible	Dosage reduction or discontinuation; clozapine as alternative; and/or botulinum toxin, tetrabenzaine, or reserpine
Neuroleptic malignant syndrome (NMS)	Catatonia, muscle ("lead pipe") rigidity, stupor, hyperpyrexia, altered mental status, autonomic instability	For hyperpyrexia, benzodiazepines and rapid physical cooling; bromocriptine (dopamine [DA]), dantrolene (for muscle rigidity); anticholinergics not recommended; withdraw neuroleptics for minimum of 14 days; clozapine as alternative

and affects 90% of patients, although it is reversible at lower doses. It is important to note that the risk of developing a tardive disorder, and the likelihood that it will become irreversible, increases with the duration of treatment and with the total cumulative dose of neuroleptic drug. As neuroleptic agent is withdrawn, dopamine activity increases, and tardive dyskinesia emerges. The effect is probably due to upregulation of dopamine receptors in the corpus striatum, especially with chronic neuroleptic treatment. *Neuroleptic malignant syndrome* (NMS) is a potentially fatal idiosyncratic complication occurring in 2% of patients on antipsychotic therapy. It is an extreme EPS reaction resulting from excessive antidopaminergic activity. Precipitating factors include abrupt withdrawal of antiparkinson dopamine agonist drugs (L-dopa) or anticholinergic medication, or a rapid increase in the dose of neuroleptics. Meningitis and anticholinergic poisoning mimic NMS and must be ruled out.

17.4.5 Clinical Management of Acute Overdose

As with TCA toxicity, cardiovascular and respiratory adverse effects with antipsychotics are monitored for 12 h and up to 6 days. The treatment for reversing neuroleptic-induced hypotension and quinidine-like effects, especially ventricular dysrhythmias, is similarly managed with lidocaine and sodium bicarbonate, as needed. Convulsions or hyperactivity are controlled with pentobarbital or diazepam.

Although ostensibly dangerous, extrapyramidal adverse effects are not fatal* and are best treated with anticholinergics or benzodiazepines. Several days of treatment are necessary to reverse acute dystonic reactions. Prophylactic anticholinergic therapy is useful in the prevention of EPS. In severe situations, consideration should be given to discontinuation of the neuroleptic, since dystonic reactions usually subside within 24 to 48 hours after drug cessation. In contrast, tardive dyskinesia is difficult to manage, especially if symptoms appear at doses where the neuroleptic is therapeutically effective. Drug discontinuation and substitution with an atypical neuroleptic, such as clozapine, is usually considered, as these agents have a lower incidence of EPS. Clozapine may also be effective in reversing phenothiazine-induced tardive dyskinesia.

17.4.6 Tolerance and Withdrawal

As with the TCA antidepressants, tolerance to the anticholinergic effects develops with continued use of antipsychotic agents. Occasionally, patients will show physical or psychic dependence with the antidepressant effects. Prolonged use of antipsychotic therapy is not associated with tolerance to the desirable effects.

17.4.7 Methods of Detection

Techniques used in the separation, detection, and confirmation of phenothiazines and related major tranquilizers are similar to those described for TCA. Gas chromatography/mass spectrometry (GC-MS) and gas chromatography/flame ionization detection (GC-FID) are the prevailing methods used in clinical and forensic analysis.

17.5 SELECTIVE SEROTONIN REUPTAKE INHIBITORS (SSRIS)

17.5.1 Classification and Clinical Use

The advent of SSRIs in 1987 revolutionized the treatment of depression. The drugs are indicated for the management of manifestations of psychiatric disorders, including the management of clinical depression, obsessive compulsive disorder (OCD), major depressive disorder (MDD), bipolar disorder, panic disorders, and syndromes that affect mood and behavior. Although their effectiveness for the former is similar

* Antipsychotic agents have therapeutic indexes (TI). For instance, the TI for chlorpromazine extends up to 5000 mg; the adult daily average dose range = 30 to 800 mg, with 100 mg daily as an average daily outpatient effective dose; for haloperidol, the TI is up to 30 mg. The adult daily average dose range = 1 to 17 mg, with 2 mg as an average daily effective dose for ambulatory patients.

to TCAs, they have replaced the tricyclics principally because of their higher selectivity and fewer adverse effects, particularly the cardiac toxicity. Table 17.4 outlines the different SSRIs along with the variety of antidepressants that interact with norepinephrine (NE) and serotonin (5-HT) reuptake or CNS neuronal activation.

17.5.2 Pharmacology and Receptor Activity

Figure 17.1 illustrates the synthesis of 5-hydroxytryptamine (5-HT, serotonin), which occurs in the terminal presynaptic neurons. Neurons that synthesize 5-HT are known as *serotonergic neurons*. Whereas the TCAs inhibit both NE and 5-HT transporter systems for reuptake into the presynaptic bulb, the SSRIs are more selective for 5-HT transporters. As a result, synaptic serotonin concentrations are elevated in the synaptic cleft, causing an increase in 5-HT receptor binding and activation.*

Serotonin receptors are located throughout the brain and vasculature and are responsible for a variety of autonomic, central, and behavioral effects. Table 17.5 outlines the different serotonin receptors according to their classes and subtypes,

TABLE 17.4
Categories, Representative Compounds, and Proprietary Names of Currently Available SSRIs and Other Related Antidepressants

Category	Compound	Proprietary Name	Block Reuptake of
SSRIs	Citalopram	Celexa	5-HT transporter (selective)
	Escitalopram	Lexapro	
	Fluoxetine	Prozac	
	Fluvoxamine	Luvox	
	Paroxetine	Paxil	
	Sertraline	Zoloft	
5-HT/NE reuptake inhibitors	Duloxetine	Cymbalta	5-HT transporter/NE transporter
	Venlafaxine	Effexor	
Atypical antidepressants	Bupropion	Wellbutrin	DA/5-HT transporter
	Mirtazapine	Remeron	↑ 5-HT/NE transmission; blocks postsynaptic 5-HT_2 receptors
	Nefazodone	Serzone	5-HT transporter, weak; block post-synaptic 5-HT_{2A} receptors
	Trazodone	Desyrel	

5-HT: 5-hydroxytryptamine (serotonin); DA: dopamine; NE: norepinephrine.

* Interestingly, 5-HT and NE require separate reuptake transporter systems to incorporate the substrates back into the presynaptic neuron, accounting for the ability to design drugs with selective inhibitory targets. The VMAT (vesicular monoamine transporter), responsible for packaging newly synthesized cytoplasmic NE and 5-HT into synaptic vesicles, is the same in both serotonergic and adrenergic neuronal systems.

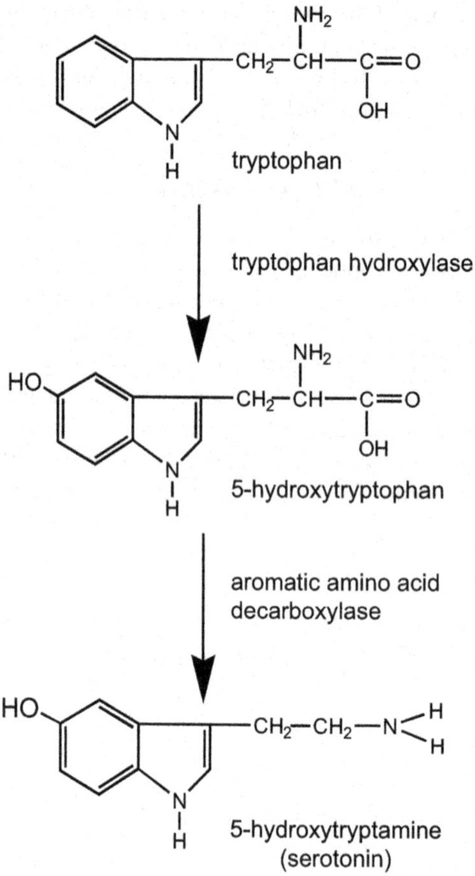

FIGURE 17.1 Synthesis of serotonin (5-hydroxytryptamine).

anatomic location, and effector organs, and the purported responses triggered by their activation. The binding of a ligand to serotonin receptors activates the G-protein coupled signal transduction pathway.* Depending on the G-protein subtype that is activated, the intracellular cascade reaction stimulates (or inhibits) adenylyl cyclase, which under normal physiological conditions, catalyzes the formation of cAMP (a second messenger). The subsequent signaling pathway singularly or simultaneously activates many downstream effector molecules that act as second messengers (e.g., Ca^{2+}, phospholipases, and protein kinases), the net result of which is an integrated cellular response.

17.5.3 Toxicokinetics

As with the TCAs, SSRIs have similar toxicokinetic properties. The drugs are lipophilic, have a relatively long plasma half-life (16 to 36 hours), and are well absorbed

* The 5-HT_3 receptor is an Na^+/K^+-gated (ligand-gated) ion channel, not a G-protein coupled receptor.

TABLE 17.5
Serotonin Receptor Classification, Anatomic Location, and Effector Responses

Main Serotonin Receptor Class	Serotonin Receptor Subtype	G-protein Coupled Effector	Major Anatomic Locations	Net Biological Effect
$5\text{-}HT_1$	$5\text{-}HT_{1A}$	G_i, ↓ cAMP	Limbic system	Sleep, cognition, mood, behavior, sensory (pain) perception
	$5\text{-}HT_{1B}$	G_i, ↓ cAMP	Corpus striatum, basal ganglia	
	$5\text{-}HT_{1D}$	G_i, ↓ cAMP[a]	Throughout cerebrum, midbrain	
	$5\text{-}HT_{1E}$	G_i, ↓ cAMP	Throughout cerebrum, midbrain	
	$5\text{-}HT_{1F}$	G_i, ↓ cAMP	Throughout cerebrum, midbrain	
$5\text{-}HT_2$	$5\text{-}HT_{2A}$	G_q, ↑ IP_3, DAG	Platelets, skeletal/smooth muscle	Platelet aggregation, smooth muscle contraction
	$5\text{-}HT_{2B}$	G_q, ↑ IP_3, DAG	Stomach	Smooth muscle contraction
$5\text{-}HT_3$	$5\text{-}HT_3$	Na^+/K^+-gated ion channel	Enteric nerves, vagal reflex, medulla (CTZ)	Smooth muscle contraction, nausea/vomiting reflex
$5\text{-}HT_{4,6,7}$	$5\text{-}HT_4$	G_s, ↑ cAMP	Mesenteric nerves, smooth muscle	Smooth muscle contraction
	$5\text{-}HT_6$	G_s, ↑ cAMP	Mesenteric nerves, smooth muscle	Smooth muscle contraction
	$5\text{-}HT_7$	G_s, ↑ cAMP	Mesenteric nerves, smooth muscle	Smooth muscle contraction

5-HT: 5-hydroxytryptamine (serotonin); cAMP: cyclic AMP; CTZ: chemoreceptor trigger zone; DA: dopamine; DAG: diacyl glycerol; G_i G_q G_s: G-protein coupled inhibitory, activating, stimulatory, respectively; IP_3: inositol trisphosphate; NE: norepinephrine.

[a] 5-HT1D are presynaptic autoreceptors important for feedback inhibition.

orally (peak levels are demonstrated after 2 to 8 hours). Their V_d ranges between 15 and 30 l/kg. With some exceptions, the drugs are metabolized by the same cytochrome P450 isoenzyme class (2D6) as TCAs. Thus, the competition for the same metabolic enzyme category may be responsible for possible drug interactions with other antidepressants, altering effective blood concentrations or increasing the risk of toxic phenomena.

17.5.4 Side Effects and Adverse Reactions

Although at pharmacologic doses SSRIs show selective affinity for their respective receptors and are considered safer than TCAs, their affinities for nonserotonergic receptor sites, enzymes, or transporter targets (e.g., α_1- α_2-, and β-adrenergic receptors) increase with higher doses. Nevertheless, their TI and acceptability are greater than those for TCAs.

At therapeutic doses, SSRIs display moderate to significant autonomic and cognitive CNS reactions, including sedation, sexual dysfunction, headache, sweating, agitation, sleep disturbances, and changes in body weight. Parasympathetic, vagal reflex reactions, such as nausea, vomiting, and diarrhea, occur at therapeutic doses, along with weakness and fatigue. The drugs do not show any significant binding to other autonomic muscarinic, histaminic, or dopaminergic receptors. Their lack of cardiac toxicity is a particularly desirable attribute of this class.

Also, at concentrations within normal target range, SSRIs can produce dangerously elevated 5-HT plasma levels when administered concomitantly with monoamine oxidase (MAO) inhibitors or other SSRIs. The resulting *serotonin syndrome* is characterized by hyperthermia, muscle rigidity, sweating, myoclonus,* changes in mental status, and alteration of vital signs. *Discontinuation syndrome* results from abrupt withdrawal or termination of SSRI therapy. Signs and symptoms are mostly related to behavioral effects and include headache, malaise, irritability, agitation, nervousness, and sleep pattern disturbances. The event is probably due to sudden rebound decrease of serotonin activity. Lastly, SSRIs have been reported to cause a *switch* phenomenon characterized by behavioral transformation from depression to mania in vulnerable patients. The difficulty, however, in identifying a *switch* event is the ability to distinguish between the developments of a true manic episode due to the patient's pathologic condition and the toxicologic reaction to the chemical.

17.5.5 Signs and Symptoms of Acute Toxicity

As with the TCAs, at higher doses, the SSRIs are capable of inducing cardiac abnormalities, attributable to their toxicologic, nonselective binding to adrenergic receptors. The cardiac aberrations manifest initially as sinus tachycardia (palpitations), sweating, tremors, vomiting, and diarrhea, followed by hypertension and arrhythmias. At toxic concentrations, the drugs also lower the seizure threshold, thus increasing the risk for mild to moderate tremors, dystonic reactions, and convulsions.

17.5.6 Clinical Management of Acute Overdose (O.D.)

As noted earlier, the SSRIs have a higher therapeutic ratio than TCAs, and O.D. due solely to SSRIs is rare. Yet, the more serious of the acute toxic reactions, either idiosyncratic or due to excessive quantities, are the *serotonin syndrome* and development of seizures. Treatments for these conditions are similar and approached as with O.D. from TCAs. Cardiac monitoring, administration of activated charcoal (1 g/kg orally), and sodium bicarbonate (used for electrocardiogram QRS prolongation) are featured modalities. Patients should be observed for complications, focusing particularly on cardiac integrity and vital signs, for at least 6 h post ingestion. In addition, the *serotonin syndrome* is managed with cyproheptadine, an antihistaminic drug used as

* Described as repetitive, rhythmic muscle twitching.

an antiserotonergic agent in this situation. A dose of 4 to 8 mg orally is repeated every 2 h until a favorable response is noted, followed by 4 mg orally every 6 h for 48 h. Necessarily, all other antidepressants are discontinued, and supportive care is optimized. Benzodiazepines, such as diazepam, are used to relieve muscle rigidity and pain. *Discontinuation syndrome* is managed primarily by avoiding abrupt discontinuation or withdrawal of the medication. The *serotonin syndrome*, as well as complications and adverse reactions resulting from interactions with other drugs of similar classes, can be avoided by incorporating a *wash-out* period of at least 1 to 2 weeks. Because of the SSRIs' long half-life, this practice of using a wash-out period prevents possible competition of the different classes for the same metabolic enzymes, thus alleviating unpredictable shifts in drug plasma concentrations.

17.5.7 Tolerance and Withdrawal

As with the TCAs, some tolerance to the anticholinergic effects of SSRIs develops with continued use of the agents. Although patients receiving SSRIs will show some psychic dependence on the antidepressant effects, prolonged therapy is not associated with tolerance, addiction, or withdrawal to their desirable effects as is seen with drugs of abuse.

17.5.8 Methods of Detection

Various studies indicate that patients at risk of toxicity from SSRI or other antidepressant overdose may be identified by neurological, cardiovascular, and electrocardiography status. This is particularly true because SSRIs that are ingested alone are rarely fatal. Patients who exhibit overdoses of combinations of SSRIs with TCAs or monoamine oxidase (MAO) inhibitors may produce a serotonin syndrome with a severe or fatal outcome. In such situations, it may be useful to chemically detect the presence of the individual drug components. Fluorescence immunoassay appears to be useful for generating rapid quantitative plasma drug concentrations for a variety of TCAs and SSRIs. It is important to note, however, that while quantitative chemical identification of these drugs following overdose is helpful in confirming the diagnosis, it is not a forensic or regulatory requirement. Careful clinical observation appears to provide a more comprehensive therapeutic picture for postingestion management.

REFERENCES

SUGGESTED READINGS

Alvarez, J.G., Chapter 14. Clinical and pharmaceutical applications of gas chromatography, in *Modern Practice of Gas Chromatography*, 4th ed., Grob, R.L. and Barry, E.F. (Eds.), John Wiley & Sons, New York,2004.

DeLas Cuevas, C. and Sanz, E.J., Safety of selective serotonin reuptake inhibitors in pregnancy. *Curr. Drug. Saf.* 1, 17, 2006.

Dyer, K.S., Anticholinergic poisoning, in *Toxicology Secrets*, Ling, L.J., Clark, R.F., Erickson, T.B., and Trestrail, J.H. III (Eds.), Hanley & Belfus, Philadelphia, PA, 2001, Chapter 10.

Frascogna, N., Physostigmine: is there a role for this antidote in pediatric poisonings? *Curr. Opin. Pediatr.* 19, 201, 2007.

Kovacova-Hanuskova, E., Buday, T., Gavliakova, S., and Plevkova, J., Histamine, histamine intoxication and intolerance. *Allergol. Immunopathol (Madr.)* 43, 498, 2015.

Leurs, R., Church, M.K., and Taglialatela, M., H1-antihistamines: inverse agonism, anti-inflammatory actions and cardiac effects, *Clin. Exp. Allergy* 32, 489, 2002.

Levin, T.T., Cortes-Ladino, A., Weiss, M., and Palomba, M.L., Life-threatening serotonin toxicity due to a citalopram-fluconazole drug interaction: case reports and discussion. *Gen. Hosp. Psychiatry* 30, 372, 2008.

MacGlashan, D. Jr., Histamine: a mediator of inflammation. *J. Allergy Clin. Immunol.* 112, S53, 2003.

Mercadamte, S., Death rattle: critical review and research agenda. *Support Care Cancer* 22, 571, 2014.

McIntyre, R.S. Panjwani, Z.D., Nguyen, H.T., Woldeyohannes, H.O., Alsuwaidan, M., Soczynska, J.K., Lourenco, M.T., Konarski, J.Z., and Kennedy, S.H., The hepatic safety profile of duloxetine: a review. *Expert Opin. Drug Metab. Toxicol.* 4, 281, 2008.

Moeller, K.E., Lee, K.C., and Kissack, J.C., Urine drug screening: practical guide for clinicians. *Mayo Clin. Proc.* 83, 66, 2008.

Roberts, D.M. and Buckley, N.A., Pharmacokinetic considerations in clinical toxicology: clinical applications. *Clin. Pharmacokinet.* 46, 897, 2007.

Wille, S.M., Cooreman, S.G., Neels, H.M., and Lambert, W.E., Relevant issues in the monitoring and the toxicology of antidepressants. *Crit. Rev. Clin. Lab. Sci.* 45, 25, 2008.

REVIEW ARTICLES

Berchou, R.C., Maximizing the benefit of pharmacotherapy in Parkinson's disease. *Pharmacotherapy* 20, 33S, 2000.

Camelo-Nunes, I.C., New antihistamines: a critical view. *J. Pediatr. (Rio J.)* 82, S173, 2006.

Catterson, M.L. and Martin, R.L., Anticholinergic toxicity masquerading as neuroleptic malignant syndrome: a case report and review. *Ann. Clin. Psychiatry* 6, 267, 1994.

Ellard, B. and Galdo, J.A., Determination of anticholinergic medication use in patients prescribed medications for the treatment of dementia. *Consult. Pharm.* 32, 764, 2017.

Forrester, M.B., Selective serotonin re-uptake inhibitor warnings and trends in exposures reported to poison control centers in Texas. *Public Health* 122, 1356, 2008.

Geddes, J.R., Freemantle, N., Mason, J., Eccles, M.P., and Boynton, J., SSRIs versus other antidepressants for depressive disorder. *Cochrane Database Syst. Rev.* 2, CD001851, 2000.

Gill, H.S., DeVane, C.L., and Risch, S.C., Extrapyramidal symptoms associated with cyclic antidepressant treatment: a review of the literature and consolidating hypotheses. *J. Clin. Psychopharmacol.* 17, 377, 1997.

Gillman, P.K., A systematic review of the serotonergic effects of mirtazapine in humans: implications for its dual action status. *Hum. Psychopharmacol.* 21, 117, 2006.

Glick, A.R., The role of serotonin in impulsive aggression, suicide, and homicide in adolescents and adults: a literature review. *Int. J. Adolesc. Med. Health* 27, 143, 2015.

Gumnick, J.F. and Nemeroff, C.B., Problems with currently available antidepressants. *J. Clin. Psychiatry* 61, 5, 2000.

Henry, J.A., Epidemiology and relative toxicity of antidepressant drugs in overdose. *Drug Saf.* 16, 374, 1997.

Kelly, C.A., Dhaun, N., Laing, W.J., Strachan, F.E., Good, A.M., and Bateman, D.N., Comparative toxicity of citalopram and the newer antidepressants after overdose. *J. Toxicol. Clin. Toxicol.* 42, 67, 2004.

Kwakye, G.F., Jiménez, J., Jiménez, J.A., and Aschner, M., *Atropa belladonna* neurotoxicity: implications to neurological disorders. *Food Chem. Toxicol.* 116(Pt B), 346, 2018.

Lane, R.M., SSRI-induced extrapyramidal side-effects and akathisia: implications for treatment. *J. Psychopharmacol.* 12, 192, 1998.

Maintz, L. and Novak, N., Histamine and histamine intolerance. *Am. J. Clin. Nutr.* 85, 1185, 2007.

Mitchell, P.B., Therapeutic drug monitoring of psychotropic medications. *Br. J. Clin. Pharmacol.* 49, 303, 2000.

Mitchell, P.B., Therapeutic drug monitoring of non-tricyclic antidepressant drugs. *Clin. Chem. Lab. Med.* 42, 1212, 2004.

Power, B.M., Hackett, L.P., Dusci, L.J., and Ilett, K.F., Antidepressant toxicity and the need for identification and concentration monitoring in overdose, *Clin. Pharmacokinet.* 29, 154, 1995.

Sachdev, P.S., The current status of tardive dyskinesia, *Aust. N. Z. J. Psychiatry* 34, 355, 2000.

Sarko, J., Antidepressants, old and new. A review of their adverse effects and toxicity in overdose. *Emerg. Med. Clin. North Am.* 18, 637, 2000.

Sen, S. and Sanacora, G., Major depression: emerging therapeutics. *Mt. Sinai J. Med.* 75, 204, 2008.

Triantafylidis, L.K., Clemons, J.S., Peron, E.P., Roefaro, J., and Zimmerman, K.M., Brain over bladder: a systematic review of dual cholinesterase inhibitor and urinary anticholinergic use. *Drugs Aging* 35, 27, 2018.

Tune, L.E., Anticholinergic effects of medication in elderly patients. *J. Clin. Psychiatry* 62, 11, 2001.

Yildirim, B.O. and Derksen, J.J., Systematic review, structural analysis, and new theoretical perspectives on the role of serotonin and associated genes in the etiology of psychopathy and sociopathy. *Neurosci. Biobehav. Rev.* 37, 1254, 2013.

Żmudzka, E., Sałaciak, K., Sapa, J., and Pytka, K., Serotonin receptors in depression and anxiety: insights from animal studies. *Life Sci.* pii: S0024-3205(18)30504-6, 2018.

18 Acetaminophen, Salicylates, and Nonsteroidal Anti-Inflammatory Drugs (NSAIDs)

18.1 HISTORY AND DESCRIPTION

Acetaminophen, salicylates (aspirin), and nonsteroidal anti-inflammatory agents are found alone and in combination with other analgesic/antipyretic* compounds, in hundreds of over-the-counter (OTC) and prescription formulas. As such, these agents are routinely reported as products responsible for acute and chronic overdose, with acetaminophen accounting for most poisoning cases within this category. Packaging laws and public service education campaigns have reduced the frequency of accidental poisoning, especially in children and teenagers. Because the drugs are available OTC, however, it suggests that these agents are innocuous, although they are marketed with clearly marked warnings and suggestions for use.

In general, analgesic, antipyretic, and anti-inflammatory agents are used in the *household* management of nonnarcotic relief of mild to moderate pain, for inflammation associated with a variety of rheumatic conditions, and for reduction of fever.

18.2 ACETAMINOPHEN (N-ACETYL-PARA-AMINOPHENOL, APAP, PARACETAMOL)

18.2.1 INCIDENCE

In 2013, analgesics were first among the substances most frequently involved in human adult exposures, accounting for 12.5% of all chemical substance poisoning, and third among pediatric cases (7.2%). Acetaminophen "in combination" was fourth among the top 25 substance categories with the highest number of fatalities (Mowry et al., 2013 Annual Report of the AAPCC National Poison Data System, 2014).

* Mild to moderate pain relief and fever-reducing properties, respectively.

Poisoning cases with APAP exceed those of all other agents in this category (five times greater incidence than with aspirin). It is the most common drug involved in overdose (O.D.), registering approximately 60% of all analgesic exposures, and the second most common cause of liver failure in the United States. The U.S. Food and Drug Administration (FDA) reports that over 56,000 emergency room visits per year are due to acetaminophen O.D. as a sole agent, resulting in about 100 deaths per year, one-fourth of which are intentional. Due to lack of hospital reporting, the numbers may greatly underestimate the extent of the problem.*

18.2.2 Medicinal Chemistry and Pharmacology

Acetaminophen is the major hydroxylation metabolite of two potent analgesic parent compounds, acetanilid and phenacetin (Figure 18.1). The antipyretic activity of the molecules resides in the aminobenzene structure.

APAP reduces fever by a direct action on the heat-regulating centers in the hypothalamus, dissipating heat via vasodilation and increased sweating. Its analgesic and antipyretic properties are equivalent to those of aspirin. Its inhibition of central prostaglandin synthetase is more effective than its peripheral action, rendering it a weak anti-inflammatory agent compared with aspirin. The site and mechanism of its analgesic action are, to this day, unclear.

18.2.3 Clinical Use

APAP is recommended as an analgesic/antipyretic in the presence of aspirin allergy, in patients who demonstrate blood coagulation disorders, in children, and in patients who receive oral anticoagulants or who demonstrate upper gastrointestinal disease. It is useful in a variety of arthritic and rheumatic conditions, including musculoskeletal disorders, headache, and other minor pain, and for the management of fever associated with bacterial and viral infections.

18.2.4 Metabolism and Mechanism of Toxicity

Acetaminophen is rapidly absorbed from the gastrointestinal tract and uniformly distributed, with peak plasma levels achieved by 0.5 to 2.0 h. Hepatic glucuronide and sulfate conjugation (Reactions 1 and 2, Figure 18.1) produce the inactive corresponding conjugates, which account for 95% of metabolism and elimination in urine. In addition, at therapeutic acute doses, the remaining 4 to 5% of the product is detoxified and eliminated in the minor cytochrome P450 oxidase pathway (Reaction 3). The result is the production of the reactive intermediate, N-acetyl-*p*-benzoquinoneimine (NAPQI) metabolite. Further conjugation by cellular glutathione results in the production of mercapturic acid and cysteine conjugates (Reaction 4).

* Recently, Parker et al. (2017) cited an association of the increase in the prevalence of autism in babies and small children with a rise in exposure to acetaminophen, whose neurotoxicity is exacerbated by inflammation and oxidative stress. However, of the 195 references cited by the authors, only two citations note a clinical association of APAP with autism spectrum disorder (ASD), both of which note that evidence from larger studies is lacking.

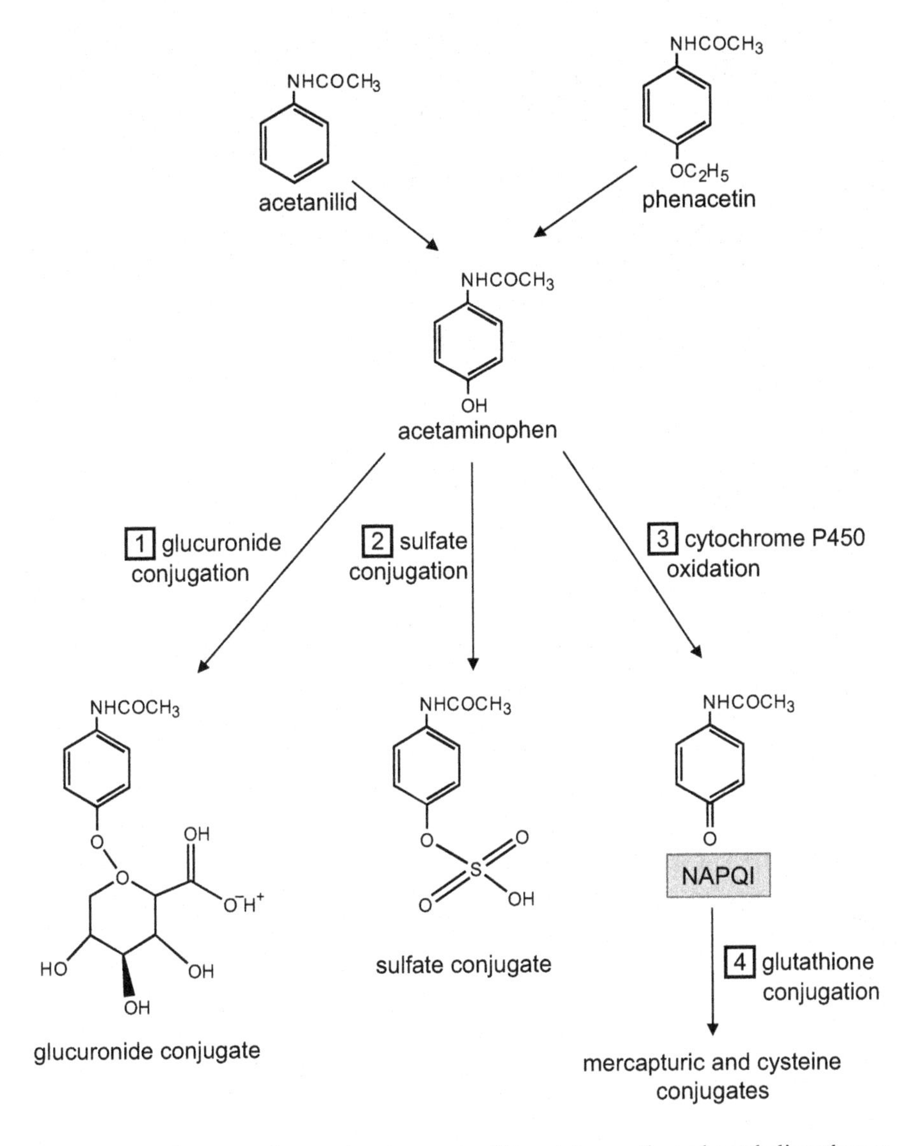

FIGURE 18.1 Structure of acetaminophen, acetanilid, and phenacetin, and metabolic pathways.

With chronic use or with large doses,* the glucuronide and sulfate conjugation metabolic routes (1 and 2) are saturated, and more importantly, glutathione stores

* Over a 24 hour period, not more than 4 g or 90 mg/kg total is recommended in adults and children, respectively. In addition, the recommended length of therapy should not exceed 10 days for adults or 5 days for children, and not more than 3 days for fever (adults and children). Interestingly, children are more tolerant to acetaminophen-induced hepatotoxicity, perhaps due to an increase in glutathione turnover. The potential hepatotoxicity of APAP may be increased in chronic alcoholics, in malnutrition, and in HIV patients. Large doses or prolonged administration of other hepatic microsomal enzyme inducers, such as phenytoin, barbiturates, isoniazid, and rifampin, also increase the hepatotoxic risk.

are depleted (4). This leaves the cytochrome P450 oxidase pathway (Reaction 3) to accumulate the toxic NAPQI metabolite. The binding of NAPQI to hepatocyte membranes and sulfhydryl proteins accounts for the hepatotoxic sequelae.

18.2.5 Signs and Symptoms of Acute Toxicity

In general, delayed acute toxicity is manifested by the lack of apparent initial signs and symptoms for about 24 h, after which symptoms may appear quickly. However, the potential for latent permanent hepatic failure and death exists even with the progression to an apparent recovery stage. Table 18.1 outlines the stages of acetaminophen toxicity.

Initial symptoms are nonspecific, and a careful history and an acetaminophen plasma level should be performed so that impending toxicity is not missed. This affords the patient the opportunity of early treatment and prevention of potential irreversible hepatic damage. A quiescent recovery period (Stage 2) is characterized by clinical improvement after 24 to 72 h post ingestion. Treatment at this stage is generally based on laboratory findings and acetaminophen plasma levels. Liver function tests are monitored throughout the critical periods until complete recovery is ensured. It is important to note that the acetaminophen plasma levels based on the Rumack–Matthew nomogram are intended for single acute overdose and do not apply in suspected toxicity with chronic or prolonged ingestion of therapeutic doses. Monitoring these patients involves the same criteria as noted in Table 18.1, especially the determination of drug concentrations every 2 to 4 h to establish baseline toxicity criteria. Continued observation and repeat levels are warranted.

18.2.6 Clinical Management of Acute Overdose

Early treatment and careful evaluation of clinical history leading to the emergency event are paramount in treatment. Activated charcoal is beneficial if administered to an individual who presents within 1 to 2 h post ingestion. An acetaminophen level obtained 4 h later determines follow-up treatment with the antidote. At 8 h post ingestion, activated charcoal, emetics, or gastric lavage are not necessary.

Rates of poison center calls with acetaminophen-related exposures decreased from 49 to 43 per 1000 calls from 2009 to 2012, respectively. Rates of emergency department (ED) visits for unintentional acetaminophen-related adverse events decreased from 58 to 50 per 1000 adverse drug events necessitating ED visits from 2009 to 2012, respectively. Rates of hospital inpatient discharges with APAP-related poisoning also decreased from 2009 to 2011, while ED visit rates remained stable. Prior to 2009, the rates of APAP-related hospitalizations had been slowly increasing (Major et al., 2016).

The Rumack–Mathew treatment nomogram is the primary tool used to guide treatment after acute ingestion of APAP (Table 18.1). Although essential in the management of APAP exposure, its apparent margins limit the nomogram's clinical use. Consequently, a thorough understanding of the guidelines in treatment management is of importance. Acute APAP poisoning is generally regarded as a single ingestion of APAP over less than a 4 h time period. Regardless of whether APAP exposure

TABLE 18.1
Stages of Acetaminophen Toxicity

Stage	Time Post Ingestion (h)	APAP Plasma Concentration (mg/dl)[a]	Signs and Symptoms	Laboratory Findings
1. Preclinical toxic effects	4 8 12 24	≥150 ≥75 ≥35 ≥5	Anorexia, nausea, vomiting, pallor, diaphoresis, malaise, confusion, hypotension, arrhythmias	Hepatic transaminases (AST, ALT) rising
2. Hepatic injury	24–72	≥1 (at 72 h)	Clinical improvement, right upper quadrant pain; hepatic and renal function deterioration	Transaminases peaking; bilirubin and PT elevated
3. Hepatic failure	72–96	≥1	Hepatic centrilobular necrosis: jaundice, coagulopathy, encephalopathy (coma); reappearance of nausea and vomiting, arrhythmias, acute renal failure, death	Peak levels of AST (20,000 U/ml) and ALT
4. Recovery	4–14 days	≥1	Resolution of hepatic dysfunction and recovery if liver damage is reversible	Return to baseline levels

[a] Based on the Rumack–Matthew nomogram; the current practice in the United States. is to treat all patients who have acetaminophen levels above these values as at "possible risk" for toxicity; the nomogram assumes that the absorption of a single overdose is complete by 4 h and that the half-life, based on elimination toxicokinetics, is 4 h.

ALT: alanine aminotransferase; AST: aspartate aminotransferase; PT: prothrombin time.

Other laboratory tests include blood urea nitrogen (BUN), creatinine, electrolytes, and blood glucose.

occurs as a result of a single O.D. or after repeated supratherapeutic ingestion, however, the progression of acetaminophen poisoning is categorized into four stages: 1. preclinical toxic effects (a normal serum alanine aminotransferase concentration); 2. hepatic injury (an elevated alanine aminotransferase concentration); 3. hepatic failure (hepatic injury with hepatic encephalopathy; and 4. recovery. Also, the Rumack–Mathew nomogram can only be used to determine the need for N-acetyl cysteine (NAC) treatment after an acute ingestion; thus, all other APAP ingestions are not applicable to the nomogram. Typically, a toxic dose is about 150 mg/kg, or 15 g in an otherwise normal 70 kg adult, corresponding to a potentially toxic level (Stage 1, Table 18.1) on the nomogram.

NAC is the antidote for acetaminophen poisoning. In its conversion to cysteine, NAC restores glutathione reserves by providing sulfhydryl donors for the eventual detoxification of NAPQI. In addition, NAC increases sulfate conjugation, thereby preventing excess NAPQI production. NAC also acts as an antioxidant, enhancing

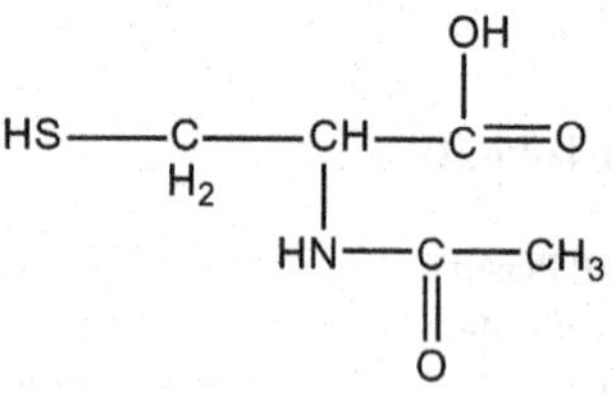

FIGURE 18.2 Structural formula for N-acetylcysteine.

oxygen use; this effect may be of benefit in patients with fulminant hepatic failure (Figure 18.2).

NAC is administered as a 10% or 20% solution for oral administration or through nasogastric (NG) instillation. It should be delivered within 8 h of ingestion and whenever a potentially toxic APAP concentration is measured above the toxic level. Protection against hepatotoxicity is 100% within 8 h of APAP ingestion. Efficacy decreases, however, when administered beyond 8 h, although NAC therapy may be beneficial even 36 h post ingestion.

Currently, the protocol is a 140 mg/kg oral loading dose followed by 17 doses of 70 mg/kg every 4 h for a total of 1330 mg/kg over 72 h. The dose is continuous over the 72 h until the acetaminophen assay reveals a nontoxic level. If the patient vomits the loading or maintenance dose within 1 h of administration, the dose is repeated. Antiemetics, such as metoclopramide (Reglan®), may be helpful in retaining the NAC. Among the possible adverse effects associated with large doses of NAC are hypersensitivity, gastrointestinal disturbances, urticaria, pruritus, angioedema, bronchospasm, tachycardia, and hypotension. Most of the serious reactions are ascribed to intravenous administration.

18.3 SALICYLATES AND ACETYLSALICYLIC ACID (ASPIRIN, ASA)

18.3.1 Incidence and Clinical Use

Prompted in part by the search for effective substitutes for quinine, early British researchers at the end of the nineteenth century synthesized and identified sodium salicylate for use as an antipyretic (for the treatment of rheumatic fever) and analgesic (for the treatment of gout). By 1815, the inception of the aspirin tablet established ASA as the most popular compound of the times.

Today, salicylates and related analgesics are one of the major sources of pediatric drug product poisoning in the United States. Although accidental poisoning in children has decreased with improved packaging safety requirements, aspirin and salicylate toxicity continue to be problems primarily because of the increased risk to children with fevers. At a pediatric toxic dose level of 150 to 200 mg/kg, as few as 10 to 12 adult 325 mg tablets of aspirin produce mild toxicity in an average 5-year-old (20 kg) child.

Aspirin and salicylic acid products are the active ingredients of hundreds of therapeutic prescription and *OTC* products. A recent review of U.S. poison control center data showed that over 40,000 exposures were a result of ingestion of salicylate-containing products, most of which were unintentional exposures (63%) or involved children (44%) (Gummin et al., 2018). ASA products are used as analgesics, antipyretics, and anti-inflammatory (arthritis) agents and in cough/cold, antihistamine, and decongestant formulations. Oil of wintergreen (betula oil) is an older product traditionally used as an analgesic liniment for the relief of sore muscles and stiff joints. It is a methyl salicylate concentrate (530 mg/ml) that produces severe toxicity in children with a 5 ml dose.* Finally, aspirin has shown some success as an antiplatelet agent in patients with thromboembolic disease, although this is an unofficial indication.

Consequently, the ubiquitous nature and ready availability of aspirin products have created the impression that aspirin is generally innocuous, which further contributes to the compound's persistent appearance in poisoning events.

18.3.2 Toxicokinetics

Salicylates are rapidly and completely absorbed but distributed unevenly throughout body tissues after oral use.† Metabolism follows first-order kinetics (dose dependent) to form oxidized and conjugated metabolites. Renal clearance accounts for most of the compound's elimination and is enhanced from 2% to more than 80% as pH, and ionization, increases.

18.3.3 Mechanism of Toxicity

Figure 18.3 outlines the sequential steps in the mechanism of aspirin-induced metabolic acidosis. In coordination with Figure 18.3, Table 18.2 summarizes the pathological mechanisms, metabolic consequences, and clinical pathology of acute aspirin toxicity. The primary result of high serum concentrations of salicylic acid is interference with acid–base balance. The stability of serum pH (7.4) depends on the maintenance of a delicate ratio of bicarbonate ion to carbonic acid (HCO_3^-/H_2CO_3: 20/1).‡ As salicylic acid levels rise and pH decreases, the medullary respiratory center is stimulated, resulting in an increase in ventilatory rate (hyperventilation). With a rise in respirations per minute, the victim expels more CO_2. As noted in the equilibrium equation for the bicarbonate buffering system (Equation 18.1), there is a shift to the left due to the decreasing concentrations of CO_2. Accordingly, this triggers a temporary but significant response from renal compensatory mechanisms. Intercalated cells located in the distal convoluted tubules of the kidney nephron have the primary responsibility of maintaining circulatory balance between HCO_3^- and H_2CO_3. Thus,

* Pepto-Bismol®, an antacid and antidiarrheal preparation, contains 262 mg of bismuth subsalicylate per 15 ml (one tablespoonful or one-half ounce), or about 130 mg total salicylate. The recommended 8 ounce dose contains the equivalent of six to seven 325 mg tablets of adult aspirin. Of the 40,405 human exposure cases to salicylate in 2004, 30% involved children.

† Aspirin is partially hydrolyzed to salicylic acid during absorption and distributed unevenly to all body fluids and tissues. *Vd* = 0.15–0.20 l/kg, which is equivalent to the extracellular space.

‡ This ratio is predicated on the physiological attempt to maintain a neutral pH. It is the kidneys' responsibility, in part, to accomplish this through reabsorption of filtered bicarbonate and/or hydrogen ions. This equilibrium is, thus, appropriately represented in Equation 18.1.

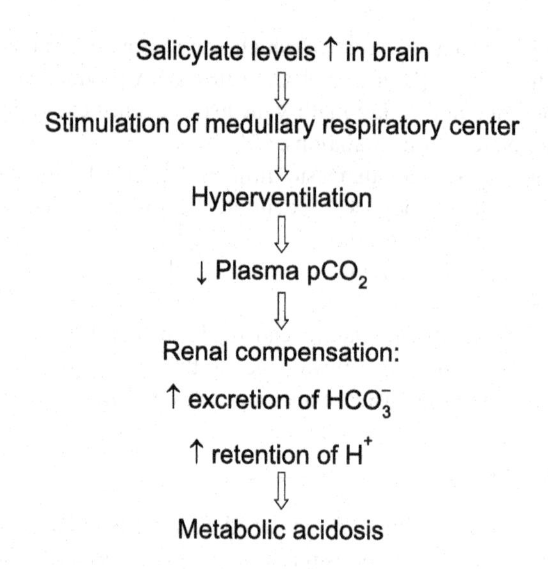

FIGURE 18.3 Schematic of mechanism of aspirin toxicity.

chemoreceptors in these cells detect a decrease in the bicarbonate:carbonic acid ratio resulting from respiratory elimination of CO_2. The cells then attempt to stabilize higher HCO_3^- concentrations by *retaining* H^+ and *eliminating* HCO_3^-, initiating a decrease in serum pH and a *compensatory* metabolic acidosis.

$$H_2O + CO_2 \leftrightarrow H_2CO_3 \leftrightarrow H^+ + HCO_3^- \qquad (18.1)$$

Toxicity associated with oxidative uncoupling of the electron transport chain (Figure 18.4) is similar to that seen with cyanide and 2,3-dinitrophenol poisoning. Salicylates reduce the effective shuttling of $NADH^+$ into the electron transport chain, preventing a number of adenosine triphosphate (ATP)–dependent reactions. As less ATP is produced, glycolysis is enhanced but is less efficient, triggering an increase in lipid metabolism and releasing excessive free fatty acids and ketones. In addition, oxygen consumption and glucose use are increased from the uncoupling effect, resulting in excess body heat production.

18.3.4 Signs and Symptoms of Acute Toxicity

The onset of symptoms is usually 1 to 2 h after ingestion but may be delayed to 4 to 6 h due to the absorption of sustained-release preparations or the formation of gastric concretions. The severity of symptoms peaks between 12 and 24 h. At low doses, the elimination half-life is constant at 3 to 4 h and follows first-order kinetics (dose *dependent*). With higher doses, or with chronic ingestion, elimination half-life shifts from first-order to zero-order kinetics (dose *independent*) as metabolic enzymes become saturated, thus contributing to the severity of chronic, low-dose toxicity.

More severe ASA toxicity, especially the development of metabolic acidosis, is associated with chronic ingestion and with children under 5 years old. The

TABLE 18.2
Pathology and Mechanism of ASA Toxicity

Mechanism of Toxicity	Pathological Consequence	Metabolic Compensation	Signs and Symptoms
Elevated ASA serum concentration (acidic substance)	Decreases serum pH	Contributes to metabolic acidosis; alters platelet function (hypoprothrombinemia)	Increased bleeding time
Stimulation of medullary respiratory center	Hyperventilation	Decreased plasma PCO_2 with respiratory alkalosis	Tachypnea, pulmonary edema, tachycardia, dehydration
Renal compensation for respiratory alkalosis	Kidneys excrete more bicarbonate ions; retain more hydrogen ions	Contributes to compensatory metabolic acidosis; CNS toxicity	Irritability, restlessness, tinnitus, dehydration, seizures, coma
Inhibition of Krebs cycle enzymes	Accumulation of organic acids (oxaloacetate)	Contributes to metabolic acidosis and lactic acidosis	Gastric irritation, nausea, vomiting
Oxidative uncoupling of electron transport chain	Prevents combination of phosphate with ADP	Decreased formation of ATP, enhanced glycolysis, lactic acid, pyruvic acid; contributes to metabolic acidosis	Hyperthermia, tachycardia, dehydration, cardiovascular collapse,
	Increases peripheral demand for glucose	Stimulates lipid metabolism, releases fatty acids; contributes to metabolic acidosis	hypoglycemia

progression is dose related. Although a single large dose of aspirin results in the signs and symptoms described in Table 18.2, death is more likely in patients with chronic poisoning and is usually a result of pulmonary or cerebral edema. In fact, the ability of ASA to traverse the blood–brain barrier is delayed due to its highly polar nature. Consequently, chronic ingestion affords the polar molecule more time for serum levels to equilibrate with central nervous system (CNS) concentrations. Interestingly, unlike acetaminophen concentrations, ASA blood levels (Done nomogram) do not correlate well with acute or chronic signs and symptoms. Treatment decisions therefore rely on empirical observations and clinical laboratory function tests.

18.3.5 Clinical Management of Toxicity

Although serum salicylate concentrations are not dependable for guiding treatment, the change in serum concentrations—that is, a drop of 10% every 3 to 4 h—is a good indicator of recovery. In addition, lack of symptoms 6 h after ingestion is associated with mild to moderate (less than severe) complications. In general, serum salicylate

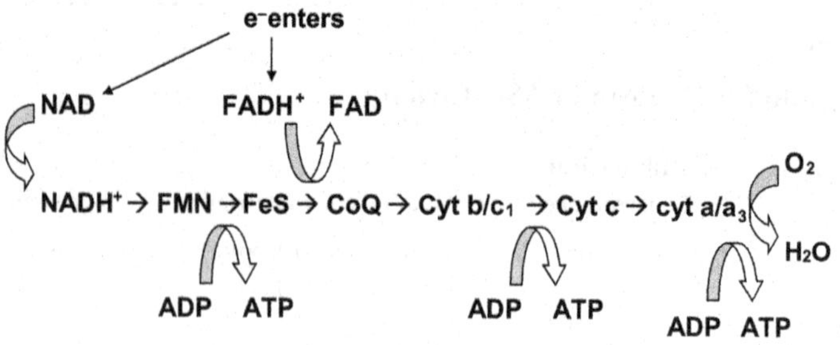

FIGURE 18.4 Electron transport and oxidative phosphorylation.

levels are used as a gauge for moderate (50 mg/dl), severe (75 mg/dl), and potentially lethal (100 mg/dl) poisoning. Complete clinical laboratory tests, especially for blood glucose, serum electrolytes (determination of the anion gap),* and liver function, are important in calculating the risk of metabolic acidosis.

The American Association of Poison Control Centers guideline (AAPCC; Chyka, et al., 2007) lists graded treatment recommendations for victims of salicylate toxicity. The recommendations are tabulated in Table 18.3 according to likely clinical use. Unless otherwise noted and as justified previously, most E.D. referrals do not rely on serum salicylate levels as a monitor for salicylate toxicity. For instance, acute ingestion of more than 150 mg/kg or 6.5 g of aspirin equivalent warrants referral to an E.D. Also, ingestion of greater than a taste (children under 6 years) and more than 4 ml (children over 6 years of age) of oil of wintergreen could cause systemic salicylate toxicity and warrants referral to the E.D. (Grade C). Routine referral to an E.D. for immediate care is not required in women in the last trimester of pregnancy who ingest below the dose for E.D. referral and who do not present with other referral conditions. Instead, these patients should be directed to a nonemergency health care facility for the evaluation of maternal and fetal risk. Asymptomatic patients with dermal exposures should be thoroughly washed with soap and water. Eyes should be irrigated with tepid tap water for 15 minutes following ocular exposure to methyl salicylate or salicylic acid.

The administration of emetics is not recommended, as vomiting can induce aspiration of stomach contents and corrosion of the esophagus and will not remove all stomach contents or sustained-release preparations. Activated charcoal and cathartics are recommended to effectively bind salicylates and to prevent intestinal obstruction due to concretions, respectively. Sodium bicarbonate administration enhances ASA elimination by alkalinizing the urine (it maintains the salicylic acid as a polar molecule) while simultaneously reversing metabolic acidosis. Other supportive measures include correcting the dehydration, maintaining kidney function (forcing fluids), rectifying electrolyte imbalance (especially potassium for hypokalemia, which may result from bicarbonate infusion), and instituting supportive measures for hyperthermia, seizure control, and pulmonary edema. Hemodialysis may be useful but is only indicated in salicylate-poisoned patients who present with severe acidosis

* See Chapter 3 for discussion of metabolic acidosis and determination of the anion gap.

TABLE 18.3
AAPCC Guidelines and Recommendations for Treatment of Acute Aspirin Toxicity

Evaluation of Toxicity	Treatment Referral or Evaluation	Grades of Recommendation[a]
Suspected self-harm or victims of potentially malicious administration	Immediate E.D. referral	"D"
Presence of typical symptoms: hematemesis, tachypnea, hyperpnea, dyspnea, tinnitus, deafness, lethargy, seizures, unexplained lethargy, confusion	E.D. for evaluation	"C"
Presence of nonspecific symptoms: unexplained lethargy, confusion, dyspnea	E.D. for evaluation—suggests development of chronic toxicity	"C"
No evidence of self-harm	E.D. for evaluation including dose, time of ingestion, symptoms, history, co-ingestants	"C"

Grade "C" (level 4) based on case series, single case reports (and poor-quality cohort and case control studies); Grade "D" (level 5) based on expert opinion without explicit critical appraisal, or based on physiology or bench research. See review articles for further details.

[a] Grades of Recommendation are based on level-of-evidence scores as developed by the Centre for Evidence-Based Medicine at Oxford University.

E.D.: emergency department; N.A.: not applicable or action taken regardless of dose.

or hypotension refractory to optimal supportive care, evidence of end organ injury (i.e., seizures, pulmonary edema), renal failure, or high serum aspirin concentrations (>100 mg/dl) despite a relatively stable metabolic picture. Chronic or severe acute toxicity is potentially fatal. Consequently, cardiovascular collapse, seizures, hyperthermia, and pulmonary edema are closely attended.

Specific antidotes are not available for salicylate poisoning.

18.4 NONSTEROIDAL ANTI-INFLAMMATORY AGENTS (NSAIDS)

18.4.1 History and Description

The search for potent anti-inflammatory agents similar to aspirin led to the discovery in 1849 of phenylbutazone, a synthetic analog and solubilizing agent for aminopyrine. It was initially used as an anti-inflammatory agent and subsequently for its analgesic and antipyretic actions. Because of the serious adverse hematologic reactions associated with phenylbutazone (agranulocytosis), the search for safer drugs with anti-inflammatory properties began. The acetic acid derivatives, including indomethacin, were introduced for the treatment of rheumatoid arthritis and related disorders. Several groups of aspirin-like drugs soon followed, many with significant advantages over aspirin but with a variety of accompanying adverse drug reactions (ADRs).

18.4.2 Classification, Pharmacology, and Clinical Use

Table 18.4 lists selected and currently available NSAIDs. NSAIDs exhibit antipyretic, analgesic, and anti-inflammatory activities resulting from the inhibition of prostaglandin synthesis. In particular, NSAIDS inhibit cyclooxygenase (COX), an enzyme that catalyzes the synthesis of prostaglandins from arachidonic acid (Figure 18.5). Two COX isoenzymes are present mainly in platelets, endothelial cells, gastrointestinal tract, and kidney. The selective functions of the enzymes, and their general inhibition by the NSAIDs, explains in part the toxic effects of the compounds. COX-1 isoenzyme activity produces prostaglandins that are important in maintaining platelet aggregation, regulation of kidney and gastric blood flow, and gastric mucus secretion. The COX-2 isoenzyme is expressed during pain and inflammation. While most NSAIDs nonselectively inhibit both enzymes, preferential inhibition of either contributes to both beneficial and adverse effects. The inhibition of COX-1 has antiplatelet activity, an effect which may be of benefit in the prevention of cardiovascular disease. However, the removal of the protective gastric mucosal secretory action exposes the stomach lining to GI toxicity. The newer COX-2 inhibitors avoid the GI upset associated with nonselective or selective COX-1 inhibitors. However, the antiplatelet activity and myocardial protective effects are precluded with the COX-2 inhibitors.

NSAIDs are indicated in the treatment of patients with osteoarthritis, rheumatoid arthritis, and arthritic conditions in patients with high risk of gastric or duodenal ulcers. Other indications include concomitant therapy with other arthritic treatments, mild to moderate pain of minor surgical procedures, and primary dysmenorrhea. Guidelines to assist in selecting the appropriate agent are mostly empirical and based on experience, the occurrence of ADRs, convenience, and cost. The major adverse reactions that limit NSAID usefulness are listed in Table 18.4. Other than the adverse reactions common to most of the NSAIDs, the highest incidence of ADRs includes GI disturbances (gastritis, heartburn); cardiovascular (hypertension, peripheral edema); CNS (dizziness, psychic disturbances); dermatologic (rash); hematologic (decreased hemoglobin and hematocrit); hepatic (elevated liver enzymes); renal (interference with renal circulation and urinary tract infection); and, respiratory disturbances (dyspnea and upper respiratory infection). Concomitant use with salicylates is not recommended.

18.4.3 Signs and Symptoms of Acute Toxicity

Most overdose exposures of NSAIDs are asymptomatic or produce self-limiting (24 h) lethargy. Nausea, vomiting, drowsiness, and potential renal ischemia and renal failure are possible, however, as a result of decrease in renal prostaglandin. Some consequences involve gastric erosion, CNS toxicity, and hemorrhage, although this toxicity requires up to 20 times the therapeutic blood concentrations.

18.4.4 Clinical Management of Acute Overdose

The ABCs of supportive care are generally warranted. Gastric lavage and administration of emetics are of benefit, as well as forced oral fluids and renal function tests.

TABLE 18.4
Representative and Currently Available Anti-Inflammatory Drugs

Classification	Compound	Proprietary Name	Enzyme Inhibition[a]	Major Toxic Effect[b]
Indole and indene acetic acids	Indomethacin	Indocin	Selective for COX-1	High incidence of peptic ulcers; CNS depression; hypersensitivity reactions; blood dyscrasias
	Sulindac	Clinoril		
Heteroaryl acetic acids	Tolmetin	Tolectin	Slightly selective for COX-1	High incidence of peptic ulcers; blood dyscrasias; hypersensitivity
	Diclofenac	Voltaren		
	Ketorolac	Toradol		
Arylpropionic acetic acids	Ibuprofen	Motrin, Advil	Slightly selective for COX-1	Gastric irritation, hepatic function impairment
	Naproxen	Naprosyn, Aleve		
	Flurbiprofen	Ansaid		
	Ketoprofen	Orudis		
	Oxaprozin	Daypro		
Anthranilic acids	Mefenamic acid	Ponstel	Slightly selective for COX-1	Gastric irritation, peptic ulcers, hepatic function impairment
	Meclofenamate	various		
Enolic acids	Piroxicam	Feldene	Selective for COX-1	High incidence of peptic ulcers; hypersensitivity
	Meloxicam	Mobic		
	Phenylbutazone[c]	Butalzolidin		
Alkanones	Nabumetone	Relafen	Slightly selective for COX-2	High incidence of peptic ulcers
Pyranocarboxylic acid	Etodolac	Lodine	Selective for COX-2	High incidence of peptic ulcers; hypersensitivity
COX-2 inhibitors	Celecoxib	Celebrex	Selective for COX-2	Gastric irritation; hypersensitivity
	Rofecoxib[d]	Vioxx		
	Valdecoxib[d]	Bextra		

[a] Action of anti-inflammatory drugs generally involves inhibition of both COX-1 and COX-2 enzymes—predominant inhibition of either enzyme is noted.
[b] Although major toxic reactions are listed categorically, some agents within the specific groups may preferentially display one or more of the toxic effects.
[c] Removed from the U.S. market principally because of high incidence of blood dyscrasias.
[d] Removed from the U.S. market because of high incidence of cardiovascular and cerebrovascular complications in patients with a predisposition to those conditions.

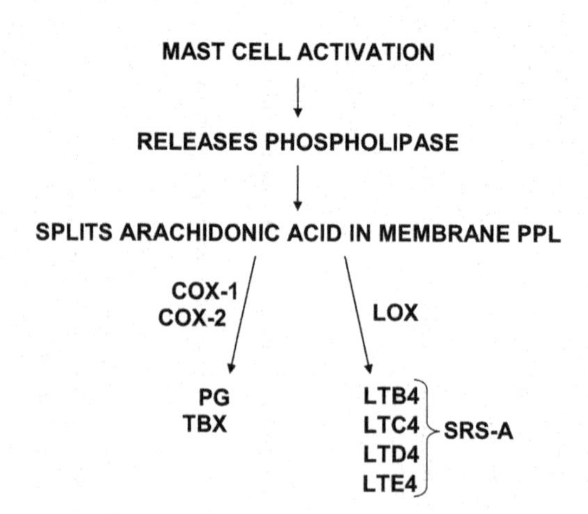

FIGURE 18.5 Schematic outline of synthesis of endoperoxides (mediators of inflammation) from arachidonic acid as a mechanism of the effector phase of type I antibody-mediated reactions.

18.5 METHODS OF DETECTION

Salicylates (acidic compounds) and acetaminophen (a neutral compound) are routinely monitored in suspected poisoning or cases involving toxicity. The methods of choice for both drugs include urine screening color test, gas chromatography, capillary electrophoresis, and solid-phase extraction (SPE) by a methylcellulose-immobilized-strong anion-exchanger (MC-SAX) HPLC system. In addition, testing for acetaminophen is routinely performed if codeine, oxycodone, hydrocodone, or propoxyphene is detected on initial urine screening.

Tables 3.7 and 3.8 show the schematic approach to testing of salicylates and APAP in biological specimens.

REFERENCES

SUGGESTED READINGS

Bastiaan-Vliegenthart, A.D., Antoine, D.J., and Dear, J.W., Target biomarker profile for the clinical management of paracetamol overdose. *Br. J. Clin. Pharmacol.* 80, 351, 2015.

Grosser, T., Smyth, E.M., and FitzGerald, G.A., Pharmacotherapy of inflammation, fever, pain and gout, in *Goodman & Gilman's the Pharmacological Basis of Therapeutics*, Brubton, L.L., Hilal-Dandan, R., and Knollman, B.C. (Eds.), 13th ed., McGraw-Hill, New York, 2018, chapter 38.

Gummin, D.D., Mowry, J.B., Spyker, D.A., Brooks, D.E., Osterthaler, K.M., and Banner, W., 2017 Annual Report of the American Association of Poison Control Centers' National Poison Data System (NPDS): 35th Annual Report. *Clin. Toxicol (Phila.).* 56, 1213, 2018.

Heard, K.J., Acetylcysteine for acetaminophen poisoning. *N. Engl. J. Med.* 359, 285, 2008.

Ingrasciotta, Y., Sultana, J., Giorgianni, F., Fontana, A., Santangelo, A., Tari, D.U., Santoro, D., Arcoraci, V., Perrotta, M., Ibanez, L., and Trifirò, G. Association of individual non-steroidal anti-inflammatory drugs and chronic kidney disease: a population-based case control study. *PLoS One*, 10, e0122899, 2015.

Mowry, J.B., Spyker, D.A., Cantilena, L.R. Jr., McMillan, N., and Ford, M., 2013 Annual Report of the American Association of Poison Control Centers' National Poison Data System (NPDS): 31st Annual Report. *Clin. Toxicol. (Phila.)*. 52, 1032, 2014.

Parker, W., Hornik, C.D., Bilbo, S., Holzknecht, Z.E., Gentry, L., Rao, R., Lin, S.S., Herbert, M.R., and Nevison, C.D., The role of oxidative stress, inflammation and acetaminophen exposure from birth to early childhood in the induction of autism. *J. Int. Med. Res.* 45, 407, 2017.

Rowden, A.K., Norvell, J., Eldridge, D.L., and Kirk, M.A., Updates on acetaminophen toxicity. *Med. Clin. N. Am.* 89, 1145, 2005.

Rowden, A.K., Norvell, J., Eldridge, D.L., and Kirk, M.A., Acetaminophen poisoning. *Clin. Lab. Med.* 26, 49, 2006.

REVIEW ARTICLES

Bello, A.E. and Holt, R.J., Cardiovascular risk with non-steroidal anti-inflammatory drugs: clinical implications. *Drug Saf.* 37, 897, 2014.

Chyka, P.A. Erdman, A.R., Christianson, G., Wax, P.M., Booze, L.L., Manoguerra, A.S., Caravati, E.M., Nelson, L.S., Olson, K.R., Cobaugh, D.J., Scharman, E.J., Woolf, A.D., and Troutman, W.G., Salicylate poisoning: an evidence-based consensus guideline for out-of-hospital management. *Clin. Toxicol. (Phila.)*. 45, 95, 2007.

Cranswick, N. and Coghlan, D., Paracetamol efficacy and safety in children: the first 40 years. *Am. J. Ther.* 7, 135, 2000.

Dawson, A.H. and Whyte, I.M., Therapeutic drug monitoring in drug overdose. *Br. J. Clin. Pharmacol.* 52, 97S, 2001.

Ghanem, C.I., Pérez, M.J., Manautou, J.E., and Mottino, A.D., Acetaminophen; from liver to brain: new insights into drug pharmacological action and toxicity. *Pharmacol. Res.* 109, 119, 2016.

Ishihara, K. and Szerlip, H.M., Anion gap acidosis. *Semin. Nephrol.* 19, 83, 1998.

Johnson, A.G., Seideman, P., and Day, R.O., Adverse drug interactions with nonsteroidal anti-inflammatory drugs (NSAIDs). Recognition, management and avoidance. *Drug Saf.* 8, 99, 1993.

Karsh, J., Adverse reactions and interactions with aspirin. Considerations in the treatment of the elderly patient. *Drug Saf.* 5, 317, 1990.

Lane, J.E., Belson, M.G., Brown, D.K., and Scheetz, A., Chronic acetaminophen toxicity: a case report and review of the literature. *J. Emerg. Med.* 23, 253, 2002.

Major, J.M.,Zhou, E.H., Wong, H.L., Trinidad, J.P., Pham, T.M., Mehta, H., Ding, Y., Staffa, J.A., Iyasu, S., Wang, C., and Willy, M.E., Trends in rates of acetaminophen-related adverse events in the United States. *Pharmacoepidemiol. Drug Saf.* 25, 590, 2016.

Moncada, S. and Vane, J.R., Mode of action of aspirin-like drugs. *Adv. Intern. Med.* 24, 1, 1979.

O'Malley, G.F., Emergency department management of the salicylate-poisoned patient. *Emerg. Med. Clin. North Am.* 25, 333, 2007.

Prescott, L.F., Paracetamol: past, present, and future. *Am. J. Ther.* 7, 143, 2000.

Riordan, S.M. and Williams, R., Alcohol exposure and paracetamol-induced hepatotoxicity. *Addict. Biol.* 7, 191, 2002.

Shah, A.D., Wood, D.M., and Dargan, P.I., Understanding lactic acidosis in paracetamol (acetaminophen) poisoning. *Br. J. Clin. Pharmacol.* 71, 20, 2011.

Sheen, C.L., Dillon, J.F., Bateman, D.N., Simpson, K.J., and MacDonald, T.M., Paracetamol toxicity: epidemiology, prevention and costs to the health-care system. *Q.J.M.* 95, 609, 2002.

Stein, R.B. and Hanauer, S.B., Comparative tolerability of treatments for inflammatory bowel disease. *Drug Saf.* 23, 429, 2000.

Vane, J.R. and Botting, R.M., Anti-inflammatory drugs and their mechanism of action. *Inflamm. Res.* 47, S78, 1998.

19 Anabolic–Androgenic Steroids

19.1 THE ENDOCRINE SYSTEM

The endocrine (ductless) glands secrete metabolically active chemical mediators into the vascular circulation, which in turn, are distributed to all organs.* The secreted hormones function as chemical mediators between endocrine glands and specific effector organs, regulating metabolic processes and physiological growth and development. In contrast to the neurochemicals of the nervous system, endocrine hormones affect all somatic and autonomic cells, require a longer onset of action, have prolonged duration, and are delivered systemically rather than locally in response to nerve impulses.

19.2 NEUROENDOCRINE PHYSIOLOGY

19.2.1 Description

The endocrine system is a conglomeration of physiological organs working interdependently and follows a complex series of feedback reactions. Normal endocrine function depends on cyclical events with stimuli originating both externally and internally. The goal of the endocrine organs, in coordination with the nervous system, is to maintain homeostasis (equilibrium) throughout the body. The nervous system coordinates the stimulation or inhibition of the release of hormones in synchronization with the endocrine glands, thus forming a communicating network of signals in response to physiological requirements. This pathway allows the neuroendocrine superstructure to regulate growth and development in adolescence as well as adulthood while maintaining physiological homeostasis.

The hypothalamic–pituitary system, through auto-feedback somatosensory control centers, regulates the secretion of hormones in response to physiological, emotional, or physical stress.† The hypothalamus synthesizes and secretes at least nine different hormones. The hormones travel from the hypothalamus through the hypothalamo-hypophyseal portal system, or neuronal tracts, to the anterior pituitary (adenohypophysis) or posterior pituitary (neurohypophysis), respectively. The hormones stimulate the respective areas of the pituitary gland to release any of seven distinct hormones, which respond to the initial stimulus in an attempt to counteract

* In some instances, mixed glands, such as the pancreas and liver, also serve exocrine functions, since they secrete chemical mediators into the lumen of organs through duct systems.

† It is important to note that the term *physiological stress* implies the existence of normal stimuli (nonpathologic) that challenge the maintenance of homeostasis. These stressors include conditions such as cold, exercise, pain, tiredness, and hunger as well as daily *routine* emotional and behavioral pressures.

or balance that force. The regulatory negative or positive feedback system ultimately adjusts all aspects of homeostasis, growth, development, and metabolism.

Disorders of the endocrine system are expressed in the form of excessive (hyperfunction) or diminished (hypofunction) activity. In general, disorders develop as a result of lack of hypothalamic–pituitary secretory control or because of inadequate or dysfunctional hormone receptors. The agents described in section 19.3.2 have therapeutic usefulness in alleviating or correcting endocrine imbalances. In addition, estrogens and progestins are clinically useful drugs for a variety of metabolic disorders, including the prevention of fertilization and implantation associated with birth control. The ubiquitous illicit use of androgens and anabolic steroids, however, taken in doses beyond their therapeutic value for nontherapeutic reasons, has contributed to an array of chronic toxic syndromes previously unrecognized in routine treatment.

19.3 ANABOLIC–ANDROGENIC STEROIDS (AAS)

19.3.1 Definition and Incidence

Androgens are hormones that stimulate the development and activity of accessory male sex organs (primary sex characteristics) and encourage the development of male sex characteristics (secondary sex characteristics). Androgens prevent changes in the latter that follow a decrease in the hormone levels, such as with castration. *Anabolic steroids* are familiar as synthetic derivatives of the androgens that promote skeletal muscle growth (anabolic effect) and the development of sex characteristics (androgenic effect). Since its discovery in 1935, testosterone, the principal male androgen, and its numerous derivatives have been used as an important component of the treatment of endocrine-related disorders. As the compound's ability to facilitate the growth of skeletal muscle in laboratory animals was understood, testosterone and related agents were incorporated into the lifestyle of bodybuilders, weightlifters, and other athletes, professional or otherwise. Since the 1950s, athletes have ingested AAS* to increase muscle size, reduce body fat, and improve general performance in sports.

Today, steroid abuse is widespread, especially among adolescents. Roughly 0.6% to 1.1% of 8th-, 10th-, and 12th-graders have taken anabolic steroids at least once in their lives. In addition, lifetime steroid use among 10th- and 12th-graders slightly decreased from 2016 to 2017. Significant reductions have been noted since 2001 for lifetime and past-year use among all grades, as well as for past-month use among 8th- and 10th-graders. However, changes in drug misuse have relatively insignificant since 2014 (NIDA, 2017).

Individuals who have experienced physical or sexual abuse state that the reasons for abusing steroids are to increase muscle size for self-protection. Some adolescents abuse steroids as part of a pattern of high-risk behaviors. Furthermore, muscle dysmorphia is a behavioral syndrome characterized by a distorted, unrealistic

* The term *steroids* will be subsequently used in this chapter to refer collectively to the androgenic–anabolic steroid compounds.

self-image. Men and women with the condition use steroids in an attempt to overcome their perception of inadequate physical appearance.

19.3.2 Pharmacology and Clinical Use

Table 19.1 shows the structure of testosterone and lists some commonly abused steroids. More than 100 different steroids have been developed and, with the exception of the dietary supplements, are classified as federal C-III controlled substances under the Anabolic Steroids Act of 1990. Dietary supplements, such as dehydroepiandrosterone (DHEA) and androstenedione (ANDRO), are not controlled substances and can be purchased legally without a prescription. Interestingly, these agents are not food products but are converted to testosterone, although the quantities available for conversion are probably insufficient to produce the desirable anabolic effects. Their adverse reactions and potential for toxicity, however, remain to be determined.

Testosterone is produced and secreted by the interstitial cells (Leydig cells) of the testes. The compound is converted at effector organs to dihydrotestosterone, which then binds to target cytosolic protein receptors. The effects of testosterone include growth and maturation of male sex organs (prostate, seminal vesicles, penis, and scrotum), development of male hair distribution, laryngeal enlargement, and alterations in body musculature and fat distribution. Androgens increase protein anabolism.

Androgenic steroids are indicated for a variety of endocrine disorders, including the treatment of hypogonadism, delayed puberty in males, and metastatic cancer in females. Androgens stimulate adolescent growth and precipitate termination of linear growth by fusion of skeletal epiphyseal growth centers. Exogenous androgen administration inhibits endogenous testosterone release through a pituitary luteinizing hormone (LH) feedback pathway.

Anabolic steroids are closely related to testosterone and have high anabolic, low androgenic properties. These compounds are indicated for the treatment of certain types of anemia, hereditary angioedema, and metastatic breast cancer. Anabolic steroids promote tissue-building and reverse tissue-depleting processes.

19.3.3 Adverse Reactions

AAS affect many organ systems as a result of suppression of normal endocrine function as well as their direct actions on target cell processes. Steroids cause serious disturbances of growth and sexual development in children and suppress gonadotropic functions of the pituitary. In particular, virilization is the most common undesirable side effect, especially in women and prepubertal males.

Table 19.1 summarizes the potential adverse reactions exhibited with prolonged* illicit use of AAS. In addition, females may experience irreversible virilization and clitoral enlargement. Females may also develop other sexual characteristics generally associated with male development, such as hirsutism (increased body hair

* Several months of frequent use is regarded as prolonged administration of steroids and frequently results in irreversible changes.

TABLE 19.1
Structure of Testosterone, Classification of Selected, Commonly Used Androgenic-Anabolic Testosterone Derivatives and Their Adverse Drug Reactions (ADRs)

Testosterone

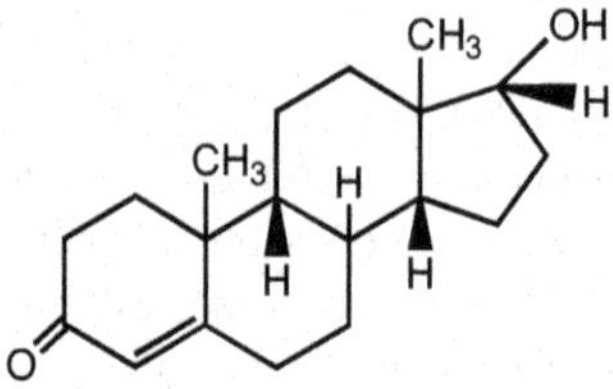

Classification[a]	Compound	Proprietary Name	ADRs
Oral steroids	Oxymethalone Oxandrolone Methadrostenolone Stanozolol	Anadrol-50 Oxandrin Dianabol Winstrol	*CNS*: headache, anxiety, depression, paresthesias; *dermatologic*: acne, male-pattern baldness, hirsutism; *GI*: nausea, hepatocellular neoplasms, abnormal liver function tests, cholestatic jaundice; *GU*: frequent and prolonged erections, gynecomastia, oligospermia; *metabolic*: electrolyte imbalance, increased serum cholesterol
Injectables	Nandrolone decanoate Nandrolone phenpropionate Testosterone cypionate Boldenone undecylenate	Deca-Durabolin Durabolin Depo-testosterone Equipoise	Same as for oral steroids plus pain and inflammation at injection site, seborrhea
Transdermal	Testosterone transdermal system	Testoderm, Androderm	Same as for oral steroids plus *CNS*: pain, dizziness, vertigo; *dermatologic*: itching, erythema, burning, pruritis, blisters at site of application; *GU*: prostate disorders

(Continued)

TABLE 19.1 (CONTINUED)
Structure of Testosterone, Classification of Selected, Commonly Used Androgenic-Anabolic Testosterone Derivatives and Their Adverse Drug Reactions (ADRs)

Classification[a]	Compound	Proprietary Name	ADRs
Implants	Testosterone pellets	Testopel	Same as for oral steroids plus pain and inflammation at injection site
Topical	Testosterone gel	Androgel 1%	Same as for oral steroids plus *CV*: hypertension, peripheral edema; *CNS*: asthenia, emotional instability; *dermatologic*: application site reactions; *GI*: none; *GU*: impaired urination, prostate disorders; *metabolic*: abnormal clinical laboratory tests

[a] The oral steroids and the nandrolone derivatives are further classified as high anabolic–low androgenic steroids, although the dissociation of these activities varies; the remaining compounds have predominantly androgenic properties.

ADRs: adverse drug reactions; CNS: central nervous system; CV: cardiovascular; GI: gastrointestinal; GU: genitourinary.

growth) and deepening of the voice. Menstrual irregularities, male-pattern baldness, acne, and changes in libido are reversible.

More severe warnings associated with steroid use include the development of *peliosis hepatis*—a liver condition characterized by the replacement of liver cells with blood-filled cysts. The disorder is reversible on discontinuation of drug use. Peliosis hepatis usually presents with minimal, unrecognized hepatic dysfunction until life-threatening liver failure or intra-abdominal hemorrhage occurs. Benign, androgen-dependent hepatic tumors have been noted. The vascularized tumors are relatively uncommon neoplastic occurrences in humans in the absence of steroid use. With continuous misuse of steroids, the tumors develop undetected until life-threatening intra-abdominal hemorrhage ensues. Increased risk of atherosclerosis and coronary artery disease is associated with chronic abuse resulting from blood lipid changes (decreased high-density lipoproteins). Hypopituitarism results from AAS abuse and is characterized by diminution or cessation of function of the adenohypophysis. Major consequences occur as a result of variable deficiency of gonadotropins (sex hormones), somatotropins (growth hormones), thyrotropin (thyroid hormones), and corticotropin (corticosteroid hormones). Table 19.2 summarizes general and age- and gender-specific effects of AAS abuse.

Steroid abusers often share contaminated needles or use steroid preparations manufactured illegally under unsanitary conditions. Such practices predispose persons to viral and bacterial infections such as bacterial endocarditis, human immunodeficiency virus (HIV), and hepatitis B.

Serious consequences of steroid administration in children and teenagers include premature epiphyseal maturation and closure without compensatory gain in linear growth. The result is a disproportionate advancement in bone maturation and termination of the growth process. This effect may continue up to 6 months following the cessation of therapeutic regimens. Ingestion or dermal contact with AAS is contraindicated during pregnancy (category X) because of potential fetal masculinization.

In general, the ability of these agents to improve athletic performance is questionable. For bodybuilders and weightlifters, the motivation to use steroids relies primarily on the increased muscle mass, decreased muscle recovery, and healing time experienced with steroid regimens. The regimens most often used include *stacking*, *pyramiding*, and *cycling*.* The goal of the regimens is to minimize side effects and drug detection while maintaining peak performance.

TABLE 19.2
General and Age- and Gender-Specific Effects of AAS Abuse

Category	Adverse Effects
Men	Shrinking of the testicles, oligospermia[a], infertility, alopecia[b], gynecomastia[c], increased risk for prostate cancer
Women	Hirsutism[d], male-pattern baldness, amenorrhea or oligomenorrhea[e], virilization[f], enlargement of the clitoris
Adolescents	Hypopituitarism, premature skeletal maturation and accelerated puberty changes
General	Hepatic damage, jaundice, edema, hypertension, hypercholesterolemia, renal failure, severe acne, tremors, and nervousness

[a] Reduced number of spermatozoa in the semen.
[b] Hair loss.
[c] Excessive growth of the male mammary glands, including development to the stage at which milk is produced.
[d] Abnormal hair growth, especially an adult male pattern of hair distribution in women.
[e] Absence or abnormal stoppage of the menses, respectively.
[f] Masculinization, especially that occurring inappropriately in a female or prepubertal male.

* *Stacking* refers to the concurrent and simultaneous use of two or more agents while varying the dosage and dosage forms over a period of 6 to 18 weeks; *pyramiding* is similar to stacking but involves the use of a single agent; *cycling* requires the scheduling of a drug-free period so that peak performance and undetectable levels will coincide with an upcoming event.

19.3.4 Addiction and Withdrawal Syndrome

An addiction syndrome is now recognized with long-term use, especially among bodybuilders, weightlifters, and professional athletes. The syndrome is characterized by physical, behavioral, and psychological changes (especially aggressiveness). The condition is not unlike the addictive and withdrawal behavior, drug craving, and signs and symptoms described for opioids, cocaine, and alcohol. Depression is the most dangerous of the withdrawal symptoms because of its association with suicidal tendencies. If left untreated, the condition persists for a year or more after discontinuation. Concomitant drug use, such as ingestion of diuretics to counteract sodium and fluid retention (nonprescription or thiazide and loop diuretics), often indicates chronic abuse.

19.3.5 Treatment of the Consequences of Chronic Steroid Use

Behavioral, supportive, and educational programs are indicated to prevent and manage the growing adolescent problem of steroid abuse. Medications used for treating steroid withdrawal aim at restoring hormonal balance, while others target specific withdrawal symptoms. Recommendations include antidepressants for depression and analgesics for headaches and muscle and joint pain.

19.4 ESTROGEN AND PROGESTINS

19.4.1 Physiology

Estrogens are secreted by the developing ovarian follicular cells as cholesterol derivatives in response to follicle stimulating hormone (FSH) and luteinizing hormone (LH). FSH and LH are synthesized and released from the anterior pituitary in response to gonadotropin releasing hormone (GnRH) secreted by the hypothalamus. This feedback system, balanced on the circulating levels of estrogens and progestins, maintains the rhythm of the normal monthly reproductive cycle.

The three principal estrogens in female plasma* are 19-β-estradiol, estrone, and estriol. Physiologically, estrogens promote the development and maintenance of female reproductive structures, including the vagina, uterus, fallopian tubes, and mammary glands. Estrogens are necessary for the maturation of secondary female sex characteristics.† As with testosterone, estrogens work in conjunction with human growth hormone (hGH) to increase protein anabolism for muscle and bone growth. This action indirectly contributes to the shaping of the skeleton and epiphyses of the long bones that allows the pubertal growth spurt and termination. The involvement of estrogens in lowering coronary heart disease in females under 50 years old is attributable to its ability to decrease blood cholesterol.

* It is important to note that estradiol also circulates in human male plasma and is usually undetectable, since most of the free precursor is used to synthesize testosterone.

† The characteristics include distribution of adipose tissue in the breasts, hips, abdomen, and mons pubis and the development of a broad pelvis, all of which prepare for pregnancy. In addition, estrogens regulate the voice pitch and hair growth pattern peculiar to females.

Progesterone, the principal progestin, is secreted by the corpus luteum after ovulation (about day 14 of a typical menstrual cycle). It acts synergistically with estrogen to repair, stimulate proliferation, and prepare the endometrium for implantation of the fertilized ovum. Consequently, it encourages the transformation of proliferative endometrium into secretory endometrium. Progesterone maintains pregnancy once implantation occurs, preventing spontaneous uterine smooth muscle contractions. Increased levels of estrogen and/or progesterone inhibit the secretion of GnRH and LH through the negative feedback pathway, which is the basis of the pharmacological activity of oral contraceptives (OCs; see next subsection). Table 19.3 summarizes the uses, proprietary names, and structures of the estrogens and progestins.

19.4.2 Pharmacology and Clinical Use

Estrogens are most commonly used in combination with progestins as OCs or as replacement therapy in postmenopausal women. Concerning the latter, estrogens are indicated for vasomotor symptoms associated with menopause—that is, for the relief of signs and symptoms associated with the decline of ovarian function. Symptoms include hot flushes, variable periods of amenorrhea, endometrial atrophy, and decrease of myometrial and vaginal epithelial mass. Estrogen replacement therapy is of value for the treatment of atrophic vaginitis and for the prevention of osteoporosis and coronary heart disease in postmenopausal women. Estrogens are also indicated as palliative therapy for advanced prostatic carcinoma in men and breast cancer in women. Unlike the AAS, estrogens and progestins do not demonstrate illicit abuse potential.

The principal clinical uses of progestational agents are mostly related to the maintenance of pregnancy or regulation of the second half of the menstrual cycle. Progestins are indicated for amenorrhea, abnormal uterine bleeding, endometriosis,* AIDS wasting syndrome (appetite enhancement), infertility, and breast or endometrial carcinoma.

For the prevention of pregnancy, oral and injectable contraceptives contain estrogen/progesterone combinations or progesterone alone. In addition, the combinations are used for emergency contraception, for long-term (1 and 3 month) contraception, and in the treatment of acne vulgaris in females 15 years of age or older†.

19.4.3 Clinical Toxicity of Prolonged Estrogen and/or Progestin Administration

Decades of clinical use of estrogens, progestins, and combinations have resulted in the accumulation of a vast amount of data related to adverse reactions of this class of substances, summarized in Table 19.2. In particular, the use of estrogens in

---

* An abnormal gynecologic condition characterized by ectopic growth and function of endometrial tissue. Serious consequences occur when abnormal growth interferes with pelvic organs.

† Interestingly, OCs are not associated with an increase in the risk of developing breast or cervical cancer. In fact, there may be a protective effect, as users appear half as likely to develop ovarian and endometrial cancer as women who have never used OCs.

TABLE 19.3
Structure, Classification, and Properties of Selected, Commonly Used Estrogen and Progesterone Derivatives

Testosterone

Classification	Compound	Proprietary Name	Indications	ADRs
Estrogens	Estradiol	Estrace	Moderate to severe vasomotor symptoms associated with menopause, primary ovarian failure (various causes), atrophic vaginitis, and other	Induction of malignant neoplasms (see text), thromboembolic disease, elevated BP, gall bladder disease, hypercalcemia, fluid retention, endometrial hyperplasia, depression, glucose intolerance, psychiatric, dermatologic, GI and GU; pregnancy— category X[a]
	Estrone	Kestrone 5		
	Estradiol transdermal systems	Climara, Estraderm		
	Estradiol valerate in oil	Delestrogen		
	Conjugated estrogens	Premarin		
	Esterified estrogens	Estratab		
	Vaginal estrogens	Premarin		
	Estrogen ring insert	Estring		
Progestins	Progesterone	Prometrium, Progesterone in Oil, Progesterone Gel	Amenorrhea, abnormal uterine bleeding, AIDS wasting syndrome, infertility, endometriosis; gel is used to support embryo implantation; megestrol is used in the palliative treatment of advanced breast or endometrial carcinoma	Congenital anomalies when administered during first 4 months of pregnancy, ophthalmic effects; rest of ADRs similar to estrogens
	Medroxy-progesterone	Provera		
	Norethindrone	Aygestin		
	Megestrol	Megace		

(Continued)

TABLE 19.3 (CONTINUED)
Structure, Classification, and Properties of Selected, Commonly Used Estrogen and Progesterone Derivatives

Classification	Compound	Proprietary Name	Indications	ADRs
Combinations	Estrogen/progesterone tablets and patches	Prempro, Premphase	As with individual components above	As above
		Alesse, Ovral, and various others	Oral contraceptives	As above

[a] Category X: should not be used during pregnancy.
BP: blood pressure; GI: gastrointestinal; GU: genitourinary.

replacement therapy is associated with an increased risk of endometrial carcinoma in postmenopausal women. In fact, recent studies report that routine replacement therapy may not be as beneficial for the prevention of osteoporosis as previously believed. In addition, long-term estrogen ingestion produces a variety of unwarranted metabolic, endocrine, and hematologic effects. Similarly, progestins are accompanied by serious metabolic side effects. Progestins have been associated with congenital anomalies because of unnecessary use during the first trimester of pregnancy for the prevention of habitual or threatened abortions. Finally, the most significant deleterious effects of continuous use of OCs appear to be the elevated risk of cardiovascular (CV) adverse reactions. The risk from CV mortality increases about 10-fold when OCs are used concomitantly with cigarette smoking and increasing age (over 35). In general, women who use OCs should refrain from smoking.

REFERENCES

SUGGESTED READINGS

Almeida, M. Laurent, M.R., Dubois, V., Claessens, F., O'Brien, C.A., Bouillon, R., Vanderschueren, D., and Manolagas, S.C., Estrogens and androgens in skeletal physiology and pathophysiology. *Physiol. Rev.* 97, 135, 2017.

Baggish, A.L. et al., Cardiovascular toxicity of illicit anabolic-androgenic steroid use. *Circulation* 135, 1991, 2017.

Dluzen, D.E. and McDermott, J.L., Estrogen, testosterone, and methamphetamine toxicity. *Ann. N.Y. Acad. Sci.* 1074, 282, 2006.

Frati, P., Busardò, F.P., Cipolloni, L., Dominicis, E.D., and Fineschi, V., Anabolic androgenic steroid (AAS) related deaths: autoptic, histopathological and toxicological findings. *Curr. Neuropharmacol.* 13, 146, 2015.

Furlanello, F., Serdoz, L.V., Cappato, R., and De Ambroggi, L., Illicit drugs and cardiac arrhythmias in athletes. *Eur. J. Cardiovasc. Prev. Rehabil.* 14, 487, 2007.

Ricke, W.A., Wang, Y., and Cunha, G.R., Steroid hormones and carcinogenesis of the prostate: the role of estrogens. *Differentiation* 75, 871, 2007.

Sanderson, J.T., The steroid hormone biosynthesis pathway as a target for endocrine-disrupting chemicals. *Toxicol. Sci.* 94, 3, 2006.

REVIEW ARTICLES

Bailey, K., Yazdi, T., Masharani, U., Tyrrell, B., Butch, A., and Schaufele, F., Advantages and limitations of androgen receptor-based methods for detecting anabolic androgenic steroid abuse as performance enhancing drugs. *PLoS One* 11(3), e0151860, 2016.

Crandall, C., Combination treatment of osteoporosis: a clinical review. *J. Womens Health Gend. Based Med.* 11, 211, 2002.

Cristoforo Pomara, C., Neri, M., Bello, S., Fiore, C., Riezzo, I., and Turillazzi, E., Neurotoxicity by synthetic androgen steroids: oxidative stress, apoptosis, and neuropathology: a review. *Curr. Neuropharmacol.* 13, 132, 2015.

Di Luigi, L., Romanelli, F., and Lenzi, A., Androgenic-anabolic steroids abuse in males. *J. Endocrinol. Invest.* 28, 81, 2005.

Dossing, M. and Sonne, J., Drug-induced hepatic disorders. Incidence, management and avoidance. *Drug Saf.* 9, 441, 1993.

Hall, R.C., Abuse of supraphysiologic doses of anabolic steroids. *South Med. J.* 98, 550, 2005.

Hoyer, P.B., Reproductive toxicology: current and future directions. *Biochem. Pharmacol.* 62, 1557, 2001.

Joosten, H.F., van Acker, F.A., van den Dobbelsteen, D.J., Horbach, G.J., and Krajnc, E.I., Genotoxicity of hormonal steroids. *Toxicol. Lett.*, 151, 113, 2004.

Jordan, A., Toxicology of progestogens of implantable contraceptives for women. *Contraception* 65, 3, 2002.

Kanayama, G. and Pope, H.G. Jr. Illicit use of androgens and other hormones: recent advances. *Curr. Opin. Endocrinol. Diabetes. Obes.* 19(3), 211–219, 2012.

Kicman, A.T., Pharmacology of anabolic steroids. *Br. J. Pharmacol.* 154, 502, 2008.

Lane, J.R. and Connor, J.D., The influence of endogenous and exogenous sex hormones in adolescents with attention to oral contraceptives and anabolic steroids. *J. Adolesc. Health* 15, 630, 1994.

NIDA Research Report Series, Anabolic Steroid Abuse, National Institute on Drug Abuse, DHHS, Public Health Service, 2017. www.nida.nih.gov/infofacts/steroids, last viewed March 2019.

Simon, J.A., Safety of estrogen/androgen regimens. *J. Reprod. Med.* 46, 281, 2001.

Wood, R.I., Anabolic-androgenic steroid dependence? Insights from animals and humans. *Front. Neuroendocrinol.* 29, 490, 2008.

20 Cardiovascular Toxicology

20.1 EPIDEMIOLOGY

About 735,000 persons in the United States suffer myocardial infarction (MI; acute heart attack) every year, of whom 525,000 experience it for the first time. An estimated 610,000 American deaths occur due to one or more forms of cardiovascular (CV) disease (CVD) every year, accounting for one in four deaths (U.S. CDC, 2019). CVD includes coronary heart disease (CHD), congestive heart failure (CHF), hypertension (HT), stroke, and congenital CV defects. According to the American Heart Association (2018), 720,000 Americans annually will have a new coronary event and approximately 335,000 will have a recurrent event.* However, the mortality associated with CHD has dropped by 34.4% from 2005 to 2015. Median survival of those over 45 years of age after first MI is 8.4 years for white males, 5.6 years for white females, 7.0 years for black males, and 5.5 years for black females (Mozafarian et al., American Heart Association, 2015). In 2013–2014, the annual direct and indirect cost of CVD and stroke in the United States was an estimated $329.7 billion.

20.2 CARDIOVASCULAR (CV) PHYSIOLOGY

20.2.1 CV Functions

The important functions of the CV system include

1. Maintenance of the cardiac pump, including the rhythmic nature of electrical signals, force of contraction, and magnitude of discharge pressure;
2. Preservation of vascular integrity, including muscular tone and structural integrity of vessel walls, and tight regulation of blood pressure;
3. Conservation of blood volume and composition, including water and electrolyte balance, lipid composition, and coagulation.

The extensive network of blood flow is initiated through the vasculature that extends between the heart and the peripheral tissues. The vasculature is divided into a pulmonary circuit, which transports blood through the gas exchange surfaces of the lungs, and a systemic circuit, which carries blood throughout the rest of the body. To maintain the complete circuit continuously from the pulmonary to the systemic circulation, venous blood returns to the heart and must complete the pulmonary circuit to become oxygenated before reentering the systemic circuit.

* Defined as first hospitalized MI or CHD.

20.2.2 Cardiac Circulation

To maintain CV integrity, the pulmonary and systemic circulations are connected through the pumping action of the heart. The heart fulfills this function by virtue of the presence of four muscular chambers: the right and left atria and the right and left ventricles. Venous blood from the systemic circulation enters the right atrium and right ventricle, from which it is then pumped into the lungs to become oxygenated. The left atrium and ventricle then receive the oxygenated blood via pulmonary veins and eject it to the systemic circulation via the aorta. In addition, the heart receives the oxygenated blood from the coronary arteries, which begin at the root of the aorta.

20.2.3 Electrophysiology

Each cardiac contraction cycle consists of an orchestrated series of events that generate the contraction of the cardiac muscle. Normal electrophysiological properties of cardiac muscles contribute to the coordinated action of the atria and ventricles. Two types of cardiac muscle cells are involved in a single contraction:

1. *Contractile* cells responsible for initiating blood circulation.
2. *Specialized* neuronal cells of the conducting system that control and coordinate the activities of the contractile muscle cells.

The functionality of heart muscle, including the conductivity and contractility of cardiac muscle cells, is regulated through membrane transport of various ions, which in turn, forms the basis of an action potential. As illustrated in Figure 20.1, the action potential is a physiological event that coordinates electrical conductivity through the heart muscle, as initiated by passage of ions through cellular membranes, with the contraction of muscle chambers. The action potential consists of four characteristic phases:

1. *Resting.* The membrane potential in a resting cell is about –90 mV (resting membrane potential, RMP). There is no membrane polarization on the muscle cell at this stage, and the muscle is not contracting.
2. *Phase 1.* At the start of an action potential, voltage-gated Na^+ channels open, leading to a rapid influx of Na^+; this results in an intracellular depolarization from approximately –90 mV to greater than 0 mV.
3. *Phase 2.* Rapid repolarization immediately follows Phase 1 and is initiated by closure of the Na^+ channels and opening of voltage-gated transient outward K^+ channels. During this phase, voltage-gated Ca^{2+} channels open, and Ca^{2+} ions begin a slower but prolonged influx at about –30 mV, producing an inward Ca^{2+} current. As the K^+ current dissipates, Ca^{2+} continues to enter the cell, and Phase 2 reaches a plateau.
4. *Phase 3.* Final repolarization of the cell results from closure of Ca^{2+} channels followed by restoration of resting K^+ currents; the resting membrane potential (RMP) is reestablished.

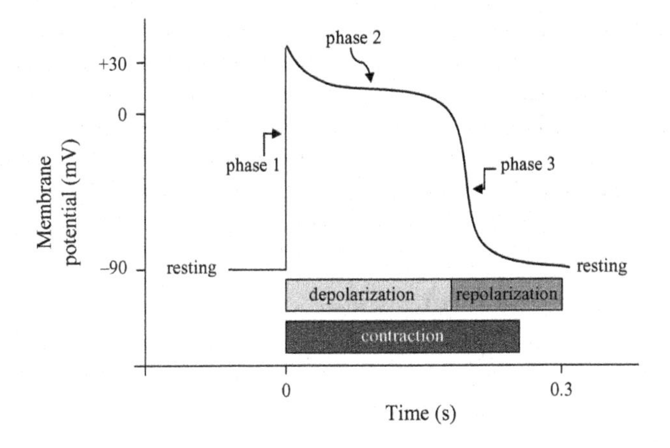

FIGURE 20.1 Action potential in cardiac muscle cells.

Following the action potential, normalization of the Na^+ and K^+ concentration gradients is achieved through the action of sarcolemma Na^+-K^+-ATPase. In addition, Ca^{2+} must be effluxed from the cellular cytoplasm or sequestered into intracellular stores, primarily through the sarcoplasmic reticulum (SR).

The appearance of an action potential in cardiac muscle cell membrane is coupled with a muscle contraction. This process occurs in two steps:

1. Ca^{2+} enters the cell during Phase 2 of the action potential and provides approximately 20% of Ca^{2+} required for contraction.
2. The influx of extracellular Ca^{2+} triggers the release of additional Ca^{2+} from the SR (i.e., Ca^{2+}-induced Ca^{2+} release). As a result, muscle cell contraction continues through Phases 2 and 3 (depolarization; Figure 20.1). Intracellular Ca^{2+} is then sequestered by the SR or pumped out of the cell, leading to cardiac muscle relaxation (repolarization; Figure 20.1).

20.2.4 The Conducting System

The cardiac conducting system is a network of specialized cardiac muscle cells that initiates and distributes electrical impulses. The conducting system consists of

1. The sinoatrial (SA) node, located in the wall of the right atrium;
2. The atrioventricular (AV) node, located at the junction between the atria and the ventricles; and
3. The conducting cells that interconnect the two nodes and distribute the contractile stimulus throughout the myocardium.

Conducting cells in the atria are found in the internodal pathways, which distribute the contractile stimulus to atrial muscle cells as the impulse travels from the SA node to the AV node. The ventricular conducting cells include those in the AV bundle and the bundle branches as well as the Purkinje fibers, which distribute the stimulus to the ventricular myocardium. The cells in the conducting system share an important feature: they cannot maintain a stable resting potential. Each time repolarization occurs, the membrane potential again gradually drifts toward a threshold. The rate of spontaneous depolarization varies in the different portions of the conducting system. It is fastest at the SA node, principally because of the constant leaking of Na^+ ions in these specialized cells, giving the SA node the distinctive title of *cardiac pacemaker.*

20.2.5 Electrocardiography

Electrical activities occurring in the heart are powerful enough to be detected by electrodes placed on the body surface. A recording of these electrical activities constitutes an electrocardiogram (ECG) (Figure 20.2). The ECG has the following components:

1. The small *P wave* results from simultaneous depolarization of the right and left atria.
2. The *QRS complex* appears with ventricular depolarization.
3. The small *T wave* indicates ventricular repolarization.

The ECG is an important tool for diagnosis of cardiac function and is particularly useful in detecting and diagnosing cardiac arrhythmias. Cardiac toxicity of therapeutic agents, especially CV drugs, is often reflected by changes in ECG.

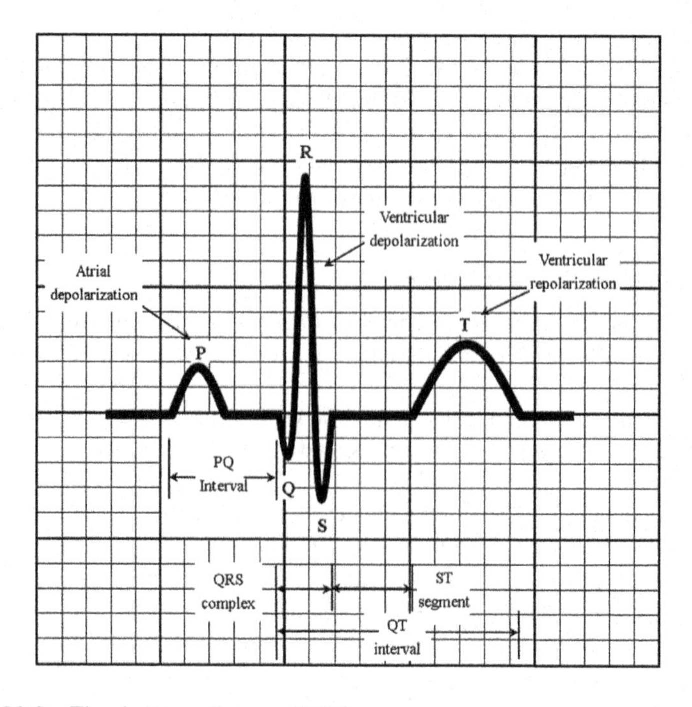

FIGURE 20.2 The electrocardiogram (ECG).

20.3 DIGITALIS GLYCOSIDES

Digitalis glycosides (DGs) have played a prominent role in the therapy of congestive heart failure since William Withering codified their use in his late eighteenth-century monograph on the efficacy of the leaves of the common foxglove plant (*Digitalis purpurea*). Digoxin in particular remains widely used today in the face of increasing rates of heart failure and atrial fibrillation and despite the emergence of newer medications. Its hemodynamic, neurohormonal, and electrophysiologic actions make it a suitable adjunctive, evidence-based therapy for these conditions. Its narrow therapeutic index and toxicity, however, have become more relevant as aging, comorbid diseases, and multiple drug ingestion increase patient population vulnerability.

20.3.1 MEDICINAL CHEMISTRY

DGs possess a common molecular motif—a steroid nucleus containing an unsaturated lactone at the C17 position and three glycoside residues at the C3 position. Digoxin (Figure 20.3) differs from digitoxin (Figure 20.4) only by the presence of a hydroxyl group at C12, which makes digitoxin more lipophilic than digoxin. Although digitoxin and ouabain are still available, digoxin is the most widely prescribed drug of this class in the United States. This is principally because the techniques for measuring serum levels are readily available, it possesses flexible routes of administration and an intermediate duration of action, and the drug has enjoyed a long history of therapeutic use.

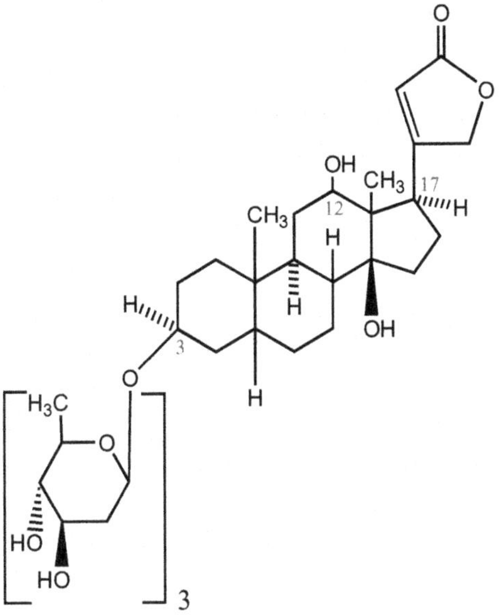

FIGURE 20.3 Structure of digoxin.

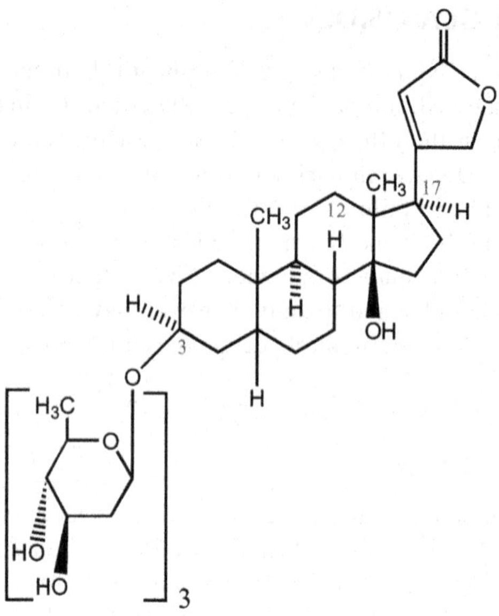

FIGURE 20.4 Structure of digitoxin.

20.3.2 Pharmacology and Clinical Use

DGs have been widely used for more than two centuries as therapeutic agents for congestive heart failure. The drugs have a positive inotropic effect on failing myocardium and consequently increase the ability to control ventricular rate in response to atrial fibrillation. DGs at therapeutic doses selectively bind and partially inhibit the intrinsic membrane protein Na^+-K^+-ATPase. This partial inhibition of the Na^+-K^+-ATPase leads to an increase in the intracellular Na^+ concentration, which is sufficient to reduce Ca^{2+} extrusion via the Na^+/Ca^{2+} exchange system in the sarcolemma. The resulting increase in cytosolic and sarcoplasmic Ca^{2+} concentrations enhances the force of cardiac muscle contraction. While the direct inhibition of Na^+-K^+-ATPase is the principal mechanism of action for the positive inotropic effects of DGs, it was recently recognized that DGs are also able to modulate sympathetic nerve system activity, an additional mechanism contributing to their efficacy in the treatment of patients with heart failure.

20.3.3 Toxicokinetics

The absorption of DGs occurs in the gastrointestinal (GI) tract. Because digoxin has greater polarity than digitoxin, GI absorption of digoxin is less rapid and less complete than in the case of digitoxin. DGs bind to plasma proteins, especially albumin. As a result, highly protein-bound substances displace DGs from circulating albumin and increase the levels of free DGs in blood, thus potentially increasing their toxicity. Once absorbed into the systemic circulation, DGs are widely distributed throughout the body, with the highest concentrations in muscular tissues. For instance, concentrations of DGs in the myocardium are about 30 times higher than those in blood. DGs are mainly metabolized in the liver and in the GI tract. Routes of elimination include urinary and bile. The half-life of digoxin is about 36 hours, whereas that of digitoxin is 5–7 days, further confirming the differences in their ionized state in serum. Consequently, digitoxin's high lipid solubility and recirculation through the enterohepatic pathway are partially responsible for its long half-life. Because the liver and kidneys are significantly involved in the metabolism and excretion of DGs, the half-life of DGs may be significantly prolonged with an increased likelihood of intoxication in patients with impaired renal and hepatic function. These features contribute to its narrow therapeutic index (~1.2–1.7 ng/ml), thus creating a slim division between effective therapeutic concentrations and toxic manifestations.

20.3.4 Clinical Manifestations of Toxicity

Because the signs and symptoms of digoxin toxicity are mostly nonspecific, a high index of suspicion is crucial for early recognition and appropriate management. Acute poisoning frequently occurs in adults following either an intentional ingestion of a large dose or from an unintentional ingestion in children. Chronic poisoning is more common in elderly or heart failure patients receiving DGs and usually results from either inadvertent medication error or faulty patient compliance. The incidence and severity of DG toxicity have been substantially reduced in the past two decades, due partially to the development of alternative drugs for the treatment of heart failure as well as to the improved management of DG intoxication.

The manifestations of DG toxicity involve the CV, the central nervous system (CNS), and the GI tract. The hallmark of cardiac toxicity is an increased automaticity coupled with concomitant conduction delay. Although no single dysrhythmia is frequently present, certain aberrations such as premature ventricular beats, bradyrhythmias, paroxysmal atrial tachycardia with block, junctional tachycardia, and bidirectional ventricular tachycardia are common. DG toxicity is also manifest as dysfunction of the CNS, including delirium, fatigue, malaise, confusion, dizziness, abnormal dreams, blurred or yellow vision, and halos. Disturbances of color vision are frequently reported. A patient's complaint of a "whitish, yellowish halo vision" is a suspicious sign of digoxin intoxication. GI disturbances may include anorexia, nausea, vomiting, and abdominal pain.

20.3.5 Mechanisms of Toxicity

The toxic effects of DGs are, at least partially, the extensions of their pharmacological actions. The electrophysiological effects of DGs are due to their direct actions on cardiac contractile and conducting cells (i.e., inhibition of sarcolemmal Na^+-K^+-ATPase) and indirect effects on the autonomic nervous system. At therapeutic, nontoxic doses, digoxin increases vagal tone and decreases sympathetic activity, resulting in decreased SA and AV node automaticity. At higher concentrations, toxicity progresses to induction of bradycardia, prolongation of AV conduction, or heart block. It is also noteworthy that toxic doses of digoxin can increase both intracellular Ca^{2+} loading and sympathetic nervous system activation, precipitating atrial and ventricular dysrhythmias, including life-threatening ventricular fibrillation. In addition, DGs may cause vasoconstriction as a result of increased intracellular Ca^{2+} in smooth muscle of the vessel wall.

CNS toxicity of DGs appears to be related to the inactivation of Na^{+}-K^{+}-ATPase, resulting in altered ionic transport across excitable neuronal membranes with consequent membrane irritability and instability.

20.3.6 Clinical Management of Intoxication

Successful management of DG intoxication depends on early recognition. Treatment relies on clinical conditions rather than serum drug levels. Management varies from temporary withdrawal of the medication, general supportive care, and proper treatment of DG-induced cardiac arrhythmias to administration of digoxin-specific *Fab* fragments (i.e., Digibind®) for life-threatening CV compromise.

Digoxin-specific *Fab* fragments are antigen-binding fragments of anti-digoxin antibody derived from immunized sheep. Its high volume of distribution (V_d), which promotes entry into extracellular spaces, coupled with an elevated half-life (15–20 hours) and the antibody's significantly higher competitive affinity for digoxin,* fashions a molecule that competitively binds and removes circulating and receptor-bound digoxin. Antitoxin intervention is indicated when serum digoxin concentrations are greater than 10 ng/ml (about 10 mg total adult oral dose, 4 mg in children) and serum potassium concentrations reach 5 meq/L or higher.

The usual dose of Digibind® is calculated according to the number of digoxin tablets ingested. In general, adults receive 20 vials (10 vials for children) of 40 mg/vial intravenously (I.V.), and the dose is repeated as needed until the signs and symptoms or serum digoxin levels fall. Chronic intoxication requires about six vials of Digibind® administered in the same way.

The therapeutic effect is revealed within minutes. Each 40 mg binds 0.6 mg digoxin; thus, the number of Digibind® vials (N) required for neutralization of DG toxicity is calculated based on the total amount of digoxin, or number of digoxin tablets, ingested (multiplied by 80% for incomplete absorption).† Also, the number of vials required can be calculated from the known serum digoxin concentration (SDC) according to the following:

$$N = \left[\text{SDC}(\text{ng/ml}) \times \text{patient weight}(\text{kg})\right] / 100$$

Since Digibind® is a biological product of immune origin, its major adverse drug reactions are consistent with therapeutic agents of that class, predominantly manifested as erythema, rash, and urticaria.

---

* The antibody possesses up to 10^{10}/M greater attraction for digoxin than the drug's affinity for the ATPase enzyme.

† Each vial of Digibind® (40 mg) binds 0.6 mg digoxin. The number of vials (N) required for treating an acute digoxin intoxication is calculated as: N = digoxin body load (mg)/0.6 mg.

20.3.7 Methods of Detection

Detection of the serum levels of DGs is of critical importance for the diagnosis of DG intoxication. Drug-specific radioimmunoassays are the most widely used methods for detecting DGs in serum, with relatively new immunoassays currently available (Digoxin III®, Abbott Laboratories, IL). Enzyme, chemiluminescence, and fluorescence polarization immunoassays are also available.

20.4 BETA-ADRENERGIC RECEPTOR ANTAGONISTS

20.4.1 β-Adrenergic Receptor System

β-Adrenergic receptors are members of the superfamily of G-protein-coupled receptors. There are at least three β-adrenergic receptor subtypes:

1. β1, found in heart and coronary blood vessels predominantly, but also present in liver, kidney, and adipose tissues;
2. β2, found in lungs, muscle tissue, and most other sympathetic target organs; and
3. β3, located primarily in adipose tissue.

On β-receptor occupation, G-proteins undergo a conformational change that activates adenylyl cyclase, leading to increased levels of intracellular cAMP. cAMP stimulates protein kinase A, which phosphorylates Ca^{2+} channels and allows Ca^{2+} to enter the cell. The influx of Ca^{2+} triggers additional Ca^{2+}-stimulated Ca^{2+} release from the SR. Stimulation of β1 receptors results in increased chronotropy* and inotropy in the heart, as well as increased renin secretion by the kidneys, both of which result in increased blood pressure. Stimulation of β2 receptors activates relaxation of smooth muscle cells in blood vessels and bronchial tree. β3 receptor stimulation appears to be involved in the increased metabolism of lipid. In view of the relative specific actions mediated by each β-receptor subtype, selective modulation by pharmacological agents has proved to be of clinical significance.

20.4.2 Pharmacology and Clinical Use

β-Adrenergic receptor antagonists (β-*blockers*) inhibit sympathetic stimulation. As a result, heart rate and blood pressure decrease in response to the adrenergic receptor inhibition, accompanied by a secondary increase in force of contraction. Consequently, the drugs are among the primary pharmacological modalities used in the treatment of hypertension, ischemic heart disease, CHF, and certain arrhythmias. Non-CV and off-label uses include the treatment of essential tremor, pheochromocytoma, glaucoma, anxiety, and migraine headaches. Due to their broad clinical applications and availability, β-blocker overdoses and intoxications are commonly encountered. Currently, three classes of β-blockers are available for clinical use, as outlined in Table 20.1.

---

* Chronotropy refers to rate of rhythmic movement, whereas inotropy implies contractility (force of contraction) of muscular movement.

TABLE 20.1
Classification of β-Adrenergic Receptor Antagonists (β-Blockers or Class II Antiarrhythmics)

Pharmacological Classification	Compound	Mechanism of Action
Nonselective β-blockers	Propranolol, nadolol, timolol, pindolol, labetalol	Inhibit β1 receptor activation—net effect: decreased cardiac chronotropy and inotropy; inhibit β2 receptor activation—net effect: blocks bronchial relaxation, resulting in bronchoconstriction
Selective β1-blockers	Metoprolol, atenolol, esmolol, acebutolol, bisoprolol	Predominantly inhibit β1 receptor activation—net effect: decreased cardiac chronotropy and inotropy
β-blockers with vasodilation activity	Carvedilol, bucindolol, nebivolol	In addition to β-receptor antagonism, produce significant vasodilation with decrease in blood pressure

20.4.3 Clinical Manifestations of Toxicity

The toxicology of β-blocker overdose is an extension of the pharmacology—that is, the drugs cause deleterious effects on the CV, CNS, and pulmonary systems through excessive inhibition of β-adrenergic receptors. Cardiac toxicity presents as bradycardia, conduction delay, and decreased cardiac contractibility with systemic hypotension. The cardiotoxicity of β-blockers, however, may also be mediated through disruption of ion transport and homeostasis in cardiac muscle cells.

The manifestations of CNS toxicity in severe intoxication include psychosis, loss of consciousness, and seizures. The mechanism underlying CNS toxicity is unclear, but it may be associated with cellular hypoxia resulting from suppressed cardiac output, or direct neuronal toxicity.

Blocking β2 receptors in bronchial smooth muscle may also precipitate life-threatening bronchoconstriction in patients with a predisposition to pulmonary disease. Acidosis and hypoglycemia are known to aggravate β-blocker intoxication.

20.4.4 Clinical Management of Intoxication

The goal of clinical management of β-blocker intoxication is to restore perfusion to critical organ systems by improving myocardial contractility or increasing heart rate, or both. General measures include supportive care and GI decontamination. Pharmacotherapy includes the use of glucagon, β-adrenergic receptor agonists, phosphodiesterase inhibitors, and atropine. In particular, glucagon enhances cardiac performance by increasing intracellular cAMP through its action on distinct glucagon receptors on cardiac muscle cells, thus restoring suppressed cardiac function.

20.5 CALCIUM CHANNEL ANTAGONISTS (CA^{2+} CHANNEL BLOCKERS)

The effect of Ca^{2+} channel antagonists is similar to induced Ca^{2+} depletion in cardiac muscle—that is, the drugs diminish the contractile force without affecting the action potential, thus inducing excitation–contraction uncoupling. Verapamil and phenylamines, together with a number of other drugs that inhibit excitation–contraction coupling (Table 20.2), have been designated Ca^{2+} antagonists. The major action of these drugs is to diminish the inward movement of Ca^{2+} through the L-type voltage-dependent Ca^{2+} channels located in the sarcolemma.

20.5.1 Pharmacology and Clinical Use

Ca^{2+} channel antagonists affect the contractility of both smooth and cardiac muscle cells. Physiologically, three distinct mechanisms have been suggested to account for increased levels of cytosolic Ca^{2+} and subsequent contraction of smooth muscle:

1. Extracellular Ca^{2+} enters the muscle cell through voltage-sensitive Ca^{2+} channels in response to depolarization of the membrane.
2. Inositol triphosphate (a second messenger) mediates release of Ca^{2+} from SR.
3. Influx of extracellular Ca^{2+} via receptor-operated Ca^{2+} channels occurs in response to receptor occupancy.

Pharmacologically, Ca^{2+} channel antagonists inhibit the voltage-dependent Ca^{2+} channels in vascular smooth muscle cells at much lower concentrations than those required to interfere with the release of intracellular Ca^{2+} from the SR or to block receptor-operated Ca^{2+} channels. All of the Ca^{2+} channel antagonists are capable of inducing relaxation of vascular smooth muscle, resulting in vasodilation. In addition, by blocking the L-type voltage-dependent Ca^{2+} channels, Ca^{2+} channel antagonists

TABLE 20.2
Classification of Ca^{2+} Channel Antagonists (Class IV Antiarrhythmics)

Chemical Classification	Compound	Mechanism of Action
Phenylalkylamines	Verapamil	Diminish the inward movement of Ca^{2+} through the L-type voltage-dependent Ca^{2+} channels located in the sarcolemma
Benzothiazepines	Diltiazem	Similar mechanism as with the phenylalkylamines
Dihydropyridines	Amlodipine,felodipine,isradipine,nicardipine,nifedipine,nimodipine,nisoldipine	In addition to the previous mechanism, this class possesses a greater degree of peripheral vasodilation
Diarylaminopropylamine esters	Bepridil	Similar mechanism as with the phenylalkylamines

exert a negative inotropic effect on myocardium. Although this is true of all classes of Ca^{2+} channel antagonists, the greater degree of peripheral vasodilation seen with the dihydropyridines is accompanied by sufficient baroreflex-mediated increase in sympathetic activity to overcome the negative inotropic effect. The depolarization of SA and AV nodes is also largely dependent on the influx of extracellular Ca^{2+} through the L-type channels. Therefore, Ca^{2+} channel antagonists have the potential to depress the rate of the SN pacemaker and to slow AV conduction. In addition, Ca^{2+} channel antagonists are able to decrease coronary vascular resistance and thereby increase coronary blood flow. In view of these pharmacological effects, the drugs are efficacious in the treatment of various types of CV disorders, including hypertension, angina pectoris, MI, and cardiac arrhythmias.

20.5.2 Clinical Manifestations of Toxicity

The most common toxic effects caused by the Ca^{2+} channel antagonists, particularly the dihydropyridines, are due to excessive vasodilation. These effects may be manifest as dizziness, hypotension, headache, flushing, digital dysesthesia,* and nausea. Patients may also experience constipation, peripheral edema, coughing, wheezing, and pulmonary edema. At therapeutic and moderately toxic doses, dihydropyridines are well recognized to produce reflex increases in heart rate with an increase in left ventricular stroke volume, leading to an increase in cardiac output. With severe overdoses that result in dramatic Ca^{2+} channel blockage, all Ca^{2+} channel antagonists exert a negative inotropic effect with depressed cardiac contraction, conduction blockage, hypotension, and shock. Other overdose effects may present as metabolic acidosis with hyperglycemia. The mechanism of hyperglycemia is likely related to the suppressive effect of the drugs on pancreatic β-cell insulin release.

20.5.3 Clinical Management of Intoxication

Patients with unexplained hypotension and conduction abnormalities, followed by careful history, may suggest overdose with Ca^{2+} channel blockers. As the toxicity produces significant morbidity and mortality, general management revolves around providing supportive care, decreasing drug absorption, and augmenting myocardial function with cardiotonic agents. I.V. calcium salts are the first-line treatment of Ca^{2+} channel antagonist overdoses. As described earlier, other cardiotonic drugs may include glucagon, atropine, and catecholamines.

20.6 OTHER CV DRUGS

Other classes of important CV drugs used in the treatment of a variety of CV diseases, including hypertension, arrhythmias, and ischemic heart disease, are further described in the following and are listed in Table 20.3.

---

* Anesthetic (tingling) sensation in the fingers due to loss of neuronal conduction.

TABLE 20.3
Classification of Other CV Drugs

Pharmacological Classification	Compound	Mechanism of Action
ACE inhibitors	Benazepril, captopril, enalapril, enalaprilat, fosinopril sodium, lisinopril, moexipril, perinopril, quinapril, ramipril, trandolapril	Block angiotensin-converting enzyme, thereby decreasing the formation of angiotensin II—net effect: vasodilation
Direct vasodilators	Hydralazine	Activates guanylate cyclase, resulting in vascular smooth muscle relaxation and vasodilation
	Minoxidil	Metabolite activates ATP-sensitive K^+ channels and hyperpolarizes arterial smooth muscle cells, resulting in vasodilation
	Diazoxide	Similar to minoxidil
	Nitroprusside	Metabolized to nitric oxide, which produces vasodilation as per hydralazine
Antiarrhythmic agents	*Class IA*: disopyramide, procainamide, quinidine	Block Na^+ & K^+ channels; depress rapid action potential upstroke; decrease conduction velocity; prolong repolarization
	Class IB: lidocaine, mexiletine, moricizine(?), tocainide	Weakly block Na^+ channels; depress rapid action potential upstroke; decrease conduction velocity; prolong repolarization
	Class IC: flecainide, propafenone	Strongly block Na^+ channels; depress rapid action potential upstroke; no effect on conduction velocity, repolarization, or K^+ channels
	Class III: amiodarone, bretylium, sotalol[a]	Block K^+ channels; depress repolarization; no effect on Na^+ channels

[a] Sotalol is also a β-adrenergic receptor antagonist; Class II (β-blockers) and Class IV (Ca2+) channel blockers are noted in Tables 21.1 and 21.2, respectively).

20.6.1 Angiotensin-Converting Enzyme (ACE) Inhibitors

ACE inhibitors block angiotensin-converting enzyme, thereby decreasing the formation of angiotensin II, a potent vasopressor, which is critically involved in raising systemic blood pressure. As such, ACE inhibitors are commonly used in the treatment of hypertension. These drugs are also efficacious in the treatment of CHF, left ventricular systolic dysfunction, acute MI, and chronic renal disease.

Hypotension is the most common manifestation in patients with ACE inhibitor overdoses. Adverse effects reported at therapeutic doses include first-dose hypotension, headache, cough, hyperkalemia, dermatitis, renal dysfunction, and angioedema.

The drugs may also cause adverse fetal effects; thus, this class of drugs is contraindicated in pregnancy. Treatment of ACE inhibitor overdose is largely supportive.

20.6.2 Direct Vasodilators

Direct vasodilators represent another class of antihypertensive drugs (Table 20.3) also used in the management of angina pectoris, CHF, and peripheral vascular disease that relax vascular smooth muscle independently of innervation or known pharmacological receptors. Among these, the most potentially toxic agent capable of inducing both arterial and venous vasodilation is nitroprusside.

Hydralazine activates guanylate cyclase, which precipitates an increase in cGMP in arterial vascular smooth muscle, resulting in vascular smooth muscle relaxation and vasodilation. The formation of cGMP results in decreased levels of cytosolic Ca^{2+} in smooth muscle. Minoxidil undergoes hepatic biotransformation, producing the active N-O sulfate metabolite. Minoxidil sulfate is able to activate the ATP-sensitive K^+ channels, which causes vasodilation by hyperpolarizing arterial smooth muscle cells. The drug has proved to be efficacious in patients with the most severe and drug-resistant forms of hypertension. Diazoxide also produces vasodilation via the activation of ATP-sensitive K^+ channels. This drug is clinically used in the treatment of hypertensive emergencies. Nitroprusside is metabolized by the vessel wall to form nitric oxide, which activates guanylyl cyclase, increases levels of cGMP, and produces subsequent vasodilation. Nitroprusside is used mainly in the treatment of hypertensive emergencies.

Two types of adverse effects have been observed with hydralazine intoxication:

1. Toxicity due to extensions of pharmacological effects of the drug, including hypotension, headache, nausea, flushing, palpitation, dizziness, tachycardia, and angina pectoris; the last occurs as a result of baroreflex-induced stimulation of the sympathetic nervous system;
2. Hydralazine-induced autoimmune reactions, including lupus syndrome, vasculitis, serum sickness, hemolytic anemia, and rapidly progressive glomerulonephritis. The reactions probably result from T-cell autoreactivity.

Clinical manifestations of minoxidil intoxication may include edema, CV compromise, hypertrichosis,* hypotension, tachycardia, and lethargy. Tachycardia is caused by the baroreceptor-mediated increase of the sympathetic tone. Hypertrichosis occurs in all patients who receive minoxidil for an extended period of time and is probably a consequence of K^+ channel activation.

The most common manifestations of diazoxide intoxication are myocardial ischemia, peripheral and systemic edema, and hyperglycemia. Myocardial ischemia may be precipitated or aggravated by diazoxide, and it results from reflex adrenergic stimulation of the heart and from increased blood flow to nonischemic regions. Hyperglycemia appears to result from its inhibition of the secretion of insulin from pancreatic β-cells.

---

* Hirsutism or excessive growth of hair.

The short-term toxic effects of nitroprusside are caused by excessive vasodilation and ensuing hypotension. Toxicity may also result from the conversion of nitroprusside to cyanide and thiocyanate, the mechanisms of which are discussed in Chapter 23. Nitroprusside-induced cyanide poisoning is a result of the development of an anion-gap metabolic acidosis. Cardiac failure, asystole, and ventricular dysrhythmias are serious CV terminal events, initially presenting as restlessness and agitation and possibly progressing to convulsion. Encephalopathy, coma, and cerebral death often occur simultaneously with the terminal CV event.

The management of vasodilator intoxication includes general supportive measures and correction of hypotension and cardiac arrhythmias. Ca^{2+} channel antagonists and β-adrenergic receptor antagonists may be useful in the treatment of myocardial ischemia caused by the vasodilators. Discontinuation of nitroprusside administration, followed by oxygen supplementation, is essential for suspected cyanide toxicity due to nitroprusside. As with cyanide toxicity, sodium nitrite and sodium thiosulfate should be given immediately to enhance the transulfuration of cyanide to thiocyanate.

20.6.3 Antiarrhythmic Drugs

Antiarrhythmic drugs are used in the treatment of cardiac arrhythmias and have selective classification based on the mechanisms of action (MOA). The drugs, classification, and MOA are tabulated in Table 20.3. Briefly stated:

Class I: depress myocardial Na^+ and K^+ channels;
Class II: possess sympatholytic activities, such as the β-adrenergic receptor antagonists (not true antiarrhythmics and not listed in Table 20.3);
Class III: prolong action potential duration and refractoriness;
Class IV: Ca^{+2} channel antagonists (not true antiarrhythmics and not listed in Table 20.3).

Class IA drugs have a low toxic-to-therapeutic ratio, and their use is associated with serious adverse effects during long-term therapy and life-threatening sequelae following acute overdoses. The most severe manifestation of intoxication is CV compromise, including sinus tachycardia, cardiac arrhythmia with ventricular tachycardia (*torsades de pointes*), and fatal ventricular fibrillation. Depressed myocardial contractility frequently manifests as vasodilation and hypotension. CNS toxicity presents as lethargy, confusion, coma, respiratory depression, and seizure. Quinidine intoxication causes cinchonism, a symptom complex that includes headache, tinnitus, vertigo, and blurred vision.* Diarrhea is the most common adverse effect during quinidine therapy. Severe immunological reactions of the lupus type have also occurred. Disopyramide also has strong anticholinergic activity, which can precipitate glaucoma, constipation, dry mouth, and urinary retention.

* "Quinidine syncope" refers to a transient loss of consciousness due to paroxysmal ventricular tachycardia.

Class IB and IC drug toxicity is marked by similar cardiac effects as occur with class IA drugs. CNS toxicity is more common and may be manifest as confusion, coma, seizure, nystagmus (an early sign of lidocaine toxicity), tremor, and nausea.

Toxicity with *Class III* drugs is not explained by blocking of K^+ channels only. For example, sotalol also has marked β-adenergic receptor antagonist activity, which explains most of the CV compromise in overdosage. Sotalol intoxication may cause Q-Tc prolongation, bradycardia, and hypotension. Coma, respiratory depression, seizure, and ventricular dysrhythmia (*torsades de pointes*) occur in severe sotalol overdoses.

Acute amiodarone toxicity following overdose is rare but may include hypotension. Although the mechanism is not understood, pulmonary fibrosis is a known complication of chronic amiodarone therapy—there is currently no effective treatment for the condition, and it carries a poor prognosis.

Clinical management of intoxication with Class I and III antiarrhythmic drugs follows general supportive measures and institution of the ABCs of emergency care. The slow absorption of amiodarone allows late GI decontamination in cases of significant ingestion, while sodium bicarbonate is efficacious in the treatment of cardiotoxic effects of Class IA and IC drugs.

REFERENCES

SUGGESTED READINGS

Brunton, L.L., Hilal-Dandan, R., and Knollmann, B.C., *Goodman & Gilman's The Pharmacological Basis of Therapeutics*, 13th ed., McGraw-Hill, New York, 2018, chapter 30.

Buchtel, L. and Ventura, H.O., Lunar Society and the discovery of digitalis, *J. La. State Med. Soc.* 158, 26, 2006.

Dasgupta, A., Risin, S.A., Reyes, M., and Actor, J.K., Rapid detection of oleander poisoning by Digoxin III, a new digoxin assay: impact on serum digoxin measurement. *Am. J. Clin. Pathol.* 129, 548, 2008.

Ford, M.D., Delaney, K.A., Ling, L.J., and Erickson, T., *Clinical Toxicology*, W.B. Saunders, Philadelphia, PA, 2001, chapters 41, 42, 43, and 44.

Magoo, R. Choudhury, A., Malik, V., Sharma, R., and Kapoor, P.M., Pharmacological update: new drugs in cardiac practice: a critical appraisal. *Ann. Card Anaesth.* 20 (Suppl 1), S49–S56, 2017.

Martini, F.H., Nath, J.L., and Bartholomew, E.F., *Fundamentals of Anatomy and Physiology*, 11th ed., Pearson, England, 2018, chapters 20 and 21.

Mozaffarian, D., Benjamin, E.J., Go, A.S., Arnett, D.K., Blaha, M.J., Cushman, M., de Ferranti, S., Després, J.P., Fullerton, H.J., Howard, V.J., Huffman, M.D., Judd, S.E., Kissela, B.M., Lackland, D.T., Lichtman, J.H., Lisabeth, L.D., Liu, S., Mackey, R.H., Matchar, D.B., McGuire, D.K., Mohler, E.R. 3rd, Moy, C.S., Muntner, P., Mussolino, M.E., Nasir, K., Neumar, R.W., Nichol, G., Palaniappan, L., Pandey, D.K., Reeves, M.J., Rodriguez, C.J., Sorlie, P.D., Stein, J., Towfighi, A., Turan, T.N., Virani, S.S., Willey, J.Z., Woo, D., Yeh, R.W., and Turner, M.B., American Heart Association Statistics Committee and Stroke Statistics Subcommittee, *Circulation* 131, e29, 2015.

Schmiedl, S., Rottenkolber, M., Szymanski, J., Hasford, J., and Thuermann, P.A., Declining public health burden of digoxin toxicity: decreased use or safer prescribing? *Clin. Pharmacol. Ther.* 85, 143, 2009.

U.S. Centers for Disease Control and Prevention (U.S. CDC), Heart disease facts. From: https://www.cdc.gov/heartdisease/facts.htm, last viewed March 2019.
Zhang, C., Qi, H., and Zhang, M., Homogeneous electrogenerated chemiluminescence immunoassay for the determination of digoxin employing Ru(bpy)(2)(dcbpy)NHS and carrier protein. *Luminescence* 22, 53, 2007.

REVIEW ARTICLES

Bara, V., Digoxin overdose: clinical features and management. *Emerg. Nurs.* 9, 16, 2001.
Benjamin, E.J., Virani, S.S., Callaway, C.W., Chamberlain, A.M., Chang, A.R., Cheng, S., Chiuve, S.E., Cushman, M., Delling, F.N., Deo, R., de Ferranti, S.D., Ferguson, J.F., Fornage, M., Gillespie, C., Isasi, C.R., Jiménez, M.C., Jordan, L.C., Judd, S.E., Lackland, D., Lichtman, J.H., Lisabeth, L., Liu, S., Longenecker, C.T., Lutsey, P.L., Mackey, J.S., Matchar, D.B., Matsushita, K., Mussolino, M.E., Nasir, K., O'Flaherty, M., Palaniappan, L.P., Pandey, A., Pandey, D.K., Reeves, M.J., Ritchey, M.D., Rodriguez, C.J., Roth, G.A., Rosamond, W.D., Sampson, U.K.A., Satou, G.M., Shah, S.H., Spartano, N.L., Tirschwell, D.L., Tsao, C.W., Voeks, J.H., Willey, J.Z., Wilkins, J.T., Wu, J.H., Alger, H.M., Wong, S.S., and Muntner, P.,Heart disease and stroke statistics—2018 update—a report from the American Heart Association. *Circulation* 67, 492, 2018.
Bristow, M.R., β-Adrenergic receptor blockade in chronic heart failure. *Circulation* 101, 558, 2000.
Connolly, S.J., Evidence-based analysis of amiodarone efficacy and safety. *Circulation* 100, 2025, 1999.
Hauptman, P.J. and Kelly, R.A., Digitalis. *Circulation* 99, 1265, 1999.
Kerns, W., Kline, J., and Ford, M.D., Beta-blocker and calcium channel blocker toxicity. *Emerg. Med. Clin. North Am.* 12, 365, 1994.
Kjeldsen, K. and Gheorghiade, A.N.M., Myocardial Na,K-ATPase: the molecular basis for the hemodynamic effect of digoxin therapy in congestive heart failure. *Cardiovasc. Res.* 55, 710, 2002.
Kolecki, P.F. and Curry, S.C., Poisoning by sodium channel blocking agents. *Crit. Care Clin.* 13, 829, 1997.
Mladěnka, P., Applová, L., Patočka, J., Costa, V.M., Remiao, F., Pourová, J., Mladěnka, A., Karlíčková, J., Jahodář, L., Vopršalová, M., Varner, K.J., and Štěrba, M., Comprehensive review of cardiovascular toxicity of drugs and related agents. *Med. Res. Rev.* 38, 1332, 2018.
Sangha, S., Uber, P.A., and Mehra, M.R., Difficult cases in heart failure: amiodarone lung injury: another heart failure mimic? *Congest. Heart Fail.* 8, 93, 2002.
Vivo, R.P., Krim, S.R., Perez, J., Inklab, M., Tenner, T. Jr., and Hodgson, J., Digoxin: current use and approach to toxicity. *Am. J. Med. Sci.* 336, 423, 2008.

21 Antineoplastic Agents

21.1 DESCRIPTION

Antineoplastic agents, traditionally referred to as *chemotherapeutic agents*, are used therapeutically in the treatment of cancer.* The drugs include a wide variety of diverse chemicals whose mechanism of action is targeted toward arresting the aberrant cell proliferation generally associated with cancerous cell growth. Although the concept that chemical agents could interfere with cell proliferation was known by about the turn of the 20th century, it was not until the end of World War I that the biological and chemical actions of alkylating agents, such as the nitrogen mustards, were understood. By the end of World War II, the conviction that such drugs could be used in the treatment of cancer prompted the search for antineoplastic drugs that would forever impact the therapeutic intervention of neoplastic disease.

21.2 REVIEW OF THE CELL CYCLE

It is to some extent ironic that most antineoplastic agents used in cancer chemotherapy are also carcinogenic, mutagenic, or teratogenic. In fact, the adverse reactions and toxicities associated with these substances often limit their usefulness and complicate the course of treatment. Consequently, a better understanding of the toxicology and pharmacology depends on the drugs' influence on cell cycle kinetics, which in turn, explains the adverse reactions and mechanisms of action, respectively.

Figure 21.1 illustrates the four successive phases of the cell cycle. Normal, proliferating somatic cells spend the majority of their existence in *interphase*. Cells also perform most of their maintenance functions in *interphase*. As a cell prepares to divide, it enters the *G_1 phase*, the period between mitosis (*M phase*) and DNA synthesis (*S phase*). The *G_1 phase* lasts from several hours to days and is characterized by synthesis of cell organelles and centriole replication (the *G_0 phase* represents a resting stage or sub-phase of G_1). The *S phase* begins when DNA synthesis starts and the chromosomes have replicated. The *G_2 phase* starts when DNA replication is complete and the content of the nucleus has doubled. The G_2

* Cancer is a general term referring to the uncontrolled growth of cells or tissue, characterized by the formation of a neoplasm, that is, a mass of cells or tissue (tumor). Cancers are classified according to their cellular origin and their benign or malignant state of histogenic differentiation. In general, benign tumors are designated by attaching the suffix *-oma* to the cell of origin (a *fibroma* is a tumor of fibrous origin; adenomas and papillomas are of benign epithelial origin). Malignant tumors are classified according to their germ cell layers and organ of derivation. Malignant tumor nomenclature follows that of benign tumors, with some exceptions: carcinoma (neoplasm of epithelial cell origin); sarcoma (tumor of mesenchymal—connective tissue—and endothelial cell origin); melanoma (melanocyte origin); lymphoma (lymphoid tissue origin); leukemia (hematopoietic cell origin); and tumors arising from the nervous system.

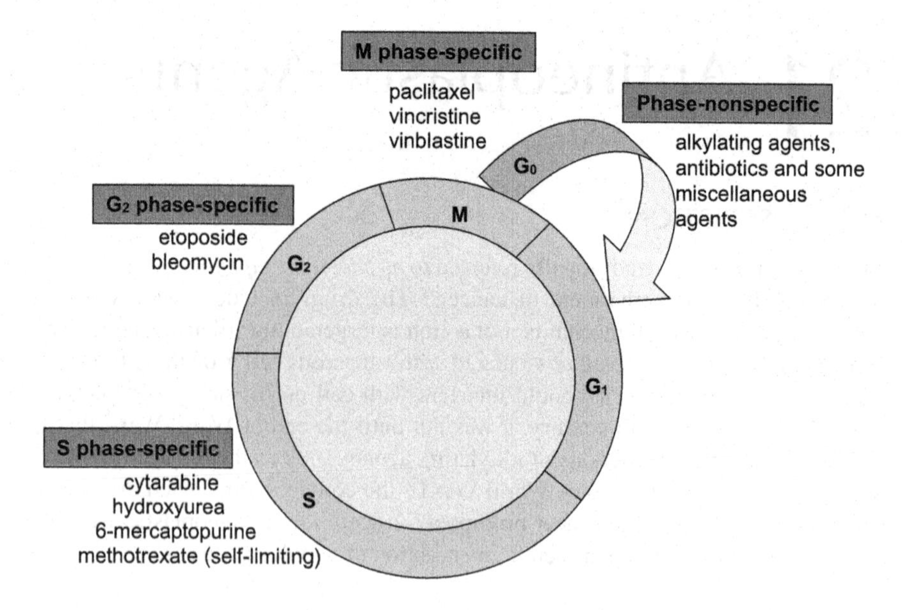

FIGURE 21.1 The four successive phases of the cell cycle.

phase ends when mitosis starts. The extended part of interphase then continues in the *M phase* with the stages of cell division— prophase, metaphase, anaphase, and telophase. During this time, the *M phase* begins with mitosis (nuclear division) and ends with cytokinesis (cytoplasmic) division. Cell division terminates with the completion of cytokinesis.

Antineoplastic agents suppress the proliferation of neoplasms through interaction at one or more phases of cell replication. *Cell cycle–specific agents*, some of which are indicated in Figure 21.1, interfere with at least one specific phase of cellular replication. *Cell cycle–nonspecific agents* act on both proliferating and resting cells.

21.3 ANTIMETABOLITES

21.3.1 Pharmacology and Clinical Use

Antimetabolites alter normal cellular functions by substituting for components within key metabolic processes. An agent may be incorporated as a substrate or may inhibit the normal functioning of an enzyme. The net effect of interference with cellular biochemical reactions is disruption of nucleic acid synthesis. Thus, antimetabolites are effective on rapidly proliferating cells, mostly during the *S phase* (DNA synthesis phase) of nucleic acid synthesis. Consequently, they are indicated for the induction and maintenance of remission of leukemias, metastatic breast cancer, and colon, rectal, stomach, and pancreatic carcinomas. Antimetabolites, alone and in combination, are also used in the management of other proliferative disorders such as psoriasis and rheumatoid arthritis (methotrexate).

21.3.2 Acute Toxicity

The toxic effects of the antimetabolites are related to their mechanisms of action and are summarized in Table 21.1. The most common, and potentially serious, adverse reactions involve organ system toxicity. The toxicities are seen at any dose and include gastrointestinal (GI) (nausea, vomiting, and diarrhea), dermatologic (alopecia), hematologic (bone marrow suppression), metabolic (hepatotoxicity), and central nervous system (CNS) (headaches and blurred vision) effects.

Extravasation* occurs with any of the parenterally administered antimetabolites. Prevention of extravasation requires optimum administration technique. However, should it occur, several steps are required to alleviate the situation, including discontinuation of the medication; administration of an antidote where appropriate; and surgical debridement with skin grafting. The antibiotic antineoplastics are generally associated with severe, local, necrotic (vesicant) reactions.

21.3.3 Clinical Management of Acute Overdose

Leucovorin (citrovorum factor, folinic acid) is a specific antidote for MTX overdose, especially as a therapeutic modality—that is, to overcome the resistance to MTX, an antifolate drug, high doses are administered concomitantly with leucovorin. This *rescue procedure* permits the intracellular accumulation of MTX in concentrations sufficient to inactivate dihydrofolate reductase while diminishing hematologic toxicity. The combination is effectively used, along with other chemotherapeutic agents, in patients with nonmetastatic osteosarcoma.

TABLE 21.1
Classification of Selected Antimetabolite Antineoplastic Agents and Predominant Adverse Drug Reactions (ADRs)

Category	Nonproprietary Name	Predominant Target of Inhibition	Predominant ADRs
Folic acid analogs	Methotrexate (MTX)	Dihydrofolate reductase	CNS, GI, GU, dermatologic, hematologic, metabolic toxicity
Purine analogs	6-TG, 6-MP	Purine ring synthesis	
Pyrimidine analogs	5-FU	TMP synthesis	
	Cytarabine	DNA synthesis	
Substituted urea	Hydroxyurea	Ribonucleotide reductase	

5-FU: 5-fluorouracil; 6-MP: 6-mercaptopurine; 6-TG: 6-thioguanine; CNS: central nervous system; CV: cardiovascular; GI: gastrointestinal; GU: genitourinary; TMP: thymidine monophosphate.

* Treatment schedules with some antineoplastic agents are generally prolonged and continuous. Consequently, the intravenous (I.V.) fluid containing the medication may leak into interstitial tissue at the site of administration, causing complications ranging from painful erythematous swelling to deep necrotic injury of underlying skin.

21.4 ALKYLATING AGENTS

21.4.1 Pharmacology and Clinical Use

Alkylating agents are strong electrophiles that form highly reactive carbonium ions. They replace hydrogen atoms with alkyl radicals, resulting in cross-linking and abnormal base-pairing in nuclear DNA. These radicals also covalently bind to sulfhydryl, phosphate, and amine groups, inhibiting cellular metabolic processes that terminate reactions in proliferating and stationary cells. The net effect is the production of defective DNA and abortive reproduction of cells. Although alkylating agents are most effective on rapidly proliferating cells, they are not *phase specific* and may act on cells at any stage of the cell cycle. The drugs are used concurrently or sequentially with other antineoplastics and are indicated, in general, in the palliative* treatment of malignancies susceptible to the drugs. The conditions include malignant lymphomas, leukemias, germ cell and progressive testicular, ovarian, and prostate cancers, myelomas, and brain tumors. Other conditions amenable to alkylating agents are metastatic carcinomas of the stomach, colon, and bladder, neuroblastomas, and nonmalignant nephritic syndrome in children (cyclophosphamide, Cytoxan®).

21.4.2 Acute Toxicity

As with the antimetabolites, the most common and serious toxic effects of the alkylating agents are related to their alkylating properties in rapidly dividing cells (Table 21.2). Adverse reactions are seen at any dose and include GI, dermatologic, hematologic (myelosuppression, especially with busulfan), metabolic, and CNS

TABLE 21.2
Classification of Selected Alkylating Agents and Predominant Adverse Drug Reactions (ADRs)

Category	Nonproprietary Name	Predominant Target of Inhibition	Predominant ADRs
Nitrogen mustards (NM)	Cyclophosphamide, Melphalan, Chlorambucil	Interfere with DNA integrity by formation of DNA adducts; inhibit key enzymes by carbomylation of amino acids in proteins	GI, GU, renal, hematologic, hepatic, CNS, dermatologic toxicity; secondary carcinogenesis; ototoxicity (Pt), anaphylactic reactions (Pt)
Nitrosoureas	Carmustine (BCNU), Lomustine (CCNU), Streptozotocin		
Triazenes	Dacarbazine (DTIC)		
Pt coordination complexes	Cisplatin, Carboplatin	Form interstrand DNA cross-links	
Alkyl sulfonates	Busulfan	Granulocytes	

CNS: central nervous system; CV: cardiovascular; GI: gastrointestinal; GU: genitourinary; Pt: platinum.

* Palliative therapy is designed to relieve or reduce the intensity of signs and symptoms rather than to produce a cure.

reactions. Although their cytotoxicities have become more manageable recently, experience dictates an adequately established diagnosis, recognition of signs of adverse drug reactions (ADRs), and supportive care.

21.4.3 Clinical Management of ADRs

The management of patients treated with alkylating agents involves monitoring for signs and symptoms of toxicity, including clinical measurements of hematological, renal, and metabolic functions, as well as general supportive treatment. Adjustment of dosage, discontinuation of the regimen, and replacement with another chemotherapeutic agent are some of the recourse actions taken to alleviate drug toxicity.

21.5 MISCELLANEOUS CHEMOTHERAPEUTIC DRUGS

21.5.1 Natural Products

These antineoplastic agents include a diverse group of naturally derived compounds of plant or prokaryotic origin with unique mechanisms of action and toxicity (Table 21.3).

The vinca alkaloids, named after the beneficial effects derived from the parent periwinkle plant,* are cell cycle-specific agents (*M phase*). Vinca alkaloids, such as vinblastine (Velban®) and vincristine (Vincasar PFS®), specifically bind to tubulin, the protein subunits of the microtubules that polymerize to form the mitotic spindle. Thus, vinca alkaloids interrupt microtubular assembly and produce metaphase arrest. Normal and malignant cells exposed to the plant alkaloids express apoptotic changes. The characteristic neurotoxicity displayed by patients receiving the drugs may be explained, in part, by the disruption of microtubules normally involved in intracellular movement, axonal transport, and phagocytosis. Considerable irritation results from improper positioning of the I.V. needle or catheter. These compounds are used by injection only.

Paclitaxel (Taxol®) is a naturally-derived *M phase–specific* antimitotic product.† Paclitaxel promotes the assembly of microtubules by stimulating tubulin polymerization, rendering stable, nonfunctioning microtubules. The resulting network "bundles" of microtubules are incapable of dynamic reorganization, thus inhibiting cellular mitosis and proliferation. ADRs associated with paclitaxel are similar to those of the vinca alkaloids.

Etoposide (VePesid®) is a semi-synthetic derivative of podophyllotoxin‡ with selective activity at the G_2 *phase* of the cell cycle. At high concentrations, the

* The anticancer potential of the family of *Catharanthus roseus*, formerly *Vinca rosea*, was recognized early in 1958. The plant's ability for producing peripheral granulocytopenia and bone marrow depression in rodents then prompted the subsequent isolation of a large number of alkaloids.

† Paclitaxel was first isolated in 1971 from the bark of the Western yew tree.

‡ Podophyllotoxin is derived from the rhizomes and root of the mandrake plant, *Podophyllum peltatum* (mayapple), which should not be confused with mandragora. The latter is referred to as *mandrake* in ancient Greek and Asian literature. The plant was used by the American Indians and introduced to colonial settlers for its emetic and cathartic properties. It was later developed as a caustic agent in a 25% dispersion in compound benzoin tincture and used topically for the treatment of papillomas.

TABLE 21.3
Classification of Selected Natural Product Antineoplastic Agents and Predominant Adverse Drug Reactions (ADRs)

Category	Nonproprietary Name	Natural Source	Predominant Target of Inhibition	Predominant ADRs
Vinca alkaloids	Vinblastine, Vincristine (V)	*Catharanthus roseus*, periwinkle plant (formerly *Vinca rosea*)	Inhibit microtubular assembly (V); enhance tubulin polymerization (P); mitotic inhibitors	CNS, CV, GI, dermatologic, hematologic, metabolic, I.V. extravasation and hypersensitivity reactions
Taxoids	Paclitaxel (P)	Western yew tree		
Epidophyllotoxins	Etoposide (E), Teniposide (T)	*Podophyllum peltatum* (mandrake plant)	Inhibits DNA synthesis at G_2 (E); inhibits type II topoisomerase (T)	
Camptothecins	CPT, Topotecan	*Camptotheca acuminata*	DNA strand break; bind to topoisomerase I	Myelosuppression, hemorrhagic cystitis
Enzymes	L-asparaginase	Bacterial (e.g., *Escherichia coli*)	H+ asparagine synthetase; inhibits protein synthesis	Hypersensitivity, GI, BM suppression
Antibiotics	Actinomycin D	*Streptomyces parvullus*	DNA intercalation; inhibit DNA/RNA synthesis; antimitotics	Extremely corrosive to soft tissue; myocardial toxicity
	Daunorubicin	*Streptomyces peucetius*		
	Doxorubicin	*Streptomyces peucetius*		
	Bleomycin	*Streptomyces verticillus*	DNA lesions at G_2, inhibits DNA repair	Pulmonary fibrosis
	Mitomycin C	*Streptomyces caespitosus*	Formation of DNA adducts	Myelosuppression, hemolytic uremic syndrome
	Pentostatin	*Streptomyces antibioticus*	Adenosine deaminase	CNS, liver, renal, pulmonary toxicity

BM: bone marrow; CNS: central nervous system; CV: cardiovascular; GI: gastrointestinal; GU: genitourinary.

compound causes cytolysis of cells entering mitosis, while at low concentrations, it prevents the cells from entering prophase. Teniposide (Vumon®) causes single- and double-stranded DNA breaks by inhibiting type II topoisomerase activity but does not intercalate DNA. Severe myelosuppression and hypersensitivity reactions are noted with both drugs.

The enzyme L-asparaginase produces tumor cell death through the activation of apoptosis. In particular, the drug deprives the cell of asparagine. This amino acid is crucial for protein synthesis because the enzyme catalyzes the hydrolysis of circulating asparagine to aspartic acid and ammonia. Consequently, L-asparaginase stops the progression of cells through the cell cycle. Its unique mechanism of action supports its inclusion with other antitumor drugs as part of a chemotherapeutic program, especially in the treatment of acute lymphoblastic leukemia and other lymphoid malignancies. The drug is conveniently administered I.V. or intramuscularly (I.M.) in a variety of regimens and schedules but poses a serious toxicity threat due to its bacterial origin—that is, antigenicity. Hypersensitivity and anaphylactic reactions occur in 5 to 20% of patients, most notably when administered intermittently. Other toxicities related to its mechanism of action include hyperglycemia, coagulopathies, immunosuppression, and hemorrhage, all of which are potentially fatal.

Unlike the antibacterial, antifungal, and antiviral antibiotic drugs, antibiotic antineoplastics are capable of disrupting DNA-dependent RNA synthesis and inhibiting mitosis. This class of natural products is derived from a variety of eubacterial and actinomycete genera.* The drugs act on rapidly proliferating normal and neoplastic cells, and with one exception (belomycin), are *phase nonspecific*. In addition to the GI, hematologic, dermatologic, and metabolic toxicities, some of the more peculiar ADRs associated with these antibiotics are listed in Table 21.3. For example, hemolytic uremic syndrome, mostly observed with mitomycin C, is characterized by microangiopathic hemolytic anemia (anemia within small blood vessels), thrombocytopenia (decrease in platelet count), and irreversible renal failure.

Pentostatin is often classified as an antimetabolite, although it is derived from microorganisms. It acts as a transition state analog and interferes with adenosine deaminase. The enzyme-specific inhibition results in accumulation of ATP and deoxyadenosine nucleotides, blockage of ribonucleotide reductase, and suspension of DNA synthesis. Interestingly, although most of the antineoplastic antibiotics have antibacterial and antifungal properties, their toxicities preclude their routine use in the treatment of infectious diseases.

21.5.2 Hormones and Antagonists

The basis of hormonal anticancer therapy is the interaction of synthetic nonsteroidal chemicals with nuclear and cell membrane growth-stimulatory receptor proteins. The drugs have the advantage of greater specificity for tissues responsive to

* Compounds and their species of derivation include actinomycin D (dactinomycin, Cosmegen®) from *Streptomyces parvullus*; pentostatin (Nipent®) from *S. antibioticus*; anthracycline antibiotics (daunorubicin, Cerubidine®, and doxorubicin, Adriamycin®) from *S. peucetius*; bleomycin (Blenoxane®) from *S. verticillus*; and mitomycin C (Mutamycin®) from *S. caespitosus*.

their counterbalancing effects, such as breast, endometrium, testes, and prostate. Consequently, they are capable of occupying nuclear or membrane steroid receptors, redirecting the feedback mechanism, and inhibiting the normal or abnormal function of the tissue. The therapeutic advantage lies in their proclivity to inhibit uncontrolled cell proliferation without significant direct cytotoxic effects. The major ADRs are listed in Table 21.4. Antineoplastic hormones are used initially or as adjunctive therapy in the palliative treatment of malignancies susceptible to the drugs, as noted in Table 21.5.

21.5.3 Platinum Coordination Complexes

Carboplatin (Paraplatin®) and cisplatin (Platinol AQ®) are platinum coordination compounds whose action is similar to that of the alkylating agents. Platinum coordination complexes are phase-nonspecific drugs that produce interstrand DNA cross-links rather than DNA–protein intercalation. Their toxicities (Table 21.4) and net effect (interference with DNA synthesis and integrity) are similar to those of the alkylating agents. The compounds are used in the initial and secondary treatment of ovarian carcinoma, testicular tumors, and advanced bladder cancer.

21.5.4 Substituted Urea

Hydroxyurea (Hydrea®) is a substituted urea with antimetabolite actions. It is *S phase specific* and interferes with the conversion of ribonucleotides to deoxyribonucleotides by selectively blocking the enzyme ribonucleotide reductase. In addition, it inhibits the incorporation of thymidine into DNA. The drug is indicated in the treatment of melanomas and leukemias. It is also used concomitantly with irradiation therapy in the management of squamous cell carcinoma of the head and neck.

TABLE 21.4
Classification of Selected Hormonal Antineoplastic Agents and Predominant Adverse Drug Reactions (ADRs)

Category	Nonproprietary Name	Predominant Target of Inhibition	Predominant ADRs
Progestins	Medroxyprogesterone, megestrol acetate	Interact with cell membrane and nuclear growth-stimulatory receptor proteins in tissues responsive to their counterbalancing effects	Weight gain, thromboembolism
Estrogens	DES, ethinyl estradiol		Thromboembolism
Androgens	Testolactone		CNS, GI
Antiestrogens	Tamoxifen		Thromboembolism, hot flashes
Antiandrogens	Flutamide, bicalutamide		Liver, GI, blood gynecomastia
GRH analog	Leuprolide (LHRH agonist)		CV, CNS, GI, GU, gynecomastia

CNS: central nervous system; CV: cardiovascular; GI: gastrointestinal; GU: genitourinary.

TABLE 21.5
Hormonal Antineoplastic Agents and Indicated Malignancies

Nonproprietary Name	Proprietary Name	Indicated Malignancy
Medroxyprogesterone	Depo-Provera®	Endometrial or renal carcinoma
Megestrol acetate	Megace®	Breast or endometrial carcinoma;appetite enhancement for AIDS patients
DES, estradiol	Various	Progressing prostatic cancer
Testolactone	Teslac®	Advanced breast carcinoma
Tamoxifen	Nolvadex®	Breast cancer, treatment and preventive
Flutamide	Eulexin®	Prostate carcinoma
Bicalutamide	Casodex®	
Leuprolide (LHRH agonist)	Lupron®	Advanced prostatic carcinoma

DES: diethylstilbestrol.

REFERENCES

SUGGESTED READINGS

Conti, M., Targeting ion channels for new strategies in cancer diagnosis and therapy. *Curr. Clin. Pharmacol.* 2, 135, 2007.

Douer, D., Is asparaginase a critical component in the treatment of acute lymphoblastic leukemia? *Best Pract. Res. Clin. Haematol.* 21, 647, 2008.

Eckhardt, S., Recent progress in the development of anticancer agents. *Curr. Med. Chem. Anti-Canc. Agents.* 2, 422, 2002.

Fimognari, C., Lenzi, M., and Hrelia, P., Apoptosis induction by sulfur-containing compounds in malignant and nonmalignant human cells. *Environ. Mol. Mutagen.* 50, 171, 2009.

García-Gómez, R., Bustelo, X.R., and Crespo, P., Protein-protein interactions: emerging oncotargets in the RAS-ERK pathway. *Trends Cancer* 4, 616, 2018.

Hansten, P.D., Horn, J.R., and Porter, R.S., *Drug Facts and Comparisons*, Lippincott Williams & Wilkins, 2017, chapter 13.

Huang, R.S. and Ratain, M.J., Pharmacogenetics and pharmacogenomics of anticancer agents. *CA Cancer J. Clin.* 59, 42, 2009.

Iwamoto, T., Clinical application of drug delivery systems in cancer chemotherapy: review of the efficacy and side effects of approved drugs. *Biol Pharm Bull.* 36, 715, 2013.

Levêque, D., Off-label use of anticancer drugs. *Lancet Oncol.* 9, 1102, 2008.

Wu, C.P., Calcagno, A.M., and Ambudkar, S.V., Reversal of ABC drug transporter-mediated multidrug resistance in cancer cells: evaluation of current strategies, *Curr. Mol. Pharmacol.* 1, 93, 2008.

REVIEW ARTICLES

Blankenship, J.R., Steinbach, W.J., Perfect, J.R., Heitman, J., Teaching old drugs new tricks: reincarnating immunosuppressants as antifungal drugs. *Curr. Opin. Investig. Drugs* 4, 222, 2003.

Bodo, A., Bakos, E., Szeri, F., Váradi, A., and Sarkadi, B., The role of multidrug transporters in drug availability, metabolism and toxicity. *Toxicol. Lett.* 140, 133, 2003.
Chabner, B.A., Cytotoxic agents in the era of molecular targets and genomics. *Oncologist* 7, 34, 2002.
Chargari, C., Toillon, R.A., Macdermed, D., Castadot, P., and Magné, N., Concurrent hormone and radiation therapy in patients with breast cancer: what is the rationale? *Lancet Oncol.* 10, 53, 2009.
Dianat-Moghadam, H., Rokni, M., Marofi, F., Panahi, Y., and Yousefi, M., Natural killer cell-based immunotherapy: from transplantation toward targeting cancer stem cells. *J. Cell Physiol.* 2018.
Dy, G.K., Haluska, P., and Adjei, A.A., Novel pharmacological agents in clinical development for solid tumours. *Expert Opin. Investig. Drugs* 10, 2059, 2001.
Elsayed, Y.A. and Sausville, E.A., Selected novel anticancer treatments targeting cell signaling proteins. *Oncologist* 6, 517, 2001.
Furst, A., Can nutrition affect chemical toxicity? *Int. J. Toxicol.* 21, 422, 2002.
Gerber, D.E. and Chan, T.A., Recent advances in radiation therapy, *Am. Fam. Physician* 78, 1254, 2008.
Gilligan, T. and Kantoff, P.W., Chemotherapy for prostate cancer. *Urology*, 60, 94, 2002.
Gold, K.A., Lee, H.Y., and Kim, E.S., Targeted therapies in squamous cell carcinoma of the head and neck. *Cancer*, Epub ahead of print, 2009.
Hogberg, T., Glimelius, B., and Nygren, P., A systematic overview of chemotherapy effects in ovarian cancer. *Acta Oncol.* 40, 340, 2001.
Iqbal, S. and Lenz, H.J., Targeted therapy and pharmacogenomic programs. *Cancer* 97, 2076, 2003.
Langebrake, C., Reinhardt, D., and Ritter, J., Minimising the long-term adverse effects of childhood leukaemia therapy. *Drug Saf.* 25, 1057, 2002.
Le Saux, O. and Falandry, C., Toxicity of cancer therapies in older patients. *Curr. Oncol. Rep.* 20, 64, 2018.
Li, L., Mok, H., Jhaveri, P., Bonnen, M.D., Sikora, A.G., Eissa, N.T., Komaki, R.U., and Ghebre, Y.T., Anticancer therapy and lung injury: molecular mechanisms. *Expert Rev. Anticancer Ther.* 23, 1, 2018.
Maeda, H., Bharate, G.Y., and Daruwalla, J., Polymeric drugs for efficient tumor-targeted drug delivery based on EPR-effect. *Eur. J. Pharm. Biopharm.* 71, 409, 2009.
Martino, E., Casamassima, G., Castiglione, S., Cellupica, E., Pantalone, S., Papagni, F., Rui, M., Siciliano, A.M., and Collina, S.,Vinca alkaloids and analogues as anti-cancer agents: looking back, peering ahead. *Bioorg. Med. Chem. Lett.* 28, 2816, 2018.
Medina, P.J., Sipols, J.M., and George, J.N., Drug-associated thrombotic thrombocytopenic purpura-hemolytic uremic syndrome. *Curr. Opin. Hematol.* 8, 286, 2001.
Mitchell, R.A., Luwor, R.B., and Burgess, A.W., Epidermal growth factor receptor: structure-function informing the design of anticancer therapeutics. *Exp. Cell Res.* 2018. pii: S0014-4827(18)30635-9.
Nabholtz, J.M. and Riva, A., Taxane/anthracycline combinations: setting a new standard in breast cancer? *Oncologist* 6, 5, 2001.
Nishikawa, G., Luo, J., and Prasad, V., A comprehensive review of exceptional responders to anticancer drugs in the biomedical literature. *Eur. J. Cancer* 101, 143, 2018.
Prinz, H., Recent advances in the field of tubulin polymerization inhibitors. *Expert Rev. Anticancer. Ther.* 2, 695, 2002.
Reiche, E.M., Nunes, S.O., and Morimoto, H.K., Stress, depression, the immune system, and cancer. *Lancet Oncol.* 5, 617, 2004.
Royce, M.E., Hoff, P.M., and Pazdur, R., Novel oral chemotherapy agents. *Curr. Oncol. Rep.* 2, 31, 2000.

Tan, S.H., Lee, S.C., Goh, B.C., and Wong, J., Pharmacogenetics in breast cancer therapy. *Clin. Cancer Res.* 14, 8027, 2008.

Tanaka, S. and Arii, S., Molecularly targeted therapy for hepatocellular carcinoma. *Cancer Sci.* 100, 1, 2009.

Tew, W.P. and Lichtman, S.M., Ovarian cancer in older women. *Semin. Oncol.* 35, 582, 2008.

Tomaszewski, J.E., Multi-species toxicology approaches for oncology drugs: the US perspective. *Eur. J. Cancer* 40, 907, 2004.

Zucchi, R. and Danesi, R., Cardiac toxicity of antineoplastic anthracyclines. *Curr. Med. Chem. Anti-Canc. Agents* 3, 151, 2003.

22 Vitamins

22.1 INTRODUCTION

Vitamins, as the Latin derivation of the name implies,* are essential for the maintenance of adequate health and life. They are diverse organic substances provided in small quantities in the diet and are found in a variety of chemical forms and structures. Vitamins have assorted essential biochemical roles in contributing to the maintenance of health and have unique therapeutic places in the treatment of related disorders. Originally, most vitamins were used for the prevention of deficiency syndromes associated with inadequate nutritional intake. For instance, in the presence of a deficiency of a particular nutrient and the subsequent induction of the corresponding condition, adequate replacement of the deficient substance resulted in a cure. And until recently, it was reasonable to conclude that for a person following a balanced diet, no additional vitamin supplementation was necessary. This assumption was valid for normal healthy children, adolescents, and adults in the absence of deficiency (with the exception of pregnancy and lactation). This message has not been heeded in recent years, however, as evidenced by the growing consumption (*megadoses*) and infatuation with vitamin supplementation in the United States. This was encouraged by the misinformation that if vitamins are important for the maintenance of health, then greater ingestion must necessarily provide better health. This belief has been defended by the equivocal results of clinical and scientific studies interpreted out of context. In particular, such studies have concluded that clinical benefits, such as prevention of aging, can be obtained from the beneficial effects of antioxidants against factors that contribute to oxidative stress.

The Food and Nutrition Board† publishes the Dietary Reference Intakes (DRIs) as standards for essential nutrients. The standards are based on periodic review of the scientific evidence for most healthy persons under normal daily stresses. Almost all of these nutrients are provided by a well-balanced dietary intake plan that includes the four basic food groups. In addition, RDA information is included in individual vitamin monographs. Table 22.1 outlines an abbreviated listing of DRIs of vitamins and minerals for children (aged 0 to 8 years) and adult males and females (aged 9 to 70+ years). The values are based on normal, healthy, height- and weight-adjusted individuals and have changed dramatically over the last 5 years.

In general, vitamin supplementation is warranted in situations where a suspected vitamin deficiency contributes to the condition. In the United States, the *relative risk* for the development of vitamin deficiencies may be higher in the following individuals: 1. those living below the poverty level; 2. patients with malabsorption syndromes; 3. patients undergoing treatment with antibiotics that alter normal

---

* *Vita (L.)*, vital, for life.

† U.S. National Research Council, Institute of Medicine of the National Academy of Sciences; DRI, Food and Nutrition Board, Dietary Reference Intakes, Office of Dietary Supplements, NIH.

TABLE 22.1
Dietary Reference Intakes (DRIs), Daily Allowances of Vitamins and Minerals

Nutrient	Infants	Children	Males	Females	Pregnant/lactating
Protein (g)	9.1–11	13–19	34–56	34–46	71
		Fat-soluble Vitamins			
Vitamin A (μg)	400–500	300–400	600–900	600–700	750–1300
Vitamin D (μg)	5	5	5–15	5–15	5
Vitamin E (mg)[a]	4–5	6–7	11–15	11–15	15–19
Vitamin K (μg)	2–2.5	30–50	60–120	60–90	75–90
		Water-Soluble Vitamins			
Vitamin C (mg)	40–50	15–25	45–90	45–75	80–120
Thiamine (B_1, mg)	0.2–0.3	0.5–0.6	0.9–1.2	0.9–1.1	1.4
Riboflavin (B_2, mg)	0.3–0.4	0.5–0.6	0.9–1.3	0.9–1.1	1.4–1.6
Niacin (B_3, mg)	2–4	6–8	12–16	12–14	18/17
Pyridoxine (B_6, mg)	0.1–0.3	0.5–0.6	1.0–1.7	1.0–1.5	1.9–2.0
Folic Acid (μg)	65–80	150–200	300–400	300–400	600/500
Vitamin B_{12}[b] (μg)	0.4–0.5	0.9–1.2	1.8–2.4	1.8–2.4	2.6–2.8
		Minerals			
Biotin (μg)	5–6	8–12	20–30	20–30	30/35
Calcium (mg)	210–270	500–800	1300/1200	1300/200	1300/1200
Choline (mg)	125–150	200–250	375–550	375–425	450/550
Pantothenic acid (mg)	1.7–3.0	2–3	4–5	4–5	6/7
Phosphorus (mg)	100–275	460–500	1250–700	1250–700	1250–700
Magnesium (mg)	30–75	80–130	240–420	240–320	400–320
Iron (mg)	0.27–11	7–10	8–11	8–18	27–9
Zinc (mg)	2–3	3–5	8–11	8–9	11–13
Iodine (μg)	110–130	90	120–150	120–150	220–290
Selenium (μg)	15–20	20–30	40–55	40–55	60–70

Age ranges (life stage group), in years: infants, 0–6 mo., 6–12 mo.; children, 1–3, 4–8; males and females, 9–70+. Ranges of daily allowances are determined by age, height, and weight (not shown) within group; in general, requirements increase with age, except with some vitamins (e.g., vitamins D, B_1, B_2, B_3, calcium, and phosphorus), for which the requirement drops for adult males and females over 25 years old. Requirements are usually higher during the first 6 months of lactation than during the second 6 months.

[a] As α-tocopherol: 1 mg d-α-tocopherol (α-TE) = 1.49 IU vitamin E.

[b] Cyanocobalamin.

Adapted and condensed from *Dietary Reference Intakes (DRIs): Recommended Intakes for Individuals*, Vitamins Food and Nutrition Board, Institute of Medicine, National Academies, National Academy Press, Washington DC, 2004. From: https://www.ncbi.nlm.nih.gov/books/NBK56068/table/summarytables.t2/?report=objectonly,last_viewed_March_2019.

vitamin-synthesizing bacterial flora; 4. individuals with behavioral problems that encourage poor dietary intake (e.g., anorexia nervosa); and 5. people with inadequate intake of nutritionally rich foods, which leads to decreased ingestion or poor absorption of vitamins (poorly planned vegetarian diets, reducing diets, and among alcoholics and the elderly).

Vitamin toxicity due to overdose is documented in clinical situations but relies on experimental scientific evidence for an adequate description of such conditions. More recently, the adverse drug reactions (ADRs) of retinoic acid derivatives of vitamin A have been well documented, and their toxicology and pharmacology will be discussed later in this chapter.

Vitamins are classified according to their biological inclination for distribution and elimination from the body. The *water-soluble* vitamins are readily eliminated in the urine and consequently, have a low incidence of toxic effects. Biochemically, they act as cofactors in most physiologically important reactions. In contrast, the *fat-soluble* vitamins are readily distributed throughout the body, with a propensity for storage in lipid-rich tissue. As a result, they tend to accumulate in physiologic compartments, risking potential cumulative toxicity. Biochemically, they elicit hormone-like activity at physiologic targets.

This chapter emphasizes the toxicity and deficiencies associated with vitamins and vitamin supplementation whose adverse reactions have been historically and therapeutically documented. It is important to note that natural diets are rarely deficient in single nutrients, and it is rare for clinical manifestations to develop from single causes. Consequently, the detection and diagnosis of vitamin deficiency, or chronic toxicity, are not without ambiguity.

22.2 FAT-SOLUBLE VITAMINS

22.2.1 Vitamin A and Retinoic Acid Derivatives

Vitamin A, and its derivatives, is essential for proper maintenance of visual acuity, dental development, skeletal muscle and bone growth, corticosteroid synthesis, and embryonic development and reproduction. Vitamin A also influences the differentiation of epithelial membranes, particularly the corneal, gastrointestinal (GI), and genitourinary epithelia. Derivatives of vitamin A include retinol, retinal, retinoic acid, and β-carotene (a precursor). These physiologically important factors support proper functioning of the reproductive cycle, visual acuity, somatic growth and differentiation, and visual adaptation to darkness, respectively. β-Carotene is a naturally occurring vitamin A precursor, found primarily in dark green and yellow-orange vegetables, and possesses antioxidant and immunostimulating (anticancer) activity. A balanced diet including fish, liver, meat, carrots, and dairy products provides the RDA for vitamin A.

Oral or parenteral administration of vitamin A (Aquasol A® injection, various tablet and capsule formulations) is indicated for the treatment of conditions associated with, but not exclusive to, vitamin A deficiency states. Deficiency syndromes have been associated with kwashiorkor,* xerophthalmia,† keratomalacia,‡ and malabsorption syndromes.

* A disease of very young African natives as a result of severe protein deficiency. It is characterized by anemia, edema, swollen abdomen, skin and hair depigmentation, and hypoalbuminemia.

† Excessive conjunctival and corneal dryness.

‡ The cornea becomes softened and vulnerable to ulceration and infection, which may lead to blindness.

Hypervitaminosis A syndrome develops with chronic ingestion of megadoses of the vitamin or in the presence of hepatic disease.* Table 22.2 lists the toxic manifestations of hypervitaminosis syndromes as well as the deficiency states. In general, acute doses of 25,000 IU/kg result in toxicity. Prolonged excessive consumption of vitamin A for 6 to 15 months (4000 IU/kg) also produces a toxic syndrome, although chronic adverse reactions are more insidious and protracted. Most symptoms resolve with prompt withdrawal of vitamin supplementation, although increased intracranial pressure requires further palliative measures.

As noted earlier, the retinoids are derivatives of retinoic acid that promote epithelial cell differentiation, keratinization, and local inflammation. These properties are advantageous in the topical or oral treatment of acne vulgaris or psoriasis, attributable to the compounds' ability to normalize follicular epithelial cell differentiation. Therapeutically, the desired effect results in the reduction of the severity and formation of the microcomedones† characteristic of acne. The adverse effects of most of the topical products are related to transient or temporary local inflammatory responses (Table 22.2). Oral retinoids, such as isotretinoin, are reserved for severe recalcitrant nodular acne. The retinoids, however, are contraindicated in women of childbearing potential because of teratogenic risk. Major human fetal abnormalities have been reported with oral isotretinoin ingestion, including skeletal, neurological, and cardiovascular anomalies. In addition, a variety of adverse effects appear with routine therapeutic use of oral isotretinoin.

22.2.2 Vitamin D

Vitamin D is a fat-soluble vitamin with hormone-like activity. Calcitriol is the most active form of vitamin D_3.‡ In conjunction with parathyroid hormone (PTH), calcitriol maintains blood calcium levels by promoting GI absorption of calcium, phosphorus, and magnesium. Calcitriol stimulates renal tubular reabsorption of calcium and mobilizes calcium from bone. Negative feedback of blood calcium controls secretion of calcitonin from the thyroid gland and PTH from the parathyroid gland. Vitamin D_2 is a plant vitamin used to fortify milk and cereal products. The role of the parathyroid glands and their interaction with calcium, vitamin D, and PTH secretion are discussed in some recent reviews (Fleet 2017; Khundmiri et al., 2016).

Oral or parenteral administration of the active forms of vitamin D is indicated in the management of hypocalcemia. The condition is prevalent in renal dialysis patients and individuals with hypoparathyroidism (treated with calcitriol), refractory *rickets*, and hypophosphatemia (treated with vitamin D_2). In addition, the

* Although mostly innocuous, the consumption of megadoses (greater than 180 mg/day of β-carotene) is manifested by yellow-orange skin discoloration.

† Dilated hair follicles filled with keratin and sebum (pustules).

‡ Calcitriol, also known as 1,25-dihydroxy vitamin D_3, is converted from 7-dehydrocholesterol to cholecalciferol (vitamin D_3) by the initial action of UV light on skin. Intermediate metabolic reactions in the liver and kidney follow. Similarly, the major transport form of vitamin D, 1,25-dihydroxyergocalciferol (vitamin D_2), is converted from provitamin D_2 by exposure of skin to UV light followed by liver and kidney metabolism.

TABLE 22.2
Fat-Soluble Vitamins and Toxic Manifestations of Excessive Consumption or Administration

Vitamin	Derivatives	Proprietary Name	Deficiency State	Common ADRs
Vitamin A	Retinol, retinal, retinoic acid	Palmitate A 5000 tablets, Aquasol A injection, various tablets and capsules	Kwashiorkor, xerophthalmia, keratomalacia	Acute: anorexia, nervousness, increased intracranial pressure, blurred vision; chronic: hepatomegaly, alopecia, dry scaly skin, bone pain and demineralization; cheilitis[a]
	β-Carotene	Various capsules		Yellow-orange skin discoloration
Retinoids	Adapalene	Differen gel	None	Erythema, scaling, dryness, pruritus, burning, photosensitivity, skin discoloration
	Tretinoin	Retin-A gels, creams		
	Tazarotene	Tazorac gel, cream		
	Isotretinoin[b]	Accutane capsules	None	Major human fetal malformations involving neurological, skeletal, muscular, hormonal, and CV systems; ADRs include pancreatitis, psychiatric disorders, visual impairment, dermatologic, CNS, GI, GU, IBS
Vitamin D	Calcitriol (D_3)	Rocaltrol	Rickets (children); osteomalacia (adults)	*Acute*: weakness, headache, NV, constipation, bone pain; *chronic*: anorexia, polydipsia, polyuria, nephrocalcinosis, vascular calcification; *chronic O.D.*: hypercalcemia, hypercalciuria, hyperphosphatemia
	Cholecalciferol (D_3)	Delta-D		
	Ergocalciferol (D_2)	Drisdol, Calciferol		
Vitamin E	*dl*-α-tocopherols	Aquavit-E, various products and dosage forms	Anemia, thrombocytosis, increased platelet aggregation (uncommon)	Hypervitaminosis E: NVD, weakness, thrombophlebitis, increased lipids and hormone levels, breast tumors, gynecomastia, altered immunity
Vitamin K	Phytonadione (K_1)	Mephyton, AquaMEPHYTON	Hemorrhagic disorders (uncommon)	Severe allergic reactions, especially on injection; transient flushing, pain at site; hemolytic anemia in infants; scleroderma-like lesions, hyperbilirubinemia

[a] Cheilitis: inflammation of the lips.
[b] Category X: should not be used during pregnancy or in women with the potential for becoming pregnant.
CNS: central nervous system; CV: cardiovascular; GI: gastrointestinal; GU: genitourinary; IBS: inflammatory bowel syndrome; NVD: nausea, vomiting, diarrhea; O.D.: overdose.

compounds are used as dietary supplements or in the management of vitamin D deficiency (vitamin D_3).

Deficiency states of vitamin D result in *rickets* in growing children and *osteomalacia* in adults, the causes and effects of which are summarized in Table 22.3. In both cases, limited exposure to sunlight is probably the most important reason for the development of a deficiency state leading to hypocalcemia. Hypocalcemia, in turn, stimulates PTH secretion, which promotes renal α1-hydroxylase activity (responsible for conversion of precursors to D_2 and D_3 in the kidney). The net effect is the enhancement of GI calcium absorption, an increase in calcium mobilization from bone, and stimulation of renal excretion of phosphate. Thus, although calcium blood levels may return to normal, loss of renal phosphate and hypophosphatemia further imperil bone mineralization. Persistent failure of bone formation in adults risks eventual loss of skeletal mass (*osteopenia*),* producing weak, fracture-sensitive long bones and vertebrae. In children, gross skeletal changes depend on age and the severity of the deficiency. In neonates, abnormalities are manifested as softened, flattened, buckled bones resulting from pressure stress. Deformation of cranial bones, pelvic bones, and the chest cavity (*pigeon breast deformity*) completes the presentation. In older children, deformities are more likely to affect the spine, pelvis, and long bones (*lumbar lordosis* and *bowing of the legs*).

Administration of vitamin D in excess of daily requirements may cause clinical signs of acute or chronic overdosage (*hypervitaminosis D syndrome*), most of which are related to elevated calcium levels (Table 22.2)[†]. Concomitant high intake of calcium and phosphate may cause the development of similar abnormalities. Treatment of accidental overdose requires general supportive measures, whereas treatment of hypervitaminosis D with hypercalcemia consists of prompt

TABLE 22.3
Causes and Effects of Rickets in Children, and Osteomalacia in Adults, Resulting from Vitamin D Deficiency

Metabolic, Physiologic, or Behavioral Cause	Pathologic or Net Effect on Vitamin D
Inadequate exposure to sunlight; descendants of African or other dark-skinned races	Reduced endogenous synthesis
Poor or inadequate nutrition (poor diet)	Poor GI absorption
Cytochrome P-450–inducing drugs[a]	Enhanced degradation
Liver, renal disease	Impaired synthesis
Chronic antacid ingestion; renal disease	Phosphate depletion

[a] Isoniazid, phenobarbital.

* Although *osteopenia* is an osteomalacia that is histologically similar to other forms of osteoporosis, the hallmark characteristic is an excess of persistent unmineralized osteoid, the protein matrix of the bone.

[†] Interestingly, prolonged exposure to sunlight does not produce excess vitamin D.

withdrawal of vitamin D supplements, a low-calcium diet, laxatives, and attention to serum electrolyte imbalances and cardiac function. Major blood vessels, myocardium, and kidneys are at risk of developing ectopic calcification. Hypercalcemic crisis with dehydration, stupor, and coma requires more immediate attention, such as prompt hydration, diuretics, short-term hemodialysis, corticosteroids, and urine acidification.

22.2.3 VITAMIN E

The most biologically active form of vitamin E, a fat-soluble vitamin, is *dl*-α-tocopherol. The antioxidant properties of vitamin E help to protect cells from oxidative processes. For example, it preserves the integrity of the red blood cell membrane, stimulates the production of cofactors in steroid metabolism, and suppresses prostaglandin synthesis and platelet aggregation. Vitamin E has numerous unlabeled uses, including the prevention of cancer, preservation of ageless skin, and stimulation of sex drive. However, it is only indicated for the treatment of nutritional states related to vitamin E deficiency. It is widely distributed in nature, predominantly found along with polyunsaturated fatty acids in vegetables and their oils, meat, dairy, wheat, soy, and nuts. Its ubiquitous presence, coupled with a low RDA, makes clinical deficiency of vitamin E a rare phenomenon. Premature infants and patients with malabsorption syndromes, however, are more likely to develop deficiency, defined as having tocopherol levels less than 0.5 mg/dl. Signs and symptoms of deficiency, particularly in premature infants, include hemolytic anemia, thrombocytosis, and increased platelet aggregation.

Symptoms associated with hypervitaminosis E are summarized in Table 22.2 and are usually experienced at doses greater than 3000 IU.

22.2.4 VITAMIN K

Vitamin K, and its equally active synthetic analog, K_1 (phytonadione), is a fat-soluble vitamin. Vitamin K promotes the hepatic synthesis of coagulation factors II (prothrombin), VII (proconvertin), IX (plasma thromboplastin), and X (Stuart factor). The vitamin acts as a cofactor for hepatic posttranslation carboxylation of glutamic acid residues to the active γ-carboxyl glutamate moieties of the clotting proteins as well as anticoagulant proteins C, S, and osteocalcin (bone matrix).

Phytonadione is indicated in the treatment of coagulation disorders related to vitamin K–associated decreased synthesis of coagulation factors. It is also valuable in the treatment of prothrombin deficiency (hypoprothrombinemia) induced by oral or parenteral anticoagulants, salicylates, or antibiotics. Sources of the vitamin are provided in various foods, particularly green vegetables, dairy products, meats, grains, and fruits. In addition, it is synthesized by normal intestinal bacterial flora.

As with vitamin E, vitamin K deficiency is uncommon except in premature and breastfed newborns, patients with malabsorption syndromes, and patients receiving chronic, anticoagulant, or broad-spectrum antibiotic therapy.

Hypervitaminosis K is rare principally because the drug is not available over the counter (OTC) as a dietary supplement.

22.3 WATER-SOLUBLE VITAMINS

22.3.1 THIAMINE

Thiamine (vitamin B_1) is essential for normal aerobic metabolism and tissue development, proper transmission of nerve impulses, and synthesis of acetylcholine. It combines with adenosine triphosphate (ATP) to form thiamine pyrophosphate, the active form of thiamine. As such, it serves as a coenzyme in many α-ketoacid decarboxylation reactions (such as with pyruvate) and transketolation (pentose-phosphate pathway) reactions.

Thiamine is indicated in the oral or parenteral treatment of B_1 deficiency syndromes (*beriberi*) and in neuritis of pregnancy. Sources of the vitamin include brewer's yeast, meats, nuts, grains, and dairy products.

Beriberi is the classic deficiency syndrome frequently associated with chronic alcoholism, renal dialysis patients, and individuals with liver and biliary dysfunction or poor diets (polished rice). The syndrome is characterized by peripheral neuritis, muscle wasting (*dry beriberi*), and cardiac failure and edema (*wet beriberi*). Other conditions due to thiamine deficiency include lactic acidosis and *Wernicke–Korsakoff* syndrome.*

Hypersensitivity reactions, anaphylactic shock, and hypervitaminosis syndrome are rare with oral or intravenous doses. The toxic manifestations, proprietary names, and deficiency states of the water-soluble vitamins are summarized in Table 22.4.

22.3.2 RIBOFLAVIN

Vitamin B_2 is a component of flavin mononucleotide (FMN) and flavin adenine dinucleotide (FAD). These two coenzymes catalyze a variety of oxidation–reduction reactions, including glucose oxidation and amino acid deamination. The vitamin is found in fish, poultry, dairy products, and leafy vegetables. Poor intestinal absorption of riboflavin limits its toxicity. The deficiency state (ariboflavinosis) is rare and often masked by other nutritional insufficiencies. Early symptoms include cheilosis (fissures in the mouth), glossitis (inflammation of the tongue), interstitial ophthalmic keratosis, and dermatitis.

22.3.3 NIACIN

Vitamin B_3† (nicotinic acid) is a component of nicotinamide adenine dinucleotide (NAD, coenzyme I) and nicotinamide adenine dinucleotide phosphate (NADP, coenzyme II). The two coenzymes catalyze oxidation–reduction reactions that act as electron acceptors and hydrogenases. Niacin is derived from niacinamide or tryptophan and occurs naturally in most red meats, fish, poultry, dairy products, nuts, and vegetables. Niacin deficiency (*pellagra*, Ital. *rough skin*) is characterized by dermatitis, dementia, and diarrhea (*3Ds*). It is often seen in chronic alcoholism,

---

* The syndrome is characterized by encephalopathy and psychosis, manifested primarily by ophthalmic nerve and muscle paralysis.

† The active forms, niacinamide and nicotinamide, are used synonymously.

TABLE 22.4
Water-Soluble Vitamins, Deficiency States, and Toxic Manifestations of Excessive Consumption or Administration

Vitamin	Compound	Proprietary Name	Deficiency State	ADRs Associated with Excessive Ingestion
Vitamin B_1	Thiamine	Thiamilate, various	Beriberi, lactic acidosis, Wernicke–Korsakoff	Pruritus, weakness, nausea, urticaria, hypersensitivity, anaphylactic shock
Vitamin B_2	Riboflavin	Various	Ariboflavinosis (rare)	No toxicity noted, limited absorption
Vitamin B_3	Niacin	Various	Pellagra	Dermatologic: cutaneous flushing, rash, dry skin; GI: ulcer, NVD, abdominal pain; hepatoxicity, hyperglycemia
Vitamin B_6	Pyridoxine HCl	Various	Dermatologic, hematologic, CNS	Sensory neuropathy, ataxia, digital and perioral numbness, paresthesias
Folic acid	Pteroylglutamic acid (folate)	Folvite, various	Megaloblastic anemia	CNS: altered sleep patterns, irritability, confusion; GI: N, anorexia, abdominal distention
Vitamin B_{12}	Cyanocobalamin	Various (oral)	Pernicious anemia	Nontoxic
		Nascobal (I.N.)		Paresthesia, headache, glossitis, nausea
		LA-12 (injection)		Pulmonary edema, CHF, itching, anaphylactic shock
Vitamin C	Ascorbic acid	Various	Scurvy	GI disturbances, poor wound healing, urinary calculi

CHF: congestive heart failure; CNS: central nervous system; GI: gastrointestinal; GU: genitourinary; I.N.: intranasal; NVD: nausea, vomiting, diarrhea.

in malabsorption syndrome, and in patients receiving isoniazid (an antituberculosis antibiotic). Pharmacologically, nicotinic acid, but not nicotinamide, is effective in reducing serum lipids. In addition, it triggers peripheral vasodilation and increases blood flow by stimulating histamine release. These properties are distinct from its nutrient role. Consequently, nicotinic acid is useful in the treatment of hypercholesterolemia and hyperlipidemia and in the management of niacin deficiency and pellagra.

22.3.4 Folic Acid

Pteroylglutamic acid (folate) is found in leafy vegetables, organ meats, and yeast. Physiologically, folic acid is required for nucleoprotein synthesis and the maintenance of hematopoiesis. It is converted intracellularly to tetrahydrofolic acid, which acts as a cofactor in the biosynthesis of purines and thymidylates of nucleic acids. Consequently, deficiency of folic acid (along with vitamin B_{12}) is often seen during pregnancy and in malabsorption conditions and is responsible for defective DNA synthesis. The condition is manifested by the production of enlarged immature red cells characteristic of megaloblastic anemia. Although folic acid is relatively nontoxic, some allergic, central nervous system (CNS), and GI reactions have been noted with large doses (Table 22.4).

22.3.5 Cyanocobalamin (Vitamin B_{12})

Dietary deficiency of vitamin B_{12} is rare, since the nutrient is found in all meats and dairy products, but it is absent from plant foods. Strict vegetarians and individuals with malabsorption syndrome, especially those experiencing atrophic gastritis who cannot absorb B_{12}, are prone to developing pernicious anemia. The presence of intestinal parasitic infections has been noted to induce B_{12} deficiency.* Signs and symptoms of the syndrome, however, may not be evident for 3 to 5 years. Pernicious anemia develops principally because of the lack of secretion of intrinsic factor in the stomach, which is necessary for absorption of cyanocobalamin in the ileum.

Vitamin B_{12} is essential for cell growth, hematopoiesis, and nucleic acid and neuronal myelin synthesis. Several preparations of cyanocobalamin are available and are indicated for the management of B_{12} deficiency (all dosage forms) and pernicious anemia (intranasal gel and injection only).

22.3.6 Ascorbic Acid (Vitamin C)

Vitamin C is abundant in citrus fruits, strawberries, tomatoes, and leafy vegetables. It is involved in numerous oxidation–reduction reactions, including the synthesis of connective tissue components, such as chondroitin sulfate and collagen. Ascorbic acid supports the hydroxylation of proline to hydroxyproline in collagen. In addition,

* *Diphyllobothrium latum*, the broad or fish tapeworm, extracts vitamin B_{12} from the intestinal lumen, preventing its normal absorption.

it improves the absorption of iron from the GI tract, promotes the synthesis of catecholamine neurotransmitters, and is an important cofactor in the wound healing process. Because of this last role of vitamin C, and its antioxidant properties, it was one of the first nutritional supplements to be sensationalized for these properties. It has been proclaimed to thwart many conditions, from the common cold to cancer, none of which have been substantiated.

Deficiency of ascorbic acid (*scurvy*) is well documented and occurs in individuals with inadequate dietary intake, including chronic alcoholics and the elderly. Some suppression of ascorbate blood levels appears in situations such as extreme cold, heat, fever, physical trauma, tuberculosis, and cigarette smoking and in women taking oral contraceptives. Scurvy is characterized by impaired wound healing, petechial hemorrhage (pinpoint, minute spots in the skin), and perifollicular hemorrhage (surrounding the hair follicles). It is also demonstrated by inflammatory bleeding gums, loss of teeth, arrested skeletal development (in children), dry skin, joint pain, and increased susceptibility to infections and fatigue.

Although the incidence is low, long-term effects of doses greater than the DRI are associated with cataracts and coronary artery disease, increased iron absorption, and development of renal oxalate stones. Interestingly, sudden curtailing of vitamin C ingestion after chronic megadose administration may precipitate signs of scurvy, although this evidence now appears to be circumstantial.

22.3.7 Pyridoxine (Vitamin B_6)

Found primarily in vegetables, peanuts, eggs, soy, and cereals, vitamin B_6 (pyridoxal, pyridoxamine) appears to play a significant role in neuronal development. It is particularly important in the formation of pyridoxal-dependent decarboxylase, necessary for the synthesis of the neurotransmitter γ-aminobutyric acid (GABA). In fact, pyridoxine deficiencies, although rare, are characterized by unremarkable features of dermatologic, hematologic, and nervous system anomalies. Vitamin B_6 deficiencies occur in individuals who are also susceptible to deficiencies of other B-vitamins, including chronic alcoholism and malabsorption syndrome. Patients receiving isoniazid, cycloserine (antituberculosis antibiotics), hydralazine (antihypertensive), and oral contraceptives are also prone to B_6 deficiency. Pyridoxine is indicated solely for the management of pyridoxine deficiency states. Sensory neuropathies, ataxia, and numbness are possible in subjects receiving megadoses for several months.

22.3.8 Pantothenic Acid (Vitamin B_5)

Vitamin B_5 is important as a cofactor in enzyme-catalyzed reactions involving carbohydrates, gluconeogenesis, fatty acids, and steroids. It is distributed in many food products, it is nontoxic, and deficiencies are seen only in association with severe multiple B-complex malnutrition.

Toxicity associated with minerals and trace metals is discussed in detail in Chapter 26.

REFERENCES

SUGGESTED READINGS

Brunton, L.L., Hilal-Dandan, R., and Knollmann, B.C., *Goodman & Gilman's The Pharmacological Basis of Therapeutics*, 13th ed., McGraw-Hill, New York, 2018, Section IX: Special Systems Pharmacology.

González, M.J. Miranda-Missari, J.RE., More, E.M., Guzman, A., Riordan, N.H., Riordan, H.D., Casciari, J.J., Jackson, J.A., and Roam-Franco, A., Orthomolecular oncology review: ascorbic acid and cancer 25 years later. *Integr. Cancer Ther.* 4, 32, 2005.

Hansten, P.D., Horn, J.R., and Porter, R.S.,*Drug Facts and Comparisons*, Lippincott Williams and Wilkins, Walters Kluwer, Philadelphia, PA, 2017; chapter 1.

Kidd, P.M., Omega-3 DHA and EPA for cognition, behavior, and mood: clinical findings and structural-functional synergies with cell membrane phospholipids. *Altern. Med. Rev.* 12, 207, 2007.

Shapiro, S.S., Seiberg, M., and Cole, C.A., Vitamin A and its derivatives in experimental photocarcinogenesis: preventive effects and relevance to humans. *J. Drugs Dermatol.* 12, 458, 2013.

Strayer, D.S., Chapter 8. Environmental and nutritional pathology, in *Rubin's Pathology, Clinicopathologic Foundations of Medicine*, 7th ed., Rubin, E. (Ed.-in-Chief), Lippincott Williams and Wilkins, 2015.

REVIEW ARTICLES

Backstrand, J.R., The history and future of food fortification in the United States: a public health perspective, *Nutr. Rev.* 60, 15, 2002.

Bast, A., Haenen, G.R., van den Berg, R., and van den Berg, H., Antioxidant effects of carotenoids. *Int. J. Vitam. Nutr. Res.* 68, 399, 1998.

Bates, C.J., Diagnosis and detection of vitamin deficiencies. *Br. Med. Bull.* 55, 643, 1999.

Blot, W.J., Vitamin/mineral supplementation and cancer risk: international chemoprevention trials. *Proc. Soc. Exp. Biol. Med.* 216, 291, 1997.

Cannell J.J., Vieth, R., Willett, W., Zasloff, M., Hathcock, J.N., White, J.H., Tanumihardjo, S.A., Larson-Meyer, D.E., Bischoff-Ferrari, H.A., Lamberg-Allardt, C.J., Lappe, J.M., Norman, A.W., Zittermann, A., Whiting, S.J., Grant, W.B., Hollis, B.W., and Giovannucci, E.Cod liver oil, vitamin A toxicity, frequent respiratory infections, and the vitamin D deficiency epidemic. *Ann. Otol. Rhinol. Laryngol.* 117, 864, 2008.

Fairfield, K.M. and Fletcher, R.H., Vitamins for chronic disease prevention in adults: scientific review. *JAMA* 287, 3116, 2002.

Fleet, J.C., The role of vitamin D in the endocrinology controlling calcium homeostasis. *Mol Cell Endocrinol.* 453, 36, 2017.

Fletcher, R.H. and Fairfield, K.M., Vitamins for chronic disease prevention in adults: clinical applications. *JAMA* 287, 3127, 2002.

Fukugawa, M. Kurokawa, K., Calcium homeostasis and imbalance, *Nephron* 92, 41, 2002.

Garewal, H.S. Diplock, A.T., How "safe" are antioxidant vitamins? *Drug Saf.* 13, 8, 1995.

Haenen, G.R. Bast, A., The use of vitamin supplements in self-medication. *Therapie* 57, 119, 2002.

Hathcock, J.N., Vitamins and minerals: efficacy and safety. *Am. J. Clin. Nutr.* 66, 427, 1997.

Jones, G., Pharmacokinetics of vitamin D toxicity. *Am. J. Clin. Nutr.* 88, 582S, 2008.

Khundmiri, S.J., Murray, R.D., and Lederer, E., PTH and Vitamin D. *Compr. Physiol.* 6, 561, 2016.

Marks, J., The safety of the vitamins: an overview. *Int. J. Vitam. Nutr. Res. Suppl.* 30, 12, 1989.
Mawer, E.B. and Davies, M., Vitamin D nutrition and bone disease in adults. *Rev. Endocr. Metab. Disord.* 2, 153, 2001.
Meyers, D.G., Maloley, P.A., and Weeks, D., Safety of antioxidant vitamins. *Arch. Intern. Med.* 156, 925, 1996.
Nagai, K., Hosaka, H., Kubo, S., Nakabayashi, T., Amagasaki, Y., Nakamura, N., Vitamin A toxicity secondary to excessive intake of yellow-green vegetables, liver and laver. *J. Hepatol.* 31, 142, 1999.
Renwick, A.G. and Walker, R., Risk assessment of micronutrients. *Toxicol. Lett.* 180, 123, 2008.
Roux, C., Bischoff-Ferrari, H.A., Papapoulos, S.E., de Papp, A.E., West, J.A., Bouillon, R., New insights into the role of vitamin D and calcium in osteoporosis management: an expert roundtable discussion. *Curr. Med. Res. Opin.* 24, 1363, 2008.
Swain, R. Kaplan, B., Vitamins as therapy in the 1990s. *J. Am. Board Fam. Pract.* 8, 206. 1995.
U.S. National Research Council, Institute of Medicine of the National Academy of Sciences; DRI, Food and Nutrition Board, Dietary Reference Intakes, Office of Dietary Supplements, NIH. From: https://ods.od.nih.gov/Health_Information/Dietary_Reference_Intakes.aspx, last viewed March 2019.
Vieth, R., Vitamin D toxicity, policy, and science. *J. Bone Miner. Res.* 22, V64, 2007.

23 Herbal Remedies

23.1 INTRODUCTION

The popularity of self-medication or prescribed therapeutic intervention with herbal products has burgeoned. These "natural" supplements* have been traditionally promoted and used for centuries in Asian and Indian medicine and later, in American folk medicine. The belief that carefully measured amounts of the leaves, flowers, bark, stems, or seeds of botanicals can treat or prevent diseases has encouraged generations to trust the beneficial effects of ingredients contained within nature's packages. As society questions the limitations, abilities, efficacy, and safety of Western medicine, this creates an opportunity for alternative therapies to demonstrate their potential. The platform for the introduction of alternative medicines into the Western medical establishment is based particularly on the prevention of diseases and the avoidance of adverse reactions associated with Western therapeutic modalities.

It is interesting how the practice of pharmacy has come full circle, particularly since the days of the medieval European apothecaries. Pharmacy's first specific and peculiar contribution to the art of healing was the recognition that botanicals contained substances that could prevent or treat disease. American folk medicine and the early colonists relied on the American Indian botanicals to help the ailing. One of the earliest American folk medicine publications is Peter Smith's *Indian Doctor's Dispensatory, Being Father Smith's Advice Respecting Diseases and Their Cure* (1812). It documented a series of diseases commonly accepted as therapeutic categories of the times and recommended herbs for their remedy. In time, detailed descriptions of plants, flowers, and shrubs were systematically categorized and incorporated as part of the discipline of *pharmacognosy*.† As the practice of medicine and pharmacy matured from the era of empiricism, the emphasis shifted from handling of the bulk plant to procedures for isolating the crude drug. Eventually, identification and isolation of the active constituents, and structure–activity relationships defined the modern age of pharmaceutical development. Another driving force for modernization was the recognition that plants and their derivatives were as likely to be poisonous as they were to be beneficial. Thus, systematic scientific studies of the unseen chemicals contained within the powdered compounds forged the path toward modern pharmacology, toxicology, and

* The terms *natural/dietary supplements*, *botanicals*, and *herbal remedies/medicines* are used interchangeably. The use of herbal products, particularly in Chinese medicine, is considered to be part of the total approach to maintaining the body in a state of well-being and thus, is incorporated into the practice of "alternative" medicine. The latter also includes the Asian practices and procedures of acupuncture, meditation, *tai chi chuan*, massage, and aromatherapies.

† The art of pharmacognosy, which means "knowledge of pharmaceuticals," developed from ancient civilizations. These societies used parts of plants and animals to concoct healing potions to prevent or eliminate disease.

therapeutics. Today, botanical remedies are desirable as a way of returning to the holistic art of healing. Thus, it has taken over 500 years for Western medicine to complete the "circle" and reacquaint itself with the power that nature always possessed. Ultimately, the goal is, perhaps, to complement the tremendous advances achieved by Western medicine.

The use of herbal remedies incorporates numerous botanical products, each claimed to contain curative or preventive compounds for a variety of disease states. The effectiveness of the herbs, however, is mostly based on empirical experience gathered over time. Documentation of the history of the medicines as recorded in Chinese and Indian pharmaceutical compendia has also contributed to the empirical database. An important Chinese herbal manuscript, for instance, dates back to the third millennium BC and the emperor Shen Nung. He was deified by the people as the "God of Agriculture" and is credited with the authorship of the first Chinese pharmacopoeia, *Pen Ts'ao* ("Great Herbal"). The manuscript is a vast Chinese medical volume containing descriptions of vegetable, animal, and mineral medicines. Huang Ti (the Yellow Emperor, circa 2800 BC), the father of Chinese internal medicine, systematically classified diseases into therapeutic categories, thus advancing the practice of treatment with herbal medicines. Today, Chinese compendia list over 5800 natural products and their ingredients. However, rigorous scrutiny, as defined by Western scientific investigations, has shown that herbal product claims are not firm or final. This understanding of the limitations of herbal claims has impeded the smooth acceptance of alternative medicine into Western society. In addition, as their use increases and information accumulates, it is apparent that the products carry significant adverse reactions, hazards, contraindications, precautions, drug interactions, and consequences of overdose. Several documented instances of adulterated or mislabeled natural products have surfaced, resulting in serious poisoning from unsuspected reactions to mislabeled botanicals. Also, hypersensitivity reactions have been attributed to the ingestion of plant components resulting in anaphylactic responses.

Much of the information relied on in the United States is based on current summaries categorized by the German Regulatory Authority's herbal watchdog agency, "Commission E." The Commission's assessment of the peer-reviewed literature encompasses over 300 herbs and botanicals. The Commission compares the quality of the clinical evidence regarding their indications, effectiveness, and toxicity. This chapter, therefore, summarizes the currently accepted knowledge, attributes, and therapeutic usefulness associated with some frequently encountered herbal products and the reported toxicities observed with their use.

23.2 NOMENCLATURE AND CLASSIFICATION

23.2.1 Nomenclature

Herbal products are listed according to their common, scientific (biological), and brand names. Common names are readily recognizable and generally accepted. The scientific name refers to the taxonomic botanical designation of the plant. Commercial preparations of the product are assigned trade names.

23.2.2 Therapeutic Category

Many of the botanical products were originally categorized in ancient compendia and folk medicine publications according to the ailments for which they were prescribed. Similarly, today, herbal products are listed according to standard accepted treatment categories—that is, organs that are affected by a disease and amenable to treatment with an herbal product. It should be noted, however, that in the United States, herbal products are marketed under the provision of the Dietary Supplement and Health Education Act of 1994. This legislation defines herbs, vitamins, minerals, amino acids, and biochemical components of tissues and organs as food supplements rather than drugs, thus relieving them from many of the efficacy and safety requirements necessary for Food and Drug Administration (FDA) drug approval. In addition, therapeutic claims of herbal products are restricted to the enhancement of general health that affects normal physiologic processes. The compounds, however, are not permitted to claim or warrant efficacy for the diagnosis, treatment, cure, or prevention of any disease.

23.3 INDICATIONS

Indications for which an herbal product is deemed useful are listed under several major categories. The primary indications are those approved by Commission E. Homeopathic indications refer to some disease symptoms that could be treated by very small doses of medicine. Asian indications are based on traditional Chinese and Indian compendia and may or may not be recognized as primary indications by Western medicine. Finally, there are many unproven uses for which a botanical has empirically been used but lacks sufficient data for proof of efficacy.

23.4 OTHER THERAPEUTIC AND TOXICOLOGIC INFORMATION ON HERBAL PRODUCTS

Other information that is available for most common herbal products includes the actions and pharmacology of the known active ingredients; their dosages, adverse drug reactions, and side effects; drug–herb interactions; precautions; and where available, treatment of acute intoxication.

Table 23.1 lists accepted and well-documented herbal preparations according to their common, scientific, and proprietary names. Botanicals used primarily as culinary spices, flavoring agents, fruits, or vegetables, or that lack documented Chinese, Indian, or Western therapeutic uses or toxicity, are not included. The principal components or parts of the botanical used in the preparation of the medicinal formulation are listed. The major active ingredients contained within the herb are outlined, most of which are structurally classified as aromatic or polycyclic glycosides.

Table 23.2 summarizes the therapeutic effects, indications, and adverse reactions of the botanicals listed in Table 23.1. The most common indications are 1. those approved by Commission E; and 2. those documented in traditional Chinese or Indian medicine compendia or in American folklore. Many other historical

TABLE 23.1
Names and Major Active Ingredients of Various Herbal Products

Common Name	Scientific Name[a]	Other Common Names	Herbal Component	Major Active Ingredients
Acacia	*Acacia arabica*	Babul bark	Bark	Tannins
Agrimony	*Agrimonia eupatoria*	Stickwort, sticklewort	Leaves	Catechin tannins
Aloe	*Aloe vera*	Various oral and topical preparations	Topical products	Anthracene and flavonoid derivatives
Amaranth	*Amaranthus hypochondriacus*	Velvet flower, "Lady bleeding"	Plant	Saponins, β-cyanins
American bittersweet	*Celastrus scandens*	Waxwork	Root, bark	Tannins, celastrol
Anise	*Pimpinella anisum*	Aniseed oil	Oil of fruit	Anetholes, flavonoids
Arnica	*Arnica montana*	Wolfsbane	Oil of flowers	Sesquiterpine lactones
Belladonna	*Atropa belladonna*	Deadly nightshade, poison black cherry	Leaves, root	(-)-Hyoscyamine, atropine, scopolamine
Benzoin	*Styrax benzoin*	Benjamin tree	Resin	Benzoate esters
Betel nut	*Piper betle*	Betel	Dried leaves	Betel phenol, eugenol
Brewer's yeast	*Saccharomyces cerevisiae*	Various preparations	Dried yeast cells	B-complex vitamins, polysaccharides, protein
Calamus	*Acorus calamus*	Sweet flag, myrtle flag	Rhizome, oil	Asarone
Camphor tree	*Cinnamomum camphora*	Gum camphor	Oil	D(+)-camphor
Cardamom	*Elettaria cardamomum*	Various preparations	Oil, fruit, seeds	Cineol
Cascara sagrada	*Rhamnus purshiana*	Sacred bark, dogwood bark	Dried bark	Anthracene
Castor oil plant	*Ricinus communis*	Castor bean, Palma Christi	Oil, seeds	Ricin D, ricinoleic acid
Cat's claw	*Unicaria tomentosa*	*Una de gato*	Root bark	Strictosidine alkaloids, triterpenes
Cayenne	*Capsicum annuum*	Capsicum, chili pepper, paprika	Dried fruit	Capsaicinoids
Clove	*Syzygium aromaticum*	Various preparations	Oil, leaves	Eugenol
Coca	*Erythroxylon coca*	Cocaine	Leaves	Cocaine
Colchicum	*Colchicum autumnale*	Meadow saffron, upstart	Flowers, seeds	Colchicine
Devil's Claw	*Harpagophytum procumbens*	Grapple plant, wood spider	Roots, tubers	Harpagoside
Digitalis	*Digitalis lanata, D. purpurea*	Foxglove, dog's finger	Leaves	Digitoxigenin glycosides
Dong Quai	*Angelica sinensis*	Chinese angelica, dang gui, tang kuei	Root	Vitamins: B-complex, E, folic acid

(Continued)

TABLE 23.1 (CONTINUED)
Names and Major Active Ingredients of Various Herbal Products

Common Name	Scientific Name[a]	Other Common Names	Herbal Component	Major Active Ingredients
Echinacea	*Echinacea angustifolia, E. purpurea*	Black Sampson, purple coneflower	Root, leaves	Arabinoxylan polysaccharides
Ergot	*Claviceps purpurea*	Cockspur rye, hornseed	Fungal sclerotium	Indole (ergot) alkaloids
Eucalyptus	*Eucalyptus globulus*	Blue gum, fever tree	Oil, dried leaves	1,8-cineol, limonene, camphene
Flax	*Linum usitatissimum*	Flaxseed, linseed	Stem, oil, seeds, flowers	Arabinoxylans (mucilages), fatty acids
Garlic	*Allium sativum*	Stinking rose, poor man's treacle	Oil, fresh bulb	Alliins (sulfoxides)
Ginger	*Zingiber officinale*	Various preparations	Root	Zingiberene, farnesene, camphor, gingerols
Ginkgo	*Ginkgo biloba*	Maidenhair tree	Leaves, seeds	Quercetin, myristicins
Ginseng	*Panax ginseng*	American, Chinese, and Korean ginseng	Root	Ginsenoside (aglycone)
Goldenseal	*Hydrastis canadensis*	Orange root, yellow root	Rhizome	Hydrastine, berberine
Gotu kola	*Centella asiatica*	Indian pennywort	Leaves, stems	Asiatocides, asiatic acid
Green tea	*Camellia sinensis*	Black tea, Chinese tea	Leaves	Methyl xanthines, theaflavins
Guaiac	*Guaiacum officinale*	Pockwood	Resin	Oleanolic acid, guaiaretic acids
Hemlock	*Conium maculatum*	Cicuta, poison parsley	Dried leaves	Piperidine alkaloids (conine)
Henbane	*Hyoscyamus niger*	Devil's eye, stinking nightshade	Dried leaves	Tropane alkaloids (hyoscyamine)
Hibiscus	*Hibiscus sabdariffa*	Guinea sorrel, red sorrel	Flowers	Fruit acids, anthocyanins
Ipecac	*Cephaelis ipecacuanha*	Ipecacuanha, *matto grosso*	Dried root	Emetine, cephalin
Jimson weed	*Datura stramonium*	Jamestown weed, devil's apple	Dried leaves	Tropane alkaloids (hyoscyamine)
Juniper	*Juniperus communis*	Juniper berry, *enebro*	Oil, berry cones	α-pinene, β-myrcene, limonene
Kava kava	*Piper methysticum*	Ava pepper, *tonga*	Dried rhizome	Kavain, methysticin
Khat	*Catha edulis*	None	Leaves	Khatamine (alkyl amine)
Licorice	*Glycyrrhiza glabra*	Sweet root, sweet wort	Dried root	Glycyrrhetic acid, glycocoumarin, glabridin
Ma-huang	*Ephedra sinica*	Ephedrine, desert herb	Cane, root	Ephedrine, pseudoephedrine

(*Continued*)

TABLE 23.1 (CONTINUED)
Names and Major Active Ingredients of Various Herbal Products

Common Name	Scientific Name[a]	Other Common Names	Herbal Component	Major Active Ingredients
Mandrake	*Mandragora officinarum*	Mandragora, Satan's apple	Dried root, fresh herb	Tropane alkaloids (hyoscyamine)
Marijuana	*Cannabis sativa*	Cannabis, pot, bhang, grass, weed	Flowering twigs	Δ^9-tetrahydrocannabinol (THC)
Morning glory	*Ipomoea hederacea*	None	Seeds, root	Ergoline alkaloids (lysergol, chanoclavine)
Nutmeg	*Myristica fragrans*	Mace	Seeds, oil	Sabinene, α-pinene, myristicin, fatty acids
Nux vomica	*Strychnos nux vomica*	Poison nut, Quaker buttons	Dried seeds, bark	Strychnine, brucine
Periwinkle	*Vinca minor*	None	Dried leaves	Vincamine
Peyote[b]	*Lophophora williamsii*	Mescal buttons, sacred mushroom	Dried shoot, fresh plant	Mescaline
Poison ivy	*Rhus toxicodendron*	Poison oak, poison vine	Dried leaves shoots	Urushiol (alkyl phenol)
Poke	*Phytolacca americana*	Pokeweed, American nightshade	Dried root, berries	Phytolaccoside, lectins, histamine
Poppy seed	*Papaver somniferum*	Garden poppy, opium poppy	Seed capsule extract	Morphine, codeine, papaverine, narcotine, thebaine
Psyllium	*Plantago ovata, P. afra*	Plantain, fleaseed	Husk, seed	Arabinoxylans
Pyrethrum	*Chrysanthemum cinerariifolium*	Dalmatian insect flowers	Flower	Pyrethrines I, II; cinerines I, II
Quinine	*Cinchona pubescens*	Peruvian bark, *cinchona*	Dried bark	Quinine, quinidine, cinchonine
Rauwolfia	*Rauwolfia serpentina*	Various proprietary names	Dried root	Reserpine, serpentine
Sarsaparilla	*Smilax* sp.	Various proprietary names	Dried root	Sarsaparilloside
Saw palmetto	*Serenoa repens*	Various proprietary names	Dried ripe fruit	β-Sitosterol
Senna	*Cassia senna*	India senna, Alexandria senna	Leaves, fruit, flowers	Sennosides A, B, C, D (anthracenes)
Soybean	*Glycine soja, G. max*	Soya, tofu, miso	Bean, seed, legume, oil	Phospholipids (PC, PE, PI)
Squill	*Urginea maritime*	Scilla	Bulbs	Cardiac glycosides (scillarene A)

(*Continued*)

TABLE 23.1 (CONTINUED)
Names and Major Active Ingredients of Various Herbal Products

Common Name	Scientific Name[a]	Other Common Names	Herbal Component	Major Active Ingredients
St. John's Wort	*Hypericum perforatum*	Amber, goatweed	Buds, flowers	Xanthones, hypericin, napthodianthrone, procyanidines
Thyme	*Thymus vulgaris*	Various proprietary names	Oil, dried leaves	Thymol, *p*-cymene, carvacrol
Tobacco	*Nicotiana tabacum*	Various formulations	Dried leaves	Nicotine
Tragacanth	*Astragalus gummifer*	Gum dragon	Exudate from branch	Tragacanthine
Uva-ursi	*Arctostaphylos uva-ursi*	Arberry, mountain cranberry	Dried leaves	Arbutin (hydroquinone glycoside), tannins
Valerian	*Valeriana officinalis*	Amantilla, Vandal root	Dried root	Valepotriates, isovaltrate, valerenate
White mustard	*Sinapis alba*	Mustard	Dried seeds	Sinalbin (*p*-hydroxy-benzylglucosinolates)
Witch hazel	*Hamamelis virginiana*	Hazel nut, tobacco wood	Plant distillates	Hamamelitannin
Yerba santa	*Eriodyctyon californicum*	Bear's weed, sacred herb	Dried leaves	Eriodictyonin
Yohimbe bark	*Pausinystalia yohimbe*	Various proprietary names	Bark	Yohimbine

[a] Botanicals may incur several scientific names for different species of the genus. The most popular designations are noted.
[b] Because of toxicity and unreliable control of dose/effect, peyote is no longer recommended for use as an herbal preparation (see Effects, Table 23.2).
PC: phosphatidylcholine; PE: phosphatidylethanolamine; PI: phosphatidylinositol.

TABLE 23.2
Therapeutic Effects, Indications, and Adverse Reactions of Common Botanicals

Common Name	Desirable Therapeutic Effects	Indications	Adverse Reactions
Acacia	Astringent	Diarrhea, inflammation	Gastric disturbances
Agrimony	Astringent	Diarrhea, inflammation	Gastric disturbances
Aloe	Laxative, antibacterial, antineoplastic, anti-inflammatory, analgesic	Constipation, antibacterial, inflammation	Gastric and electrolyte disturbances, hypersensitivity, possible malignancy
Amaranth	Astringent	Diarrhea, inflammation	No significant ADRs
American bittersweet	Diuretic	Inflammation, menstrual disorders	No significant ADRs; botanical is obsolete
Anise	Expectorant, antispasmodic, antibacterial	Cough, colds, fever, inflammation, rheumatism	Allergic reactions
Arnica	Analgesic, antiseptic, ↓ respiratory secretions	Cough, colds, fever, inflammation, rheumatism	Allergic reactions, ↑ skin sensitivity
Belladonna	Anticholinergic, GI antispasmodic, positive chronotropy	Liver and gall bladder disorders	Mydriasis, dryness of mouth, perspiration, arrhythmias
Benzoin	Expectorant	Respiratory cold, stroke, syncope	No significant ADRs
Betel nut	Antibacterial, anti-inflammatory	Cough, asthma, bronchitis	None; compound is obsolete
Brewer's yeast	Antibacterial, immunostimulant	Dyspepsia, eczema	Gastric bloating, allergic reactions
Calamus	Aromatic, cholinergic, antispasmodic, sedative	Dyspepsia, gastritis, rheumatism	Possible carcinogen with chronic use
Camphor tree	Hyperemic, antispasmodic, ↑ bronchial secretions	Bronchitis, rheumatism, arrhythmias	Dermal irritation, delirium, spasms, dyspnea
Cardamom	Antimicrobial, cholinergic	Cough, colds, fever, inflammation	Gastric colic
Cascara sagrada	Laxative, cholinergic	Constipation	Gastric spasms, edema, electrolyte disturbances, nephropathy
Castor oil plant	Laxative, cholinergic	Constipation, dyspepsia, dermal inflammation	NVD, electrolyte disturbances, gastroenteritis, shock, fatal with O.D.
Cat's claw	Anti-inflammatory, immunostimulant, ↓ platelet aggregation, anti-HT, contraceptive	Inflammation, antiviral	Interference with gonadotropin levels
Cayenne	Hyperemic, antibacterial, thrombolytic, antineoplastic	Rheumatism, muscular tension	Diarrhea, dermal ulcers, anaphylaxis, hypocoagulation

(Continued)

TABLE 23.2 (CONTINUED)
Therapeutic Effects, Indications, and Adverse Reactions of Common Botanicals

Common Name	Desirable Therapeutic Effects	Indications	Adverse Reactions
Clove	Antiseptic, antispasmodic, local anesthetic	Oral inflammation, dental analgesia	Allergic reactions, local irritation
Coca	Local anesthetic, CNS stimulant	Local anesthesia	Hallucinations, depression, addictive, embryotoxic
Colchicum	Anti-inflammatory, inhibits phagocytizing lymphocytes	Gout, Mediterranean fever	NVD, hemorrhage, kidney, liver, CNS, BM pathology
Devil's Claw	Cholinergic, anti-inflammatory	Dyspepsia, anorexia, rheumatism	Allergic reactions
Digitalis	Cardioactive, positive inotropy, negative chronotropy	Cardiac insufficiency[a]	NVD, HA, anorexia, arrhythmias, visual disorders, confusion[a]
Dong Quai	Relaxes uterine smooth muscle, coronary vasodilator	Menopause, irregular menstruation, Raynaud's syndrome	Photosensitization, abortifacient, anticoagulant
Echinacea	Anti-inflammatory, promotes wound healing, antibacterial, immunostimulant	Cough, colds, fever, inflammation, UTI, rheumatism	NV, fever, hypersensitivity, dyspnea, hypotension
Ergot	Smooth muscle contraction	Hemorrhage, migraine, uterine and muscle spasm[a]	NVD, muscle pain, multiple systemic pathology[a]
Eucalyptus	Expectorant, diuretic, aromatic	Chronic bronchitis, rheumatism	NVD, asthmatic attacks with facial application
Flax	Bulk laxative, hypolipidemic	Constipation, dermal inflammation	Delayed GI absorption
Garlic	Antibacterial, hypolipidemic, antioxidant, antiplatelet	Hypertension, hypercholesterolemia	Gastric irritation, allergic reactions, bleeding disorders
Ginger	Antiemetic, anti-inflammatory	Dyspepsia, anorexia, motion sickness	Gastric irritation, allergic reactions
Ginkgo	Antiplatelet, antioxidant, improves cognitive function, thrombolytic	Inflammation, tinnitus, intermittent claudication, vertigo	Gastric irritation, allergic reactions
Ginseng	Improves cognitive function, antineoplastic, antioxidant, antiplatelet, hypolipidemic	Malaise	HT, insomnia, edema
Goldenseal	Antiparasitic, antiseptic, antibacterial	Infection (antiseptic, cold sores), acute diarrhea	Gastric disorders, constipation, hallucinations

(Continued)

TABLE 23.2 (CONTINUED)
Therapeutic Effects, Indications, and Adverse Reactions of Common Botanicals

Common Name	Desirable Therapeutic Effects	Indications	Adverse Reactions
Gotu kola	Anti-inflammatory, antineoplastic, promotes wound healing, anti-HT	Inflammation, neoplasms, gastric ulcers	Allergic contact dermatitis
Green tea	CNS stimulation, diuresis, positive inotropic, anticancer	Cancer prevention	Gastric irritation, reduction of appetite
Guaiac	Antifungal	Rheumatism	Diarrhea, gastroenteritis
Hemlock	Autonomic stimulant	Rheumatism, cramps, bronchial spasms[a]	Tonic contractions, hypotension, cramps, paralysis[a]
Henbane	Anticholinergic, smooth muscle relaxation	Dyspepsia	Mydriasis, dryness of mouth, perspiration, arrhythmias
Hibiscus	Hypotensive, uterine relaxation	Anorexia, colds, constipation, dermal inflammation	No significant ADRs
Ipecac	Emetic, gastric irritation	Asthma, bronchitis, cough, colds, poisoning	Allergic reactions, myopathy
Jimson weed	Anticholinergic, parasympatholytic	Stomach complaints, colic, dyspnea	Restlessness, irritability, hallucinations, delirium
Juniper	Antihypertensive, diuretic, antidiabetic	Anorexia, dyspepsia	Kidney irritation
Kava kava	Anticonvulsant, antispasmodic, sedative/hypnotic	Nervousness, spasms, insomnia	GI upset, allergic reactions
Khat	CNS stimulant, sympathomimetic	Depression, headache, asthma, fever, obesity[a]	Irritability, hypertonia nervousness, insomnia[a]
Licorice	Anti-inflammatory, antiplatelet, antifungal, mineralocorticoid	Cough, bronchitis, gastritis	Hypokalemia, hypernatremia, edema, HT
Ma-huang	Sympathomimetic	Cough, bronchitis, asthma, edema, fever	Irritability, HT, arrhythmias, restlessness, HA
Mandrake	Anticholinergic	Stomach complaints, colic, whooping cough	Drowsiness, delirium, excitation, tachycardia, hallucinations
Marijuana	Psychotropic, antiemetic, analgesic, anticonvulsive, CV stimulant, ↓ IOP	Anorexia, depression, insomnia, vomiting	Alterations of perception, sensory disturbances
Morning glory	Laxative	Constipation, intestinal parasitic infections	Abdominal cramps
Nutmeg	Antidiarrheal, anti-inflammatory	Diarrhea, vomiting, stomach cramps	Allergic contact dermatitis

(*Continued*)

TABLE 23.2 (CONTINUED)
Therapeutic Effects, Indications, and Adverse Reactions of Common Botanicals

Common Name	Desirable Therapeutic Effects	Indications	Adverse Reactions
Nux vomica	↑ CNS and muscular excitability	Anorexia, depression, CV stimulant, pain, respiratory disorders	Anxiety, tonic-clonic spasms and convulsions, nervousness
Periwinkle	Hypotension, hypoglycemia, sympatholytic	Circulatory disorders, memory loss	GI upset, skin flushing, hypotension
Peyote	Hallucinogenic	Used as an hallucinogen	Visual, auditory, sensory disturbances[a]
Poison ivy	Dermal immune stimulant	Rheumatism, muscle strain, inflammation	Skin irritation, swelling, blisters
Poke	Immune stimulant, antiedemic, emetic	Rheumatism, dermal ulcers, inflammation	NV, diarrhea, hypotension, tachycardia
Poppyseed	Analgesic, sedative, euphoric, antitussive	Pain, nervousness, muscle spasms, cough, diarrhea	Constipation, weakness, dizziness, clonic twitching
Psyllium	Laxative, antidiarrheal (swelling)	Constipation, diarrhea, hypercholesterolemia, hemorrhoids	Allergic reactions, GI obstruction
Pyrethrum	Neurotoxic, contact insecticide	Mite and lice infestation (scabies, pediculosis)	NV, paresthesia, tinnitus, HA
Quinine	Anti-inflammatory, appetite stimulant	Anorexia, dyspepsia, malaria, fever	Eczema, itching, arrhythmias, tinnitus, NV
Rauwolfia	Sympatholytic, hypotensive effect	Hypertension, nervousness, insomnia	Nasal congestion, depression, erectile dysfunction
Sarsaparilla	Diuretic, diaphoretic	Dermal inflammation, rheumatism, edema	GI upset
Saw palmetto	Antiandrogenic, antiestrogenic	Prostate and urinary complaints, irritable bladder	GI upset
Senna	Stimulant laxative	Constipation	Spastic GI upset, electrolyte and CV abnormalities
Soybean	Lipid-lowering effects	Hypercholesterolemia	GI upset
Squill	Positive inotropy, negative chronotropy	Cardiac insufficiency, arrhythmias, venous conditions	Arrhythmias, depression, confusion
St. John's Wort	Antidepressant, sedative, anxiolytic	Anxiety, depression, dermal inflammation, wounds, burns	Photosensitization, drug interactions (see text)
Thyme	Relaxes bronchial smooth muscle	Cough, bronchitis	No significant ADRs

(Continued)

TABLE 23.2 (CONTINUED)
Therapeutic Effects, Indications, and Adverse Reactions of Common Botanicals

Common Name	Desirable Therapeutic Effects	Indications	Adverse Reactions
Tobacco	CV, respiratory, and CNS stimulant	Dental pain, angina pectoris, diarrhea, skin parasites	Habituation, nicotine poisoning
Tragacanth	Laxative	Constipation	No significant ADRs
Uva-ursi	Diuretic, antibacterial	UTI, biliary tract disorders	GI irritation, NV
Valerian	Sedative, anxiolytic, muscle relaxant, spasmolytic	Nervousness, insomnia	Headache, restlessness, mydriasis
White mustard	Dermal hyperemia, decongestant, swelling	Cough, bronchitis, cold, rheumatism	Skin necrosis, nerve damage
Witch hazel	Astringent, hemostatic, anti-inflammatory	Inflammation, wounds, burns, hemorrhoids	No significant ADRs
Yerba santa	Diuretic	Asthma, masks bitter taste	No significant ADRs
Yohimbe bark	Aphrodisiac, analgesic, sympatholytic (α-2 antagonist)	Impotence, sexual dysfunction	Anxiety, HT, mydriasis

[a] Because of toxicity and unreliable control of dose/effect, digitalis is not recommended for use as an herbal preparation.

BM: bone marrow; CV: cardiovascular; GI: gastrointestinal; HA: headache; HT: hypertension; IOP: intraocular pressure; NVD: nausea, vomiting, diarrhea; O.D.: overdose; UTI: urinary tract infection.

TABLE 23.3
Some Significant Drug–Drug Interactions Associated with Therapeutic Use of Common Botanicals

Botanical	Generic Therapeutic Drug	Drug–Drug Interaction: Potential Influence of Botanical on Drug
Belladonna	Anticholinergic agents (e.g., tricyclic antidepressants, atropine)	↑ Anticholinergic effects
Castor oil plant	Cholinergic agents, oral medications, diuretics	↑ Cholinergic effects, ↑ GI motility (produces diarrhea), delays absorption of oral medications, ↑ potential for electrolyte disturbances (diuretics)
Digitalis	Cardioactive agents	Counteracts effectiveness of cardioactive agents; potentiates effect of cardiac stimulants
Dong Quai	Oral anticoagulants	Enhances anticoagulant effects
Echinacea	Immunosuppressant agents	Reduces immunosuppression activity
Flax	Oral medications	Delays absorption and oral bioavailability
Garlic	Oral anticoagulants	Enhances anticoagulant effects, ↑ risk of bleeding episodes
Ginkgo	Oral anticoagulants	Enhances anticoagulant effects, ↑ risk of bleeding episodes
Ginseng	Oral anticoagulants; oral hypoglycemics	Enhances anticoagulant effects, ↑ risk of bleeding episodes; enhances hypoglycemic effect, raises risk of diabetic shock
Jimson weed	Anticholinergic agents (e.g., tricyclic drugs, atropine)	↑ Anticholinergic effects
Kava kava	Anesthetics, opioids, sedative/hypnotics	Enhances sedative effects
Ma-huang	CNS stimulants, sympathomimetics, MAOI	Enhances CNS stimulation, potential for HT, vascular complications, arrhythmias, hemodynamic instability
Rauwolfia	Anti-HT agents, antidepressants	Enhances anti-HT effect (syncope, dizziness, shock); may counteract effect of antidepressants
Saw palmetto	Oral anticoagulants	Enhances anticoagulant effects, increases risk of bleeding episodes
St. John's Wort	Drugs requiring cytochrome P_{450} inactivation	Cytochrome P_{450} enzyme inducer, ↓ bioavailability of many therapeutic drugs
Valerian	Anesthetic, opioids, sedative/hypnotics	Enhances sedative effects
Yohimbe bark	CNS stimulants, sympathomimetics, MAOI	Enhances CNS stimulation, potential for HT, vascular complications, arrhythmias, hemodynamic instability

MAOI: monoamine oxidase inhibitor.

applications for which the compounds have claimed usefulness are not included. Among the most common adverse reactions noted with ingestion or application of herbal products is the propensity for development of allergic reactions. In particular, the lectin, glycosidic, or peptide components of the herbs behave as haptens. Haptens, by definition, are incapable of inducing inflammatory reactions alone but can stimulate an antigenic response by binding to circulating proteins. Toxicity of some herbs has also resulted from contamination of the product with metals and pharmaceutical agents. Because of the lenient federal oversight for these products, some poisonings have resulted from misidentification, mislabeling, and inadequate purification of herbal components. In addition, the toxic effects of herbal compounds generally result from ingestion of higher doses than recommended or recommended doses for extended periods.

As more information is gathered with the increasing popularity of these agents, notable drug interactions are surfacing, some of which are summarized in Table 23.3. In general, several classes of drugs have the potential for undesirable interactions with botanical preparations. These include anticoagulants and other medications requiring therapeutic drug monitoring, drugs that require P450 biotransformation, drugs that sensitize the myocardium, and compounds that increase the risk of systemic allergic reactions.

REFERENCES

SUGGESTED READINGS

Blumenthal, M., (Ed.), *The Complete German Commission E Monographs: Therapeutic Guide to Herbal Medicines*, Austin, TX, American Botanical Council, 1999, p. 218.

Dasgupta, A., Herbal supplements and therapeutic drug monitoring: focus on digoxin immunoassays and interactions with St. John's wort. *Ther. Drug Monit.* 30, 212, 2008.

Dietary Supplement Health and Education Act of 1994, Public Law 103-417, U.S. Code 2006, Title 21: Food and Drugs, Supplement 1, January 4, 2007 to January 8, 2008, U.S. Food and Drug Administration.

Haller, C., Chapter 58. Herbal preparations, in *Toxicology Secrets*, Ling, L.J., Clark, R.F., Erickson, T.B., and Trestrail J.H. (Eds.), Hanley and Belfus, 2001.

Keys, J.D., *Chinese Herbs: Their Botany, Chemistry, and Pharmacodynamics*, Charles E. Tuttle Co., Rutland, VT, 1993.

Mandlekar, S., Hong, J.L., and Kong, A.N., Modulation of metabolic enzymes by dietary phytochemicals: a review of mechanisms underlying beneficial versus unfavorable effects. *Curr. Drug Metab.* 7, 661, 2006.

Meyer, C., *American Folk Medicine*, Plume Books, New American Library, Times Mirror, 1973.

PDR Physicians' Desk Reference for Nonprescription Drugs, Dietary Supplements and Herbs, Physicians' Desk Reference Inc., 35th edition, 2014.

Phillipson, J.D., Phytochemistry and pharmacognosy. *Phytochemistry* 68, 2960, 2007.

Smith, P., *Indian Doctor's Dispensatory: Being Father Smith's Advice Respecting Diseases and Their Cure,* 1812.

Wu, K.M., Ghantous, H., and Birnkrant, D.B., Current regulatory toxicology perspectives on the development of herbal medicines to prescription drug products in the United States. *Food Chem. Toxicol.* 46, 2606, 2008.

Zhou, S.F., Xue, C.C., Yu, X.Q., and Wang, G., Metabolic activation of herbal and dietary constituents and its clinical and toxicological implications: an update. *Curr. Drug Metab.* 8, 526, 2007.

REVIEW ARTICLES

Aiello, E., Russo, R., Cristiano, C., and Calignano, A., The safety assessment of herbals with a new and ethical approach. *Nat. Prod. Res.* 32, 1838, 2018.

Barnes, J., McLachlan, A.J., Sherwin, C.M., and Enioutina, E.Y., Herbal medicines: challenges in the modern world. Part 1. Australia and New Zealand. *Expert Rev. Clin. Pharmacol.* 9, 905, 2016.

Chiu, J., Yau, T., and Epstein, R.J., Complications of traditional Chinese/herbal medicines (TCM)—a guide for perplexed oncologists and other cancer caregivers. *Support Care Cancer* 17, 231, 2009.

Di, Y.M., Li, C.G., Xue, C.C., and Zhou, S.F., Clinical drugs that interact with St. John's wort and implication in drug development. *Curr. Pharm. Des.* 14, 1723, 2008.

Durante, K.M., Whitmore, B., Jones, C.A., and Campbell, N.R., Use of vitamins, minerals and herbs: a survey of patients attending family practice clinics. *Clin. Invest. Med.* 24, 242, 2001.

Ernst, E., Heavy metals in traditional Indian remedies. *Eur. J. Clin. Pharmacol.* 57, 891, 2002.

Ervin, R.B., Wright, J.D., and Kennedy-Stephenson, J., Use of dietary supplements in the United States, 1988–94. *Vital Health Stat.* 11, 1, 1999.

Haller, C.A., Anderson, I.B., Kim, S.Y., and Blanc, P.D., An evaluation of selected herbal reference texts and comparison to published reports of adverse herbal events. *Adverse Drug React. Toxicol. Rev.* 21, 143, 2002.

Jordan, S.A., Cunningham, D.G., and Marles, R.J., Assessment of herbal medicinal products: challenges, and opportunities to increase the knowledge base for safety assessment. *Toxicol. Appl. Pharmacol.* 243, 198, 2010.

Lee, K.H., Research and future trends in the pharmaceutical development of medicinal herbs from Chinese medicine. *Public Health Nutr.* 3, 515, 2000.

Nelson, L. and Perrone, J., Herbal and alternative medicine. *Emerg. Med. Clin. North Am.* 18, 709, 2000.

Palmer, M.E., Haller, C., McKinney, P.E., Klein-Schwartz, W., Tschirgi, A., Smolinske, S.C., Woolf, A., Sprague, B.M., Ko, R., Everson, G., Nelson, L.S., Dodd-Butera, T., Bartlett, W.D., Landzberg, B.R. Adverse events associated with dietary supplements: an observational study. *Lancet* 361, 101, 2003.

Perharic, L., Shaw, D., Colbridge, M., House, I., Leon, C., and Murray, V., Toxicological problems resulting from exposure to traditional remedies and food supplements. *Drug Saf.* 11, 284, 1994.

Perharic, L., Shaw, D., and Murray, V., Toxic effects of herbal medicines and food supplements. *Lancet* 342, 180, 1993.

Roe, A.L., Paine, M.F., Gurley, B.J., Brouwer, K.R., Jordan, S., and Griffiths, J.C., Assessing natural product-drug interactions: an end-to-end safety framework. *Regul. Toxicol. Pharmacol.* 76, 1, 2016.

Sammons, H.M,. Gubarev, M.I., Krepkova, L.V., Bortnikova, V.V., Corrick, F., Job, K.M., Sherwin, C.M., and Enioutina, E.Y., Herbal medicines: challenges in the modern world. Part 2. European Union and Russia. *Expert Rev. Clin. Pharmacol.* 9, 1117, 2016.

Shaw, D., Leon, C., Kolev, S., and Murray, V., Traditional remedies and food supplements. A 5-year toxicological study (1991–1995). *Drug Saf.* 17, 342, 1997.

Wang, J.F., Wei, D.Q., and Chou, K.C., Drug candidates from traditional Chinese medicines. *Curr. Top. Med. Chem.* 8, 1656, 2008.

Wu, K.M., Ghantous, H., and Birnkrant, D.B., Current regulatory toxicology perspectives on the development of herbal medicines to prescription drug products in the United States. *Food Chem. Toxicol.* 46, 2606, 2008.

Yeung, K.S., Gubili, J., and Cassileth, B., Evidence-based botanical research: applications and challenges. *Hematol. Oncol. Clin. North Am.* 22, 661, 2008.

Section III

Toxicity of Nontherapeutic Agents

24 Alcohols and Aldehydes

Alcohols and aldehydes are carbon compounds containing hydroxyl (H–R–OH) or carbonyl (H–R=O) groups. The chemicals affect biological processes directly or indirectly. Direct action of these compounds involves interaction of the parent molecules with the substrates of biological pathways. Indirect action interferes with metabolic products of the parent chemical. Ethanol, methanol, and isopropanol are important alcohols primarily because of their commercial availability, importance in a variety of essential chemical reactions, and involvement in intermediate chemical and biological pathways. Among the aldehydes, formaldehyde is ubiquitously distributed in the environment. Human exposure to this chemical has received considerable attention due to its human carcinogenic potential and developmental toxicity in laboratory animals.

24.1 ETHANOL

Ethanol (ethyl alcohol) has been produced from fermented grain, fruit juice, and honey for thousands of years. The presence of ethanol in wine, beer, and liquor and the use of it as a common solvent make it widely available to adults. The consumption and abuse of ethanol are not only a serious public health problem but also one of the major social problems, especially in Western countries.

24.1.1 Incidence and Occurrence

Alcohol-mediated psychomotor disruption and subsequent deaths stand third among the leading preventable causes of death in the United States, causing an estimated 88,000 annual deaths. Ethanol is a socially accepted recreational drug due to its legal approval and ubiquitous availability. According to the National Survey on Drug Use and Health (U.S. National Institute on Drug Abuse [NIDA], NSDUH, 2018), 2.5% of children aged 12–17 years and 6.2% aged 18 years and above in the United States suffered with alcohol use disorder. Furthermore, studies in pregnant women demonstrate about a 10% incidence of alcohol consumption during pregnancy.

24.1.2 Chemical Characteristics

Although a polar molecule, ethanol is a colorless aliphatic hydroxy-hydrocarbon molecule, possessing both water and lipid solubility. Ethanol diffuses across cell membranes easily and is absorbed from the gastrointestinal (GI) tract rapidly. The average apparent volume of distribution (V_d) of ethanol is about 0.6 l/kg, nearly equivalent to that of water. It is thus capable of readily distributing throughout physiologic compartments; particularly, it is able to penetrate the blood–brain barrier and placenta. Because the oxidation of ethanol yields 7.1 kcal/g, sufficient calories can be obtained from ethanol alone for chronic drinkers if the daily intake of ethanol exceeds 5 g/kg of body weight.

24.1.3 Toxicokinetics

Ethanol is rapidly absorbed from the GI tract within 30 to 60 minutes after ingestion. The stomach extracts about 20%, primarily by diffusion, with the bulk of absorption occurring in the small intestine. Several factors may delay absorption, including concomitant intake with food, drugs, and medical conditions that inhibit gastric emptying. After it enters the portal vein, ethanol first passes through the liver before it distributes systemically. More than 90% of ingested ethanol is oxidized to acetaldehyde by the liver and gastric mucosal cells; 5% to 10% is excreted unchanged by the kidneys and lungs and through dermal pores.

Figure 24.1 illustrates the ethanol metabolic pathway. The primary pathway for ethanol oxidation occurs predominantly in the liver, where alcohol dehydrogenase (ADH) catalyzes the reaction of the alcohol to acetaldehyde. Also, CYP2E1 is a foremost hepatic cytochrome P450 isoform that catalyzes an analogous oxidation reaction in the confines of the endoplasmic reticulum, whereas catalase promotes oxidation of ethanol in peroxisomes. However it is produced, the acetaldehyde product is further converted to acetate via the action of aldehyde dehydrogenase (ALDH) present in liver mitochondria (Figure 24.1).

Generally, women have a higher peak ethanol concentration if exposed to the same amount of the alcohol due to their lower body water content and lower level of ADH in gastric mucosal cells. ADH, ALDH, and CYP2E1 exhibit genetic polymorphisms that subsequently may lead to alterations in ethanol elimination rate. For example, the facial flushing reaction following alcohol consumption in some Asian populations is due to the lower ALDH efficiency and accumulated blood acetaldehyde. The average rate of ethanol metabolism in adults is 100–124 mg/kg/hour in occasional drinkers and can be up to 175 mg/kg/hour in chronic drinkers. On average, blood alcohol concentrations (BAC) fall by 15–20 mg/dl/h in adults.

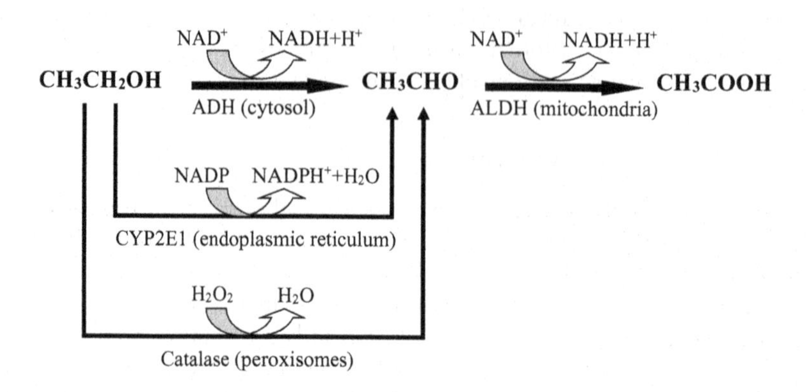

FIGURE 24.1 Metabolic pathways of ethanol.

24.1.4 Calculation of Blood Alcohol Concentrations (BAC)

Estimation of the blood concentration of ethanol provides useful information regarding the severity of intoxication. BAC is calculated according to the following equation:

$$\text{BAC}(\text{mg/dL}) = [\text{Et}](\text{mg})/V_d(\text{L/kg})] \times \text{b.w.}(\text{kg}) \times 10$$

where
- V_d is the apparent volume of distribution
- [Et] is the amount of ethanol ingested
- b.w. is the body weight

Alternatively, BAC is traditionally calculated as outlined according to Widmark's factor:

$$A = \frac{Wr\text{CT}}{0.8}$$

where
- A = ml of absolute ethanol
- W = body weight (g)
- r = compartmental distribution of ethanol, based on [% whole body alcohol concentration]/%BAC; $r = 0.68$ for men, 0.55 for women
- CT= BAC (decimal units)
- 0.8 is the specific gravity of ethanol

Ethanol elimination depends on the overall enzyme rate for ADH activity for the metabolism of ethanol. Figure 24.2 illustrates the plot of ADH activity versus ethanol concentration (per unit time, usually hours). At or below the maximum enzyme rate of about 23 mg/dl (V_{max}) for ADH, ethanol metabolism (and elimination) follows

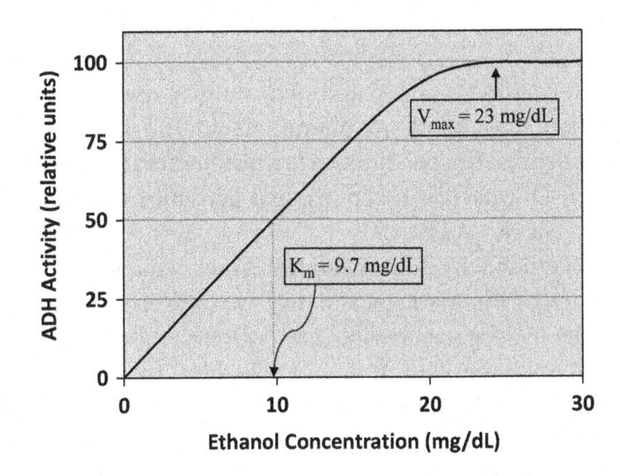

FIGURE 24.2 Ethanol and alcohol dehydrogenase (ADH) kinetics.

first-order kinetics; that is, the amount of ethanol metabolized is directly proportional to the concentration. Above the V_{max} for the enzyme, ethanol metabolism follows zero-order kinetics, whereby a further increase in ethanol concentration is not followed by a proportionate increase in enzyme activity. Thus, ethanol metabolism remains constant above V_{max} even as the concentration increases. Based on these rates, the equilibrium constant (or average enzyme rate constant, K_m) is calculated as

$$K_m = 9.7\,\text{mg/dL}$$

Practically, the K_m is usually standardized at about 10 mg/dl. For an average 70 kg male, the overall enzyme rate (V_{max}) is

$$V_{max} = 18\,\text{mg/dL/hr}$$

which computes to an elimination rate (K_{elim}) for ethanol at

$$K_{elim} = 100\,\text{mg/kg/hr}$$

Thus, the same 70 kg individual metabolizing at 100 mg/kg/h will be able to eliminate

$$7\,\text{g/hr}\,(\text{by weight})$$

$$\text{or } 9\,\text{ml/hr}\,(\text{by volume})$$

of 100% (absolute) ethanol, assuming the specific gravity of ethanol = 0.8.

24.1.5 Mechanisms of Toxicity

A multifactorial setting is responsible for the mechanisms of ethanol's toxic effects. The liver, nervous system, gastrointestinal (GI) tract, and cardiovascular (CV) system are the principal targets of ethanol toxicity. Several cellular processes have been proposed to be crucially involved in ethanol-induced toxicity:

1. Ethanol directly affects cell membrane fluidity and modifies membrane proteins, which may result in alterations in the liquid-crystal state of membranes, membrane ion transport, transmembrane signal transduction (i.e., N-methyl-D-aspartate receptor), and activities of intrinsic membrane enzymes (i.e., Na^+-K^+-ATPase).
2. Metabolism of ethanol by microsomal enzymes, especially CYP2E1, results in the formation of free radicals and reactive oxygen species (ROS), including superoxide, hydrogen peroxide. and hydroxyl radicals. These molecules are capable of oxidizing important biomolecules, including lipids, proteins, and nucleic acids, leading to oxidative cell injury.
3. Ethanol toxicity is also attributable to the formation of phosphatidylethanol (PE) and dihydroxyacetone phosphate (DAP), unique phospholipid reactive species formed intracellularly in the presence of ethanol. The formation of

PE is catalyzed by phospholipase D, an enzyme that normally catalyzes the hydrolysis of phospholipids, leading to the formation of phosphatidic acid. DAP drives cellular metabolism toward the formation of glycerol, which reverses fatty oxidation and results in accumulation of lipids. The reactions are summarized in Figure 24.3.

4. Fatty acid ethyl esters (FAEEs) are produced from the conjugation of ethanol and fatty acids in target organs such as the heart, brain, pancreas, and liver. These lipophilic molecules accumulate in cell membranes and mitochondria, resulting in impaired organelle function.
5. Ethanol changes mitochondrial membrane permeability, thus inhibiting the expression of mitochondrial electron transport chain components, such as $NADH^+$ dehydrogenase and cytochrome *c* oxidase. This effect is a consequence, initially, of overproduction of $NADH^+$ from ethanol metabolism. Increasing levels of $NADH^+$ result in a reversal of the citric acid cycle (red arrows, Figure 24.3) and shuttles acetyl-CoA toward the formation of metabolic acids (such as β-hydroxybutyrate and lactate). The residual effect is decreased formation of $NADH^+$ (reducing equivalents necessary for neutralizing the destructive activity of ROS), further accumulation of reactive oxygen and electron donors, accumulation of metabolic acids, and promotion of ROS.

The liver is the most important target organ of ethanol-induced toxicity, especially chronic ethanol intoxication. Increased NADH levels resulting from ethanol oxidation not only reduce gluconeogenesis but also accelerate triglyceride synthesis from free fatty acids, which leads to steatosis.* Ethanol stimulates the release of endotoxin by Gram-negative bacteria in the GI tract. Endotoxins then enter the liver through the hepatic portal vein and activate immune-reactive Kupffer cells, resulting in release of the cytotoxic ROS. This inflammatory reaction triggers further cytokine recruitment, aggravating hepatic injury.

The generation of reactive acetaldehyde, as a product of ethanol metabolism, leads to the formation of protein adducts and enzyme inactivation. Moreover, acetaldehyde promotes glutathione depletion, free-radical-mediated toxicity, and lipid peroxidation. Acetaldehyde has also been shown to stimulate collagen synthesis by liver stellate cells, which might be involved in the pathogenesis of ethanol-induced liver cirrhosis.

As a central nervous system (CNS) depressant, ethanol alters neurotransmission by interfering with ligand-gated, G-protein-regulated, and voltage-sensitive channels, as well as its interaction with glutamate and GABA-ergic neurons. Acute ethanol exposure augments GABA activity at ligand-gated $GABA_A$ receptors and inhibits the activation of N-methyl-D-aspartate (NMDA) glutamate receptors involved in cognition.†

Chronic ethanol intoxication induces cardiotoxicity by interfering with myocardial stores of catecholamines and by decreasing the synthesis of cardiac contractile proteins, leading to depression of myocardial contractility.

* Fatty changes, particularly in the liver parenchyma.

† GABA and glutamate are inhibitory and excitatory amino acid neurotransmitters, respectively.

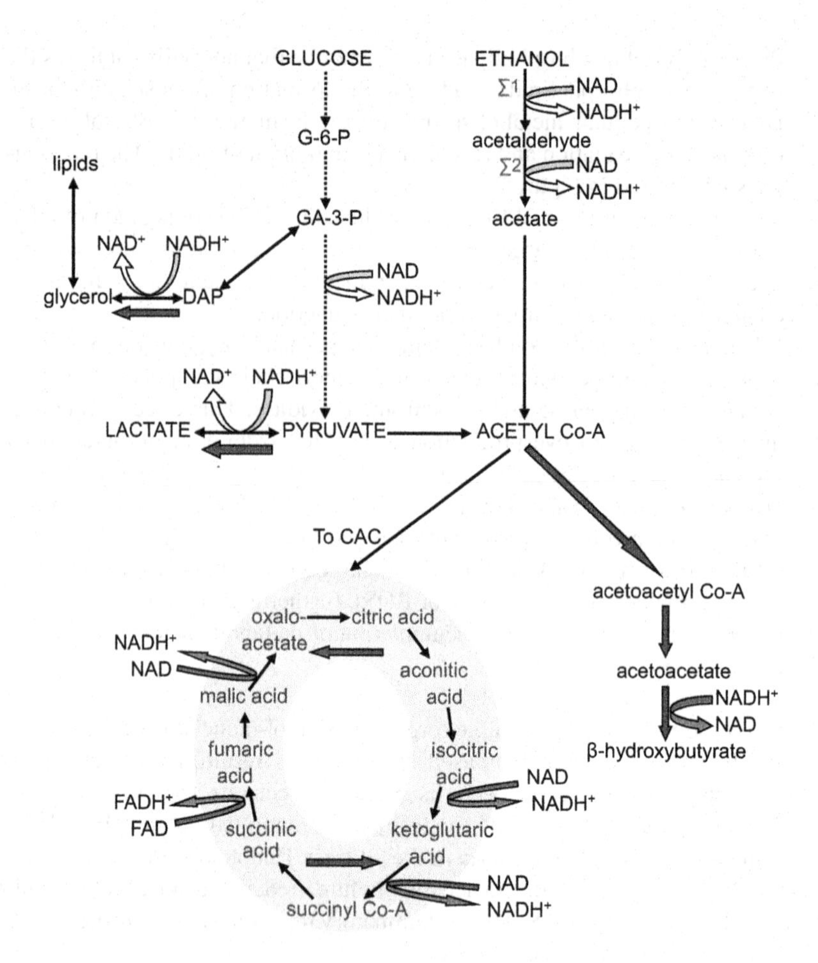

FIGURE 24.3 Influence of ethanol on energy metabolism and its pathways. Glycolysis is on left side of figure; solid thinner arrows illustrate normal pathways; shaded circle = citric acid cycle (CAC); lighter gray curved arrows = generation of NADH+; reverse block arrows = effect of ethanol and reversed CAC due to ethanol toxicity; dashed arrows = presence of intermediates (not shown in diagram). ∑1: aldehyde dehydrogenase; ∑2: acetaldehyde dehydrogenase; DAP: dihydroxyacetone phosphate; GA-3-P: glyceraldehyde-3-phosphate; G-6-P: glucose-6-phosphate.

GI effects of ethanol ingestion in acute intoxication impair the gastric mucosal barrier and increase gastric and pancreatic secretions. Commonly, ethanol irritates the GI tract and aggravates concomitant diseases of the stomach and intestinal tract. Table 24.1 categorizes and correlates the BAC with the affected CNS regions and the corresponding signs and symptoms of acute alcohol toxicity.

24.1.6 Clinical Manifestations of Acute Toxicity

In the last decade, the legal BAC level for ethanol intoxication was lowered from 100 to 80 mg/dl (0.08% or 0.08), which is now the legal limit for DUI/DWI laws in all 50

TABLE 24.1
Correlation of BAC with Affected CNS Regions and Signs and Symptoms of Acute Alcohol Intoxication

BAC	Anatomical CNS Region	Observed Toxic Effects	
		Signs	Symptoms
50–100 mg/dl (0.05–0.10%)	Frontal lobe	↓RPM, ↑HR	Nausea, vomiting, alcoholic breath, euphoria, blurred vision, ↓attention span, altered judgment, slow reaction (reflex) time, ↑ confidence
100–300 mg/dl (0.1–0.3%)	Frontal, parietal, occipital lobes; cerebellum	Diplopia[a], staggering gait, ataxia, hypoglycemia, loss of critical judgment	Muscular incoordination, blurred vision, stuporous state, emotional instability, tremors, slurred speech
300–500 mg/dl (0.3–0.5%)	Frontal, parietal, occipital lobes; cerebellum, diencephalon	Marked ataxia, disorientation, visual impairment, slow shallow respirations, convulsions, coma	Stuporous state, severe hypoglycemia, ↓pain perception
Greater than 500 mg/dl (>0.5%)	Frontal, parietal, occipital lobes; cerebellum, diencephalon, brain stem (medulla)	CV collapse, respiratory arrest, ↓HR, coma, shock, convulsions	Stuporous, anesthetic, comatose state

[a] double vision.
CV: cardiovascular.
HR: heart rate; RPM: respirations per minute.

U.S. states.* The average lethal blood concentration equals 400 mg/dl. For chronic drinkers who become tolerant to the effects of ethanol, the corresponding concentrations to elicit the same degree of toxicity might be much higher than those for casual drinkers. The clinical manifestations of acute intoxication include sedation and relief of anxiety, reduced tension and coordination, impaired concentration and reaction time, tachycardia, and more severely, slurred speech, ataxia, and altered emotions (interestingly, respiratory depression is manifested only at high concentrations).

* DUI/DWI, or "driving under the influence" and "driving while intoxicated," respectively, are part of the individual states' drunk driving laws. Interestingly, under the Impaired Driving Law in Michigan, it is a crime for a driver to have a BAC of .08 or greater if over age 21 or .02 or greater if under 21. In addition, Michigan has a high-BAC law with enhanced penalties for anyone caught driving with a BAC of 0.17 or higher.

BAC corresponding to greater than 300 mg/dl produces metabolic and toxic coma, which presents clinically as muscular hypotonia, respiratory depression, hypotension, and hypothermia. Clinical laboratory testing reveals hypoglycemia, ketosis, and electrolyte derangements in patients with severe ethanol intoxication.

24.1.7 Management of Acute Intoxication

Treatment of severe acute alcohol intoxication depends on the level, rate, and duration of the BAC. The goal of management of acute ethanol intoxication follows the ABCs of supportive therapy—that is, prevention of respiratory depression and the pulmonary aspiration of vomitus. In addition, glucose is administered for the prevention of hypoglycemia and ketosis. Electrolyte solutions are also administered to prevent the dehydration that accompanies vomiting. Moreover, chronic drinkers are at risk of developing a malnutrition status. Therefore, nutrients, such as thiamine, folate, and magnesium, accompany the course of therapy. Other strategies, including hemodialysis and gastric decontamination, have limited efficacy in the management of ethanol intoxication.

24.1.8 Clinical Manifestations of Chronic Toxicity

Long-term consumption of ethanol impairs almost all of the organ systems (Table 24.2), and chronic alcoholic addiction leads to twice the risk of death compared with nondrinkers. The higher risk is principally due to the greater incidence of alcoholic liver cirrhosis, infections, accidents, cancers, and cardiovascular diseases.

Chronic ethanol consumption affects the central (CNS) and peripheral (PNS) nervous systems. CNS damage in chronic alcoholics often presents signs and symptoms corresponding to the Wernicke–Korsakoff syndrome. This condition is classically characterized by a triad of:

1. Paralysis of external eye muscles
2. Cerebellar ataxia
3. Mental confusion

The full clinical presentation is rarely encountered, but it is believed to be associated with thiamine deficiency due to reduced thiamine (vitamin B_1) absorption in alcoholics. *Wernicke's encephalopathy* is the acute phase of the disease, with longer duration of the confusional state in comparison with acute ethanol intoxication. Administration of thiamine alleviates the ataxia, ocular difficulty, and confusion; however, a memory deficit, known as *Korsakoff psychosis*, may linger. Moreover, ethanol may cause bilateral and symmetrical visual impairment because of optic nerve degeneration.

Ethanol-associated neuropathy characterizes PNS damage in chronic ethanol intoxication. It is characterized by axonal degeneration of peripheral nerve fibers with earlier and more frequent involvement of sensory fibers of the lower limbs. It usually commences with symmetrical paresthesias of hands and feet.

TABLE 24.2
Major Systemic Effects of Chronic Ethanol Consumption

Liver	Steatosis
	Alcoholic hepatitis
	Cirrhosis
	Liver cancer
Cardiovascular System	Alcoholic cardiomyopathy
	Cardiac arrhythmias
	Hypertension
Central Nervous System	Wernicke–Korsakoff syndrome
	Dementia
	Tolerance, dependence, and withdrawal
Gastrointestinal System	Esophagitis
	Gastritis
	Malabsorption
	Pancreatitis
	Cancer of mouth, pharynx, esophagus
Endocrine and Metabolic Systems	Hypoglycemia
	Alcoholic ketoacidosis
	Hypomagnesemia
	Hypokalemia
	Malnutrition
	Gynecomastia
	Menstrual cycle abnormalities
Reproductive System	Hypogonadism
	Impotence
	Infertility
Hematologic System	Iron, folate, B_{12} deficiency anemias
	Leukopenia

As with acute intoxication, liver disease is the most frequent clinical complication of chronic ethanol abuse. About 90% of chronic drinkers have steatosis of the liver, defined by the abnormal accumulation of triglycerides in hepatocytes. While fatty liver is reversible, it may develop into alcoholic hepatitis,* cirrhosis, or liver cancer. Clinically, patients present with nausea, vomiting, jaundice, abdominal pain, and hepatosplenomegaly. Alcoholic cirrhosis, characterized by fibroblastic proliferation and the production of connective tissue in the periportal and centrilobular regions, is the most common type of cirrhosis. The development, duration, and intensity of liver diseases correlate with the amount and the duration of ethanol consumption.

Long-term ingestion of ethanol increases the occurrence of gastritis and pancreatitis and causes intestinal injury, further aggravating or complicating vitamin deficiencies, diarrhea, and weight loss.

* Alcoholic hepatitis refers to hepatocyte degeneration and necrosis.

Cardiovascular toxicity from chronic ethanol intoxication is associated with dilated cardiomyopathy with ventricular hypertrophy and fibrosis, which often precipitates atrial and ventricular arrhythmias. Interestingly, chronic consumption of ethanol is associated with an increased incidence of hypertension.

24.1.9 Management of Chronic Intoxication

The management of chronic intoxication and addiction presents significant obstacles to the treatment of ethanol-related diseases. Psychological, medical, and pharmacological interventions, combined with abstinence, are the resources available for treatment of chronic alcoholism. While the drugs are designed to interfere with ethanol metabolism, pharmacological intervention alone has limited efficacy.

Disulfiram (tetraethylthiuram) is the most commonly used drug to deter drinking. Disulfiram is an inhibitor of aldehyde dehydrogenase and causes the accumulation of acetaldehyde. If the drug is administered soon after or before ethanol ingestion, extreme discomfort, including flushing, throbbing headache, nausea, vomiting, sweating, hypotension, and confusion, develops. Disulfiram is rapidly absorbed, with peak effects at about 12 h, and its duration is 24–48 h due to slow elimination. Disulfiram is understood to inhibit the metabolism of other therapeutic agents such as phenytoin, isoniazid, and oral anticoagulants. Additionally, the drug alters liver function and associated diagnostic tests. Due to poor compliance and side effects, disulfiram is becoming less favored in alcoholism therapy.

Naltrexone, the opioid receptor antagonist, has high oral bioavailability and a long duration of action. The basis for incorporation of an opioid antagonist into the treatment of chronic alcoholism relies on the evidence that opioid receptor antagonists can reduce the cravings associated with alcohol drinking. Typically, as with disulfiram, naltrexone is part of a psychosocial therapeutic management. Adverse effects include nausea, dizziness, and headache, but an overdose of naltrexone can result in severe liver damage, acute hepatitis, and liver failure. In contrast to disulfiram, naltrexone is associated with good compliance in alcoholism therapy. Due to the potential hepatotoxicity of both drugs, the combination of naltrexone and disulfiram should be avoided.

Nalmefene has also been developed as an opioid antagonist in preliminary clinical trials. It has some advantages over naltrexone, including greater oral bioavailability, longer duration of action, and lack of dose-dependent liver toxicity. Other areas of clinical pharmacological interest include the development of drugs that alter glutamatergic, serotonergic, and dopaminergic neurotransmitter systems. Of these, a more recent addition to the drug treatment battery for chronic alcoholism is acamprosate (Campral®). The drug is a structural analog of γ-aminobutyric acid (GABA) and the amino acid taurine and is a competitive inhibitor of the NMDA (glutamate) receptor family. The pharmacological objective of acamprosate is to reestablish the balance between neuronal excitation (glutamate) and inhibition (GABA), which is upset by excessive ethanol consumption. Thus, the compound is indicated for the maintenance of abstinence from alcohol in patients with alcohol dependence who are abstinent at treatment initiation. It decreases the rate of relapse and enhances the duration of abstinence. Other drugs currently marketed as "off-label" uses for the treatment of chronic alcohol addiction include buspirone, a serotonin receptor antagonist; fluoxetine, a selective serotonin reuptake inhibitor (SSRI); and D1/D2 dopamine receptor antagonists. In addition to pharmacotherapy, magnesium, potassium phosphate, multivitamins, and folate are administrated to alcoholic patients to correct the hypomagnesium, hypophosphatemia, hypokalemia, and vitamin deficiency associated with chronic alcohol abuse, respectively.

24.1.10 Fetal Alcohol Syndrome (FAS)

FAS is the most common preventable cause of mental retardation and congenital malformation in humans. It is estimated that FAS affects 4000 infants per year in the United States, and an additional 7000 cases show fetal alcohol effects. The typical features of FAS include retarded body growth, craniofacial abnormalities, and CNS dysfunction. The mechanism underlying the teratogenic effects caused by chronic ethanol abuse remains uncertain.

24.1.11 Tolerance, Dependence, and Withdrawal

As with other drugs of abuse (see Chapter 15), the Department of Mental Health and Substance Dependence at the World Health Organization (WHO), in collaboration with the U.S. National Institute on Drug Abuse (NIDA), defines several terms that are important in understanding drug and alcohol abuse and the phenomena of tolerance and withdrawal. *Addiction* involves compulsive psychoactive substance abuse with an overwhelming involvement in the securing and using of such drugs. The withdrawal syndrome occurs as a result of sudden or abrupt discontinuation of the substance. *Compulsive substance abuse* involves the psychological need to procure and use the chemical, often referred to as *craving.* In this case, the uncontrollable drive to drink alcohol is necessary to maintain an optimum state of well-being. *Habituation* refers to psychological dependence. *Physical (physiological) dependence* involves the need for repeated ingestion to prevent withdrawal (abstinence) syndrome. The more complex phenomenon of *tolerance* requires the satisfaction of several criteria. With repeated ingestion, alcoholic individuals require greater amounts of alcohol to achieve the desired effect. Conversely, the euphoric effect is markedly diminished with continued use of the same amount of ethanol. Since various pharmacological effects on different organ systems are not uniformly distributed, tolerance is not evenly demonstrated. Increased metabolism and adjustment to the sedative and euphoric effects are proposed as possible mechanisms for the development of tolerance—that is, the physiological drive to achieve homeostasis. Unlike other substance abuse, the development of tolerance to ethanol involves induction of alcohol-metabolizing enzymes, downregulation of the GABA-mediated response, and upregulation of NMDA receptor function, all of which appear to account for the CNS hyperexcitability in ethanol withdrawal.

The alcohol withdrawal syndrome is characterized by three phases, the intensity of which depends on the extent of addiction:

1. The *tremulous phase* occurs shortly after the last consumption and is confirmed by the psychological compulsion to obtain a drink (*craving*), nausea, mydriasis, sleep disturbances, tachycardia, sweating, and hypertension.

2. The *seizure phase* continues about 12 to 18 hours after the last consumption and is mainly demonstrated by the presence of seizures.
3. The *delirium tremens* (DTs) begins within 3 days after the last consumption and is characterized by mental confusion, hallucinations, severe agitation, and fever; the patient risks cardiovascular collapse if the condition is not managed.

Treatment of withdrawal involves reassurance, restoration of fluid and electrolyte balance, and correction of nutritional deficiencies, particularly the inadequacies of vitamin C and B-complex that accompany chronic alcohol intake. For pharmacologically induced sedation, benzodiazepines are the mainstay of therapy. Short-acting barbiturates are not dependable except for phenobarbital, a long-acting barbiturate. Phenothiazines, including chlorpromazine, and anticonvulsants, such as phenytoin, are not especially useful for the DTs or seizures, respectively, as the former may lower the seizure threshold. Patients with repeated seizures respond well to diazepam.

24.1.12 Methods of Detection

To detect and quantify ethanol, many methods have been established. Blood alcohol levels can be determined accurately by immunoassay or gas chromatography in most hospitals, but it takes longer to obtain these results than by using other methods. An electrochemical meter has been applied to test alcohol concentration in venous blood, and this method has high sensitivity but poor specificity.

Breath alcohol analyzers (*Breathalyzers*) are widely used as alcohol screening tools, especially by law-enforcement agencies. The calculation of the BAC depends on the constant vapor pressure (volatility) of ethanol and the distribution of ethanol between blood and pulmonary air spaces, which compute to a ratio of the concentration of ethanol in capillary blood to that in alveolar air of about 2100. This test is performed by microprocessors and infrared spectral analysis with good accuracy and precision. The test relies on the oxidation of any alcohol with oxidizing substrates and the production of a green reaction product according to the following equation:

$$K_2Cr_2O_7 + CH_3 - CH_2 - OH + H_2SO_4 \rightarrow CH_3 - COOH + K_2SO_4 + Cr_2\left(SO_4\right)_3 + H_2O$$

where

$K_2Cr_2O_7$ = potassium dichromate
CH_3-CH_2-OH = ethanol
H_2SO_4 = sulfuric acid
CH_3-COOH = acetic acid
K_2SO_4 = potassium sulfate
$Cr_2(SO_4)_3$ = green dichromium sulfate

The absorbance of the colored product is monitored against a reference standard, which is then quantified and converted to BAC units, yielding a measurable, legally enforced concentration value. The presence of reducing agents in the test sample can interfere with the accuracy of the test, especially the incorporation of acetaldehyde, methanol, or other alcohols.

Although not technically difficult, determination of BAC "at the scene" is complicated by the cooperation and the physical condition of the suspect (e.g., the individual may be comatose), and the sample is not available for re-examination.

24.2 METHANOL

Methanol (methyl alcohol, wood alcohol, Columbian spirits) is readily available at various concentrations in numerous industrial and household products, including windshield washer fluid and paint thinner. It is also used in the manufacture of formaldehyde and methyl *t*-butyl ether. Exposure to methanol commonly occurs in the workplace and in the household, and it is often intentionally ingested by alcoholics.

24.2.1 Incidence and Occurrence

The consumption of adulterated counterfeit or informally produced spirit drinks causes outbreaks of methanol poisoning. Many countries, including Cambodia, the Czech Republic, Ecuador, Estonia, India, Indonesia, Kenya, Libya, Nicaragua, Norway, Pakistan, Turkey, and Uganda, have seen numerous outbreaks in recent years. According to the fact sheet published by WHO in July 2014, fatality rates were over 30%, and the size of these outbreaks ranged from 20 to 800 victims (Agency for Toxic Substances and Disease Registry [U.S. ATSDR], Atlanta, Georgia, U.S. Department of Health and Human Services).

24.2.2 Toxicokinetics

Methanol is readily absorbed via the skin, GI, and respiratory routes. When ingested, peak methanol blood levels occur within 30 to 60 minutes. Methanol (CH_3OH) is oxidized by ADH in the liver to formaldehyde (CH_2O), which in turn is converted to formic acid (formate, CHOH=O) via the action of formaldehyde dehydrogenase (also an ALDH) according to the following formula:

$$CH_3{-}OH \xrightarrow{\text{ADH}} CH_2{=}O \xrightarrow{\text{ALDH}} HO{-}CH{=}O \xrightarrow{\text{THFR}} CO_2 + H_2O$$

The formation of formic acid is largely responsible for the toxicity of methanol. Formate is further eliminated by the action of formyl-tetrahydrofolate-reductase (formyl-THFR) by combination with tetrahydrofolate (THF) to form 10-formyl-THF (not shown). The product is then converted to carbon dioxide (and water) by the catalytic action of formyl-THF-dehydrogenase (F-THF-DH). The elimination half-life of formate is about 3.5 h.

24.2.3 Mechanism of Toxicity

Most of the toxic effects of methanol are attributed to the formation of formic acid. Formic acid inhibits the mitochondrial cytochrome *c* oxidase complex and concomitantly decreases ATP production, resulting in increased anaerobic glycolysis and production of lactate, in a manner similar to the inhibition of the citric acid cycle by ethanol (Figure 24.3). The accumulation of formate and lactate causes systemic acidosis, which facilitates the formation of nonionized formate, leading to its increased cellular accumulation and cytotoxic effects. In addition, the production of hydroxyl radicals and induction of lipid peroxidation may also be implicated in the cellular damage induced by methanol intoxication. Ocular tissues such as optic nerve and retina are more susceptible to the toxicity of formate because the retina is able to metabolize methanol to formate, resulting in retinal edema and optic nerve damage.

24.2.4 Clinical Manifestations of Acute Intoxication

Methanol primarily affects the nervous, ocular, and GI systems. While intoxication symptoms appear within a few hours after methanol intake, they can be delayed for more than 30 hours due to varied individual response and/or co-ingestion of ethanol. Among the symptoms induced by methanol, the most typical clinical presentation is visual disturbance, including blurred vision, visual hallucination, and loss of vision. On eye examination, retinal edema, visual field constriction, and nonreactive pupils are sequelae. Nonspecific neurological manifestations, such as headache, dizziness, bradycardia, impaired consciousness, seizures, and coma, are nervous system consequences of methanol intoxication. Abdominal pain, diarrhea, GI hemorrhage, and pancreatitis represent signs and symptoms of GI complications. Finally, systemic metabolic acidosis induced by methanol intoxication may also increase the respiratory rate.

24.2.5 Management of Acute Intoxication

The general ABCs of supportive care are part of immediate emergency support. Diazepam or phenobarbital are administrated to patients with seizures.

Fomepizole and ethanol are specific antidotes for methanol intoxication. Fomepizole is an ADH inhibitor that effectively blocks the metabolism of methanol and reduces the conversion of the alcohol to formaldehyde and formic acid. In addition, the administration of ethanol also forces the competition of the alcohols for ADH, thus diverting the metabolism away from methanol. As such, fomepizole should be given intravenously (I.V.)* to all patients with methanol intoxication. Caution is in order for ethanol administration, since it may cause CNS depression and hypoglycemia, as the dose is not easily manipulated.† However, fomepizole has great efficacy and fewer side effects than ethanol; hence, it is the favored antidote in the treatment of methanol intoxication.

24.3 ISOPROPANOL

Isopropanol (isopropyl alcohol) is a colorless and volatile alcohol with a fruity odor and a slightly bitter taste. It is present in rubbing alcohol, industrial solvents, paints,

---

* 15 mg/kg body weight I.V. stat, followed by 10 mg/kg every 12 hours.

† Initial ethanol loading dose of 0.7 g/kg body weight, infused over 1 hour, followed by 0.1–0.2 g/kg/h to achieve an ethanol maintenance level (BAC) of 100 mg/dl.

disinfectants, and parenteral drugs. It may be ingested accidentally by nonalcoholics and intentionally by alcoholics.

24.3.1 Toxicokinetics

Isopropanol enters biological compartments by ingestion, by inhalation of vapors, or through the skin. Isopropanol reaches peak blood levels 30 minutes after ingestion. It is metabolized by hepatic ADH, by which 80% is converted to acetone, as shown:

$$H_3C-\underset{|}{\overset{HO}{C}}H_2-CH_3 \xrightarrow[ADH]{} H_3C-\overset{O}{\overset{\|}{C}}-CH_3$$

Acetone is excreted primarily through the renal route (20% unchanged) and a small amount through lungs, saliva, and gastric juices. The range of elimination half-life is 2.5 to 6.6 hours for isopropanol and 10 to 31 hours for acetone.

24.3.2 Mechanisms of Toxicity

Because acetone cannot be further converted to an acid, isopropanol intoxication is not considered to cause metabolic acidosis. The metabolite, however, is a potent CNS depressant and about twice as toxic as ethanol.

24.3.3 Clinical Manifestations of Acute Toxicity

The presentation of CNS depression with isopropanol intoxication includes inebriation with drowsiness, incoordination, staggering gait, and slurred speech as well as sweating, stupor, coma, and death due to respiratory depression. GI responses such as nausea, vomiting, and abdominal pain and hemorrhage into the bronchi and the chest cavity may occur. In addition, the patient displays an acetone odor to the breath. In severe intoxication, CV compromise may occur, including myocardial depression and severe hypotension. For young children after isopropanol ingestion, irritability, hypotonia, and seizures may be indicators of intoxication.

24.3.4 Management of Acute Intoxication

Respiratory support and cardiac monitoring are required in the treatment of isopropanol toxicity. Patients who demonstrate respiratory depression require ventilatory support, gastric lavage, and hemodialysis.

24.4 FORMALDEHYDE

Formaldehyde is a nearly colorless gas with a pungent irritating odor. It dissolves easily in water and is used to formulate formalin* and methanol (wood alcohol).

* A 10% formaldehyde solution in water or buffer, also known as neutral buffered formalin (NBF).

Formaldehyde is used as a preservative, a hardening and reducing agent, a corrosion inhibitor, a sterilizing agent, a histological preservative, and an embalming fluid. It is also found in glues, pressed wood products, foam insulation, and a wide variety of molded or extruded plastic items. Indoor sources include permanent-press fabrics, carpets, pesticide formulations, and cardboard and paper products. Outdoor sources include emissions from fuel combustion, oil refining processes, and environmental tobacco smoke.

24.4.1 Incidence and Occurrence

Building and furnishing materials such as chipboard, textiles, and cosmetics contribute to the major route of exposure of formaldehyde; that is, inhalation. Between 10% and 25% of total indoor exposure can be attributed to environmental tobacco smoke. Exposure to formaldehyde by ingestion is limited because formaldehyde is not available in free form when ingested. Dermal exposure to formaldehyde has also been reported with the use of cosmetics such as hand cream or suntan lotion. Inhalation remains a major route of exposure to formaldehyde. Studies have shown that exposure to formaldehyde can occur during commuting in gasoline-powered vehicles. Occupational exposure is also reported in the literature. From 1981 to 1983, the National Occupational Exposure Survey (NOES) posted the results of a survey, which showed that 1,329,332 workers employed in various professions were potentially exposed to formaldehyde in the United States; 45% of 2 million workers from 112,000 firms were found to have been exposed to formaldehyde. The majority of these workers belonged to the garment industry. Firefighters are another group of workers exposed to high concentrations of formaldehyde. Exposure levels as high as 8 ppm can occur depending on the type of fire control exercise. Exposure levels of 0.3 ppm were found inside the firefighters' self-contained breathing apparatus. Children are also at risk due to exposure.

24.4.2 Toxicokinetics

Formaldehyde can enter the body by inhalation, ingestion, or skin contact. Formaldehyde is readily absorbed via the respiratory and GI routes. Dermal absorption of formaldehyde appears to be very slight. In the systemic circulation, formaldehyde (CH_2O) is rapidly metabolized to formate (CHOOH) by a glutathione-dependent formaldehyde dehydrogenase (ALDH), the product of which is further converted to carbon dioxide and water as described in Section 24.2.1.

$$CH_2{=}O \xrightarrow{\text{ALDH}} HO{-}CH{=}O \xrightarrow{\text{THFR}} CO_2 + H_2O$$

Alternatively, formic acid may enter the one-carbon cycle and be incorporated as a methyl group into nucleic acids and proteins. Metabolites of formaldehyde are mainly excreted through respiratory and renal routes.

24.4.3 Mechanisms of Toxicity

Formaldehyde toxicity causes multi-organ damage on either acute or chronic exposure. Acute toxicity is largely attributable to its irritating and corrosive properties. The mechanisms underlying formaldehyde-induced chronic toxicity remain to be elucidated. The binding of formaldehyde to endogenous proteins may result in the formation of neoantigens, which can elicit an inflammatory immune response. This mechanism may account for the occurrence of asthma and other respiratory health complaints associated with formaldehyde exposure. Consequently, long-term exposure to formaldehyde is associated with the formation of formaldehyde–albumin adducts, autoantibodies, and immune activation in occupationally exposed patients or residents of mobile homes or homes containing particleboard subflooring. In addition, formaldehyde is a known carcinogen in animals and a suspected carcinogen in humans. The ability of formaldehyde to increase the frequency of DNA–protein crosslinks (DPCs) and sister chromatid exchanges (SCEs) in humans has been reported as its potential carcinogenic mechanism.

24.4.4 Clinical Manifestations of Acute Intoxication

Formaldehyde primarily affects the mucous membranes of the upper airways and eyes. The manifestations include watery eyes, burning sensations in the eyes and throat, nausea, wheezing, coughing, chest tightness, and difficulty in breathing at elevated levels (above 0.1 ppm). As little as 30 ml of oral exposure to formalin results in severe burns to the throat and stomach, and is fatal.

24.4.5 Management of Acute Intoxication

There is no antidote for formaldehyde poisoning. Supportive care constitutes the primary management of patients with acute formaldehyde intoxication.

REFERENCES

SUGGESTED READINGS

Abrahao, K.P., Salinas, A.G., and Lovinger, D.M., Alcohol and the brain: neuronal molecular targets, synapses, and circuits. *Neuron* 96, 1223, 2017.

Brunton, L.L., Hilal-Dandan, R., and Knollmann, B.C., *Goodman & Gilman's The Pharmacological Basis of Therapeutics*, 13th ed., McGraw-Hill, New York, 2018, Chapter 23.

"CDC—Fact Sheets—Alcohol". www.cdc.gov/alcohol/fact-sheets.htm, last viewed March 2019.

Clemens, D.L., Use of cultured cells to study alcohol metabolism. *Alcohol Res. Health* 29, 291, 2006.

Caprara, D.L., Nash, K., Greenbaum, R., Rovet, J., and Koren, G., Novel approaches to the diagnosis of fetal alcohol spectrum disorder. *Neurosci. Biobehav. Rev.* 31, 254, 2007.

Crabb, D.W. and Liangpunsakul, S., Acetaldehyde generating enzyme systems: roles of alcohol dehydrogenase, CYP2E1 and catalase, and speculations on the role of other enzymes and processes. *Novartis Found. Symp.* 285, 4, 2007.

Crews, F.T., Walter, T.J., Coleman, L.G. Jr., Vetreno, R.P., Toll-like receptor signaling and stages of addiction. *Psychopharmacology (Berl.)* 234, 1483, 2017.
Hack, J.B., Goldlust, E.J., Ferrante, D., and Zink, B.J., Performance of the Hack's impairment index score: a novel tool to assess impairment from alcohol in emergency department patients. *Acad. Emerg. Med.* 24, 1193, 2017.
LoVecchio, F., Sawyers, B., Thole, D., Beuler, M.C., Winchell, J., and Curry, S.C., Outcomes following abuse of methanol-containing carburetor cleaners. *Hum. Exp. Toxicol.* 23, 473, 2004.
Martin, T.L., Solbeck, P.A., Mayers, D.J., Langille, R.M., Buczek, Y., and Pelletier, M.R., A review of alcohol-impaired driving: the role of blood alcohol concentration and complexity of the driving task. *J. Forensic Sci.* 58, 1238, 2013.
Perry, P.J., Doroudgar, S., and Van Dyke, P., Ethanol forensic toxicology. *J. Am. Acad. Psychiatry Law* 45, 429, 2017.
Pizon, A.F., Becker, C.E., and Bikin, D., The clinical significance of variations in ethanol toxicokinetics. *J. Med. Toxicol.* 3, 63, 2007.
Preedy, V.R., Crabb, D.W., Farrés, J., and Emery, P.W., Alcoholic myopathy and acetaldehyde. *Novartis Found. Symp.* 285, 158, 2007.
Roberts, S.M., James, R.C., and Williams, P.L., *Principles of Toxicology: Environmental and Industrial Applications*, 3rd ed., John Wiley & Sons, New York, 2015, chapter 18.
Seitz, H.K. and Becker, P., Alcohol metabolism and cancer risk. *Alcohol Res. Health* 30, 38, 2007.
U.S. National Institute on Drug Abuse (NIDA), National Institutes of Health (NIH), U.S., Natuional Survey of Drug Use and Health (NSDUH), Monitoring the Future Study, 2018. From: www.monitoringthefuture.org, last viewed March 2019.
Vasiliou, V., Pappa, A., and Estey, T., Role of human aldehyde dehydrogenases in endobiotic and xenobiotic metabolism. *Drug Metab. Rev.* 36, 279, 2004.

REVIEW ARTICLES

Agarwal, D.P., Genetic polymorphisms of alcohol metabolizing enzymes. *Pathol. Biol.* 49, 703, 2001.
Baker, R.C. and Kramer, R.E., Cytotoxicity of short-chain alcohols. *Annu. Rev. Pharmacol. Toxicol.* 39, 127, 1999.
Barceloux, D.G., Bond, G.R., Krenzelok, E.P., Cooper, H., and Vale, J.A., American Academy of Clinical Toxicology practice guidelines on the treatment of methanol poisoning. American Academy of Clinical Toxicology Ad Hoc Committee on the Treatment Guidelines for Methanol Poisoning. *J. Toxicol. Clin. Toxicol.* 40, 415, 2002.
Beauchamp, G.A., Carey, J.L., Adams, T., Wier, A., Colón, M.F., Cook, M., Cannon, R., Katz, K.D., and Greenberg, M.R.Sex differences in poisonings among older adults: an analysis of the Toxicology Investigators Consortium (ToxIC) Registry, 2010 to 2016. *Clin. Ther.* 2018.
Beckemeier, M.E. and Bora, P.S., Fatty acid ethyl esters: potentially toxic products of myocardial ethanol metabolism. *J. Mol. Cell Cardiol.* 30, 2487, 1998.
Brooks, P.J. and Theruvathu, J.A., DNA adducts from acetaldehyde: implications for alcohol-related carcinogenesis. *Alcohol* 35, 187, 2005.
Cogliano, V. Cogliano., V., Grosse, Y., Baan, R., Straif, K., Secretan, B., and El Ghissassi, F., WHO International Agency for Research on Cancer. Advice on formaldehyde and glycol ethers. *Lancet Oncol.* 5, 528, 2004.
Collins, J.J. Ness, R., Tyl, R.W., Krivanek, N., Esmen, N.A., and Hall, T.A., A review of adverse pregnancy outcomes and formaldehyde exposure in human and animal studies. *Regul. Toxicol. Pharmacol.* 34, 17, 2001.

Cowan, J.M. Jr., Weathermon, A., McCutcheon, J.R., and Oliver, R.D., Determination of volume of distribution for ethanol in male and female subjects. *J. Anal. Toxicol.* 20, 287, 1996.

Dey, A. and Cederbaum, A.I., Alcohol and oxidative liver injury. *Hepatology* 43, S63, 2006.

Garner, C.D., Lee, E.W., Terzo, T.S., and Louis-Ferdinand, R.T., Role of retinal metabolism in methanol-induced retinal toxicity. *J. Toxicol. Environ. Health* 44, 43, 1995.

Guerzoni, S., Pellesi, L., Pini, L.A., and Caputo, F., Drug-drug interactions in the treatment for alcohol use disorders: a comprehensive review. *Pharmacol. Res.* 133, 65, 2018.

Henderson, G.I., Chen, J.J., and Schenker, S., Ethanol, oxidative stress, reactive aldehydes, and the fetus. *Front. Biosci.* 4, D541, 1999.

Hipólito, L., Sánchez, M.J., Polache, A., and Granero, L., Brain metabolism of ethanol and alcoholism: an update. *Curr. Drug Metab.* 8, 716, 2007.

Hoek, J.B. and Pastorino, J.G., Ethanol, oxidative stress, and cytokine-induced liver cell injury. *Alcohol* 27, 63, 2002.

Holland, M.G. and Ferner, R.E., A systematic review of the evidence for acute tolerance to alcohol—the "Mellanby effect". *Clin. Toxicol. (Phila.)* 55, 545, 2017.

Johnson, B.A. and Ait-Daoud, N., Neuropharmacological treatments for alcoholism: scientific basis and clinical findings. *Psychopharmacology* 149, 327, 2000.

Jung, Y.C. and Namkoong, K., Alcohol: intoxication and poisoning—diagnosis and treatment. *Handb. Clin. Neurol.* 125, 115, 2014.

Kaphalia, B.S. and Ansari, G.A., Fatty acid ethyl esters and ethanol-induced pancreatitis. *Cell Mol. Biol.* 47, 173, 2001.

Lang, C.H., Kimball, S.R., Frost, R.A., and Vary, T.C., Alcohol myopathy: impairment of protein synthesis and translation initiation. *Int. J. Biochem. Cell Biol.* 33, 457, 2001.

Leo, M.A. and Lieber, C.S., Alcohol, vitamin A, and beta-carotene: adverse interactions, including hepatotoxicity and carcinogenicity. *Am. J. Clin. Nutr.* 69, 1071, 1999.

Mégarbane, B., Borron, S.W., and Baud, F.J., Current recommendations for treatment of severe toxic alcohol poisonings. *Intensive Care Med.* 31, 189, 2005.

Mullins, P.M., Mazer-Amirshahi, M., and Pines, J.M. Alcohol-related visits to US Emergency Departments, 2001–2011. *Alcohol* 52, 119, 2013.

Patel, V.B., Why, H.J., Richardson, P.J., and Preedy, V.R.,The effects of alcohol on the heart. *Adverse Drug React. Toxicol. Rev* 16, 15, 1997.

Pohanka, M., Toxicology and the biological role of methanol and ethanol: current view. *Biomed. Pap Med. Fac. Univ. Palacky Olomouc Czech Repub.* 160, 54, 2016.

Pragst, F. and Yegles, M. Determination of fatty acid ethyl esters (FAEE) and ethyl glucuronide (EtG) in hair: a promising way for retrospective detection of alcohol abuse during pregnancy? *Ther. Drug Monit.* 30, 255, 2008.

Ramchandani, V.A., Bosron, W.F., and Li, T.K., Research advances in ethanol metabolism. *Pathol. Biol.* 49, 676, 2001.

Romaguera, A., Torrens, M., Papaseit, E., Arellano, A.L., and Farré, M.,Concurrent use of cannabis and alcohol: neuropsychiatric effect consequences. *CNS Neurol. Disord. Drug Targets* 16, 592, 2017.

Sanghani, P.C., Stone, C.L., Ray, B.D., Pindel, E.V., Hurley, T.D., and Bosron, W.F., Kinetic mechanism of human glutathione-dependent formaldehyde dehydrogenase. *Biochemistry* 39, 10720, 2000.

Song, B.J., Ethanol-inducible cytochrome P450 (CYP2E1): biochemistry, molecular biology and clinical relevance: 1996 update. *Alcohol Clin. Exp. Res.* 20, 138A, 1996.

Zachman, R.D. and Grummer, M.A., The interaction of ethanol and vitamin A as a potential mechanism for the pathogenesis of fetal alcohol syndrome. *Alcohol Clin. Exp. Res.* 22, 1544, 1998.

25 Gases

25.1 INTRODUCTION

Pulmonary irritants, simple asphyxiants, toxic products of combustion, lacrimating agents, and chemical asphyxiants comprise a diverse group of toxic gases capable of causing a variety of local and pulmonary reactions. The sources of these compounds are as varied, encompassing both naturally occurring and synthetic chemicals developed over the last 50 years. In addition, exposures to the gases are encountered *accidentally*, at home or in the workplace with industrial products, as environmental hazards, or *intentionally*, as commercial sprays for individual protection, for law enforcement, or as potential bioterrorist weapons. The most common routes of exposure to these agents occur through oral ingestion, local dermal or mucous membrane contact, or deep inhalation. Because of their seemingly unrelated chemical structures and properties, the gases are classified principally according to the clinical effects and metabolic consequences of exposure.

Physiologically, the respiratory system is composed of

1. The upper respiratory tract (URT), consisting of the nasal and oral cavities, pharynx, and larynx;
2. The lower respiratory tract (LRT), namely, the trachea, bronchi, bronchioles, and lungs.

Two main functions of the respiratory system are ventilation (inspiration and expiration of air down to the level of the alveoli) and gas exchange.* In support of these functions, the mucociliary system produces mucus. This tenacious fluid captures inhaled particles and transports them to the nasal and oral mucosa by ciliary action of epithelial cells of the URT and LRT. The cough reflex then acts to expectorate the sequestered mucus. Together, these innate reflex responses cooperate in the *clearance mechanism* of pulmonary function. For effective gas exchange, therefore, ventilation and perfusion must match closely (calculated as the *V/Q ratio*). Consequently, any alteration of the V/Q ratio, or interference with the *clearance mechanism* due to the presence of toxic gases, results in altered lung mechanics. This explains the ensuing clinical effects.

25.2 PULMONARY IRRITANTS

As the label implies, pulmonary irritants are chemicals that when in contact with mucous membranes, produce an inflammatory response and cytotoxicity. The extent

* Gas exchange involves the passage of gas molecules through the respiratory membranes. Oxygen diffuses from air in the alveoli, across the alveolar membranes, and into the capillary circulation. Carbon dioxide diffuses from the capillaries to the alveolar space. Gas exchange is driven by partial pressures of the gases in inspired air.

of trauma ranges from mild to severe irritation and depends on the nature of the chemical (solubility, concentration, and pH). Table 25.1 lists commonly encountered pulmonary irritants, sources, and major pathophysiologic effects. The presentation of signs and symptoms of exposure to pulmonary irritants, summarized in Table 25.2, appears progressively. Toxicity depends on several factors: the extent of exposure (local contact or deep inhalation), the duration of exposure, the degree of remedial measures (physiologic clearance, availability of treatment), and the development of secondary complications (such as infections).

25.3 SIMPLE ASPHYXIANTS

25.3.1 Introduction

Unlike the pulmonary irritants or the chemical asphyxiants (described later), simple asphyxiants are nonirritating, chemically inert gases. Simple asphyxiants interfere with pulmonary function by overwhelming the oxygen concentration in inspired air. The net result is a lowered oxygen content of inspired air and decreased oxygen availability for gas exchange. The toxicity of these agents depends on the available oxygen concentration remaining.

25.3.2 Gaseous Agents

The characteristics and properties of the simple asphyxiants are summarized in Table 25.3. The agents are naturally occurring atmospheric components or are produced as products of industrial combustion. Toxic concentrations, however, are encountered in areas where the gases tend to accumulate. The acute effects and corresponding signs and symptoms of simple asphyxiants, therefore, result from replacing oxygen in inspired air with the gas, causing a decrease in oxygen concentration and corresponding signs and symptoms related to oxygen deprivation (Table 25.4).

Conversely, therapeutic administration of oxygen (O_2) is associated with toxicity, depending on the concentration, the duration of inhalation, and the forced pressure. Respiratory asphyxiation or irritation occurs with prolonged inhalation (>24 h) of 100% O_2 at 1 atmosphere of pressure. This concentration is capable of raising the partial pressure of oxygen (PO_2) in arterial blood to 600 mmHg.* Oxygen toxicity also occurs with the administration of hyperbaric 100% O_2 (about 8 h at 2 atmospheres of pressure or 1 h at 3 atmospheres of pressure). With hyperbaric oxygen, arterial and venous plasma are at risk of complete saturation. For instance, at 2–3 atmospheres of hyperbaric oxygen for 1 h, the volume % O_2 dissolved in plasma is 6.1%; normal breathing at sea level is 0.3%–0.5%. For these high pressures, the PO_2 increases to 1500–2000. At this saturation level, patients experience signs and symptoms related to O_2 apnea—that is, sore throat, coughing, retraction of eardrums, and obstruction of paranasal sinuses. Tracheobronchitis, pulmonary edema, and atalectasis (complete alveolar collapse) follow progressively. Other adverse reactions

* The normal PO_2 for arterial and venous blood, breathing air at sea level, is 100 and 40 mmHg, respectively.

TABLE 25.1
Pulmonary Irritants: Sources, Chemical and Clinical Properties

Agent	Source	Industrial[a] or Household Uses	Chemical Properties	Acute Effects
Acrolein	Petroleum by-product, synthetic	Plastics, metals, aquatic herbicide	WS, F liquid	Local irritation, delayed pulmonary edema
Acrylonitrile	Synthetic	Acrylics, fumigant, chemical intermediate	WS, LS, F, explosive	Cyanide toxicity, skin vesiculation, dermatitis
Ammonia	Synthetic	Plastics, refrigerants, household cleaners, petroleum products	Alkaline, WS gas	Liquefaction necrosis, hemoptysis, ARDS, RADS[b]
Arsine (AsH_3)	Petroleum by-product	Glass, enamels, herbicides, textiles, preservative	Neutral gas, slightly WS (garlic odor)	Local irritation, hemolysis, renal failure, peripheral neuropathy
Carbon disulfide	Petroleum by-product, natural gas	Electroplating, degreaser, production of rayon	F liquid	Peripheral neuropathy, euphoria, restlessness, NV
Chlorine	Synthetic chemical intermediate	Bleach, chlorination of water, rubber, plastics, disinfectant	Greenish-yellow gas, oxidizing agent	Local, URT, LRT irritation
Formaldehyde	Environmental, synthetic, wood/coal smoke	Disinfectant, histological preservative (fixative)	Reducing agent, 37% gas in water	Local, URT, LRT irritation, convulsions, coma, respiratory failure
Hydrogen fluoride	Photographic film, solvents, plastics	Insecticide, bleach, metal and glass industry	Colorless gas, WS, LS	Local, URT, LRT irritation; cardiac arrhythmias
Hydrogen sulfide	Naturally occurring, organic decomposition	Petroleum, paper industries, metallurgy	F gas (rotten egg odor)	Irritant and asphyxiant, respiratory paralysis; mechanism similar to CN
Metal fumes (zinc chloride)	Welding	Undesirable product of welding and metal industries	Lustrous metal, WS	Local, URT, LRT irritation (*metal fume fever*)
Nitrogen dioxide	Synthetic, photochemical smog	Chemical intermediate, nitration	Reddish-brown gas, oxidizing agent	Local, URT, LRT irritation; pulmonary edema and arrest

(*Continued*)

TABLE 25.1 (CONTINUED)
Pulmonary Irritants: Sources, Chemical and Clinical Properties

Agent	Source	Industrial[a] or Household Uses	Chemical Properties	Acute Effects
Nitrogen oxide	Welding, farming, explosives	Undesirable product of welding and farming industries	Oxidizing agent	Pulmonary edema, methemoglobinemia
Ozone (O_3)	Welding, sewage treatment plants, air pollution	Disinfectant, bleaching, oxidizing agent	Bluish explosive gas or liquid, oxidizing agent	Irritation, pulmonary edema, chronic respiratory disease
Phosgene	Welding	Production of solvents, plastics, pesticides	Colorless gas, poor WS	Pulmonary edema, URT irritation
Sulfur dioxide	Synthetic	Preservative, disinfectant, bleaching	Colorless gas	Local, URT irritation
Sulfur oxide	Industrial, air pollution	Bleaching, paper industry	Oxidizing agent	Bronchoconstriction, cough, chest tightness

[a] Used in manufacturing, synthesis, or as part of industrial processes.

[b] RADS (reactive airway dysfunction) develops secondary to massive, acute, accidental, high-level irritant exposure. It is manifested by persistent, nonspecific hyperreactivity.

ARDS: adult respiratory distress syndrome; F: flammable; LRT: lower respiratory tract; LS: lipid soluble; N: nausea; URT: upper respiratory tract; V: vomiting; WS: water soluble.

TABLE 25.2
Presentation of Signs and Symptoms of Pulmonary Irritants Depending on Site of Exposure

Pulmonary Target	Pathological Effects
Nasal, oral, ocular mucous membranes	Local irritation: rhinitis, conjunctivitis, sneezing, coughing, lacrimation
URT: nasal and oral cavities, pharynx, larynx	Burning throat, sinusitis, laryngitis, swelling, laryngoedema, dyspnea, epistaxis
LRT: trachea, bronchi, bronchioles, lungs	Nausea, vomiting, dyspnea, bronchospasm, chest pain, bronchitis, chemical pneumonitis, epithelial denudation, intravascular thrombosis, pulmonary edema, development of secondary infections

LRT: lower respiratory tract; URT: upper respiratory tract.

resulting from inhalation of high-concentration or hyperbaric oxygen include retrolental fibroplasia,* paresthesias, vertigo, loss of consciousness, nausea, and vomiting. The conditions are reversible on discontinuation before the onset of permanent damage.

Distinct among the simple asphyxiants is radon, a naturally occurring radioactive gas. Radon is a liquid below 211 K but is found in the atmosphere at a concentration of 6×10^{-14} parts per million (ppm) in air. Although the isotopes (^{220}Rn and ^{222}Rn) are short-lived ($t_{1/2} = 3.825$ days for ^{222}Rn), these α-emitters release ionizing radiation and are strongly adsorbed onto various surfaces. These properties account for radon's ubiquitous and resilient presence in the earth's crust and atmosphere. Radon is a known carcinogen suspected to produce lung cancer in hard rock mine workers due to occupational exposure. The Environmental Protection Agency (EPA) has established an action level of 4 pCi/l (0.02 working level) for home indoor radon, with a maximum permissible concentration of ^{222}Rn in air of 10^{-8} μCi/ml.

Normal atmospheric concentrations of carbon dioxide (CO_2) range between 0.027 and 0.036% v/v. Inhaling 2–5% carbon dioxide stimulates the medullary respiratory center in the brain stem, resulting in an increase in the respiratory minute volume (RMV). Initial local peripheral and skeletal muscle vasodilation and hypotension soon develop. In addition, since CO_2 freely diffuses between plasma and tissue, an increase in blood PCO_2 (hypercapnia) risks the development of respiratory acidosis. This development promotes medullary and peripheral chemoreceptors and sympathetic innervation. The net result is an increased tachycardia, arrhythmias, and seizure activity† (5–10% CO_2 further stimulates RMV and produces a characteristic acidic oral taste).

* Characterized by retinal detachment in premature infants administered high oxygen at birth; a cause of blindness.

† Interestingly, about 50% of patients receiving general anesthetics hypoventilate (decrease in normal respirations per minute [rpm]), a state accompanied by depressed alveolar ventilation and retention of CO_2. Thus, hypoventilation increases blood PCO_2. Hyperventilation (increase in rpm) enhances elimination of blood CO_2 and causes respiratory alkalosis. The response here is peripheral vasodilation and a fall in blood pressure. Kidney, CNS, and intestinal arteries, however, constrict.

TABLE 25.3
Simple Asphyxiants, Sources, and Chemical Properties

Agent	Source	Industrial[a] or Household Uses	Chemical Properties[b]
Carbon dioxide	Atmospheric, fermentation, mammalian metabolism, coal mines (*afterdamp*)	Carbonation, propellant, fumigation, aerosols, fire prevention	Colorless, odorless, noncombustible, heavier than air
Ethane	Petroleum by-product, constituent of natural gas	Fuel gas, refrigerant	Colorless, odorless, flammable
Helium	Constituent of natural gas	Cryogen, lasers, welding, metal processing, carrier in GC, filling balloons and airships	Colorless, odorless, nonflammable
Hydrogen	Electrolysis of water, action of HCl on Fe or Zn, hydrolysis of metal hydrides	Welding, chemical production, filling balloons and airships, coolant, thermonuclear reactors	Colorless, odorless, flammable, explosive
Methane	Constituent of natural gas, action of aluminum carbide and water, coal mines (*firedamp*)	Illuminating and cooking gas, organic synthesis	Colorless, odorless, flammable, lighter than air
Nitrogen	Atmospheric, mine gases, heating of Na azides	Inorganic synthesis, explosives, incandescent bulbs, cryogen, pharmaceutic aid	Odorless
Oxygen	Atmospheric, liquefaction of air	Welding, propellant, lighting, liquid fuels, therapeutic administration	Colorless, odorless, supports combustion
Radon (^{220}Rn)	Naturally occurring, decay product of ^{235}U	Study of chemical reactions, source of neutrons, study of radium	Radioactive isotope, α-emitter, colorless, odorless, tasteless
Xenon	Naturally occurring, distillation/liquefaction of air	Gas lamps, leak detection systems for nuclear reactors	Colorless, odorless, noncombustible, tasteless

[a] Used in manufacturing, in synthesis, or as part of industrial processes.
[b] All are gases.
GC: gas chromatography.

Inhalation of 10% CO_2 for 1 minute causes *carbon dioxide narcosis*, a condition characterized by dizziness, dyspnea, sweating, malaise, restlessness, paresthesias, coma, convulsions, and death. Because it is heavier than air, CO_2 accumulates in low-lying, enclosed compartments, such as in mines, wells, and caverns. Table 25.5 summarizes the consequences of hyper- and hypoventilation, the resulting effects on plasma CO_2 concentrations, and corresponding pathological characteristics.

TABLE 25.4
Oxygen Concentration and Corresponding Signs and Symptoms Associated with Oxygen Deprivation

Oxygen Concentration in Inspired Air	Acute Pathologic Effects
16%–21% (normal)	No signs and symptoms
12%–16%	Tachypnea, tachycardia, muscular incoordination
10%–12%	Hypoxia; confusion, fatigue, dizziness
6%–10%	Hypoxia and decreased oxygen saturation; nausea, vomiting, lethargy, unconsciousness
<6%	Convulsions, apnea, cardiac arrest

25.4 TOXIC PRODUCTS OF COMBUSTION (TCP)

25.4.1 Introduction

TCPs are by-products of combustion generated from a variety of mostly synthetic materials* that are released with the smoke. TCPs include ammonia, acids, aldehydes, cyanide and isocyanates, carbon monoxide, halogenated hydrocarbons, oxides of nitrogen and sulfur, and styrene. Smoke inhalation of these substances produces toxicity from the heat in the gases, vapors, and fumes (thermal damage), or they may act as simple asphyxiants or pulmonary irritants. As with the pulmonary irritants, URT and LRT injury depends on duration of exposure, RMV†, and solubility.

25.4.2 Clinical Toxicity

URT and LRT toxicity are distinguished on the basis of signs and symptoms. Upper airway injury is characterized primarily by local inflammation and irritation of ocular, oral, and nasal mucous membranes. Symptoms include conjunctivitis, lacrimation, rhinitis, pharyngitis, and stridor (an abnormal, high-pitched musical sound caused by an obstruction in the larynx or trachea). LRT symptoms entail wheezing, deep chest pain, and carbonaceous sputum (material coughed up by the lungs, expectorated, and accompanied by the presence of particulate residues of combustion).

Treatment of smoke inhalation victims is primarily supportive, including decontamination (removal of clothes, washing or showering of patient) and removal of the individual from the source. This is followed by endotracheal intubation and administration of humidified oxygen, β-agonists and/or steroids for bronchospasm, and epinephrine for stridor. Carbon monoxide or cyanide poisoning, which often accompanies smoke inhalation, complicates the case (see sections 25.7 and 25.8).

* The burning of cellulose acetate film, resins, paper, polystyrene, nylon, petroleum products, rubber, wool, polyurethane, acrylics, wood, and organic material generates TCPs.

† RMV is a product of number of respirations per minute (rpm) by volume of air inspired per respiration (V_T, tidal volume).

TABLE 25.5
Respiratory Rates, Plasma CO_2 Levels, and Associated Signs and Symptoms

Clinical Respiratory Rate	Arterial Plasma CO_2 levels	Consequences	Toxicological Characteristics	Pathology, Signs, and Symptoms
Hyperventilation	<40 mm Hg	Hypocapnia, respiratory alkalosis	$\uparrow$ CO_2 elimination, $\downarrow$arterial PCO_2	Similar to mild oxygen toxicity
Hypoventilation	>46 mm Hg	Hypercapnia, respiratory acidosis	$\downarrow$ CO_2 elimination, enhances CO_2 retention, $\uparrow$arterial PCO_2	Vasodilation, hypotension; dizziness
Inhalation of >atmospheric CO_2	Inhalation of 2% CO_2	Respiratory stimulation, respiratory acidosis	Stimulation of medullary and peripheral chemoreceptors, stimulates SNS peripheral and muscle vasodilation	Vasodilation, hypotension; dizziness, loss of consciousness
	3%–6% CO_2	Respiratory stimulation, respiratory acidosis	Stimulation of medullary and peripheral chemoreceptors, stimulates SNS peripheral and muscle vasodilation	Tachycardia, arrhythmias, dyspnea, seizures; acidic taste, HA, dizziness, sweating
	10% CO_2	Respiratory stimulation, respiratory acidosis	*Carbon dioxide narcosis*	Malaise, restlessness, paresthesias, CCC, lethal

CCC: cardiovascular collapse, coma, convulsions; HA: headache; SNS: sympathetic nervous system.

25.5 LACRIMATING AGENTS (TEAR GAS)

25.5.1 Introduction

Before World War I, the mechanisms of biological and chemical alkylating agents were surfacing. The search for less toxic yet severely irritating compounds was already progressing. Law enforcement and governing bodies were convinced that such chemicals could be used in domestic (personal protection, crowd control) or in military situations (war). This understanding prompted the effort to develop agents that could be effective tools for law enforcement while avoiding life-threatening force. This spawned the introduction of lacrimating agents, popularly referred to as *tear gas* or *pepper spray*. Unlike the pulmonary irritants or asphyxiants that have practical industrial and commercial applications, lacrimating agents were developed specifically to cause irritation.

25.5.2 Chemical Agents

The compounds consist mostly of chemically invariable groups of brominated or chlorinated, simple or aromatic hydrocarbons that cause severe local, upper respiratory, and lower respiratory illness. Most of the agents are highly lipid-soluble powders. They are dissolved in organic solvents to effect aerosol delivery or burned and exploded for military use. Table 25.6 summarizes the properties, chemistry, and clinical effects of popular lacrimating agents currently used for domestic and military purposes. Today, the compounds are all organically synthesized and have otherwise limited commercial or industrial utility.

25.6 CHEMICAL ASPHYXIANTS

As noted earlier, chemical asphyxiants produce toxicity through induction of cellular hypoxia or anoxia. The agents alter the oxygen-carrying capacity of hemoglobin (as with carbon monoxide) or inhibit cellular metabolic enzymes (cyanide, hydrogen sulfide), ultimately interfering with normal physiologic respiration. Among the chemical asphyxiants, carbon monoxide, cyanide, and hydrogen sulfide are the most frequently encountered chemical asphyxiants and are discussed in the following sections.

25.7 CARBON MONOXIDE (CO)

25.7.1 Incidence

Each year, nearly 500 unintentional deaths and more than 1700 suicides are related to carbon monoxide poisoning in the United States. An estimated 3000 to 5000 people are treated annually for CO poisoning in emergency departments (EDs). Thousands more are either misdiagnosed or do not seek medical care. The statistics support the conclusion that CO poisoning is a serious public health issue.

TABLE 25.6
Lacrimating Agents: Uses, Chemical and Clinical Properties

Agent	Chemical (or Common Name)	Chemical Properties[a]	Uses	Acute Clinical Effects
Benzyl bromide	Bromomethyl-benzene	*Liquid*, decomposed by water	Chemical war gas	Intense local irritation; large doses cause CNS depression
Bromoacetone	1-Bromo-2-propanone	*Liquid*, turns violet in air	Chemical war gas	Intense local irritation
α-Bromobenzyl cyanide	α-Bromobenzene acetonitrile; *camite*	Crystalline *powder*, odor of soured fruit	Chemical war gas	Intense local irritation
Chloroacetone	1-Chloro-2-propanone	*Liquid*, pungent odor, turns dark with light	Tear gas component for police and military use; insecticide; lead, perfume, and drug manufacturing	Intense local irritation
ω-Chloroacetophenone	2-Chloro-1-phenylethanone; *chemical mace*	Crystalline *powder*	Riot control agent	Intense local irritation; URT and LRT irritation, pulmonary edema
o-Chlorobenzylidenemalononitrile	[(2-Chloro-phenyl) methylene] propanedinitrile	Crystalline *solid*	Riot control agent, chemical warfare agent	Intense local irritation; URT and LRT irritation plus erythema, chest constriction, vesiculation
Chloropicrin	Trichloronitromethane; *acquinite*	Oily *liquid*	War gas, insecticide, disinfectant, fumigant	URT irritation and lacrimation, potent skin irritant, NVD (orally)

[a] At standard temperature and pressure (STP).

All the compounds are miscible or soluble in acetone, alcohol, chloroform, or ether and are poorly or slightly soluble in water.

URT and LRT symptoms are as described in Table 23.2.

D: diarrhea; LRT: lower respiratory tract; N: nausea; URT: upper respiratory tract; V: vomiting.

25.7.2 Chemical Characteristics and Sources of Exposure

CO is odorless, colorless, nonirritating, and an abundant product of industrial combustion*; thus, it is appropriately labeled as the *silent killer.*

Principal sources of the gas include commercial and passenger motor vehicle exhaust fumes (1% from new automobiles, above 10% in older models) as well as other gasoline, diesel, and propane-powered engines. Smoke from charcoal fires and organic materials, tobacco smoke (3%–6% CO), and methylene chloride account for the majority of sources. Methylene chloride is a useful industrial solvent in paint, cleaning, and food-processing industries as well as an aerosol propellant and insecticide. In fact, on ingestion, methylene chloride is metabolized by hepatic mixed function oxidases (MFO) to carbon monoxide and carbon dioxide. Because of the wide distribution of the pollutant, it is not surprising to detect normal adult blood CO levels between 0.40% and 0.55%.

25.7.3 Toxicokinetics

Although CO has low aqueous (plasma) solubility, its binding affinity, particularly for hemoglobin (Hb), is high. As for other toxic gases, the absorption and binding of CO to hemoglobin depend on the same factors that increase exposure to the substance—that is, %CO in ambient air, duration of exposure, and RMV. The degree of binding is estimated according to the following formula:

$$\mathbf{\%CoHb = RMV \times [CO] \times time}$$

where

%COHb is the percentage carboxyhemoglobin formed
RMV is the respiratory minute volume (described earlier and equal to about 6 l/min in average adults)
[CO] is the CO concentration in ambient air
time of exposure is in minutes

According to this formula, inhaling 500 ppm CO from exhaust fumes (0.05% in a typical open garage with a running motor vehicle engine) for 30 min yields a %COHb concentration in blood equal to 15%. The compound is not metabolized, and its half-life is approximately 4 to 5 h.

25.7.4 Mechanism of Toxicity

The net effect of CO toxicity is tissue hypoxia. This is mediated through its reversible, but high, affinity for ferrous ion (Fe^{+2}) in hemoglobin in the red blood cell. The binding is estimated to range from 200 to 250 times that of molecular oxygen for Hb. The strength of the binding results in the formation of a stable carboxyhemoglobin (COHb) moiety. COHb then displaces the oxygen-carrying capacity of Hb and shifts

* It is the most abundant pollutant, accounting for 0.001% of atmospheric gases.

the *oxygen–Hb dissociation curve* leftward (Figure 25.1). The diagram illustrates the normal sigmoidal relationship between Hb saturation and the partial pressure of oxygen (PO_2, mmHg) dissolved in blood at normal body temperature.* At normal atmospheric pressure, the higher the PO_2, the more oxygen combines with Hb. The curve reaches a plateau at 100 mmHg PO_2, where Hb is almost completely saturated (98%). In the presence of CO, oxygen is displaced from Hb binding sites, rendering less oxygen available for delivery to tissues. The oxygen remaining within the Hb molecule combines more tightly with Hb. At any given PO_2, in the presence of CO, Hb is more saturated with oxygen. This phenomenon is known as the *Bohr effect* (the Bohr effect also occurs in metabolic alkalosis and is stimulated by high blood pH or low blood PCO_2). In addition, CO also binds myoglobin and cytochrome oxidase enzymes with high intensity.

25.7.5 Signs and Symptoms of Acute Toxicity

The clinical presentation of CO poisoning depends on the time of exposure and the concentration of CO in the area, as noted previously. Acute, high-concentration exposure, such as might occur in an enclosed space (automobile exhaust in a closed garage), will produce more severe signs and symptoms than chronic, low-concentration exposure (as with faulty heating systems). The latter scenario may be

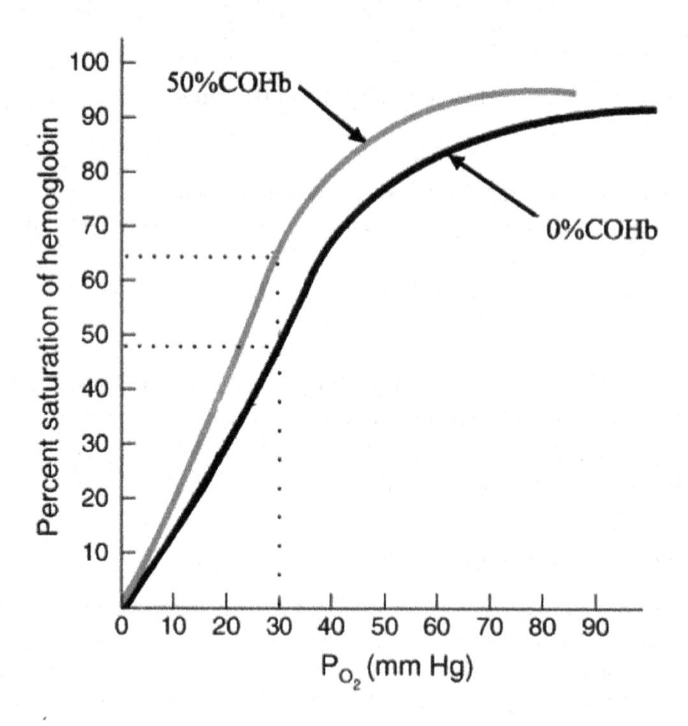

FIGURE 25.1 Oxygen–hemoglobin and carboxyhemoglobin dissociation curves.

* The percent saturation expresses the average saturation of Hb with oxygen.

misdiagnosed as mimicking a bacterial or viral infection. Symptoms from acute, mild exposure range from asymptomatic to headache, dizziness, malaise, and fatigue. Moderate exposure may present with confusion, lethargy, ataxia, syncope, and nystagmus.* Severe intoxication manifests as seizures, pulmonary edema, myocardial infarction, and coma. The classic cherry-red discoloration of the face and extremities, due to uncompensated peripheral vasodilation, is evident only in severe poisoning. Blood samples for gas analysis must be obtained immediately after exposure (using blood gas CO-oximetry). The calculation of percentage of arterial blood oxyhemoglobin (SaO_2), based on blood gas analysis, is often falsely elevated because of COHb high-affinity binding. Other routine clinical laboratory values may also lead to inaccurate conclusions.

Although recovery following nonfatal acute exposure is often complete within several days, subacute complications develop depending on the severity of exposure. The complications include persistent neurologic and myocardial dysfunction, peripheral neuropathy, aspiration pneumonitis, and ischemic skin. Approximately 10% to 30% of victims of severe acute poisoning will display delayed-onset neurobehavioral dysfunction, also known as *CO-induced delayed neuropsychiatric syndrome* (CO-DNS). The condition is characterized by impaired cognitive function, personality changes, dementia, and symptoms resembling Parkinson's disease. Individuals at greater risk for the development of complications are patients with a history of heart disease, anemia, and chronic obstructive pulmonary disease (COPD), and infants. Exposure in the presence of alcohol or respiratory depressants also increases the risk for CO-DNS.

25.7.6 Treatment of Acute Poisoning

As with any agent suspected of causing CNS depression or disrupting cardiovascular function, clinical history and evaluation should determine other etiologies, such as intoxication with alcohol or other CNS depressants. The presence of concurrent cyanide poisoning (particularly in burn victims) may aggravate the complications. The goal of treatment of CO inhalation victims, then, is to reduce the development of cerebral and cardiovascular ischemia and to increase the dissociation of COHb. Initial management includes removal of the individual from the source (while minimizing muscle and spinal movement, if possible), followed by administration of supplemental humidified oxygen soon after. Maintenance of respiration, fluid and electrolyte replacement, and clinical chemistry determination are largely supportive. Administration of 100% normobaric† oxygen reduces the half-life of 50% COHb from about 4 h in room air to approximately 50 to 60 min,‡ although longer periods may be required in high-risk patients. Treatment continues until COHb levels drop to within the normal range.

* Pendular or jerky rhythmical oscillation of the eyeballs.

† At 1 atmosphere of pressure.

‡ Although some studies have demonstrated a further reduction of the half-life to less than 40 minutes with hyperbaric oxygen (i.e., 100% oxygen at 3 atmospheres of pressure), the results from this mode of therapy are equivocal. As noted earlier, hyperbaric oxygen is associated with signs and symptoms of oxygen toxicity.

25.8 CYANIDE

25.8.1 Chemical Characteristics, Occurrence, and Uses

Cyanic acid (hydrogen cyanate, HCNO) is the starting chemical principle for the various salt forms of cyanide, including the sodium (cyanogran, NaCN), potassium (KCN), and calcium (CaCN) salts. Hydrogen cyanide (HCN, hydrocyanic acid, prussic acid) is a gas and a catalyst and is prepared from the cyanate salts. In addition, the compounds occur naturally as cyanogenic glycosides. The compounds are found from 0.01% to 14% in the seeds of various nuts, including almonds (highest concentration, 2%–14%), cherries, plums, apples, peaches, apricots, pears, plums, and rosaceous plants, as well as in bamboo sprouts and cassava. Figure 25.2 illustrates the hydrolysis of amygdalin, the most widely distributed cyanogenic glycoside. Most hydrolyzing agents, in the presence of the enzyme β-glucuronidase, are capable of producing the hydrolysis products of amygdalin—that is, mandelonitrile glucoside (an intermediate) plus glucose, benzaldehyde, and hydrocyanic acid.

Cyanide compounds are also valuable industrial chemicals used in electroplating and electropolishing, the manufacture of plastics, and the extraction of gold and silver from ores; as fumigants; in fertilizer; and in artificial nail glue removers. Therapeutically, sodium nitroprusside, a direct arterial vasodilator used in the treatment of emergency hypertension, releases five molecules of CN when metabolized, which also accumulates with fast infusion rates (see Chapter 20). As with CO poisoning, fire victims are also prone to CN intoxication.

25.8.2 Mechanism of Toxicity

Cyanide produces histotoxic anoxia by inhibiting oxidative phosphorylation, resulting in arrest of cellular respiration (see Figure 18.4). By binding to cytochrome a/a3, CN forms a CN–cytochrome oxidase–Fe^{3+} complex. The complex interferes with the transfer of electrons to O_2, the final electron acceptor. Ultimately, CN blocks the electron transport chain and inhibits metabolic respiration. It provokes a decrease in cellular oxygen use, prevents oxidative phosphorylation of ADP to ATP, and prompts an increase in venous PO_2 (arterialization of venous blood).* The decrease in aerobic respiration forces the cell to revert to anaerobic metabolism, which generates excess lactic acid, triggering metabolic acidosis.

25.8.3 Toxicokinetics

Acute lethal toxicity results within 1 h from an oral dose of 100–200 mg, while inhalation of 150–200 ppm of HCN gas is fatal (only approximately 60% of the population can smell 0.2–5.0 ppm). Trace amounts of CN are generally detoxified slowly by binding to circulating methemoglobin (methHb). The resulting cyanomethemoglobin complex prevents access to the cytochrome enzymes. Normally, circulating rhodanese enzyme (thiosulfate cyanide sulfur transferase) transfers a sulfur group

* Interestingly, the patient is not cyanotic, and the availability and binding of oxygen are not compromised. In fact, arterial PO_2 appears normal (100 mmHg).

to the cyanomethemoglobin complex, forming a relatively nontoxic thiocyanate ion that is eventually eliminated by renal excretion. Chronic, low-dose intoxication is more insidious.

25.8.4 Signs and Symptoms of Acute Toxicity

Signs and symptoms are precipitated rapidly with exposure to HCN vapors. Initially, symptoms of neurological toxicity appear, including headache, nausea, vomiting, weakness, and dizziness. The chemical stimulates chemoreceptors in the carotid artery, triggering reflex hyperpnea (increase in respirations), tachypnea (gasping for air), and pulmonary edema. Hypotension with reflex tachycardia completes the cardiovascular presentation. With high doses, the victim is stuporous yet responsive; the condition may deteriorate to hypoxic convulsions, hypotension, coma, and death.

25.8.5 Treatment of Acute Poisoning

As with CO poisoning, the initial management of patients with CN intoxication includes removal of the individual from the source, decontamination (removal of clothes, flushing with water, if necessary), and administration of activated charcoal or gastric lavage if the victim is discovered soon after ingestion. The goal of treatment is to immediately decrease CN binding to cytochrome enzymes with the specific antidote available. The Cyanide Antidote Package (various manufacturers) consists of three major components:

1. Amyl nitrite inhalant, 0.3 ml (12 aspirols);
2. Sodium nitrite, 300 mg in 10 ml (2 ampoules);
3. Sodium thiosulfate, 12.5 g in 50 ml (25% solution, 2 ampoules).

Disposable syringes, a stomach tube, a tourniquet, and instructions are also included in the kit. The primary mechanism of detoxification involves conversion of CN to the nontoxic thiocyanate ion in preparation for renal elimination. Initially, either intravenous (I.V.) sodium nitrite (300 mg over 3 to 5 min)* or amyl nitrite inhalant (1 or 2 crushed aspirols every 2 to 3 min if the I.V. route is not accessible) is administered. Nitrites convert reduced $Hb\text{-}Fe^{2+}$ ([H]) to oxidized methHb ([O]), freeing the cytochrome oxidase enzyme to resume oxidative phosphorylation. The sequence is outlined in Equation 25.1:

$$\underset{[H]}{Hb\text{-}Fe^{+2}} + \underset{[O]}{NO_2} \rightarrow \underset{methHb[O]}{Hb\text{-}Fe^{+3}} + \underset{[H]}{NO} \tag{25.1}$$

Since methHb has a greater affinity for CN than cytochrome oxidase, it induces the transfer of CN from the cytochrome enzyme complex to methHb, forming cyanomethemoglobin ($CN\text{-}Hb\text{-}Fe^{3+}$) according to Equation 25.2:

* In children, the initial dose of sodium nitrite (mg/kg) is calculated according to the patient's Hb level (g/dl).

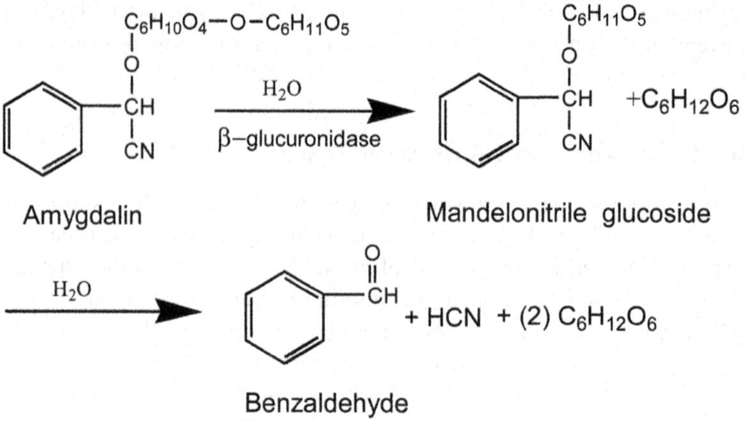

FIGURE 25.2 Cyanogenic glycosides and hydrolysis of amygdalin.

$$\text{Hb-Fe}^{+3} + \text{CN-cytochrome-Fe}^{+3} \rightarrow \text{CN-Hb-Fe}^{+3} + \text{cytochrome-Fe}^{+3} \quad (25.2)$$

Overall, the nitrites induce formation of the cyanomethemoglobin–Fe^{3+} (CN–Hb–Fe^{3+}) complex in preference to CN–cytochrome oxidase–Fe^{3+}. Peak methHb levels are reached within 30 minutes of I.V. administration in adults. Since cyanomethemoglobin is relatively unstable and reversible, the subsequent step is to force renal excretion of the CN moiety by administration of sodium thiosulfate. As mentioned earlier, this requires the rhodanese enzyme reaction, which naturally detoxifies trace amounts of circulating CN ions. This reaction is accelerated by supplying exogenous sulfur from the administration of sodium thiosulfate ($Na_2S_2O_3$) (Equation 25.3). $Na_2S_2O_3$ (12.5 g I.V. over 10 min) is administered immediately after sodium nitrite (400 mg/kg up to 12.5 g total in children). The treatment results in the formation of thiocyanate (CN-S), sodium sulfite (Na_2SO_3), and regenerated methemoglobin, respectively.

$$\text{CN-Hb-Fe}^{+3} + Na_2S_2O_3 \rightarrow \text{CN-S} + Na_2SO_3 + \text{Hb-Fe}^{+3} \quad (25.3)$$

Adverse reactions associated with nitrites involve hypotension and the risk of production of excess, life-threatening amounts of methemoglobin. In excess, methemoglobin decreases the availability of oxyhemoglobin (reduced form) necessary for oxygen transport. Other antidotes for CN poisoning, such as 4-methylaminophenol (4-DMAP), hydroxycobalamin, dicobalt-EDTA, and hyperbaric oxygen, are not Food and Drug Administration (FDA) approved or recommended.

Permanent neurological damage (Parkinson-like syndrome) is a complication of severe CN toxicity. Higher levels of thiocyanate are also implicated in the development of tobacco amblyopia (in chronic smokers) and tropical ataxic neuropathy (in diets rich in cassava).

25.9 METHODS OF DETECTION

Clinical chemistry analysis, hematology assays (including hemoglobin and hematocrit tests), and arterial blood gas determinations are not clinically useful indicators for CO poisoning. Routine blood gas analysis (pulse oximetry), used to measure changes in oxyhemoglobin content, may not be sensitive enough due to the high-affinity COHb complex. Carboxyhemoglobin blood levels are useful if performed soon after acute exposure. Automated spectrophotometric devices (CO-oximeters) provide valuable measures of carboxyhemoglobin, oxyhemoglobin, and methemoglobin, the levels of which are correlated with the severity of CO exposure. The technique simultaneously estimates total hemoglobin, % oxyhemoglobin, and % carboxyhemoglobin. The procedures are recommended for most clinical purposes. For the investigation of low-level exposure and the detection of increased hemolysis in neonates, more sensitive methods involving the release of CO and its measurement by gas chromatography are required.

As with CO, pulse oximetry may not be suitable for therapeutic management of CN poisoning. In fact, the onset and rate of CN toxicity are often too rapid to allow CN blood levels to be of any utility. Consequently, the determination of hemoglobin levels is a better indicator of the progress of CN poisoning and can be used to manage initial treatment with sodium nitrite. Elevated plasma lactate, associated with cardiovascular collapse, should also suggest cyanide intoxication. Other clinical chemistry and hematology tests can be of value as indicators of supportive measures.

REFERENCES

SUGGESTED READINGS

Becker, M. and Forrester, M., Pattern of chlorine gas exposures reported to Texas poison control centers, 2000 through 2005. *Tex. Med.* 104, 52, 2008.

Belanger, K. and Triche, E.W., Indoor combustion and asthma. *Immunol. Allergy Clin. North Am.* 28, 507,2008.

Bui, L., Cyanide and hydrogen sulfide, in *Toxicology Secrets*, Ling, L.J., Clark, R.F., Erickson, T.B., and Trestrail, J.H. (Eds.), Hanley & Belfus, Philadelphia, PA, 2001, chapter 53.

Burke, G., Planning for cyanide emergencies. *AAOHN J.*, 56, 319, 2008.

Chee, K.J. et al., Finding needles in a haystack: a case series of carbon monoxide poisoning detected using new technology in the emergency department. *Clin. Toxicol. (Phila.)* 46, 461, 2008.

Eckstein, M., Enhancing public health preparedness for a terrorist attack involving cyanide. *J. Emerg. Med.* 35, 59, 2008.

Fung, F., Simple asphyxiants and pulmonary irritants, in *Toxicology Secrets*, Ling, L.J., Clark, R.F., Erickson, T.B., and Trestrail, J.H. (Eds.), Hanley & Belfus, Philadelphia, PA, 2001, chapter 50.

Griffin, S.M., Ward, M.K., Terrell, A.R., and Stewart, D., Diesel fumes do kill: a case of fatal carbon monoxide poisoning directly attributed to diesel fuel exhaust with a 10-year retrospective case and literature review. *J. Forensic Sci.* 53, 1206, 2008.

Hartzell, G.E., Overview of combustion toxicology. *Toxicology* 115, 7, 1996.

Holmes, H.N. (Ed.), Respiratory system, in *Handbook of Pathophysiology*, Springhouse, PA, 2001, chapter 7.

Kinobe, R.T., Dercho, R.A., and Nakatsu, K., Inhibitors of the heme oxygenase–carbon monoxide system: on the doorstep of the clinic? *Can. J. Physiol. Pharmacol.* 86, 577, 2008.

Liu, G., Niu, Z., Van Niekerk, D., Xue, J., and Zheng, L., Polycyclic aromatic hydrocarbons (PAHs) from coal combustion: emissions, analysis, and toxicology. *Rev. Environ. Contam. Toxicol.* 192, 1, 2008.

The Merck Index, 14th ed., Budavari, S. (Ed.), Merck, Whitehouse Station, NJ, 2007.

Shepherd, G. and Velez, L.I., Role of hydroxocobalamin in acute cyanide poisoning. *Ann. Pharmacother.* 42, 661, 2008.

Stefanidou, M., Athanaselis, S., and Spiliopoulou, C., Health impacts of fire smoke inhalation. *Inhal. Toxicol.* 20, 761, 2008.

Streitz, M.J., Bebarta, V.S., Borys, D.J., and Morgan, D.L., Patterns of cyanide antidote use since regulatory approval of hydroxocobalamin in the United States. *Am. J. Ther.* 21, 244, 2014.

Tesseraux, I., Risk factors of jet fuel combustion products. *Toxicol. Lett.* 149, 295, 2004.

Wallace, K.L., Carbon monoxide, in *Toxicology Secrets*, Ling, L.J., Clark, R.F., Erickson, T.B., and Trestrail, J.H. (Eds.), Hanley & Belfus, Philadelphia, PA, 2001, chapter 52.

REVIEW ARTICLES

Borron, S.W. and Baud, F.J., Acute cyanide poisoning: clinical spectrum, diagnosis, and treatment. *Arh. Hig. Rada. Toksikol.* 47, 307, 1996.

Choi, A.M. and Otterbein, L.E., Emerging role of carbon monoxide in physiologic and pathophysiologic states. *Antioxid. Redox. Signal* 4, 227, 2002.

Cooper, C.E. and Brown, G.C., The inhibition of mitochondrial cytochrome oxidase by the gases carbon monoxide, nitric oxide, hydrogen cyanide and hydrogen sulfide: chemical mechanism and physiological significance. *J. Bioenerg. Biomembr.* 40, 533, 2008.

Hamad, E., Babu, K., and Bebarta, V.S., Case files of the University of Massachusetts Toxicology Fellowship: does this smoke inhalation victim require treatment with cyanide antidote? *J. Med. Toxicol.* 12, 192, 2016.

Hartzell, G.E., Overview of combustion toxicology. *Toxicology* 115, 7, 1996.

Jones, J., McMullen, M.J., and Dougherty, J., Toxic smoke inhalation: cyanide poisoning in fire victims. *Am. J. Emerg. Med.*, 5, 317, 1987.

McKay, C.A. Jr., Toxin-induced respiratory distress. *Emerg. Med. Clin. North Am.* 32, 127, 2014.

Medinsky, M.A. and Bond, J.A., Sites and mechanisms for uptake of gases and vapors in the respiratory tract. *Toxicology* 160, 165, 2001.

Mintegi, S., Clerigue, N., Tipo, V., Ponticiello, E., Lonati, D., Burillo-Putze, G., Delvau, N., and Anseeuw, K.,Pediatric cyanide poisoning by fire smoke inhalation: a European expert consensus. Toxicology Surveillance System of the Intoxications Working Group of the Spanish Society of Paediatric Emergencies. *Pediatr. Emerg. Care* 29, 1234, 2013.

Olajos, E.J. and Salem, H., Riot control agents: pharmacology, toxicology, biochemistry and chemistry. *J. Appl. Toxicol.* 21, 355, 2001.

Omaye, S.T., Metabolic modulation of carbon monoxide toxicity. *Toxicology* 180, 139, 2002.

Opresko, D.M., Opresko, D.M., Young, R.A., Faust, R.A., Talmage, S.S., Watson, A.P., Ross, R.H., Davidson, K.A., and King, J.,Chemical warfare agents: estimating oral reference doses. *Rev. Environ. Contam. Toxicol.* 156, 1, 1998.

Parker-Cote, J.L., Rizer, J., Vakkalanka, J.P., Rege, S.V., and Holstege, C.P., Challenges in the diagnosis of acute cyanide poisoning. *Clin. Toxicol. (Phila.)* 56, 609, 2018.

Piantadosi, C.A., Carbon monoxide, reactive oxygen signaling, and oxidative stress. *Free Radic. Biol. Med.* 45, 562, 2008.

Roderique, J.D., Josef, C.S., Feldman, M.J., and Spiess, B.D., A modern literature review of carbon monoxide poisoning theories, therapies, and potential targets for therapy advancement. *Toxicology* 334, 45, 2015.

Shifrin, N.S., Beck, B.D., Gauthier, T.D., Chapnick, S.D., and Goodman, G., Chemistry, toxicology, and human health risk of cyanide compounds in soils at former manufactured gas plant sites. *Regul. Toxicol. Pharmacol.* 23, 106, 1996.

Shusterman, D., Toxicology of nasal irritants. *Curr. Allergy Asthma Rep.* 3, 258, 2003.

Strickland, J.A. and Foureman, G.L., US EPA's acute reference exposure methodology for acute inhalation exposures. *Sci. Total Environ.* 288, 51, 2002.

Widdop, B., Analysis of carbon monoxide. *Ann. Clin. Biochem.* 39, 378, 2002.

26 Metals

Anirudh J. Chintalapati and Frank A. Barile

26.1 INTRODUCTION

26.1.1 BACKGROUND

Metals and their alloys are unquestionably responsible for the transition of human civilizations from their existence in the ancient world, through the Bronze and Iron Ages, to our present modern society. Chemically, elements in the periodic table are systematically grouped into metals, nonmetals, metalloids, and noble gases. Metals are electropositive elements, which readily lose electrons from the outermost shell. Physically, they are usually hard, lustrous, and exceptional conductors of heat and electricity. In general, metals are classified into ferrous (iron; Fe) and nonferrous materials. Ferrous metals, such as cast iron and carbon steel, contain iron as their core constituent in addition to other trace compounds. However, based on its physical composition and chemical properties, Fe is classified as a heavy metal. Nonferrous metals are subclassified as heavy, light, and noble. Heavy metals such as copper and zinc differ from light metals, sodium and magnesium; the former have relatively higher density and tensile strength and are relatively less reactive and soluble. Noble metals, including palladium, silver, osmium, iridium, platinum, and gold, have high resistivity to corrosion and oxidation in normal atmospheric conditions.

A few heavy metals, such as iron, copper, selenium, and zinc, possess qualities that are biologically essential for normal physiological processes, typically at low concentrations; however, these same chemicals that allow organisms to function at standard temperature and pressure (STP) are lethal beyond specific threshold concentration ranges. In contrast, lead, mercury, and cadmium have no known biological function for survival but exhibit toxic consequences on exposure.

26.1.2 EXPOSURE AND APPLICATIONS

Exposure to metals and metallic elements is a source of significant toxicity. Since metal compounds, in their elemental, inorganic, or organic states, are the most numerous of all elements, and their source of exposure is ubiquitous, their potential for toxicity is significant. Despite their various and seemingly unrelated mechanisms of toxicity, accumulation and the resulting oxidative stress are considered the central mechanistic pathway of metal-induced toxicity. Uncontrolled acute or chronic exposure to metals, such as arsenic, cadmium, cobalt, iron, mercury, and lead, extensively disrupts the production and detoxification equilibrium of reactive oxygen species (ROS), which results in membrane lipid peroxidation and loss of the membrane's structural and functional integrity. When accompanied by altered DNA repair pathways, ROS oxidize DNA and inhibit its replication, with possible irreparable mutagenic and carcinogenic transformations. In addition, the oxidation of amino acids through reactions with metals results in considerable changes in native protein structure, rendering the latter dysfunctional (Figure 26.1).

Many metallic chemicals, micronutrients, and related compounds are essential for the proper functioning of biological systems wherever they are required in trace amounts. These same biologically relevant substances are toxic with acute or chronic exposure to supraphysiological doses. The realization by early hominids that metals have properties that allow them to be crafted and manipulated proved invaluable for their survival. Today, all metals are useful in some way; some of them are indispensable for our modern social and industrial needs, while others have mostly intrinsic value (e.g., gold and silver). Consequently, in their pure form or when combined with other compatible ingredients (such as in the formation of alloys), metals have applications as diverse as fine jewelry as well as in the industries of architecture, construction, health care, and electronics. The relevance and utility of individual compounds with respect to their toxicological potential are discussed in sections 26.3 through 26.14.

26.1.3 Physiological Role of Metals

These universal compounds are associated with a variety of physiological processes. For instance, iron catalyzes normal cellular enzymatic reactions involved with oxygen transport. Na^+, K^+, and Ca^{2+} operate as essential charged molecules critical for neurotransmission and muscle contraction. Several essential metals are complexed in electron transfer reactions; toxicity occurs when the biochemical positions of the vital metals are shifted or replaced by abnormal, misfitting, detrimental metals. In addition, trace metallic chemicals can undo physiologic reactions by generating free radicals, or when combined with locally present molecules, thus perpetuating membrane degradation and organelle dysfunction. The metals listed in Table 26.1 represent the most common compounds with toxic potential.

26.2 CHELATION THERAPY

26.2.1 Description

Toxicity from metals and metallic elements depends largely on the route of exposure (inhalation, dermal, or oral), chemical species (inorganic, organic, or gaseous state),

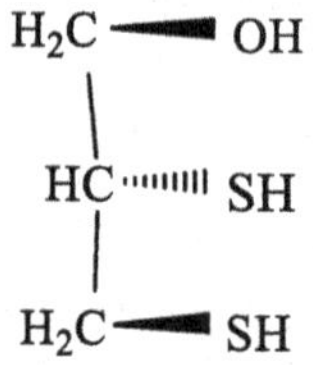

H_2C — OH
|
HC ··· SH
|
H_2C — SH

FIGURE 26.1 Chemical structure of dimercaprol (BAL).

TABLE 26.1
Chemical Characteristics and Methods of Detection of Some Metals

Metal	Chemical Symbol	Atomic #	Atomic Wt.	Appearance	Method of Detection	Clinical Sample
Antimony	Sb	51	121.76	Silvery-white, brittle crystalline metalloid	AAS, ESI-MS, C-ICP-MS	Urine, blood
Arsenic	As	33	74.9216	Gray metalloid	AAS, ICP-AES, ICP-MS	Hair, nails, blood, urine
Asbestos	–	–	Formula wt. 277.11 g	Mixture of minerals	PCM, TEM, Polarized light, and X-ray diffraction	Lung digests, bronchoalveolar lavage material
Cadmium	Cd	48	112.40	Silvery-white metal	GFAAS, ICP-MS	Urine, blood
Cobalt	Co	27	58.93	Silvery-gray	GFAAS, ICP-AES	Urine, blood, tissue
Copper	Cu	29	63.546	Brownish-red metal	ICP-MS, AAS	Blood
Iron	Fe	26	55.847	Silvery-white metal	AAS	Blood
Lead	Pb	82	207.2	Bluish-gray metal	GFAAS, radiographic techniques (lead lines)	Blood
Mercury	Hg	80	200.59	Silvery liquid metal	AAS, ICP-AES, ICP-MS	Hair, nails, blood, bone, urine
Selenium	Se	34	78.96	Brick red powder, red or gray crystals, metalloid	AAS, GC, X-ray diffraction, NAA	Blood, urine, placenta, hair, nails
Silver	Ag	47	107.86	Silvery-white	GFAAS, AAS, DCP-AES	Blood, tissue
Zinc	Zn	30	65.38	Brittle bluish-white metal	AAS, ICP-AES, ICP-MS	Urine, blood, bone, hair

AAS: atomic absorption spectroscopy; DCP-AES: direct current plasma atomic emission spectroscopy; ESI-MS: electron spray ionization mass spectroscopy; FM: formula mass; GC-ICP-MS: gas chromatography–inductively coupled mass spectroscopy; ICP-AES: inductively coupled plasma atomic emission spectroscopy; ICP-MS: inductively coupled plasma mass spectroscopy; GFAAS: graphite furnace AAS; NAA: neutron activation analysis; PCM: phase contrast microscopy; TEM: transmission electron microscopy.

concentration, frequency, and duration. The management of toxicity generally entails instituting the ABCs of emergency treatment (see Chapter 3). In the case of a few chemical agents, antidotes are available for neutralizing, eliminating, or ameliorating the effects of metal exposure. Chelation therapy is generally associated with the treatment of metal poisoning for the particular goal of decreasing the *body burden* of the chemicals that have been absorbed and distributed to physiological compartments.* Chelators have one or more binding sites for particular compounds; their affinity varies according to their structure and properties.

Ideally, chelators must have minimal adverse drug reactions (ADRs) associated with their therapeutic use. The chelator–metal complex should be water soluble to enhance elimination via the kidneys. Also, chelators have the potential to bind to essential metals or cause the movement of physiologically essential components from storage sites, thus increasing the probability of deficiency of required trace metals. Chelators can also shift metals from innocuous locations in physiological compartments to more critical areas. For instance, when combined with a chelator, lead that is sequestered in bone is released into the systemic circulation. Finally, oral administration is the most desirable route of administration, especially for the treatment of chronic metal poisoning.

Commonly used chelators are listed in Table 26.2. Their affinity, indications, and common ADRs are noted.

26.2.2 Dimercaprol

Dimercaprol (2,3-dimercapto-1-propanol, British anti-lewisite, BAL; Figure 26.1) is an effective chelating agent for heavy metal poisoning (see Table 26.1).† BAL forms intracellular metallic ligands with its constituent sulfur atoms, rendering the complex unable to traverse cell membranes in preparation for renal elimination.

26.2.3 Ethylenediaminetetraacetic Acid (EDTA)

EDTA forms four or six bonds with metal ions and chelates both transition-metal and main-group ions. The Ca^{2+} salt of EDTA (Figure 26.2) is the chelator of choice for Pb toxicity. EDTA is also used as an anticoagulant for stored blood in blood banks and in collection tubes; it prevents coagulation by sequestering Ca^{2+} ions required for clotting. As an antidote for Pb poisoning, calcium disodium EDTA (Ca^{2+}-(2)Na^{+}-EDTA) exchanges the chelated Ca^{2+} for Pb^{2+}, resulting in the formation of a Pb^{2+} chelate that is rapidly excreted in the urine. The Ca^{2+}-(2)Na^{+}-salt of EDTA, administered intravenously (I.V.), is also used in the treatment of acute cadmium and iron poisoning. As a result of the high incidence of nephrotoxicity, Ca^{2+}-(2)Na^{+}-EDTA is administered for the treatment of lead toxicity only when blood lead levels exceed 45 μg/dl; it is also used as part of the mobilization test for lead when blood lead levels

* The word *chelate* is derived from the Greek meaning "claw" (e.g., the chelicerae or claw-shaped mouthpart of *Crustacea*). The implication is that the chelator acts as a "claw" adhering to the metal and thus, hastening the elimination of the latter from the compartment.

† Dimercaprol was developed during World War I to treat the toxic effects of As-containing mustard gases (Lewisite and dichloro-(2-chlorovinyl) arsine); hence the name BAL.

TABLE 26.2
Chelators and Their Properties

Agent	Common or Proprietary Name	Metal Binding Affinity	Indications[a]	Common ADRs
Dimercaprol	British anti-lewisite (BAL)	As, Hg	Acute As toxicity; Hg-induced renal damage	HT, tachycardia, NVD, HA
Ca-(2) Na-EDTA (calcium disodium versenate)	Ethylene diamine tetraacetate	Ca^{2+}, Pb	Severe Pb toxicity	Renal damage
Penicillamine	Cuprimine®	Cu, Pb, Hg, Zn	Cu and Pb toxicity; Hg elimination;Wilson's disease	Allergic reactions
Deferoxamine	Desferal®	Fe^{2+}, Fe^{3+}	Fe toxicity	Allergic reactions
Succimer	DMSA, Chemet®	Pb	Pb toxicity	NVD, anorexia
Unithiol	DMPS	Hg	Inorganic acute and chronic Hg poisoning	

[a] Approved for use in the treatment of the listed conditions or diseases.
ADRs: adverse drug reactions; As: arsenic; Cu: copper; Fe: iron; HA: headache; Hb: hemoglobin; Hg: mercury; HT: hypertension; NVD: nausea, vomiting, diarrhea; Pb: lead; Zn: zinc.

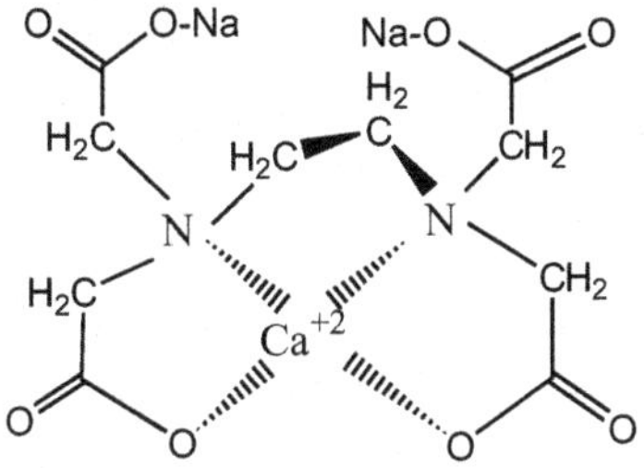

FIGURE 26.2 Chemical structure of calcium disodium versenate (ethylenediamine tetraacetate).

exceed 25 μg/dl. EDTA is used in combination with other chelators in an effort to reduce the toxic effects of each of the agents individually.

26.2.4 Penicillamine

Penicillamine (cuprimine) is a chelating agent for Hg, Pb, Fe, and Cu. It forms soluble complexes, thus decreasing toxic levels of the metals.* Penicillamine is well

* Penicillamine has been used in the treatment of Wilson's disease, a genetic disorder that results in accumulation of Cu.

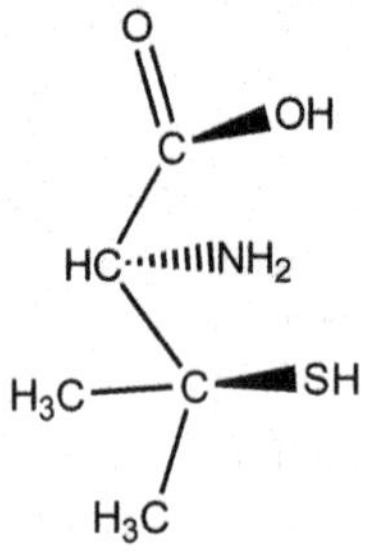

FIGURE 26.3 Chemical structure of penicillamine.

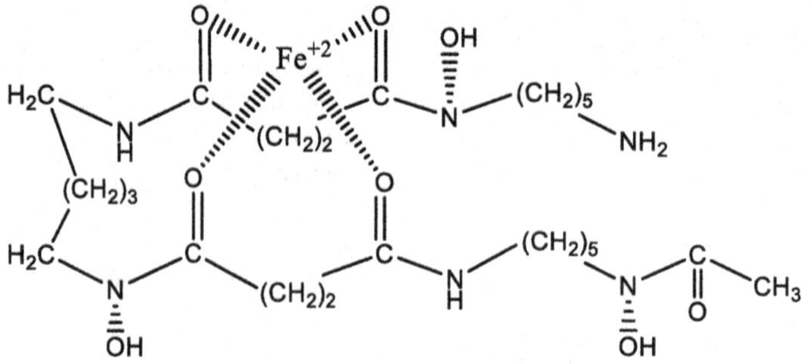

FIGURE 26.4 Chemical structure of deferoxamine with Fe in the reduced form.

absorbed from the gastrointestinal (GI tract) and excreted in urine. Food decreases the absorption of the drug by over 50%. Since the chelator is a hydrolytic product of penicillin (Figure 26.3), it should not be used in patients who are allergic to this antibiotic.

26.2.5 Deferoxamine

Deferoxamine is an aluminum (Al) and iron (Fe(II)) chelator that is beneficial in the treatment of acute and chronic Fe poisoning and for Al overload. The compound is isolated from bacteria (*Streptomyces pilosus*) and is one of the few chelators recommended for the alleviation of secondary Fe overload. It is not effective orally and requires continuous subcutaneous administration to achieve efficient Fe elimination. Figure 26.4 illustrates the structure of deferoxamine with a chelated Fe moiety. It preferentially binds both free and bound Fe from hemosiderin and ferritin but not Fe contained in hemoglobin, transferrin, or cytochromes.

26.2.6 Succimer

Succimer (dimercaptosuccinic acid, DMSA; Figure 26.5) and sodium dimercaptopropanesulfonate (DMPS) are incorporated in the secondary treatment of acute Pb poisoning in children, as they chelate excess Pb from physiological compartments.

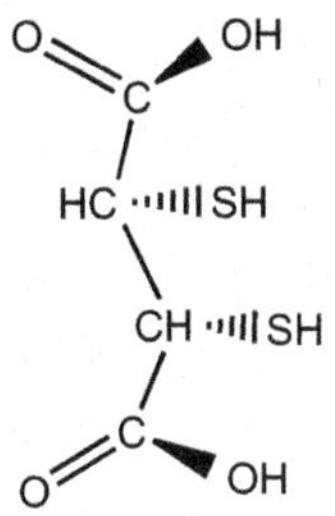

FIGURE 26.5 Chemical structure of Succimer.

Succimer combines with Pb in serum and facilitates excretion through the kidneys. By removing excess Pb, the chelator reduces damage to various organs and tissues. In healthy individuals, approximately 20% of an oral dose of DMSA is absorbed from the GI tract; about 95% of circulating DMSA binds to cysteine residues on circulating albumin, leaving one of the sulfhydryls (–SH) in DMSA available to chelate metals.

26.2.7 Unithiol

Unithiol is the water-soluble sodium salt of DMPS, derived from BAL. The World Health Organization (WHO) Expert Committee considers DMPS to be the first-line drug for inorganic acute and chronic mercury poisoning. DMPS is administered both orally and by continuous I.V. infusion. Clinical and experimental evidence suggests that DMPS is capable of removing inorganic mercury sequestered in human tissue reservoirs. Experimentally, DMPS penetrates renal nephrons and promotes accumulated mercury mobilization from renal tissues. On chelation, DMPS enhances the renal excretion of solubilized mercury, thus clinically reducing the chronic mercury body burden.

26.3 ANTIMONY

26.3.1 Physical and Chemical Properties

Sb is a silver-white, brittle metal found in different types of minerals. As a metalloid, Sb exhibits both metallic and nonmetallic characteristics. It is moderately flammable as a gas and when exposed to heat or on contact with acid, emits toxic fumes of stibine (SbH_3). It exists in the trivalent (Sb^{3+}) oxidation state as antimony trisulfide and stibine and in a pentavalent (Sb^{5+}) state as antimony pentachloride.

26.3.2 Occurrence and Uses

Geologically, Sb is most commonly present as the sulfide ore. It has demonstrated usefulness as a solder for metal sheets and piping; in the manufacture of storage battery casings; in pewter, paints, and ceramics; and in enamels for plastics and glass. Sb oxides have been added to clothing for their flame-retardant properties.

Clinically, it is an effective disinfectant and antibiotic, particularly for the treatment of parasitic infections, such as leishmaniasis and schistosomiasis.

26.3.3 Mechanism of Toxicity

Like other chemicals in this category, Sb induces toxicity by forming ligands with cellular organic constituents. Once the metalloid–organic complex is formed, the molecules lose their normal function, progressing to the disruption of affected cells. In addition, the reactive properties of Sb facilitate binding to oxygen, sulfur, and nitrogen molecules, resulting in the inactivation of essential enzymes and proteins that rely on these elements.

26.3.4 Toxicokinetics

As a function of its insolubility, Sb is poorly absorbed orally and by inhalation. GI absorption is estimated at less than 10% in humans; systemic distribution varies among species and is directly related to its valence state. The parent molecule and its derivatives are not metabolized; they bind to macromolecules via covalent reactivity with –SH and phosphate groups.

Pulmonary absorption of Sb is dependent on particle size. Pentavalent forms are detected more frequently in the liver and spleen while trivalent forms are more frequently located in neuroendocrine glands, especially the thyroid. Sb accumulates in the skeletal system and in animal fur. Trivalent Sb undergoes glutathione conjugation in the liver and is excreted in the bile; it then crosses the threshold into the enterohepatic circulation. Minimal amounts of Sb are slowly excreted in urine.

26.3.5 Signs and Symptoms of Acute Poisoning

In humans, acute poisoning has occurred as a result of accidental or suicidal ingestion of Sb compounds, with death ensuing within several hours. Initial symptoms of severe Sb poisoning include vomiting, watery diarrhea, irregular respiration, hypothermia, and circulatory collapse. Other toxicological complications following exposure include pneumoconiosis, altered EKG readings, increased blood pressure, dermatosis, ocular irritation, abdominal distress, headache, nausea, vomiting, jaundice, and anemia. Inhalation of Sb dust in occupational settings has produced GI irritation, probably as a result of Sb dust being transported via the upper respiratory tract. Information regarding the acute inhalation toxicity of Sb in humans is substantially lacking due to insufficient reporting of emergency department (ED) cases.

26.3.6 Treatment of Acute Poisoning

Treatment for acute Sb poisoning is most effectively accomplished using DMPS or dimercaprol, although chelation is indicated when blood Sb levels are detectable. BAL may be the most effective treatment for trivalent Sb after systemic exposure, although this treatment has little effect following contact with stibine gas. Dialysis is recommended for the treatment of pentavalent Sb toxicity.

26.4 ARSENIC (AS)

26.4.1 Physical and Chemical Properties

Arsenic is a crystalline metalloid, which exists as yellow, black, and gray allotropic forms. It is rapidly oxidized to arsenic trioxide (As_2O_3). Organic formulas are considered to be less toxic than inorganic states. As binds several metals and non-metals to form stable organic complexes, which are toxic on exposure. Some organic As compounds exist as gases or low-boiling liquids at STP. Highly poisonous arsine gas is produced by heating As salts or treatment with acids. More than 21 As complexes of significant toxicological concern have been identified as environmental pollutants.

26.4.2 Occurrence and Uses

A major source of As pollution results from As contamination of groundwater as a consequence of progressive leaching from rocks and sediments infiltrated with the metal. As is the 20th most common element in the earth's crust, and it is estimated that half of global production occurs in the United States.

As has historical medicinal applications. For instance, arsenic trioxide (As_2O_3) is a major ingredient of traditional Chinese medicine (TCM) and is approved by the U.S. Food and Drug Administration (U.S. FDA) for the treatment of acute promyelocytic leukemia (APL). Fowler's solution (potassium arsenite) was used for alleviating asthma. Inorganic As compounds have a wide range of applications in the industrial and agricultural sectors. Arsenic trihydride (arsine) gas is chemically incorporated into reactions for the manufacture of electronic components, for soldering, and for lead plating. Sodium arsenite and arsenic pentoxide are useful as wood preservatives and as commercial insecticides.

26.4.3 Mechanisms of Toxicity

As exerts toxicity by inhibiting enzymes involved in cellular energy pathways and nucleic acid synthesis and repair. Environmentally, the metal predominantly exists in the inorganic trivalent (As^{3+}; arsenite) and pentavalent (As^{5+}; arsenate) states. The trivalent form binds to sulfhydryl groups of several enzymes involved in cellular respiration and the detoxification of oxygen radicals (glutathione), leading to oxidative stress. Pentavalent As requires conversion to the trivalent form to elicit similar toxic effects. However, at the cellular level, the pentavalent form competes with inorganic phosphate in cellular respiratory pathways, forming stable complexes. For instance, arsenate competes with inorganic phosphate for binding to ADP, resulting in the formation of ADP–arsenate complex instead of the requisite and highly reactive ATP. As targets and accumulates within mitochondria, thus inhibiting succinic dehydrogenase activity and oxidative phosphorylation, a process that results in the disruption of all energy-dependent cellular functions.

26.4.4 Toxicokinetics

As is well absorbed through cutaneous, oral, and inhalation routes. Inorganic trivalent and pentavalent, as well as organic As, are well absorbed through the GI tract.

Upon absorption, the metal distributes to systemic compartments, preferentially localizing to muscle, bone, kidneys, and lungs. Arsenite is methylated in the liver to inactive mono-methylarsenic acid (MMA) and di-methylarsenic acid (DMA) and excreted by the kidneys. Thus, methylation is a significant detoxification mechanism for As. Since glutathione and other thiol-containing enzymes and cofactors act as methylating and reducing agents, any interaction with these compounds, or their storage organs, has the potential to increase tissue or circulatory concentrations of As. In mammals, the liver is a crucial site of As methylation, especially following hepatic first passage. The testes, kidney, and lung also contribute to arsenic methylation. Finally, significant levels of the metalloid are deposited indefinitely in hair and bone, where it replaces phosphorus. As also effectively crosses the placental barrier, resulting in fetal intrauterine developmental anomalies.

26.4.5 Signs and Symptoms of Acute Toxicity

Acute oral exposure to As produces GI signs and symptoms, including nausea, vomiting, severe abdominal pain, and profuse watery or bloody diarrhea. Acute myocardial dysfunction, cardiac arrhythmias, and cardiogenic shock are fatal cardiovascular consequences of acute As ingestion. Precipitation of cardiac ischemia caused by severe hypotension followed by renal tubular toxicity is observed in a few cases of acute As toxicity. Low-level intoxication has been shown to induce pulmonary hemorrhage, resulting in respiratory distress, particularly in newborn children. Both oral ingestion of inorganic forms and inhalation of the fumes effect moderate to severe respiratory irritation and dyspnea. Dermal exposure precipitates hair loss and produces white transverse bands of opacity on fingernail beds, referred to as *Mees lines*. Acute encephalopathy, hyperpyrexia, convulsions, and micro-hemorrhages are the neurological effects of acute As exposure. Interestingly, the toxic effects associated with acute As poisoning, such as anemia and leukopenia, are some of the same ADRs precipitated by the clinically approved version of the As-containing drug used in the treatment protocol for APL.

26.4.6 Signs and Symptoms of Chronic Toxicity

Chronic As toxicity occurring from environmental exposures and occupational settings causes changes in skin pigmentation, exemplified as hyper/hypopigmentation, plantar and palmar hyperkeratosis, and Bowen's disease. Nausea, vomiting, and diarrhea are less common than with acute exposure. In patients treated with Fowler's solution, which contains potassium arsenite, noncirrhotic portal hypertension has been demonstrated. Aplastic anemia, leucopenia, Raynaud's phenomena, and agranulocytosis are commonly reported in victims exposed for prolonged periods. Obstructive and restrictive lung disease, as well as peripheral neuropathies, has also been reported with chronic interaction with the metal (Axelson et al., 1978).

26.4.7 Treatment of Acute Poisoning

Acute arsenic toxicity is life-threatening and, in most cases, induces coma and death if vital signs are not stabilized and life support monitoring is not instituted. Delay or prevention of As absorption in cases of high-dose oral exposure is possible by the consumption of large volumes of water, commencement of gastric lavage, or administration of cathartics, if initiated in a timely manner. Chelation therapy with BAL and/or pencillamine is indicated for acute As poisoning.

26.4.8 Treatment of Chronic Poisoning

The primary course of action involves removal of the patient from the exposure source. Any evidence of arsenic in the GI tract should be managed with gastric decontamination. The patient must be frequently and continuously monitored for the appearance of, or increase in, blood arsenic levels even after the source of the contamination has been obviously eliminated.

26.4.9 Carcinogenesis

There is convincing evidence, based on clinical case studies and epidemiological reports, suggesting that acute and chronic exposure to a variety of As compounds increase the risk of cancer, particularly pulmonary and hematological neoplasms. The categorization of inorganic arsenic as a Group A known human carcinogen (U.S. EPA, 2000; Agency for Toxic Substances and Disease Registry [ATSDR], 2008) is based on clinical studies of As_2O_3 inhalation, As_2O_5 ingestion, and inorganic contamination of ground and drinking water (ref. Cu-smelting plants). The risk assessment that ingestion of inorganic As results in an increased threat of skin cancers and basal cell carcinomas has also been noted (ATSDR 2008.).

26.5 ASBESTOS

26.5.1 Physical and Chemical Properties

Asbestos is composed of a group of six different fibrous minerals (amosite, chrysotile, crocidolite, and the fibrous varieties of tremolite, actinolite, and anthophyllite) that occur naturally in the environment. There are two different silicate mineral groups to which asbestos belongs. Chrysotile belongs to the serpentine family of minerals, while the rest belong to the amphibole family. Chrysotile accounts for over 90% of the world's asbestos production. Of the five members of the amphibole group, crocidolite (blue asbestos) and amosite (brown asbestos) are widely used for commercial purposes. Amphibole fibers are more resistant to acids, and all asbestos fibers are resistant to alkalis.

26.5.2 Occurrence and Uses

The ancient Greeks spun and wove asbestos into cloth much like cotton. Until the late 1800s, when major deposits were discovered in Canada, asbestos was not widely available. Today, asbestos deposits are found throughout the world and are still mined in Australia, Canada, South Africa, and the former Soviet Union.

Asbestos was used to make thermal insulation for boilers and pipes and for fireproofing and reinforcement material. The military used asbestos extensively in ships and other applications during World Wars I and II. Commercial uses in buildings increased greatly thereafter until the 1970s, when growing concerns about health risks led to voluntary reductions. It has been used in thousands of products, mainly because it is plentiful, readily available, inexpensive, strong, fire retardant, heat resistant, chemical corrosion resistant, and a poor conductor of electricity. Products and building materials made with asbestos are often referred to as *asbestos-containing materials* (ACM) and *asbestos-containing building materials* (ACBM), respectively. Fibers mixed into asbestos cement, asphalt, and vinyl are usually firmly bound but pose potential toxicity when the fibers are released through drilling, cutting, grinding, or sanding.

26.5.3 Mechanism of Toxicity

The exact mechanism of toxicity of asbestos fibers to lung cells and pleura has not been fully elucidated. However, the generation of oxidants by fiber uptake or cellular interaction appears to be involved in the cellular response to the compound. Information about the associated health risks has come from studies of individuals exposed to levels of asbestos fibers that are greater than 5 μm in length. Inhalation of asbestos fibers may lead to the development of a slow accumulation of fibrous (scar-like) tissue in the lungs and pleura, preventing proper expansion and contraction.

Lung cancer and mesothelioma are two cancers frequently associated with asbestos exposure. The mechanism is not precisely known, but chromosomal alterations and the generation of ROS appear to be associated with the toxic consequences.

26.5.4 Toxicokinetics

Following the ingestion of asbestos, some of the fibers penetrate the GI epithelium and distribute to other tissues such as the lymphatic or systemic circulation, resulting in widespread distribution. Macrophages are probably involved in the uptake and distribution process. Most of the asbestos fibers that are deposited in the lung during inhalation are transported by mucociliary action. Only a small fraction of the inhaled fibers penetrate the epithelial layer of the lungs (the retention of asbestos fibers in the lungs of asbestos workers was estimated to have a clearance half-life of greater than 10 years).

26.5.5 Signs and Symptoms of Acute Toxicity

Temporary breathing difficulties have been reported in individuals exposed to high concentrations of asbestos dust, which may be accompanied by a local inflammatory response in the terminal bronchioles. Progressive fibrosis followed within a few weeks of the first exposure to dust.

26.5.6 Signs and Symptoms of Chronic Toxicity

Long-term inhalation of asbestos fibers by humans results in chronic, progressive pneumoconiosis (*asbestosis*). The disease is common among occupational groups directly exposed to asbestos fibers, such as insulation workers, but also extends to those working near the application or removal of asbestos. A prolonged inflammatory response caused by the presence of fibers in the lungs is characterized by fibrosis of the lung parenchyma. This can be seen radiographically 10 years after the first exposure. The main clinical symptom is shortness of breath, often accompanied by abnormal chest sounds (rales) and cough. In severe cases, disruption of respiratory function may result ultimately in death.

26.5.7 Treatment of Acute Poisoning

Bronchial irritation and congestion caused by acute inhalation of asbestos dust are managed by administering bronchodilators with supplemental oxygen therapy. Anti-inflammatory drugs are prescribed to relieve pulmonary inflammation and pain.

26.5.8 Treatment of Chronic Poisoning

Asbestosis is a life-threatening condition, which markedly increases patients' susceptibility to bacterial infections. Similarly to other chronic lung disorders, patients suffering from asbestosis have low circulating blood oxygen levels demanding supplemental oxygen and bronchodilators to relieve pulmonary congestion. Surgery is necessary if the patient's lungs show signs of scar tissue formation. Smoking can worsen asbestosis dramatically; therefore, smokers must undergo smoking cessation therapy.

26.6 CADMIUM

26.6.1 Physical and Chemical Properties

Cd exists in the form of cadmium sulfide as a trace contaminant of zinc, copper, and lead metal ores. It exists as a bluish solid at room temperature and is readily oxidized at atmospheric conditions to a divalent ion, Cd^{2+}. When combined with other metals, Cd forms alloys with relatively low melting points.

26.6.2 Occurrence and Uses

Since Cd exists as an impurity in ores, locations involving ore mining or refining activities and Cd-emitting industries or incinerators are major sources of Cd contamination. However, exposure to Cd in the general population occurs through cigarette smoke, food, and drinking water. The U.S. Environmental Protection Agency (EPA) reports that average Cd levels in food range from 2 to 40 parts per billion (ppb). Grain and cereal products, potatoes, leafy vegetables, and root vegetables contain the highest levels of cadmium. Furthermore, shellfish such as scallops and oysters are commonly considered dietary sources of Cd. On average, each person ingests about 30 μg of Cd daily, while smokers absorb an additional 1 to 3 μg per day.

Cd is a principal reagent in electroplating automobile engine parts and the production of batteries and photovoltaic cells. Therapeutically, a 1% concentrate of Cd in bath shampoo is used for treating dandruff and seborrheic dermatitis. The ability of Cd to lower the melting point when alloyed with other metals results in its use in the manufacture of fusible metals for automobile sprinkler systems, fire alarms, and electric fuses. Cd salts are also used in photography, in the manufacture of fireworks, rubber, fluorescent paints, and glass, and as shielding materials in atomic energy plants.

26.6.3 Mechanism of Toxicity

The liver is the primary organ affected by acute Cd exposure, causing hepatocellular necrosis with infiltration by inflammatory cells. Although the underlying mechanism of Cd toxicity has not been elucidated, possible factors include oxidative stress and/or lipid peroxidation of cell membranes. An alternate mechanism of Cd toxicity is its ability to substitute zinc, an essential trace element required for the functionality of several enzymes essential to maintain metabolic homeostasis.

In 1993, Cd was classified as a suspected human carcinogen. Occupational exposure is associated with lung and prostate cancers, although the carcinogenic mechanism is not known. Cd does not form stable DNA adducts but stimulates cell proliferation and inhibits DNA repair.

26.6.4 Toxicokinetics

The absorption of Cd varies considerably with the route of exposure. About 5–20% of orally ingested Cd salts are bioavailable, yet exposure to cadmium fumes results in up to 90% bioavailability. Once Cd is orally absorbed, it binds to albumin and alpha2-macroglobulin in serum and is distributed via systemic circulation, preferentially localizing in the liver and kidneys. Trace amounts are found in the pancreas, spleen, lung, kidneys, and testes. It is bound and sequestered by metallothionein (MT) produced in the liver, forming a cadmium–metallothionein complex (Cd-MT) that is released into the systemic circulation and undergoes glomerular filtration. The filtered Cd-MT is then reabsorbed and concentrated in proximal tubular cells, causing renal toxicity. Once formed, the Cd-MT complex is quite stable and is responsible for the extremely long biological half-life, estimated at 12–20 years.

26.6.5 Signs and Symptoms of Acute Toxicity

When heated at high temperatures, Cd and its compounds produce Cd oxide fumes, which when inhaled, cause Cd pneumonitis. Further inhalation occasionally progresses to respiratory insufficiency and death. Pulmonary symptoms vary from nasopharyngeal and bronchial irritation to severe pulmonary edema and death. Other common respiratory symptoms include severe cough, dyspnea, sore throat, and tracheobronchitis. Nonrespiratory effects of acute Cd toxicity include metallic taste, salivation, nausea, vomiting, and myalgia, which may linger for several weeks, and impairment of pulmonary function may persist for several months. Cd oxide inhalation can occasionally produce flu-like symptoms analogous to *metal fume fever.*

26.6.6 Signs and Symptoms of Chronic Toxicity

Occupational exposure to Cd is the most common cause of chronic toxicity, primarily affecting the kidneys, lungs, and musculoskeletal system. Cd-MT complex in liver as a function of chronic Cd exposure and accumulation results in hepatocellular injury and alters hepatic metabolism. Also, recurrent Cd exposure is linked to damage to the olfactory nerve and Parkinson's syndrome on rare occasions. In kidneys and lungs, chronic exposure is implicated in carcinogenesis and has also been associated with decreased pleural friction and emphysema. Other common consequences of chronic Cd exposure are anemia, osteomalacia, yellow discoloration of teeth, rhinitis, occasional ulceration of nasal septum, and damage to the olfactory nerve and subsequent loss of sense of smell (anosmia).

26.6.7 Treatment of Acute Poisoning

Acute oral exposure to Cd is life-threatening and must be addressed by instituting the ABCs of protocol treatment, followed by GI decontamination with activated charcoal. Chelation therapy with compounds such as EDTA, DPA, BAL, Succimer, and Unithiol may be beneficial to increase the chance of patient's recovery and survival. Airway support and appropriate oxygenation must be ensured for patients with acute exposure to Cd fumes.

26.6.8 Treatment of Chronic Poisoning

Patients reporting chronic Cd exposure are screened for renal and pulmonary functional changes to assess the extent of damage. Chelation therapy is not generally beneficial; however, the administration of dithiocarbamates is linked to a reduction in total Cd body burden. Chelation intervention may only be successful for Cd-induced dysfunction if the length, time, rate and concentrations associated with exposure are favorable enough to reverse significant or permanent damage. The age of the patient, the presence of concomitant pathological conditions and comorbidities, interactions with other pathological complications, and concomitant use of drug treatments may complicate the rate of successful recovery.

26.7 COBALT

26.7.1 Physical and Chemical Properties

Co is a light, silver-gray metal with high melting and boiling points. Inorganic Co exists in cobaltous (Co^{2+}) and cobaltic (Co^{3+}) oxidation states.

26.7.2 Occurrence and Uses

Co is a constituent of green and red leafy vegetables and plants as Co^{3+} at the center of the cyanocobalamin (Vitamin B_{12}) molecule. A dietary deficit in cyanocobalamin causes pernicious anemia, characterized by weakness, fatigue, and weight loss. Co salts, in combination with different iron salts, are used to treat anemia and associated symptoms. The first reported sources of Co exposure and toxicity were linked to pulmonary complications, referred to as *hard metal lung disease* (HMLD). Inhalation of metal fumes at industrial manufacturing sites results from the electrochemical combination of Co and tungsten carbide to form amalgamated cemented tungsten carbide. Historically, Co sulfate ($CoSO_4$) was used as an additive in beer preparation to function as a foam stabilizer. Following the consumption of Co-tainted beer, reports of cardiomyopathy and goiter (also known as *beer-drinkers' cardiomyopathy* and *cobalt-induced goiter*, respectively) were documented. Co, along with copper and aluminum, is also used in electrical and aircraft manufacturing industries and as a drying agent in the production of paints and varnishes.

26.7.3 Mechanism of Toxicity

Co exhibits multiple mechanistic toxicity; complexation with sulfhydryl groups in alpha-lipoic acid and dihydrolipoic acid is considered the primary target. This results in inhibition of acetyl Co-A production, with untoward consequences for cellular metabolism. Additionally, Co causes free radical-mediated pulmonary toxicity and inhibits tyrosine iodinase, an enzyme essential for thyroid hormone synthesis, leading to hypothyroidism. In the hematopoietic system, the metal chelates with iron-binding sites on transferrin, causing iron imbalance. Furthermore, Co competes with calcium at motor nerve terminals and inhibits neuromuscular transmission.

26.7.4 Toxicokinetics

The absorption, distribution, metabolism, and excretion (ADME) profile of Co in humans is not well documented. However, increased absorption of cobalt in the human small bowel has been linked to both iron deficiency and overload. Animal studies have demonstrated variable oral bioavailability between 5% and 45%. Co salts and oxides are primarily excreted by the kidneys, and the remainder is eliminated through feces.

26.7.5 Signs and Symptoms of Acute Toxicity

Both acute and chronic Co exposure affects multiple organ systems, including the GI, endocrine, central nervous (CNS), peripheral nervous (PNS), and cardiovascular (CVS) systems. Patients presenting with beer-drinkers' cardiomyopathy display signs and symptoms of respiratory and CV distress, including dyspnea, tachycardia, and elevated serum lactate concentration. Furthermore, acute Co exposure results in iron deficiency anemia, thyroid hyperplasia, and goiter. Oral ingestion of elemental Co or its salts causes GI distress, damage to the vestibulocochlear nerve, and impaired proprioception.

26.7.6 Signs and Symptoms of Chronic Toxicity

Respiratory irritation that slowly progresses to asthma and HMLD is a common pulmonary complication with chronic Co exposure. In addition, patients exhibiting chronic dermal exposure have the propensity to develop dermatitis as well as endocrine effects such as hypothyroidism and goiter.

26.7.7 Treatment of Acute Poisoning

GI decontamination followed by chelation therapy with Succimer, EDTA, and glutathione enhances Co elimination and is considered the first-line management strategy. However, chelation therapy is not beneficial for Co-induced cardiomyopathy and pulmonary symptoms of HMLD. Intensive supportive care must be ensured when cardiomyopathy is suspected, along with the administration of diuretics and potassium supplements as needed. Total blood counts and thyroid function tests are required after chelation. Pulmonary symptoms are also further analyzed and evaluated by chest X-ray. Symptomatic relief of asthmatic complications is achieved by administering bronchodilators.

26.7.8 Treatment of Chronic Poisoning

In an occupational work environment involving frequent, long-term Co exposure, chronic Co poisoning is a common occurrence. Shifting the workplace to a Co-free atmosphere or requiring workers to wear skin protection with emollient creams is helpful in preventing Co-associated dermatitis. However, symptomatic patients must be screened for several months for any signs of multiple–organ system toxicity.

26.8 COPPER

26.8.1 Physical and Chemical Properties

Cu is a malleable, odorless solid with a reddish-brown color. It is an excellent conductor of heat and electricity and reacts with water and carbon dioxide in the atmosphere to form greenish-colored hydrated copper carbonate ($CuCO_3$). With time, Cu oxidizes by reacting with oxygen in the air.

26.8.2 Occurrence and Uses

Cu simultaneously functions as an essential element for living organisms and displays toxicity in cumulative scenarios. It is one of the eight essential trace elements that are required cofactors for enzymes, such as Cu/Zn superoxide dismutase and nitrate reductases. Cu is used in household and industrial applications for its high electrical and thermal conducting properties. Its usefulness is manifested in electroplating and the manufacturing of insecticides, fungicides, paints, and varnishes. In ancient Greek and Roman civilizations, Cu was useful for its curative powers, largely due to its antibacterial and antifungal properties, in the treatment of wounds and skin and pulmonary diseases. Today, it is widely recognized for its effectiveness in the management of numerous internal diseases, including anemia, cancer, rheumatoid arthritis, stroke, and heart disease. For instance, Cu complexes, such as Cu aspirinate and Cu tryptophanate, markedly increase the healing rates of ulcers and skin wounds. While nonsteroidal anti-inflammatory drugs, such as ibuprofen, suppress wound healing, Cu complexes with these drugs to promote normal wound healing while retaining anti-inflammatory activity.

Cu protects against free radical damage, which affects proteins, membrane lipids, and nucleic acids. Several key reactive substrates and enzymes depend on Cu for proper physiological and metabolic function. These include enzymes of mitochondrial aerobic metabolism (cytochrome-C oxidase); enzymes that complex with connective tissue (lysyl oxidase); reactive substrates located in the CNS) (dopamine ß-hydroxylase); and Cu transport proteins (ceruloplasmin). Thus, Cu deficiencies in mammals are related to poor development of the CNS; initiate anemia, hypothermia, and bone fragility; and impair cardiac and immune function.

26.8.3 Mechanism of Toxicity

The toxicity of the metal is attributed to its ability to cause membrane lipid peroxidation and mitochondrial dysfunction. The propensity of Cu to exist in different valence states complements its participation in several redox reactions, which is responsible for the initiation of oxidative stress conditions as well as the inhibition of several metabolic factors that participate in the electron transport chain. In addition, Cu poisoning precipitates the depletion of cellular glutathione, the cofactor that neutralizes ROS. This shifts the formula toward formation of peroxidative products, particularly ROS, producing highly reactive supercharged oxygen radicals and subsequent generation of membrane damage. Cu-induced toxicity of the liver and kidneys is mediated by free radical damage as a result of ROS formation. Histopathological analysis of kidneys from patients with acute Cu poisoning shows acute tubular necrosis and the presence of hemoglobin casts.

26.8.4 Toxicokinetics

Absorption of Cu starts in the stomach following oral uptake of Cu-containing food, where low pH liberates metal ions from partially digested food particles. Cu complexes and solubilizes with amino acids and organic acids in the intestinal tract. The largest portion of ingested Cu is absorbed by the duodenum and ileum through simple diffusion. Cu-transporting ATPases discharge the charged ion into serosal capillaries, where they bind to albumin and amino acids for transport to the liver. Circulating Cu may also combine with ceruloplasmin, a copper-containing protein in the systemic circulation, from where it is returned to the liver as ceruloplasmin-bound Cu.

Cu has high affinity for metallothionein, a small cysteine-rich protein linked to Cu storage and transport. High Cu levels stimulate metallothionein synthesis. Glutathione, amino acids, ATP, and recently identified copper metallochaperones escort intracellular Cu transport. Physiological Cu-binding ligands are normally present in serum to protect against its toxicity by coordinating the cellular transport of Cu with intracellular enzymes, thereby effectively controlling and balancing its circulating levels. However, the bioavailability of Cu depends on the physiological age of the patient or victim and the amount ingested. Approximately one-half of the

Cu consumed by the average adult is absorbed in the GI tract, of which two-thirds is secreted into bile and excreted in feces, with trace amounts eliminated via urine, hair, and sweat.

26.8.5 Signs and Symptoms of Acute Toxicity

Most human cases of acute poisoning with high doses of Cu result from attempted suicides and accidental ingestion of contaminated food and beverages. The principal targets are the GI tract, liver, kidney, blood, CNS, and CVS. Symptoms such as GI irritation, abdominal pain, and vomiting are the initial manifestations of acute Cu poisoning, which are sometimes followed by GI ulcerations and perforations. Hepatic necrosis accompanied by jaundice, hypotension, tachycardia, tachypnea, nephropathy, and CNS manifestations (dizziness, headache, and convulsions) results from acute exposure. Cu-induced hemolysis is a common manifestation of acute exposure and is linked to liver damage. In cases of severe Cu poisoning, cardiovascular collapse occurs in conjunction with severe hypotension. There are few recognized effects on the muscular, integumentary, or ocular systems.

26.8.6 Signs and Symptoms of Chronic Toxicity

Chronic Cu exposures are reported with frequent intake of high-dose Cu-containing supplements and contaminated drinking water, resulting in manifestations analogous to Wilson's disease, an autosomal recessive disorder characterized by excessive copper deposition in the liver, brain, and visceral organs. Accumulation of Cu in the basal ganglia is linked to CNS symptoms such as tremor, ataxia, dystonia, and dysphagia. Chronic exposure results in hemolytic anemia secondary to hepatic necrosis. This eventually releases additional Cu into the circulation, resulting in red blood cell (RBC) destruction.

26.8.7 Treatment of Acute Poisoning

Supportive care, including restoration of body fluid, maintenance of electrolyte balance, and stabilization of blood pressure are the initial management steps for severe acute Cu poisoning. If possible, drinking 4 to 8 oz of milk or egg white prior to chelation is helpful in complexing Cu in the GI tract, thus forming Cu-albuminate, which inhibits its further absorption. Activated charcoal is also used for intestinal adsorption of Cu salts. If symptoms persist, Ca-disodium-EDTA or intramuscular BAL is administered, followed by penicillamine. Vigorous irrigation with water is helpful for ocular exposure, while topical corticosteroids counteract Cu-related dermatitis. As with any potential corrosive chemical ingestion, induction of vomiting is discouraged.

26.8.8 Treatment of Chronic Poisoning

Chelation therapy with Ca-disodium-EDTA, BAL, or penicillamine is preferred for patients with chronic Cu toxicity. Patients with severe kidney damage or end-stage renal failure should undergo hemodialysis. Liver transplantation is required for patients diagnosed with severe hepatocellular necrosis and liver failure.

26.9 IRON (FE^{2+}, FE^{3+})

26.9.1 Physical and Chemical Properties

Fe occurs as a silvery-white metal in the form of hematite (Fe_2O_3) or magnetite. Charged Fe valences exist as ferrous (2^+) or ferric (3^+) compounds; the former ion is readily oxidized to ferric valences. Ferrous sulfate (green vitriol or *cuas*), the most notable of the ferrous compounds, usually occurs as pale green crystals. Ferrous and ferric ions combine with cyanides to form complex ferro-cyanic composites.

26.9.2 Occurrence and Uses

Industrially, Fe is used in powder metallurgy and as a catalyst in chemical reactions. Steel is one of the most important alloys of iron and is incorporated into construction materials. Ferric ferrocyanide, a dark blue, amorphous solid formed by the reaction of potassium ferrocyanide with a ferric salt (*Prussian blue*), is used as a pigment in paint and in laundry bluing. Potassium ferricyanide (red prussiate of potash) is obtained from ferrous ferricyanide (*Turnbull's blue*) and is integrated into blueprint paper.

Biologically, Fe is the major component of hemoglobin, and its physiological deficiency causes hypochromic or Fe-deficiency anemia. Average adult humans store 3.9 to 4.5 g of Fe^0.* Of this, 65% is bound to hemoglobin, 20% to 30% is bound to the Fe storage proteins ferritin and hemosiderin, and the remaining 10% is a constituent of myoglobin, cytochromes, and other Fe-containing enzymes. The metal is a component of several catalytic enzymatic pathways and Fe-containing proteins.

Cytochrome enzymes rely on normal concentrations of Fe for proper functioning of many biotransformation reactions. Transferrin is a β_1-globulin responsible for transporting Fe^{3+} throughout the circulation, with the liver and bone marrow as ultimate destination targets. As a component of cytochromes and non-heme Fe proteins, it is required for oxidative phosphorylation reactions. Myeloperoxidase, a lysosomal enzyme, requires Fe for proper phagocytosis and oxidation of bacteria by neutrophils.

26.9.3 Mechanism of Toxicity

Fe toxicity results when the total iron-binding capacity (TIBC) of transferrin is exceeded, which usually occurs after a few days of 20 to 60 mg/kg of continuous administration. The mechanism of Fe toxicity is linked to its ability to exist in different oxidation states and its key role of participation in several redox reactions.

* Fe^o = elemental iron.

$$Fe^{+2} + H_2O_2 \underset{\text{reduce}}{\overset{\text{oxidize}}{\rightleftharpoons}} Fe^{+3} + \bullet OH + OH^-$$

[ferrous, reduced form] [ferric, oxidized form]

FIGURE 26.6 Fenton Reaction. Hydrogen peroxide oxidizes reduced form of iron (ferrous) to the oxidized form (ferric), thus generating hydroxy radicals.

Oxidative damage is attributed to the ability of iron to participate in the Fenton reaction in its ferrous form and the Haber–Weiss cycle in its ferric form, both of which lead to oxidative stress within the cell. Hydroxyl radicals formed by Fenton reactions and Haber–Weiss cycles (Fig. 26.6) oxidize membrane-bound lipids, resulting in progressive cell death.

Once transferrin is saturated with bound Fe, the unbound freely circulating Fe in blood is redistributed and deposited in other target organs, resulting in additional oxidative damage. Furthermore, excess Fe deposited in cells induces intracellular changes, forcing the cells to release H^+ ions, whose ultimate destination is participation in the formation of life-threatening metabolic acidosis.

Organ and cell damage arising from chronic Fe overload affects the liver, the heart, and pancreatic β-cells. Hemosiderosis is a rare condition related to excess Fe intake or improper Fe metabolism. The reason for large numbers of premature deaths in individuals with hematochromatosis, the hereditary form of hemosiderosis, is the development of complications from hepatic cirrhosis and hepatoma. Treatment with antioxidants effectively combats Fe-mediated oxidative damage. In addition, iron accumulation within cellular lysosomal compartments sensitizes the organelles to loss of integrity and rupture. The release of lysosomal enzymes into the cytoplasm of cells induces autophagocytosis, apoptosis, or necrosis.

26.9.4 Toxicokinetics

Fe metabolism is unique in operating primarily as a closed system, as Fe stores are efficiently recycled in the body. Physiologically, net Fe loss is less than 1 mg/day, but its absorption is usually poor. It is generally present in red meat products and dark green leafy vegetables in the ferric form bound to proteins and organic acids. The release of iron from these carriers is a prerequisite for absorption. While only 10% of iron ingested in the diet is absorbed, severe deficiency increases absorption to about 30%.

As the bidirectional equation outlines in Figure 26.6, the acidic pH of the stomach results in the **reverse** Fenton reaction; that is, hydrochloric acid reduces ferric iron (Fe^{3+}) to the ferrous form (Fe^{2+}), which is better absorbed by intestinal mucosal cells. Under the influence of apoferritin in the intestinal wall, Fe^{2+} is transformed back to its oxidized state, Fe^{3+} moiety, and eventually enters the plasma. Transferrin transports ferric iron to the liver, where it is bound to ferritin and hemosiderin. From these storage depots, iron is transported out as needed via transferrin to the bone marrow to produce hemoglobin and myoglobin and to other tissues for incorporation into cytochromes and non-heme iron.

26.9.5 Signs and Symptoms of Acute Toxicity

Based on its pathophysiology, acute Fe toxicity is characterized by five sequential clinical stages: 1. GI toxicity; 2. relative stability; 3. shock and acidosis; 4. hepatotoxicity; and 5. GI scarring.

1. *Stage 1.* Acute Fe intoxication is characterized by nausea, vomiting, and GI symptoms such as abdominal pain and diarrhea. Ulceration of the GI tract, inflammation of the bowel, and in severe cases, necrosis may occur.
2. *Stage 2.* Also referred to as the *latent stage,* it constitutes the time period of 6 to 24 hours after the GI symptoms resolve. This stage is relatively asymptomatic, and patients either recover or progress to Stage 3. For patients who progress to the third stage of Fe toxicity, major organ collapse occurs progressively without any observable signs and symptoms.
3. *Stage 3.* The third stage, or *shock stage,* is the result of prolonged acute systemic toxicity resulting in extensive vasodilation, poor cardiac output, and metabolic acidosis. Cardiogenic shock usually occurs 26 to 48 hours after ingestion and represents a depressant effect of Fe on myocardial cells.
4. *Stage 4.* Hepatic failure characterizes this stage of Fe toxicity. It typically occurs 2–3 days after ingestion of high levels of Fe and is a consequence of extensive oxidative damage to the liver.
5. *Stage 5.* GI scarring occurs 2 to 4 weeks after ingestion. Patients present with partial or complete bowel obstruction and formation of Fe concretions as the initial injury to the gut lumen heals following scarring and stenosis.

26.9.6 Signs and Symptoms of Chronic Toxicity

Chronic Fe toxicity is caused by hereditary hematochromatosis due to abnormal absorption of the metal from the intestinal tract, from excess dietary Fe, or from repeated blood transfusions for certain forms of anemia (*transfusional siderosis*). The symptoms of all three types are very similar and result in disturbances of liver function, diabetes mellitus, endocrine disturbances, and CV effects. Chronic Fe accumulation has been associated with infection with *Yersinia enterocolitica,* which requires Fe as its essential growth factor. Persistence of fever, bloody diarrhea, and GI spasm followed by resolution of toxicity are the common signs of *Y. enterocolitica* infection.

26.9.7 Treatment of Acute Poisoning

Traditionally, treatment for acute intoxication is guided toward removal of Fe from the GI tract by induction of vomiting and gastric lavage.* Alternatively, deferox-

* In general, the American Academy of Clinical Toxicology and the European Association of Poisons Centres and Clinical Toxicologists toxicological guidelines for the treatment of ingestion of acute oral toxins do not recommend induction of vomiting (Krenzelok et al., 1997). There is a high risk of permanent GI or esophageal scarring as a result of corrosive stomach and intestinal fluids (oxidized iron + acidic fluids) passing repeatedly through the alimentary canal.

amine is an Fe chelator and is the treatment of choice for acute Fe overload. Repeated phlebotomy has also been suggested.

26.9.8 Treatment of Chronic Poisoning

Chronic Fe poisoning causes septicemia primarily due to opportunistic *Y. enterocolitica* infection. In patients suspected with signs and symptoms of contamination, cephalosporin or fluoroquinolone classes of antibiotics are the treatment of choice along with adequate fluid and electrolyte replenishment.

26.9.9 Clinical Monitoring

Percent transferrin saturation indirectly measures Fe stores (serum Fe and Fe-binding capacity). Serum ferritin correlates directly with Fe stores, but it can also be elevated in liver disease, inflammatory conditions, and malignant neoplasms. A complete blood count (CBC) is also an indirect measure of Fe stores because the mean corpuscular volume (MCV) of the RBC is increased with Fe overload. Excessive Fe stores are calculated using histological Fe stains of bone marrow and liver biopsies. The technique is primarily performed to determine the amount of available Fe for erythropoiesis.

26.10 LEAD

26.10.1 Physical and Chemical Properties

Elemental Pb is a heavy, lustrous, silver-gray metal. At normal atmospheric pressure, Pb has a melting point of 327°C and a boiling point of 1620°C. Inorganic Pb compounds have bright colors and differential water solubility. Tetramethyl and tetraethyl Pb are examples of organic–Pb complexes. The ability of Pb to form complexes with sulfhydryl (–SH) present in biologic molecules is considered the primary toxic mechanistic cause of Pb toxicity.

26.10.2 Occurrence and Uses

Writings from ancient Babylonian, Egyptian, and Roman civilizations indicate the usefulness of lead (Pb) for its properties in the production of water pipes, solder, plaster, and pottery. Pb ranks 36th in abundance in the earth's crust. It is seldom found alone, and its compounds are widely distributed throughout the world. The three principal ores are galena (sulfide form), cerussite, and angelsite. The main use of Pb is in the fabrication of storage batteries and in sheathing electric cables. It is also useful as protective shielding from common X-rays and from radiation emitted from nuclear reactors.

Pb compounds are frequently used as pigments in paint, putty, and ceramic and as insecticides. For decades, Pb was incorporated in gasoline as an "antiknock" agent, until it was banned as an environmental pollutant in the United States in the 1970s. Recently, New York City (NYC) Department of Health and Mental Hygiene

(DHMH) described three NYC cases of adult lead poisoning associated with the ingestion of contaminated herbal remedies imported from China and India (see Suggested Readings).

26.10.3 Mechanism of Toxicity

Although Pb toxicity may affect all organ systems, the principal targets for Pb intoxication are the bone marrow and blood-forming pathways, the GI tract, the CNS, and the neuromuscular system. Pb is incorporated into Ca^{2+}-selective molecular structures and mimics its action, interfering with vital proteins. Thus, it is capable of binding to sulfhydryl, amine, phosphate, and carboxyl groups. Consequently, Pb increases intracellular levels of Ca^{2+} in brain capillaries, neurons, hepatocytes, and arteries, which triggers smooth muscle contraction, thereby inducing hypertension.

The effects of Pb on blood formation and heme biosynthesis have been extensively documented. Figure 26.7 illustrates the sequence of reactions in the formation of heme and the important enzymes involved in the assembly and expansion of glycine residues. The final heme structure is composed of four N-containing pyrrole rings linked by methylene bridges. After the incorporation of Fe^{2+}, four heme molecules are positioned, one into each of four polypeptide chains, two α and two β, of the globin molecule in hemoglobin.

Pb interferes with heme biosynthesis by meddling with ferrochelatase, ALAS (δ-aminolevulinic acid synthetase), and ALAD (δ-aminolevulinic acid dehydratase) (Figure 26.7). The net effect is accumulation of protoporphyrin in RBCs and decreased formation of hemoglobin due to defective heme assembly. Clinically, patients manifest with *basophilic stippling* (due to accretion of polyribosomes and RNA aggregates in RBCs), which is noticeable on a Gram-stain RBC smear at 100× magnification. In addition, defective heme synthesis results in increased production of reticulocytes (immature RBCs), which is demonstrated by stippled cell cytoplasm and microcytic hypochromic anemia.

The effects of Pb on heme synthesis also impact skeletal, renal, and neurological functions. In bone, Pb alters circulating levels of 1,25-dihydroxyvitamin D, affecting Ca homeostasis and osteocyte function. Pb deposits in renal nephrons, eventually revealing itself in glomerulonephritis and glomerular atrophy. In the nervous system, Pb substitutes for Ca^{2+} as a secondary messenger in neurons, blocking voltage-gated Ca^{2+} channels, inhibiting influx of Ca^{2+}, and obstructing subsequent release of neurotransmitter. The result is an inhibition of synaptic transmission. Pb inhibits glutamate uptake and glutamate synthetase activity in astroglia, thus inhibiting the regeneration of glutamate, a major excitatory neurotransmitter. These mechanisms explain the pathophysiology of Pb toxicity, described next.

26.10.4 Toxicokinetics

The bioavailability of Pb is dependent on its ingestion in the presence of food (10% or less bioavailable) or after fasting (60–80%). Immediately following ingestion, Pb is distributed widely to plasma and soft tissue and redistributes and accumulates in bone. In children, bone Pb accounts for about 73% of the total body burden, while

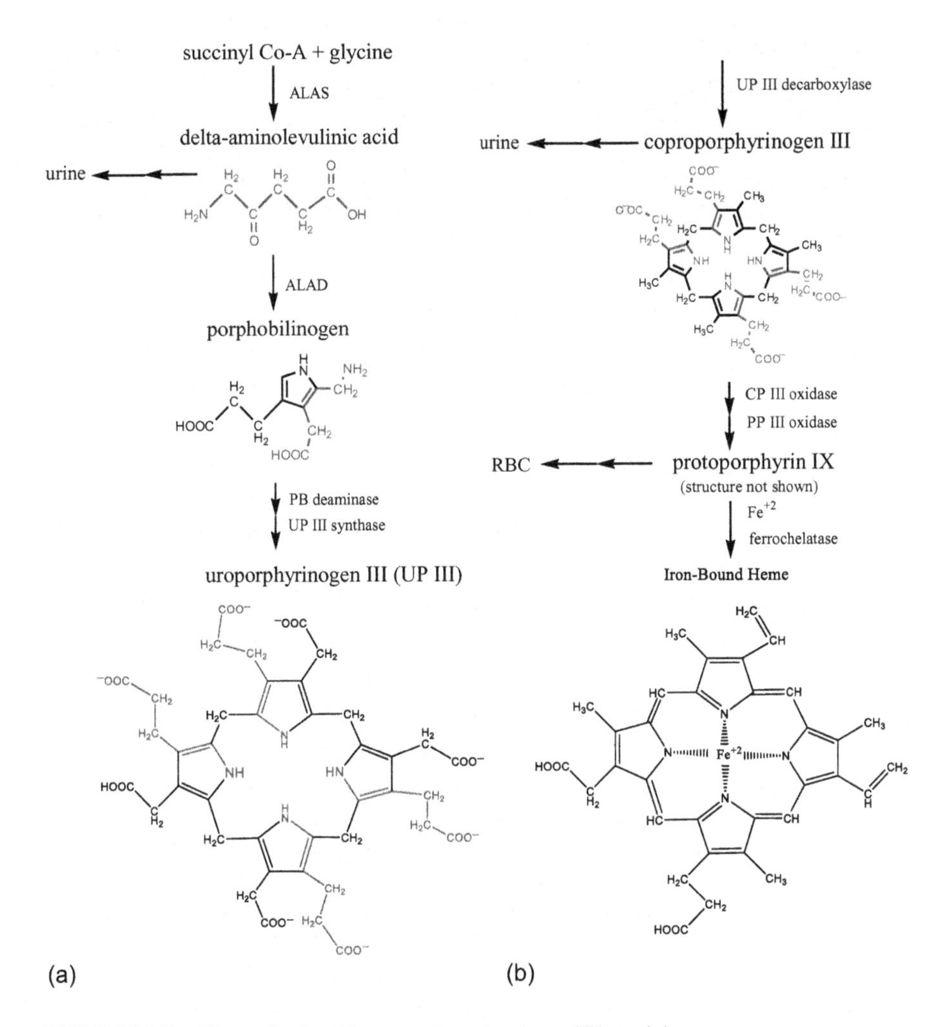

FIGURE 26.7 Biosynthesis of heme and mechanism of Pb toxicity.

in adults, it increases to 94% due to the slower turnover rate of bone with age.* The rate of disposition of inhaled inorganic Pb is about 30–50% but is largely dependent on the particle size and ventilation rate.

Pb not retained in the body is excreted primarily by the kidneys as soluble salts or through biliary clearance in the GI tract in the form of conjugates with organic compounds. Exhalation is also considered to be a major excretion route for organic Pb.

The distribution of the heavy metal in humans is well characterized. Freely circulating Pb is uncommon; thus, the measurement of blood Pb levels is often poor or inaccurate. This is principally because blood Pb is found mainly in RBCs (99%),

* Pb absorption is dependent on nutritional status. In children, Fe deficiency correlates with higher blood Pb levels, suggesting that Fe may affect Pb absorption. A similar correlation was found with Ca^{2+} levels.

which distribute chiefly to soft tissue, accounting for the high levels occurring in the liver, lung, spleen, and kidneys. Redistribution to Ca^{2+}-containing structures explains the high Pb concentrations found in bone. Interestingly, Pb does not distribute characteristically in bone but will accumulate in the form of lead phosphate in those regions undergoing active calcification at the time of exposure.

Inorganic Pb is not metabolized or biotransformed but forms complexes with a variety of protein and nonprotein ligands. Organic Pb is metabolized in the liver by an oxidative dealkylation reaction catalyzed by cytochrome P450.

26.10.5 Signs and Symptoms of Acute Toxicity

Exposure to excessive Pb via the GI tract or by inhalation is rare but may result in cramping, constipation, and colicky abdominal pain. The latter is accompanied by nausea, vomiting, and bloody black stools, with the patient complaining of an oral metallic taste. Early symptoms of Pb exposure include fatigue, apathy, and vague GI pain. Arthralgias and myalgias of the extremities may also occur. Headache, confusion, stupor, coma, seizures, and optic neuritis are all manifestations of Pb neurotoxicity. On further exposure, CNS symptoms, including insomnia, confusion, impaired concentration, and memory problems, become more pronounced.

26.10.6 Signs and Symptoms of Chronic Toxicity

Chronic Pb toxicity is more frequently seen, especially in children and young adults, and is known as *plumbism*. Although initial signs and symptoms are subtle, the child complains of fatigue, muscular weakness, and incoordination. *Lead colic* describes GI signs that present as nausea, vomiting, diarrhea, anorexia, constipation, and severe abdominal cramps. Interference with blood-forming organs shows anemia as a common sequel. Blue-gray pigmentation of the gingiva ("Burtonian lines" or "lead lines") is also a diagnostic sign. *Pb encephalopathy*, fatal in 25% of untreated or poorly treated patients, describes the cerebral and neuromuscular effects of chronic lead intoxication and is characterized by irritation, memory loss, and ataxia, followed by delirium, convulsions, and coma. Demyelination of the radial nerve of the forearm typifies *lead palsy*, demonstrated by *wrist drop*,* while demyelination of the femoral nerve results in *foot drop*.†

26.10.7 Treatment of Acute Poisoning

Treatment for chronic Pb poisoning is determined primarily by blood lead levels (BLL) and follow-up testing and referral. Table 26.3 outlines the guidelines, recommended action, and suggested chelation therapy used for treatment of childhood lead intoxication in NYC, based on BLL parameters. Parenteral administration of chelating agents, Ca-diNa-EDTA, and dimercaprol (BAL) is used to reduce the body burden of absorbed Pb. Penicillamine has been used as an oral chelating agent for

---

* An inability to freely flex the hand–wrist with a normal hinge reflex movement.

† An inability to freely flex the foot–ankle with a normal hinge reflex movement.

outpatient continuation of I.V. chelation treatment, although it is not as effective as EDTA. In patients with kidney impairment, dimercaprol is recommended, since excretion is primarily in bile rather than urine. EDTA also mobilizes Pb complexed in bone to soft tissue but may aggravate acute toxicity if not administered in conjunction with dimercaprol. DMSA (Succimer) is the only FDA-approved orally available chelating agent for treating children with blood Pb levels greater than 45 μg/dl.

26.10.8 Treatment of Chronic Poisoning

Workplace exposure is the most common cause of chronic Pb poisoning. Patients diagnosed with elevated Pb concentrations should be relocated from the worksite into well-ventilated surroundings. It is important to note that eating and drinking are prohibited inside Pb-exposed worksites. The treatment protocol for symptomatic patients is similar to that for acute poisoning (see earlier).

26.10.9 Clinical Monitoring

Because Pb redistributes and accumulates in RBCs from plasma, the method of choice for determination of Pb exposure was the erythrocyte protoporphyrin (EP) test. For two decades, this assay was used to determine the accumulation of protoporphyrin in erythrocytes. The assay, however, proved to be insensitive to Pb levels in the 10–25 μg/dl range; it is not recommended for tracking childhood Pb exposure. The evolution of more sensitive testing techniques has made measuring Pb in blood more feasible. For example, the free erythrocyte protoporphyrin (FEP) or zinc protoporphyrin (ZPP) assays are more sensitive, especially at concentrations as low as 1 μg/dl (Table 26.3). As Pb reallocates to bone after ingestion, the use of radiographic techniques also occasionally proves useful for its detection.

26.11 MERCURY (HG)

26.11.1 Physical and Chemical Properties

Elemental Hg exists as a silvery liquid. It is volatile at room temperature and has a low enough vapor pressure to vaporize. Three toxic forms of Hg are elemental (Hg^0), inorganic, and organic. Hg compounds are considered major pollutants of the biosphere, with organic mercurials being the most toxic. The chemical properties, characteristics, and pathophysiologic consequences of the different forms of Hg are summarized in Table 26.4.

26.11.2 Occurrence and Uses

Two major sources of Hg deposition in the environment are the natural degassing of the earth's crust and the leaching of sediment. This natural source of contamination is estimated at 25,000 to 150,000 tons of Hg per year, which binds to organic or inorganic particles and to sediment that has a high sulfur content. Although discharges of Hg have been strictly regulated, some industrial activities still release substantial

TABLE 26.3
Chelation Therapy for Chronic Lead Toxicity in Children in NYC DHMH

Blood Lead Level (BLL, μg/dl)	Recommended Action	Chelation Therapy
5 to <10	Provide education, testing, reporting BLL to NYC DHMH; follow-up in 3–6 months	No chelation therapy
10 to <15	Provide education, testing, risk assessment, reporting BLL to NYC DHMH; follow-up in 3 months	No chelation therapy
15 to <45	Provide education, testing, risk assessment, reporting BLL to NYC DHMH; follow-up in 1–3 months	No chelation therapy
>45	As above, also confirm BLL with venous sample, perform FEP test + medical exams; follow-up as soon as possible	Chelation therapy: Ca-EDTA for 3–5 days, followed by Ca-EDTA + BAL if BLL > 69 μg/dl

From the New York City Department of Health and Mental Hygiene, Lead poisoning: Prevention, identification and management, *City Health Information*, 2018, www1.nyc.gov/assets/doh/downloads/pdf/lead/lead-guidelines-children.pdf; last viewed March 2019.
FEP: free erythrocyte protoporphyrin; NYC DHMH: New York City Department of Health and Mental Hygiene.

TABLE 26.4
Mercury Poisoning and Its Pathophysiologic Characteristics

Characteristics	Elemental Hg (Hg°)	Inorganic Hg	Organic Hg
Form and chemical properties	Hg°, lipophilic, converted to charged cations	Salt forms, water soluble	Methyl Hg, lipophilic
Occurrence	Industrial, thermometers	Industrial, chemical laboratories, marine life	Pesticide, industrial
Absorption	Inhalation	GI tract	GI tract
Target organs	Pulmonary, CNS	Kidneys, GI	CNS
Elimination	Fecal, enterohepatic	Urinary, fecal	Fecal, enterohepatic

quantities of the metal. For example, fossil fuel contains as much as 1.0 ppm. Since 1973, approximately 5000–10,000 tons of Hg per year has been discharged from burning coal, natural gas, and the refining of petroleum products, with one-third of the atmospheric Hg due to industrial releases. The element is used in a number of products, including thermometers, barometers, electrical apparatus, paints, and pharmaceuticals.

Regardless of the source, both organic and inorganic Hg undergoes environmental transformation. The conversion of inorganic Hg to methyl Hg results in its release from sediment at a relatively fast rate and leads to its wider distribution. Inorganic Hg may be methylated and demethylated by microorganisms. Elemental Hg at ambient air temperatures volatilizes and is extremely dangerous.

26.11.3 Occupational and Environmental Exposure

Methyl Hg released in the aqueous environment bioaccumulates in plankton, algae, and seafood. With the latter, the absorption of methyl Hg is faster than inorganic Hg, and its clearance rate is slower, resulting in high methyl Hg concentrations. This is of biological as well as public health interest, since fish enter the food chain.

Pollution of the environment with Hg compounds has resulted in an increased level of neurotoxicity, referred to as *Minamata disease*, named after a historic outbreak of Hg poisoning in Japan. Two poisonings, one in Minamata Bay (1956–1975) and one in Niigata (1964), occurred as a consequence of industrial releases of Hg compounds into Minamata Bay and the Agano River. An additional outbreak occurred in Iraq (1971–1972), which resulted from eating bread made from seed grain coated with a methyl Hg fungicide.

Most human exposure to Hg is via inhalation, since the highly lipid-soluble compound readily diffuses across the alveolar membrane. This property confers on Hg high affinity for RBCs and the CNS.

26.11.4 Mechanism of Toxicity

Hg toxicity is attributable to high-affinity binding of divalent mercuric ions to thiol and –SH groups in proteins. Inactivation of various enzymes and structural proteins and alterations of cell membrane permeability contribute to severe toxic effects. Increased oxidative stress, disruption of microtubule formation, and interference with protein synthesis, DNA replication, and Ca homeostasis are purported pathways.

26.11.5 Toxicokinetics

Inhaled Hg vapor is efficiently absorbed (70–80%); absorption of liquid metallic Hg is considered insignificant. The absorption rate for inorganic mercuric salts varies greatly and is dependent largely on the chemical form. Oral absorption of organic Hg is nearly 100%.

Although all forms of Hg have the potential to accumulate in the kidney, inorganic mercuries have a propensity for kidney accumulation. Because of the high lipophilicity of metallic Hg, transfer through the placenta and the blood–brain barrier is thorough. Inorganic Hg compounds have lower lipophilicity, and although there is distribution to most organs, penetration through biomembranes is not as efficient.

Metallic Hg is oxidized to the divalent form by the catalase pathway; the divalent ion is reduced to the metallic form. Elimination occurs via urine, feces. and expired air for metallic Hg. Inorganic Hg is eliminated in urine and feces, while organic Hg is eliminated primarily in the feces. Both inorganic and organic formulae are excreted in breast milk. A small fraction of the metal is exhaled after exposure to the vapor. About 90% of methyl Hg is excreted in feces after acute or chronic exposure. Inorganic Hg is preferentially excreted via the feces and subsequently through the renal route. Renal excretion of inorganic Hg increases with time. while methyl Hg excretion is not proportional with time.

26.11.6 Signs and Symptoms of Acute Toxicity (Inhalation and Ingestion)

Hg poisoning is principally defined according to exposure to the chemical form and the target organ for accumulation. The five syndromes resulting from Hg intoxication are described accordingly in Table 26.5.

Accidental acute exposure to high concentrations of metallic Hg vapor has resulted in human fatalities. The most commonly reported symptoms of acute inhalation exposure are cough, dyspnea, tightness and burning pain in the chest. GI effects include acute inflammation of the oral cavity, abdominal pain, nausea, and vomiting. Increased heart rate and blood pressure are evidence of CV toxicity. Renal effects resulting in proteinuria, hematuria, and oliguria have been demonstrated, and severe neurotoxic reactions are noted as behavioral, motor, and sensory disruptions.

The ingestion of inorganic mercurial salts causes severe GI irritation, including pain, vomiting, diarrhea, and renal failure. Contact dermatitis, hyperkeratosis, acrodynia (*pink disease*), shock, and CV collapse are observed in patients with acute exposure to inorganic mercurial salts.

TABLE 26.5
Five Syndromes That Characterize Mercury Poisoning

Mercurial Form	Exposure Route	Pathophysiology	Description	Treatment
Hg°	Inhalation of vapors	Pulmonary—acute toxicity	Pulmonary injury, edema	None
Acute inorganic Hg salts	Oral ingestion	Gastric—acute toxicity	Gastroenteritis, dermatitis	BAL, DMSA
Chronic inorganic Hg salts	Oral ingestion	Gastrointestinal, neurological, renal toxicities	Gastroenteritis, neuromuscular abnormalities, renal abnormalities	BAL, DMSA
Methyl Hg intoxication	Oral ingestion	Gastrointestinal, neurological, pulmonary toxicities	Neurotoxicity, gastroenteritis, respiratory distress	None
Any form	All exposure routes	Acrodynia	Idiosyncratic hypersensitivity to mercury	None

BAL: British anti-lewisite, dimercaprol; DMSA: 2,3-dimercaptosuccinic acid.

26.11.7 Signs and Symptoms of Subacute or Chronic Poisoning

The major clinical symptoms of methyl Hg toxicity are neurologic and include, in order, paresthesia, ataxia, dysarthria, and deafness. A latency period of weeks or months is characterized from the time of exposure until the development of symptoms (*onset*). Some pathological features include degeneration and necrosis of neurons in focal areas of the occipital cortex and in the granular layer of the cerebellum. This particular distribution of lesions in the CNS is thought to reflect a propensity of Hg to damage small neurons in the cerebellum and visual cortex.

26.11.8 Clinical Management of Hg Poisoning

Treatment of Hg poisoning is still considered experimental. Even prompt action and monitoring may result in fatality. For dermal or ocular exposure, washing of exposed areas is suggested. Reducing absorption from the GI tract refers mostly to inorganic forms. Oral administration of a protein solution (egg yolks) is suggested to reduce absorption, based on Hg's affinity for binding to –SH groups. Administration of activated charcoal may be useful in the case of acute high-dose situations. Gastric lavage and induction of emesis are recommended as a contingency measure, although emesis is contraindicated following ingestion of mercuric oxide due to its caustic nature.

Chelation therapy is the preferred treatment to reduce body burden. The choice of chelator depends on the ionic or elemental form of Hg, the route of exposure, and possible anticipated side effects. BAL is a more effective chelator for inorganic Hg salts, while D-penicillamine is marginally useful for elemental and inorganic Hg.

26.12 SELENIUM (SE)

26.12.1 Physical and Chemical Properties

Se is a metalloid that exists in several allotropic forms with different physical characteristics, such as color and crystalline structure. The crystalline form of Se has a density of 4.5 g/cm^3. Chemically, it combines with the halogens chlorine and bromine, as well as with metals, to form complex selenides.

26.12.2 Occurrence and Uses

Chemically, Se closely resembles sulfur and is related to tellurium. Like sulfur, it exists in several different forms. The burning of fossil fuels and coal accounts for much of the Se released into the environment. Se is incorporated into photoelectric devices (*gray Se*), giving the material a flat appearance; it imparts a scarlet red color to glass and enamels (*red Se*, sodium selenide); and it has the ability to decolorize glass. Sodium selenate is included in insecticides, while Se sulfide is available as a pharmaceutical agent for the treatment of skin diseases, including itchy and flaking skin of the scalp, acne, eczema, and seborrheic dermatitis.

The amount of Se in food sources varies according to soil levels, whether consumed from plants or as meat from animals. Most Se in foods is lost during processing, such as in the production of white rice or white flour. Regulatory agencies promote the addition of Se to drinking water when it is noted to be deficient.

26.12.3 Physiological Role

Se, once classified solely as a toxic mineral, was recognized in the 1970s as an essential trace element. The metal has a variety of properties, ranging from anticancer and antioxidant to effects on the immune, reproductive, and nervous systems. Biochemically, it is a component of the enzyme glutathione peroxidase, which accounts for its antioxidant function. It is also a constituent of deiodinases and thioredoxin reductase, represented in the biologically active form as *selenocysteine.*

Two endemic diseases have been associated with Se deficiency. *Keshan's disease* is a type of heart disease characterized by cardiomegaly, congestive heart failure, abnormal ECG, and multifocal necrosis of the myocardium. The disease is prevalent in children and women of childbearing age. *Kashin–Beck disease* is an osteoarthropathic disease characterized by atrophic necrosis and degeneration of cartilage. Both conditions are satisfactorily treated with Se supplementation.

The total body concentration of Se is less than 1 mg in the average adult, most of it concentrated in the liver, kidneys, pancreas, testes, and seminal vesicles. Men have a greater need for Se, which functions in sperm production and motility. Some Se is lost through sperm as well as through urine and feces. It is absorbed fairly well from the intestinal tract, with an absorption rate of nearly 60%.

26.12.4 Mechanism of Toxicity

Long-term effects of excess Se on the hair, nails, skin, liver, and nervous system are well documented; however, the biochemical mechanisms by which Se exerts toxicity remain largely unknown.

Acute Se toxicity is the result of inactivation of sulfhydryl enzymes that are necessary for cellular respiration. High Se concentrations replace sulfur in biomolecules, especially under conditions of low sulfur, possibly resulting in toxicity. Replacement of selenomethionine for selenocysteine in protein synthesis is another purported mechanism of toxicity.

26.12.5 Toxicokinetics

Se compounds are absorbed by the GI route, ranging from 44% to 95% of the ingested dose. Absorption depends on the dose as well as the physical state of the ingested form, with soluble Se having greater absorption rates than solid. The metal is also absorbed via inhalation and distributes to all tissues, with highest concentrations in the kidneys, liver, spleen, and pancreas. Se also concentrates in RBCs and is amenable to placental transfer.

Excessive amounts of Se are excreted primarily via the urinary system, leaving trace amounts systemically. Excretion of Se occurs in feces, expired air, and sweat.

26.12.6 Signs and Symptoms of Acute Toxicity

Although clinical cases are not frequently encountered, clinical signs of acute toxicity following ingestion of high doses of Se include excessive salivation, garlic odor to the breath, shallow breathing, and diarrhea. Pulmonary edema and lung lesions have been observed in acute lethal cases. Tachycardia, abdominal pain, nausea, vomiting, and abnormal liver functions have been seen in human acute *selenosis*. Inhalation of Se or its compounds results in irritation of the mucous membranes of the respiratory tract, dyspnea, bronchial spasms, bronchitis, and chemical pneumonias.

26.12.7 Signs and Symptoms of Chronic Toxicity

Clinical signs of chronic selenosis include loss of hair and fingernails, skin lesions, and clubbing of fingers. Nervous system effects such as numbness, convulsions, paralysis, and motor disturbances have been described.

26.12.8 Clinical Management of Poisoning

There are no specific protocols recommended for treatment of acute high-dose exposure to Se via inhalation, and only supportive treatment has been suggested for oral overdose. Gastric lavage and induction of vomiting with emetics may reduce absorption, but since selenious acid is caustic, both procedures are contraindicated. The chelators EDTA and BAL have not been successfully employed and, in fact, may increase the toxic effects of Se.

26.13 SILVER (AG)

26.13.1 Physical and Chemical Properties

Ag is a relatively rare and precious metal with high thermal and electrical conductivity. It occurs as a silvery-white powder in the chemical state of silver nitrate ($AgNO_3$), silver chloride (AgCl), dark gray silver sulfide (Ag_2S), and silver oxide (Ag_2O).

26.13.2 Occurrence and Uses

Ag is naturally occurring in the earth's crust, water, and soil. Estimated low to rare levels of silver occur in soil at 0.2–0.3 ppm and in surface waters at 0.2–2.0 ppb. Consequently, oral ingestion of Ag from food and drinking water is minimal and thus, insufficient to cause significant toxicity in humans. Typically, high levels of Ag exposure generally occur in occupational settings involving the handling of Ag and its derivatives, which are useful as key components of batteries and mechanical bearings. Hazardous waste sites are also a significant source of Ag compounds. Topical Ag preparations were historically used for its antibacterial properties.

26.13.3 Toxicokinetics

Ag is absorbed in the GI tract and transported systemically, bound to plasma globulins. Following its rapid distribution, it is primarily localized in skin and liver, in addition to kidney and spleen in trace amounts. The liver is responsible for the major metabolic pathway, and excretion occurs primarily in feces.

The mechanism of toxicity and the acute and chronic symptoms of silver exposure are not well documented due to limited animal and human studies. *Argyria* is the most significant consequence of excess exposure to Ag and its related compounds.

26.13.4 Argyria

Argyria results from inhalation and oral absorption of particulate Ag as well as prolonged mechanical impregnation of skin. Chronic topical treatment with Ag salts, resulting in increased Ag absorption from the oral mucosa and conjunctiva, often leads to argyria. The condition is characterized by increased bluish to black pigmentation of skin in sun-exposed areas; this is a permanent, irreversible manifestation. However, avoidance of elemental Ag exposure and supportive therapy help to reduce its progression. The pathogenesis of argyria is hypothesized to proceed from Ag-complexed proteins that are reduced to their elemental form by photoactivation when exposed to sunlight, causing increased production of melanin. Skin biopsy confirms the diagnosis and reveals brown clusters of metallic granules. Serum and urine Ag concentration measurements are also performed to assess the extent of silver load and body burden.

Animal studies have shown that Se supplementation can counteract elevated blood Ag levels. Se complexes with Ag to form insoluble silver selenide, which renders the metal soluble and amenable to excretion. In patients with progressive skin pigmentation, the use of sunscreen and other topical protectants helps reduce further pigmentation of the underlying skin.

26.14 ZINC (ZN)

26.14.1 Physical and Chemical Properties

Zn exists naturally as a blue to white transition metal in elemental (Zn) and divalent forms (Zn^{2+}). It is highly reactive and in the presence of air (O_2, CO_2) and moisture (H_2O), forms zinc carbonate ($ZnCO_3$, former) or zinc oxide (ZnO, both).

26.14.2 Occurrence and Uses

Zn is used extensively as an industrial protective coating or galvanizer for iron and steel. Its brass alloy is used for plates in dry cells and die casting. Clinically, Zn oxide has antiseptic and astringent properties and is available in topical preparations. Zn salts also have utility in electroplating, for soldering, as rodenticides, herbicides, pigments, and wood preservatives, and as solubilizing agents (zinc insulin suspensions).

26.14.3 Physiological Role

Zn is an essential metal for the physiological functioning of a large number of metalloenzymes, including alcohol dehydrogenase, alkaline phosphatase, carbonic anhydrase, Zn-Cu superoxide dismutase, leucine peptidase, and DNA and RNA polymerase. Zn deficiency is clinically encountered more often than deficiencies of other trace metals and results in dermatitis, growth retardation, impaired immune function, and congenital malformations. The metal has a significant role in the maintenance of nucleic acid structure of genes through the formation of "Zn finger" proteins.

Zn interacts with other physiologically important metals. Zn and Cu have a reciprocal relationship—that is, a high intake of Zn precipitates Cu depletion. The metal also interacts with Ca and is necessary for proper bone calcification. Cd competes with Zn for binding to sulfhydryl groups present on macromolecules. It also induces the metal-binding protein, metallothionein, which is involved in the absorption, metabolism, and storage of both essential and nonessential metals. Recently, the aggregation of ionic zinc and copper in cortical glutamatergic synapses has been implicated in modulating the response of the NMDA receptor and the vulnerability of amyloid-β to abnormal interaction with these metal ions, leading to a possible explanation for the mechanism underlying Alzheimer's disease. This *metals hypothesis of Alzheimer's disease* is based on observations of the precipitation of amyloid-β by Zn and its radicalization by Cu. Interestingly, both metals are markedly enriched in plaques.

26.14.4 Mechanism of Toxicity

The mechanism of Zn toxicity has not been fully clarified, although it enters cells via channels that are shared by Fe and Ca. This pathway may be a prerequisite for cell injury.

26.14.5 Toxicokinetics

Under normal physiologic conditions, 20–30% of an ingested dose of Zn is absorbed via the GI route. Zn absorption is influenced by P, Ca, and dietary fiber, but once absorbed, it is widely distributed. The highest content is found in muscle, bone, GI tract, brain, skin, lung, heart, and pancreas. In the circulation, about two-thirds of Zn is bound to albumin. The principal route of excretion is in feces and to a lesser extent, through the urinary system.

26.14.6 Signs and Symptoms of Acute Toxicity

Acute toxicity varies depending on the ingested or inhaled ion. *Metal fume fever*, which is a result of inhalation of Zn oxides, causes chest pains, cough, and dyspnea.*

* Although the term *metal fume fever* is now applied to the condition resulting from inhalation of any heated or flamed metal exposure.

Zn chloride is more damaging and corrosive to mucous membranes. Bilateral diffuse infiltrates, pneumothorax, and acute pneumonitis have been described. Oral ingestion of large doses of Zn sulfate has been associated with GI distress and alterations of GI tissue, including vomiting, burning in the throat, abdominal cramps, and diarrhea.

26.14.7 Signs and Symptoms of Chronic Toxicity

Hematological changes are reported in patients with chronic exposure to Zn. Long-term administration of Zn supplements is implicated in cases of anemia.

26.14.8 Clinical Management of Poisoning

General recommendations for management of excess Zn exposure include removal of the subject from the immediate area of exposure in the case of inhalation, or irrigation with water for ocular and dermal contact. For acute oral toxicity, administration of ipecac to induce vomiting is not recommended in the presence of caustic Zn compounds. Ingestion of large amounts of milk, cheese, or egg yolks may reduce gastrointestinal Zn absorption due to the high levels of phosphorus and Ca in these products.

To reduce Zn body burden, administration of $CaNa_2$-EDTA is the treatment of choice, while BAL is also recommended.

REFERENCES

SUGGESTED READINGS

Agency for Toxic Substances and Disease Registry (ATSDR), *Toxicological Profiles for Metals*, U.S. Department of Health and Human Services, Public Health Service, Atlanta, GA. Last updates available, 2008. From: www.atsdr.cdc.gov/substances/index.asp, last viewed March 2019.

Axelson, O., Dahlgren, E., Jansson, C.D., and Rehnlund, S.O., Arsenic exposure and mortality: a case-referent study from a Swedish copper smelter. *Br. J. Ind. Med.* 35, 8, 1978.

Castoldi, A.F., Onishchenko, N., Johansson, C., Coccini, T., Roda, E., Vahter, M., Ceccatelli, S., and Manzo, L., Neurodevelopmental toxicity of methylmercury: laboratory animal data and their contribution to human risk assessment. *Regul. Toxicol. Pharmacol.* 51, 215, 2008.

Choi, A.L., Cordier, S., Weihe, P., and Grandjean, P., Negative confounding in the evaluation of toxicity: the case of methylmercury in fish and seafood. *Crit. Rev. Toxicol.* 38, 877, 2008.

Ellenhorn, M.J. and Barceloux, D.G., *Medical Toxicology. Diagnosis and Treatment of Human Poisoning*, Elsevier, New York, 1988.

Tokar, E.J., Boyd, W.A., Freedman, J.H., and Waalkes, M.P., Toxic effects of metals, in *Caserett and Doull's Toxicology: The Basic Science of Poisons*, 8th ed., Klaassen, C.D. (Ed.), McGraw Hill, New York, 2013.

Harper, M., 10th Anniversary Critical Review: naturally occurring asbestos. *J. Environ. Monit.* 10, 1394, 2008.

Hazardous Substances Data Bank, TOXNET, Toxicology Data Network, Hazardous Substances Data Bank (HSDB), National Library of Medicine, Bethesda, MD. https://toxnet.nlm.nih.gov/newtoxnet/hsdb.htm, last viewed March 2019.

Kitchin, K.T. and Wallace, K., The role of protein binding of trivalent arsenicals in arsenic carcinogenesis and toxicity. *J. Inorg. Biochem.* 102, 532, 2008.

Krenzlok, E.P., McGuigan, M., and Lheur, P., Position statement: ipecac syrup. American Academy of Clinical Toxicology; European Association of Poisons Centres and Clinical Toxicologists. *J Toxicol Clin Toxicol.* 35, 699, 1997.

Lewis's Occupational and Environmental Health and Safety, Vincoli, J.R. (Ed.), CRC, Boca Raton, FL, 2000.

The Merck Index: An Encyclopedia of Chemicals, Drugs, and Biologicals, 15th ed., O'Neil, M.J. (Ed.), Merck, Rahway, NJ, 2013.

New York City Department of Health and Mental Hygiene, *Lead Poisoning for Healthcare Providers.* www1.nyc.gov/site/doh/health/health-topics/lead-poisoning-for-healthcare-providers.page, last viewed March 2019.

New York City Department of Health and Mental Hygiene, *Contaminated Herbal Medicine Products.* www1.nyc.gov/site/doh/health/health-topics/lead-poisoning-contaminated-herbal-medicine-products.page, last viewed March 2019.

Reeves, P.G. and Chaney, R.L., Bioavailability as an issue in risk assessment and management of food cadmium: a review, *Sci. Total Environ.* 398, 13, 2008.

Rice, D.C., Overview of modifiers of methylmercury neurotoxicity: chemicals, nutrients, and the social environment. *Neurotoxicology* 29, 761, 2008.

Sundar, S. and Chakravarty, J., Antimony toxicity. *Int. J. Environ. Res Public Health* 7(12), 4267, 2010.

Thompson, J. and Bannigan, J., Cadmium: toxic effects on the reproductive system and the embryo. *Reprod. Toxicol.* 25, 304, 2008.

U.S. Environmental Protection Agency (U.S. EPA), Health Assessment Document for Inorganic Arsenic. EPA107-02-8. Environmental Criteria and Assessment Office, https://www.epa.gov/sites/production/files/2016-09/documents/arsenic-compounds.pdf, 1984; updated January 2000; last viewed March 2019.

Wild, P., Bourgkard, E., and Paris, C., Lung cancer and exposure to metals: the epidemiological evidence, *Methods Mol. Biol.* 472, 139, 2009.

World Health Organization (WHO), *Environmental Health Criteria 118. Inorganic Mercury*, International Program on Chemical Safety, Geneva, 1991. www.who.int/ipcs/publications/ehc/ehc_numerical/en/, last viewed March 2019.

Wright, R.O. and Baccarelli, A., Metals and neurotoxicology. *J. Nutr.* 137, 2809, 2007.

Yoon, S., Han, S.S., and Rana, S.V., Molecular markers of heavy metal toxicity—a new paradigm for health risk assessment. *J. Environ. Biol.* 29, 1, 2008.

Zheng, Y., Li, X.K., Wang, Y., and Cai, L., The role of zinc, copper and iron in the pathogenesis of diabetes and diabetic complications: therapeutic effects by chelators. *Hemoglobin* 32, 135, 2008.

REVIEW ARTICLES

Abdul, K.S.M., Jayasinghe, S.S., Chandana, E.P., Jayasumana, C., and De Silva, P.M.C., Arsenic and human health effects: a review. *Environ. Toxicol. Pharmacol.* 40, 828–846, 2015.

Addison, J. and McConnell, E.E., A review of carcinogenicity studies of asbestos and non-asbestos tremolite and other amphiboles. *Regul. Toxicol. Pharmacol.* 52, S187, 2008.

Baldwin, D.R. and Marshall, W.J., Heavy metal poisoning and its laboratory investigation. *Ann. Clin. Biochem.* 36, 267, 1999.

Becklake, M.R., Bagatin, E., and Neder, J.A., Asbestos-related diseases of the lungs and pleura: uses, trends and management over the last century. *Int. J. Tuberc. Lung Dis.* 11, 356, 2007. Erratum in: *Int. J. Tuberc. Lung Dis.* 12, 824, 2008.

Beyersmann, D. and Hartwig, A., Carcinogenic metal compounds: recent insight into molecular and cellular mechanisms. *Arch. Toxicol.* 82, 493, 2008.

Bush, A.I., Drug development based on the metals hypothesis of Alzheimer's disease. *J. Alzheimers Dis.* 15, 223, 2008.

Dopp, E. and Schiffman, D., Analysis of chromosomal alterations induced by asbestos and ceramic fibers. *Tox. Lett.* 96–97, 155, 1998.

Drake, P.L. and Hazelwood, K.J., Exposure-related health effects of silver and silver compounds: a review. *Ann. Occup. Hyg.* 49, 575–585, 2005.

Eaton, J.W. and Qian, M., Molecular bases of cellular Fe toxicity. *Free Rad. Bio. Med.* 32, 833, 2002.

Flora, S.J., Mittal, M., and Mehta, A., Heavy metal induced oxidative stress and its possible reversal by chelation therapy. *Indian J. Med. Res.* 128, 501, 2008.

Garelick, H., Jones, H., Dybowska, A., and Valsami-Jones, E., Arsenic pollution sources. *Rev. Environ. Contam. Toxicol.* 197, 17, 2008.

Hadrup, N. and Lam, H.R., Oral toxicity of silver ions, silver nanoparticles and colloidal silver—a review. *Regul. Toxicol. Pharmacol.* 68, 7, 2014.

Hall, A.H., Chronic arsenic poisoning. *Tox. Lett.* 128, 69, 2002.

Hershko, C., Mechanism of iron toxicity. *Food Nutr. Bull.* 28, S500, 2007.

Klaassen, C.D., Liu, J., and Choudhuri, S., Metallothionein: an intracellular protein to protect against cadmium toxicity. *Ann. Rev. Pharm. Tox.* 39, 267, 1999.

Lansdown, A.B., A pharmacological and toxicological profile of silver as an antimicrobial agent in medical devices. *Adv. Pharmacol. Sci.* 2010, 910686 , 2010.

Laovitthayanggoon, S., Henderson, C.J., McCluskey, C., MacDonald, M., Tate, R.J., Grant, M.H., and Currie, S., Cobalt Administration Causes Reduced Contractility with Parallel Increases in TRPC6 and TRPM7 Transporter Protein Expression in Adult Rat Hearts. *Cardiovasc Toxicol.* doi: 10.1007/s12012-018-9498-3, 2018. [Epub ahead of print].

Leyssens, L., Vinck, B., Van Der Straeten, C., Wuyts, F., and Maes, L., Cobalt toxicity in humans—a review of the potential sources and systemic health effects. *Toxicology* 387, 43, 2017.

Miller, A.L., Dimercaptosuccinic acid (DMSA), a non-toxic, water-soluble treatment for heavy metal toxicity. *Altern. Med. Rev.* 3, 199, 1998.

Mutter, J. and Yeter, D., Kawasaki's disease, acrodynia, and mercury. *Curr. Med. Chem.* 15, 3000, 2008.

Rahimzadeh, M.R., Rahimzadeh, M.R., Kazemi, S., and Moghadamnia, A.A., Cadmium toxicity and treatment: an update. *Caspian J. Intern. Med.* 8, 135, 2017.

Rana, S.V., Metals and apoptosis: recent developments. *J. Trace Elem. Med. Biol.* 22, 262, 2008.

Rayman, M., The importance of selenium to human health. *Lancet* 356, 233, 2000.

Rebelo, F.M. and Caldas, E.D., Arsenic, lead, mercury and cadmium: toxicity, levels in breast milk and the risks for breastfed infants. *Environ. Res.* 151, 671–688, 2016.

Sun, H.J., Rathinasabapathi, B., Wu, B., Luo, J., Pu, L.P., and Ma, L.Q., Arsenic and selenium toxicity and their interactive effects in humans. *Environ. Int.* 69, 148–158, 2014.

Tenenbein, M., Toxicokinetics and toxicodynamics of iron poisoning. *Tox. Lett.* 102, 653, 1998.

Tossavainen, A., A. Karjaleinen, and P. J. Karhunen, Retention of asbestos fibers in the body. *Environ. Health Perspect.* 102, 253, 1994.

Tvermoes, B.E., Paustenbach, D.J., Kerger, B.D., Finley, B.L., and Unice, K.M.,Review of cobalt toxicokinetics following oral dosing: Implications for health risk assessments and metal-on-metal hip implant patients. *Crit. Rev. Toxicol.* 45, 367–387, 2015.

27 Aliphatic and Aromatic Hydrocarbons

27.1 INTRODUCTION

27.1.1 ALIPHATIC AND ALICYCLIC HYDROCARBONS

Hydrocarbons (HCs) are composed of carbon and hydrogen molecules whose carbon–carbon (C–C) bonds are composed of either all single (*saturated*) bonds or combinations of single and multiple (*unsaturated*) bonds. Aliphatic HCs, the term usually pertaining to fats or oils, applies to straight, open chains of carbon atoms, rather than ring structures, the simplest of which are the saturated HCs (*alkanes*). In addition, alkanes exist as *unbranched* straight chains or *branched* (for butane or longer chains), depending on the structural isomerism. Multiple C–C bonds result when hydrogens are removed from the alkanes, yielding unsaturated HCs such as *alkenes* (double C–C bonds) and *alkynes* (triple C–C bonds). Alicyclic HCs are saturated ring structures consisting of three or more carbon atoms. Unlike the aromatic chemicals (see next subsection), cyclopropane, cyclobutane, cyclopentane, and cyclohexane, for example, have three-, four-, five-, and six-membered rings, respectively, but do not exhibit double bonds within the rings.

27.1.2 AROMATIC HCs

Aromatic HCs are a special class of unsaturated HCs that contain one or more planar rings, such as a benzene ring. Benzene is the simplest aromatic molecule that can exist alone, attached as a substituent to other HCs (as a *phenyl* group), or as a substituted benzene (*halogenated* or *methylated* benzene). Benzene also exists as part of more complex aromatic systems consisting of a number of fused benzene rings, such as naphthalene or anthracene.

The vast majority of these compounds, however, are fundamental HCs but contain additional elements (*functional groups*) classifying them as *HC derivatives*, each of which possesses characteristic chemical properties. The functional groups consist of halogens (–X, such as F^-, Cl^-, Br^-, I^-), alcohols (–OH), ethers (–O–), aldehydes (–CO–H), ketones (–CO–), carboxylic acids (–COOH), esters (–CO–O–), and amines ($-NH_2$).*

27.1.3 GENERAL SIGNS AND SYMPTOMS OF ACUTE TOXICITY

In general, exposure to HCs occurs primarily through the inhalation, oral ingestion, and dermal routes. The ubiquitous nature of HC mixtures, and their ready

* The formula –CO– represents a C=O double bond.

availability in consumer products, renders them easily accessible and a source of toxicity and fatality from accidental ingestion. In fact, in the United States, 65,000 HC exposures were recorded in 2001 that required clinical intervention, 99% of which were accidental and 65% involved children under 5 years of age.

Intentional inhalation is reported among adolescents and drug abusers for the apparent euphoric effects. In 2012, 21.7 million adolescent Americans (aged 12 and older) reported using inhalants at least once in the past year, as well as 13.1% of 8th-graders. Among young adults aged 18 to 25, past-year use of inhalants decreased from 2.1% in 2005 to 1.8% in 2006. The National Institute on Drug Abuse (NIDA)-funded *2016 Monitoring the Future Study* (NIDA, 2017) showed that 4.7% of 8th-graders, 2.3% of 10th-graders, and 1.5% of 12th-graders had abused inhalants at least once in the year prior to being surveyed. The trends in prevalence of inhalants* over a "lifetime" for ages 12 or older appear to have decreased since the last reported survey in 2006 (8.3% for ages 12 to 17, 9.8% for ages 18 to 25, and 9.0% for ages 26 or older; NIDA, 2016). Deep inhalation of volatile HCs from cigarette lighter fluids (*sniffing*), paint adhesives (model airplane glue), and paint thinners, applied to the inside of a paper bag (*bagging*) or to a fabric (*huffing*), produces intense hallucinations and delusions. The effects, however, are accompanied by undesirable agitation, stupor, and seizures and increase the risk of cardiac arrhythmias.

Aliphatic and alicyclic HC gases and liquids are generally nontoxic or of low acute toxicity, with the majority of effects described being similar to those for simple asphyxiants or pulmonary irritants (see Chapter 25). Aromatic and halogenated HCs are associated with acute, systemic toxic potential, including hematological and hepatorenal toxicity, attributable in part to their volatility and viscosity. Some substances cause symptoms such as central nervous system (CNS) depression, narcosis, loss of consciousness, hallucination, stupor, and seizures. Gastrointestinal effects involve nausea and vomiting. Hematological consequences of HC toxicity result in mild hemolysis. Cardiovascular sensitization to circulating catecholamines, and the risk of arrhythmias, is a consequence of exposure to high concentrations of aromatics and chlorinated HCs.

Depending on the concentration of the chemical, dermal exposure produces thermal burns, requiring therapeutic intervention. Typical treatments for burn injuries, as well as the use of HC ointments such as Polysorbate 80, mineral oil, and petroleum jelly, are useful in managing the dermal trauma.

Although no antidote is available for HC poisoning, life-threatening toxicity is managed using supportive care. Because of the volatility of these chemicals, the risk of aspiration pneumonitis increases with induction of vomiting, especially with administration of emetics, or with the use of gastric lavage.

---

* According to NIDA, *inhalants* refers to the various substances that can only be inhaled, including solvents (liquids that become gases at room temperature), aerosol sprays, gases, and nitrites (prescription drugs for chest pain).

27.2 PETROLEUM DISTILLATES

27.2.1 Occurrence and Uses

Petroleum and petroleum distillates represent an extremely valuable natural resource consisting of a complex mixture of thousands of aliphatic and aromatic compounds. Ninety percent of this mixture of petrochemicals is used for fuels for heating and transportation, as well as raw materials for the chemical industry. Thus, the derivatives of petroleum are used in the manufacturing of many common substances found in household products and in the industrial setting, including, but not limited to, the production of oils, waxes, cements, fuels (gasoline, kerosene, lighter fluids), polymers (for plastics), paint and furniture products, and therapeutic agents (cod liver oil, mineral oil, laxatives).

27.2.2 Mechanism of Toxicity

In general, the toxicity and reactivity of aliphatic and alicyclic HCs are low. For instance, gaseous compounds have properties similar to those of pulmonary irritants or simple asphyxiants. The degree of cyclization or unsaturation does not correlate with toxicity, and they usually produce anesthetic effects at very high ambient concentrations. As with the simple asphyxiants, therefore, their toxicity is limited to the amount of substituted oxygen.

Aliphatic and alicyclic HCs are mostly flammable or combustible gases or liquids. Their ability to form explosive mixtures with air decreases as the carbon number increases above 13. Their ominous chemical nature appears to exaggerate their toxic potential since most HCs are classified as nonpoisonous or have low acute toxicity on exposure. Table 27.1 summarizes the properties and toxicities of a variety of aliphatic and alicyclic HCs.

27.3 AROMATIC HYDROCARBONS

27.3.1 Occurrence and Uses

Aromatic HCs contain mononuclear or polynuclear benzene rings as part of their structure. They are widely distributed in petroleum and its products and are used as solvents and in the synthesis of organic chemicals.

27.3.2 Mechanism of Toxicity

As with the aliphatic and alicyclic HCs, the toxicity of aromatic HCs is generally low.* At high concentrations, the substances affect the CNS, resulting in depres-

* Although many of the polynuclear HCs are known or suspected human or animal carcinogens, the carcinogenic potential and clinical outcome of chronic exposure to these chemicals are addressed in Chapter 32.

TABLE 27.1
Properties, Sources, Uses, and Toxicity of Aliphatic and Alicyclic Hydrocarbons

Chemical	Properties	Source	Industrial[a] or Household Uses	Acute Human Toxic Effects
1,3-Butadiene	F colorless gas, aromatic odor	Petroleum distillate	Synthetic rubber, elastomers, food wrapping material	Asphyxiant; irritation, hallucinations; at high doses, narcosis; suspected carcinogen
Acetylene	F colorless gas	Petroleum distillate	Welding, illuminant, fuel, synthesis of acetylides	Simple asphyxiant; at high doses, headache, dyspnea, death
Cyclohexane	F colorless liquid	Petroleum distillate	Solvent for paints and resins, organic synthesis	Low acute toxicity, CNS depression, pulmonary irritant
Cyclohexene	F colorless liquid	Petroleum distillate, coal tar	Oil extraction, organic synthesis, gasoline stabilizer	Low acute toxicity, CNS depression, pulmonary irritant
Cyclopentadiene	F colorless liquid	Petroleum distillate	Resins, camphors, preparation of metals	Pulmonary irritant in rodents
Cyclopentane	F colorless liquid	Petroleum distillate	Solvent, shoe industry, wax extraction	Low acute toxicity, CNS and respiratory depression, pulmonary irritant
Ethane	F odorless, colorless gas	Natural gas; petroleum *cracking*[b]	Fuel, refrigerant	Nonpoisonous, simple asphyxiant
Ethylene	F colorless gas	Petroleum distillate	Major starting material for organic synthesis; anesthetic	Simple asphyxiant; at high doses, narcosis and unconsciousness
Isobutane	F colorless gas	Natural gas; petroleum *cracking*	Liquid and motor fuels, organic synthesis	Simple asphyxiant; at high doses, narcosis
Isooctane	F colorless liquid, gasoline odor	Petroleum distillate	Octane fuel reference for gasoline, solvent	Low acute toxicity, pulmonary irritant
Methane	F odorless, colorless gas	Natural gas from earth's crust and landfills	Heating fuel, production of other gases	Nonpoisonous, simple asphyxiant
n-Butane	F colorless gas	Natural gas; petroleum *cracking*	Liquid and motor fuels, propellant, organic synthesis	Simple asphyxiant; at high doses, narcosis

(*Continued*)

TABLE 27.1 (CONTINUED)
Properties, Sources, Uses, and Toxicity of Aliphatic and Alicyclic Hydrocarbons

Chemical	Properties	Source	Industrial[a] or Household Uses	Acute Human Toxic Effects
n-Octane	F colorless liquid	Petroleum *cracking*, gasoline	Solvent, organic synthesis	Low acute toxicity, CNS depression, pulmonary irritant
n-Pentane	F colorless liquid, gasoline odor	Petroleum distillate	Solvent, plastics	Pulmonary irritant, narcosis
Propane	F colorless gas	Petroleum distillate	Fuel, refrigerant, organic synthesis	Simple asphyxiant; at high doses, narcosis
Propylene	F colorless gas (burns yellow)	Gasoline refining, catalytic *cracking* of petroleum	Polypropylene (plastics)	Simple asphyxiant; at high doses, mild anesthetic

[a] Used in manufacturing, in synthesis, or as part of industrial processes.
[b] *Cracking* refers to the breaking of carbon–carbon bonds in the formation of smaller chemicals using pyrolysis (*thermal cracking*) or a catalyst (*catalytic cracking*).
F: flammable; HC: hydrocarbon.

sion (narcosis), hallucinations, and stupor. Similarly, acute toxicity appears minimal except in high concentrations. Common routes of exposure for liquids or solids include oral ingestion, inhalation, and dermal absorption. Aromatic HCs form explosive mixtures with air and are flammable, with low flash points. They react with oxidizing agents, have moderate to strong aromatic odors, and are designated as priority pollutants or hazardous waste by the U.S. Environmental Protection Agency (EPA). Table 27.2 lists the properties and toxicities of some commonly encountered aromatic HCs possessing acute or chronic human toxicity. Other HCs have carcinogenic or mutagenic potential, primarily in animals, yet have not demonstrated acute or chronic toxicity in humans. These aromatic HCs are noted in Table 27.3.

27.4 HALOGENATED HYDROCARBONS

27.4.1 Occurrence and Uses

Halogenated aromatic or halogen-substituted aliphatic HCs (halocarbons) consist mostly of mononuclear benzene rings or straight, open chains of carbon atoms, respectively, whose hydrogen atoms are partially or fully replaced by halogen atoms: fluorine, chlorine, bromine, and iodine. These liquids or solids are widely used as

TABLE 27.2
Properties, Sources, Uses, and Toxicity of Aromatic HCs

Chemical (or *Common Name*)	Properties	Source	Industrial[a] or Household Uses	Clinical Toxic Effects
Benzene	F colorless liquid	Coal tar, landfills, petroleum	Solvent, dyes, paints, organic synthesis	*Acute*: pulmonary and dermal irritant; narcosis, convulsions; *chronic*: blood, BM, CNS, dermal, and respiratory toxicity
Cumene (*cumol*)	F colorless liquid	Petroleum, gasoline	Solvent, organic synthesis	*Acute*: pulmonary and dermal irritant; narcosis; *chronic*: unknown
Naphthalene (*tar camphor, mothballs*)	White crystalline flakes	Petroleum, gasoline	Moth repellent, dyes, explosives, lubricants	*Acute*: pulmonary and dermal irritant; narcosis; hemolytic anemia (oral); *chronic*: ocular opacities
Styrene (*cinnamol*)	F yellow liquid	Petroleum, gasoline	Polystyrene plastics, rubber, resins	*Acute*: pulmonary and dermal irritant; narcosis; *chronic*: unknown
Toluene (*toluol*)	F colorless liquid	Coal tar, petroleum, gasoline	Production of benzene, TNT; solvent	*Acute*: similar to benzene; *chronic*: less toxic than benzene
Xylene (*xylol*)	F colorless liquid	Petroleum, gasoline	Solvent, dyes, paints, organic synthesis	Similar to toluene

[a] Used in manufacturing, synthesis, or as part of industrial processes.
BM: bone marrow; CNS: central nervous system; F: flammable.

solvents, as refrigerants, as fire retardants, in the organic synthesis of a variety of chemicals, and in the past, as general anesthetics.*

27.4.2 Mechanism of Toxicity

Exposure to volatile liquid and gaseous halocarbons occurs primarily through inhalation, oral ingestion, and dermal exposure. As with the aliphatic, alicyclic, and aromatic HCs, the toxicity of halocarbons is generally low. At high concentrations, the volatile liquids and gases appear to affect the CNS and intestinal tract, resulting in anesthesia, drowsiness, incoordination, nausea, and vomiting. Hepatic, renal, and cardiac toxicity can prove fatal with inhalation at high concentrations. Some

* Because of their toxicity, chlorinated halocarbons as general anesthetics have been replaced by less toxic agents.

TABLE 27.3
Properties, Sources, and Effects of Aromatic HC with Low Suspicion of Human Toxicity

Chemical (or *Common Name*)	Properties	Source	Toxic Effects
Anthracene (*green oil*)	Greenish crystals	Dyes	Low toxicity; unknown carcinogenic potential
Benzo[a]-anthracene (*tetraphene*)	Greenish-yellow fluorescent crystals	Petroleum, gasoline	Lethal in mice (10 mg/kg, intravenous only); animal carcinogen
Benzo[a]pyrene	Yellowish crystals	Coal tar, petroleum, gasoline, tobacco smoke, smog	*Acute*: low; known mutagen, procarcinogen, teratogen in test species
Dibenz[a,h]-anthracene	Colorless crystals	Dyes	Similar to benzo[a] anthracene
Phenanthrene	Blue fluorescent crystals	Dyes, explosives, pharmaceutical industry	*Acute*: low toxicity in humans; animal carcinogen
Pyrene	Colorless-yellowish crystals	Coal tar, petroleum, gasoline, tobacco smoke, smog	Pulmonary irritant in animals, no evidence of carcinogenicity or mutagenicity in animals

TABLE 27.4
Properties, Sources, Uses, and Toxicity of Halocarbons

Chemical	Properties	Industrial[a] or Household Uses	Toxic Effects
		Aliphatic Halocarbons	
1,2-Dibromoethane	NC colorless liquid (sweet odor)	Fumigant, gasoline antiknock compound, solvent	*Acute*: CNS depression, pulmonary irritant, dermal, hepatic, renal; *chronic*: respiratory, depression, HA; AC, SHC
Carbon tetrachloride	NC colorless liquid (characteristic odor)	Solvent, fire extinguishers, dry cleaning, aerosol propellants	Similar to chloroform
Chloroform	NC colorless liquid (sweet odor)	Solvent for paints, degreaser, refrigerant (former anesthetic)	*Acute*: low—anesthesia, dizziness, hallucinations, HA, fatigue; *chronic*: hepatic, renal, CNS, cardiac toxicity; AC, SHC
Ethyl chloride (muriatic ether)	F colorless liquid, gas (ether odor)	Solvent, refrigerant, topical anesthetic	*Acute*: anesthesia, narcosis, unconsciousness, respiratory irritation, cardiac arrest
Halothane (fluothane)	NC colorless liquid (sweet odor)	General clinical anesthetic	*Acute*: CNS depression, anesthesia, cardiac, ocular irritation; *chronic*: hepatic
Methyl chloride	F colorless explosive gas (sweet odor)	Refrigerant, local anesthetic, organic synthesis	*Acute*: HA, nausea, vomiting, narcosis, convulsions, coma; *chronic*: ocular, hepatic, renal, BM toxicity
Methylene chloride	NC colorless liquid	Solvent for paints, degreasing agent	*Acute*: low (inhalation)— irritation, fatigue, HA, narcosis;*chronic*: hepatic; AC, SHC
Trichloroethylene, tetrachloroethylene	NC colorless liquid (chloroform, ether odor, respectively)	Solvent, dry cleaning, degreaser, anesthetic (limited)	*Acute*: CNS depression, narcosis;*chronic*: respiratory, cardiac failure; AC, SHC (metabolized to phosgene, trichloroethanol)
Vinyl chloride	F colorless gas	Resins, plastics, refrigerant	*Acute*: low—anesthesia, dizziness, narcosis;*chronic*: hepatic, renal; AC, known HC

(*Continued*)

TABLE 27.4 (CONTINUED)
Properties, Sources, Uses, and Toxicity of Halocarbons

Chemical	Properties	Industrial[a] or Household Uses	Toxic Effects
		Aromatic Halocarbons	
Benzyl chloride	NC colorless- yellowish liquid	Dyes, solvents, resins, plastics, perfumes, lubricants	*Acute*: ocular, dermal, respiratory corrosive liquid; pulmonary edema, CNS depression; AC
Chlorobenzene	F colorless-yellowish liquid (almond-like odor)	Solvent, manufacture of phenol and aniline	*Acute*: low—anesthesia, dizziness, narcosis (in animals); AC
Hexachloronaphthalene (halowax)	NC yellow solid	Electric wire insulation, lubricant	*Acute*: chloracne, dermatitis, nausea, confusion
Monohalogenated benzenes (fluoro-, chloro-, bromo-, iodobenzene)	F liquids (aromatic odor)	Solvent, dyes, paints, organic synthesis	*Acute*: cardiac arrhythmias, anesthesia, pulmonary irritation, narcosis, convulsions; *chronic*: blood, BM, CNS, dermal, respiratory toxicity

AC: known animal carcinogen; BM: bone marrow; CNS: central nervous system; F: flammable; NC: noncombustible; SHC: suspected human carcinogen.

compounds are known animal and human carcinogens. Halocarbons range from noncombustible to highly flammable gases or liquids, the latter of which form explosive mixtures with air or when heated. Although the compounds are generally stable, they react violently with alkali metals and oxidizers. Table 27.4 outlines the properties and toxicities of common halocarbons with acute or chronic human toxicity.

27.5 METHODS OF DETECTION

Quantitative analysis of HCs is usually performed when sampling of ambient or contaminated areas is necessary. Identification of HC liquids is also performed in suspected forensic cases. Gas chromatographic (GC) techniques are sensitive to detecting HCs in air, soil, and biological samples. The volatility of the liquids renders them suitable for monitoring through a packed GC column (solid phase). For examining the composition of the flowing gas, a flame ionization or electron capture detector is required. Gas chromatography-mass spectrometry (GC-MS) or nitrogen-specific detectors have been employed when absolute specificity is a principal requirement. These techniques have also been important in validating other screening methods.

GC-MS is currently the method of choice for the detection and characterization of aliphatic and aromatic hydrocarbons (American Society of Testing and Materials [ASTM]). Since petroleum products are the most common type of organic chemical encountered in occupational, environmental, or household areas, the analytical chemist must have at least a basic knowledge of the manufacturing process. In general, petroleum products, especially those used in accelerants or paints, are flammable liquids with high vapor pressure and low flash points, the components of which are easily separated by GC-MS (although flame ionization detection [FID] was the detection method of choice until recently). Consequently, GC can separate organic hydrocarbons based on the carbon length up to about C_{27} with impressive resolution. The essential features of chromatographic separation, elution, and resolution include the following: capillary glass column with headspace analysis,* methylsilicone as the nonpolar liquid stationary phase, and helium as the gas carrier; several organic solvents are used for sample extraction (e.g., carbon tetrachloride, acetone, and hexane).

REFERENCES

SUGGESTED READINGS

American Society of Testing and Meterials (ASTM), PDF Publication: ASTM D6968 - 03(2015), Standard Test Method for Simultaneous Measurement of Sulfur Compounds and Minor Hydrocarbons in Natural Gas and Gaseous Fuels by Gas Chromatography and Atomic Emission Detection. From: https://www.astm.org/Standards/D6968.htm, last viewed March 2019.

* 30 m × 0.75 mm i.d. × 1.0 μm film column; GC-MS conditions: carrier gas helium = 6 ml/min; oven = 40 C (5 min) to 260 C at 12 C/min; injection = 260 C, detector = 270 C.

Chamberlain, J., Chapter 6. Gas chromatography, in *The Analysis of Drugs in Biological Fluids*, 2nd ed., CRC Press, Florida, 1995.

Henry, J.A., Composition and toxicity of petroleum products and their additives. *Hum. Exp. Toxicol.* 17, 111, 1998.

Kenyon, E.M., Benignus, V., Eklund, C., Highfill, J.W., Oshiro, W.M., Samsam, T.E., and Bushnell, P.J., Modeling the toxicokinetics of inhaled toluene in rats: influence of physical activity and feeding status. *J. Toxicol. Environ. Health A* 71, 249, 2008.

Kropp, K.G. and Fedorak, P.M., A review of the occurrence, toxicity, and biodegradation of condensed thiophenes found in petroleum. *Can. J. Microbiol.* 44, 605, 1998.

Li, C., Wang, L.E., and Wei, Q., DNA repair phenotype and cancer susceptibility—a mini review. *Int. J. Cancer* 1124, 999, 2009.

Mehlman, M.A., Dangerous and cancer-causing properties of products and chemicals in the oil refining and petrochemical industries. Part XXX: Causal relationship between chronic myelogenous leukemia and benzene-containing solvents. *Ann. N.Y. Acad. Sci.* 1076, 110, 2006.

NIDA (National Institute on Drug Abuse), *2016 Monitoring the Future Study: Inhalants*, National Institutes of Health, U.S. Department of Health and Human Services (HHS), 2017. www.drugabuse.gov/drugs-abuse/inhalants, last viewed March 2019.

Patnaik, P., *Comprehensive Guide to Hazardous Properties of Chemical Substances*, Van Nostrand Reinhold, Wiley Publishers, New York,1992.

Ritchie, G., Still, K., Rossi, J. 3rd, Bekkedal, M., Bobb, A., and Arfsten, D., Biological and health effects of exposure to kerosene-based jet fuels and performance additives. *J. Toxicol. Environ. Health B Crit. Rev.* 6(4), 357–451, 2003. Review.

Sawyer, T.W., Chapter 9. Cellular methods of genotoxicity and carcinogenicity, in *Introduction to In Vitro Cytotoxicology: Mechanisms and Methods*, Barile, F.A. (Ed.), CRC Press, Boca Raton, Florida, 1994.

Shiao, Y.H., Genetic signature for human risk assessment: lessons from trichloroethylene. *Environ. Mol. Mutagen.* 50, 68, 2009.

Tormoehlen, L.M., Tekulve, K.J., and Nañagas, K.A., Hydrocarbon toxicity: a review. *Clin. Toxicol. (Phila.)* 52, 479, 2014.

REVIEW ARTICLES

Ahmed, F.E., Toxicology and human health effects following exposure to oxygenated or reformulated gasoline. *Toxicol. Lett.* 123, 89, 2001.

Baudouin, C., Charveron, M., Tarroux, R., and Gall, Y., Environmental pollutants and skin cancer. *Cell Biol. Toxicol.* 18, 341, 2002.

Cooper, G.S. and Jones, S., Pentachlorophenol and cancer risk: focusing the lens on specific chlorophenols and contaminants. *Environ. Health Perspect.* 116, 1001, 2008.

Cronkite, E.P., Drew, R.T., Inoue, T., Hirabayashi, Y., and Bullis, J.E., Hematotoxicity and carcinogenicity of inhaled benzene. *Environ. Health Perspect.* 82, 97, 1989.

De Rosa, C.T., Pohl, H.R., Williams, M., Ademoyero, A.A., Chou, C.H., and Jones, D.E., Public health implications of environmental exposures. *Environ. Health Perspect.* 106, 369, 1998.

Ferguson, L.R., Natural and human-made mutagens and carcinogens in the human diet. *Toxicology* 181, 79, 2002.

Gao, P., da Silva, E., Hou, L., Denslow, N.D., Xiang, P., and Ma, L.Q., Human exposure to polycyclic aromatic hydrocarbons: metabolomics perspective. *Environ Int.* 119, 466, 2018.

Gold, L.S., De Roos, A.J., Waters, M., and Stewart, P., Systematic literature review of uses and levels of occupational exposure to tetrachloroethylene. *J. Occup. Environ. Hyg.* 5, 807, 2008.

Himmel, H.M., Mechanisms involved in cardiac sensitization by volatile anesthetics: general applicability to halogenated hydrocarbons? *Crit. Rev. Toxicol.* 38, 773, 2008.

Hughes, K., Meek, M.E., Walker, M., and Beauchamp, R., 1,3-Butadiene: exposure estimation, hazard characterization, and exposure-response analysis. *J. Toxicol. Environ. Health B. Crit. Rev.* 6, 55, 2003.

Langer, P., Persistent organochlorinated pollutants (PCB, DDE, HCB, dioxins, furans) and the thyroid. *Endocr. Regul.* 42, 79, 2008.

Linden, C.H., Volatile substances of abuse. *Emerg. Med. Clin. North Am.* 8, 559, 1990.

Lock, E.A. and Smith, L.L., The role of mode of action studies in extrapolating to human risks in toxicology. *Toxicol. Lett.* 140, 317, 2003.

Lucena, R.A., Allam, M.F., Jiménez, S.S., and Villarejo, M.L., A review of environmental exposure to persistent organochlorine residuals during the last fifty years. *Curr. Drug Saf.* 2, 163, 2007.

Moore, M.R., Vetter, W., Gaus, C., Shaw, G.R., and Müller, J.F., Trace organic compounds in the marine environment. *Mar. Pollut. Bull.* 45, 62, 2002.

Periago, J.F., Zambudio, A., and Prado, C., Evaluation of environmental levels of aromatic hydrocarbons in gasoline service stations by gas chromatography. *J. Chromatogr. A* 778, 263, 1997.

Plaa, G.L., Chlorinated methanes and liver injury: highlights of the past 50 years. *Annu. Rev. Pharmacol. Toxicol.* 40, 42, 2000.

Ritchie, G.D., Still, K.R., Alexander, W.K., Nordholm, A.F., Wilson, C.L., Rossi, J. 3rd., and Mattie, D.R., A review of the neurotoxicity risk of selected hydrocarbon fuels. *J. Toxicol. Environ. Health B. Crit. Rev.* 4, 223, 2001.

Shelton, K.L., Discriminative stimulus effects of abused inhalants. *Curr. Top Behav. Neurosci.* 39, 113, 2018.

Suvorov, A., and Takser, L., Facing the challenge of data transfer from animal models to humans: the case of persistent organohalogens. *Environ. Health* 7, 58, 2008.

Yoon, B.I., Hirabayashi, Y., Kawasaki, Y., Kodama, Y., Kaneko, T., Kim, D.Y., and Inoue, T., Mechanism of action of benzene toxicity: cell cycle suppression in hemopoietic progenitor cells (CFU-GM). *Exp. Hematol.* 29, 278, 2001.

Zapponi, G.A., Attias, L., and Marcello, I., Risk assessment of complex mixtures: some considerations on polycyclic aromatic hydrocarbons in urban areas. *J. Environ. Pathol. Toxicol. Oncol.* 16, 209, 1997.

28 Insecticides

28.1 INTRODUCTION

In the rural developing world, about 300,000 people die each year from pesticide self-poisoning where pesticides are widely used in agricultural practice. Organophosphorus and carbamate insecticides cause most acute fatalities, whereas paraquat and aluminum phosphide are major problems in some countries, where fatalities may climb to over 60%. Conversely, significant acute poisoning is much less common in industrialized countries, where the long-term effects of low-dose chronic exposure are of utmost concern. In the United States, approximately 50 million tons of pesticide in general, and insecticide applications in particular are used annually, making it the largest consumer of these products worldwide.* Consequently, there are a number of nonfatal poisonings from general pesticide use, averaging about 5000 cases annually in the United States. These cases primarily result from inadvertent ingestion of insecticide applicators and solutions for home use as well as dermal and respiratory exposure to industrial insecticides. Insecticides are also responsible for the production of acute dermal and respiratory inflammation. Chronically, these agents are implicated in the production of dermal neoplasms.

28.2 ORGANOPHOSPHORUS COMPOUNDS (ORGANOPHOSPHATES, OP)

28.2.1 Chemical Characteristics

The structural core of organophosphates (OP) contains a central phosphate nucleus to which are attached a variety of aliphatic side chains, according to the following formula:

$$[R_1O]_2\text{—}\overset{\overset{\text{O (or S)}}{\|}}{P}\text{—O (or S)—}R_2$$

where

R_1 is a methyl or ethyl group
R_2 is an aliphatic or aromatic hydrocarbon side chain

Most of these compounds are dense liquids (specific gravity above 1.25) or solids at room temperature, have low vapor pressure, and are slightly soluble or insoluble in water at 20°C. Table 28.1 outlines the physical characteristics and relative rodent toxicity of a variety of OP insecticides. The sulfur-containing compounds, such as dimethoate,

* The NCFAP (National Center for Food and Agricultural Policy) reported that over 131, 461, 183, and 210 million pounds of fungicide, herbicide, insecticide, and "other" pesticides, respectively, were applied to U.S. crops per year between 1992 and 1997.

TABLE 28.1
Names, Chemical Properties, and Rodent Toxicity of Organophosphorus Insecticides Arranged According to Rodent Oral LD_{50}

Common Name	Chemical Name or Synonym	Physical Characteristics	Median Oral Rodent LD_{50} (mg/kg)
TEPP	Tetraethyl pyrophosphate	Liq, agreeable odor, misc in water and org liq	1.1
Disulfoton	Phosphorodithioic acid O,O-diethyl ester	Colorless oil, immisc in water	2.3–6.8
Mevinphos	Phosdrin	Yellow liq, misc with water	3.7–6.1
Parathion ethyl	Paraphos, Thiophos	Pale yellow liq, practically insol in water	3.6–13
Azinophos methyl	Guthion	Crystalline solid, sol in alcohols and org liq	11
DDVP	Dichlorvos[a]	Non-flam liq, misc with alcohol and org liq	56–80
Chlorpyrifos	Dursban, Lorsban	White crystals, misc with alcohol and org liq	145
Dimethoate	Cygon, Roxion	S-containing crystals, ss in water	250
Malathion	Malamar 50, Cythion	Brown-yellow liq, ss in water, misc in org liq	1000–1375
Ronnel	Fenchlorphos, Ectoral	White powder, insol in water, misc in org liq	1250–2630

[a] Found in the commercial household product *No Pest Strip*®.
flam: flammable; immisc: immiscible; insol: insoluble; liq: liquid; misc: miscible; org: organic; sol: soluble; ss: slightly soluble.

require metabolic activation prior to binding to the enzymatic target. Figure 28.1 illustrates and compares the chemical structures of some of the OP compounds. Tetraethyl pyrophosphate (TEPP) is one of the more potent OP insecticides with the prototype structural features of the OP molecules. Substitution of the diethyl phosphate with the para-nitrophenyl substituent and a thiol replacement for the oxygen group renders the parathion molecule comparatively less potent and toxic. Further substitutions within the parathion molecule, as illustrated in Figure 28.1, result in the formation of malathion, a compound with much lower toxicity and of high commercial value.

28.2.2 Occurrence and Uses

The OP insecticides were originally developed as nerve gases for use as possible chemical warfare agents during World War II. The first of these compounds was TEPP. The biological action of the nerve gases, such as sarin, tabun, and soman, is similar to, but more toxic than, that of the OPs. Soman is not only the most toxic of the three but one of the most toxic compounds ever synthesized, with fatalities occurring with an oral dose of 10 μg/kg in humans (see Chapter 34).

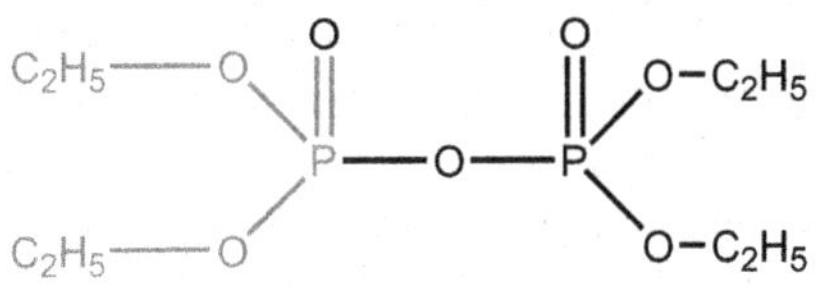

TEPP

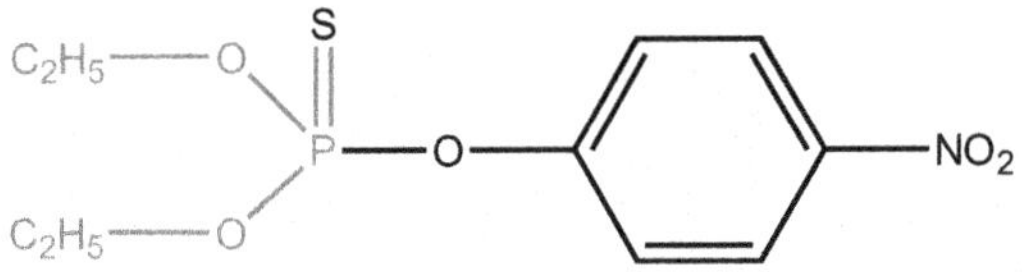

ethyl parathion

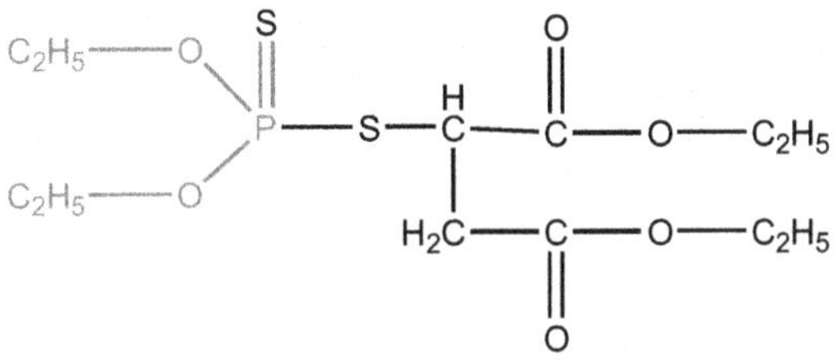

malathion

FIGURE 28.1 Structures of selective OP compounds. Structures of several selected OP molecules are shown. Blue highlight indicates structural similarities to the prototypic molecule (TEPP, tetraethylpyrophosphate).

Currently, OPs are popular chemicals used as household and agricultural insecticides. Their wide distribution as industrial chemicals allows access to the general population. They are conveniently used for suicidal attempts and are found as crop contaminants* and in accidental occupational exposure. Consequently, in the United States, they accounted for over 10,000 cases of exposure. About 1% of these cases resulted in fatalities in 2002, confirming their role as a public health hazard. As with the organic hydrocarbons (Chapter 27), exposure routes include oral ingestion, inhalation, and dermal contact. The agents are most rapidly absorbed after inhalation, especially when delivered in aromatic hydrocarbon vehicle solvents.†

28.2.3 Mechanism of Toxicity

Acetylcholine (ACh) is a neurotransmitter found throughout the central (CNS), autonomic (sympathetic and parasympathetic), and somatic nervous systems. Specifically, ACh receptors are located at autonomic preganglionic sympathetic

* *Orange-picker's flu* is a condition characterized by chronic low-level exposure to handling of fruits and vegetables that are sprayed with OP agents. The syndrome is common among farm workers in U.S.

† A recent study suggests that human offspring's risk of autism spectrum disorder increases following prenatal exposure to ambient pesticides, including malathion & chlorpyrifos; **BMJ 364, 1962, 2019.**

nicotinic and parasympathetic muscarinic synapses, autonomic postganglionic muscarinic parasympathetic synapses, somatic neuromuscular cholinergic synapses, and central cholinergic synapses (see Figure 15.1 for an illustration of autonomic receptors). ACh activity is either terminated or potentiated by alteration of metabolism at cholinergic synapses, principally through interference with reuptake or enzymatic degradation. In the enzymatic pathway, acetylcholine esterase (ACh-∑, *true* or *RBC* cholinesterase) is predominantly distributed among red blood cells (RBCs), central, somatic and autonomic neurons, gray matter of the brain, spinal cord, lung, and spleen. Pseudocholinesterase (*plasma* cholinesterase) is located in serum, plasma, white matter, liver, and pancreas and heart tissue.

ACh forms an ester link with the anionic site of the enzyme and is subsequently hydrolyzed to inactive components according to the following *reversible* reactions:

1. ACh + ACh-∑ → Ach—ACh-∑ complex
2. ACh—ACh-∑ complex → choline + acetyl—ACh-∑
3. acetyl—Ach-∑ + H_2O → acetic acid + Ach-∑

The active enzyme is subsequently regenerated.

OPs inhibit the action of ACh-∑ by phosphorylating the active esteratic site, forming an *irreversible* OP–Ach-∑ complex, rendering it incapable of hydrolyzing ACh. Thus, inhibition of the enzyme results in accumulation and over-stimulation of Ach at autonomic and somatic receptors. The irreversible nature of the complex requires days to weeks for disassembly. In addition, phosphorylation prompts the enzyme to undergo an accelerated "aging" process, whereby it loses an alkyl group, which prevents it from spontaneously regenerating. These circumstances account for the toxic manifestations of OPs.

28.2.4 Signs and Symptoms of Acute Toxicity

Low doses of OP insecticides produce mild to moderate toxicity depending on the remaining percentage of available ACh-∑ activity.* Acute initial toxic effects mimic exaggerated cholinergic muscarinic stimulation, including salivation, lacrimation, excessive sweating (diaphoresis), miosis, tachycardia, hypertension, and tightness of the chest (bronchoconstriction). Higher doses cause over-stimulation of both nicotinic and muscarinic receptors,† resulting in diarrhea, urinary incontinence, bradycardia, muscle twitching, fatigue, hyperglycemia, bronchospasm, and bronchorrhea. Increased depolarization at nicotinic neuromuscular synapses results in muscle weakness and flaccid paralysis.

CNS cholinergic stimulation suppresses central medullary centers, accounting for depressed respiration, headache, anxiety, restlessness, confusion, psychosis,

* While 20–50% of enzyme activity remains with mild toxicity and 10–20% remains with moderate toxicity, RBC cholinesterase levels normally vary within the general population. Enzyme concentrations also vary in the presence of other conditions, such as malnutrition, cirrhosis, and pregnancy, making interpretation of the levels unreliable for diagnosis or management.

† Less than 10% of normal enzyme activity is available for degrading ACh.

seizures, and coma.* Death from OP poisoning is secondary to respiratory paralysis and cardiovascular collapse.

Acute pancreatitis, myocardial dysrhythmias, and hydrocarbon pneumonitis following aspiration of the solvent vehicle are complications of OP poisoning. Although chronic effects from OP toxicity are limited, some latent pathology may develop. OP-induced delayed neuropathy (OPIDN) is a delayed neurotoxicity that may develop 1 to 3 weeks after exposure. The syndrome is characterized by muscular weakness and paralysis of extremities, especially of hand and foot muscles, progressing to a persistent spastic spinal paresis. OPIDN is associated with inhibition of the neuronal enzyme, neurotoxic esterase, and ACh-∑ "aging." The syndrome is either slowly reversible over several months or irreversible. Intermediate syndrome (IMS) is also characterized by muscular paralysis innervated by cranial nerves. It is associated with excessive exposure to OPs and results in prolonged ACh-∑ inhibition at the neuromuscular junction. The prevention, or recovery from development, of IMS is mostly accomplished by early therapeutic intervention following OP exposure.

28.2.5 Clinical Management of Acute Poisoning

Decontamination, airway stabilization, and activated charcoal are important initial supportive measures for OP intoxication. Washing dermal areas with a mild soap, removal of contaminated clothes, and rinsing the eyes are necessary for dermal or ocular exposure. Atropine and the oxime cholinesterase reactivators follow as specific OP antidotes.

Atropine is a competitive antimuscarinic cholinergic antagonist at central and peripheral autonomic receptors. It has no effect at neuromuscular or nicotinic receptors. It should be given intravenously (I.V.) (initially, adult 1–3 mg, child 0.02 mg/kg) and repeated in doubling doses at 5-minute intervals until bronchorrhea and bronchospasm are abolished and cardiovascular function is restored (blood pressure >80 mmHg, pulse >80 bpm). Once this is attained, a constant infusion of atropine is continued to sustain cardiorespiratory function. Patients with severe poisoning may require very large doses of atropine. However, the development of dry mouth, tachycardia, and dry skin is evidence of atropine intoxication.

Oximes, such as pralidoxime mesylate (P2S), pralidoxime chloride, and obidoxime, reactivate phosphorylated cholinesterases provided they are not given after "aging" has already occurred.† The oximes are quaternary amines that reactivate ACh-∑ by severing the OP—ACh-∑ covalent bond at nicotinic, muscarinic, and central cholinergic sites. The drugs also scavenge remaining OP molecules, they have few adverse reactions, and their actions are synergistic with those of atropine. An initial loading dose of 30 mg/kg body weight is administered by continuous slow I.V. infusion over 30 minutes and is followed by 8–10 mg/kg/h to maintain effective therapeutic levels of 4 μg/l. The subsequent infusion is continued until atropine has not been required for 12–24 hours.

* It is noteworthy that CNS depression is manifested more commonly in children than in adults.

† Pralidoxime is most effective when administered soon after exposure, before the poisoned enzyme complex has "aged" and become resistant to the effects of the antidote.

28.3 CARBAMATES

28.3.1 Chemical Characteristics, Occurrence, and Uses

As with the OP compounds, the carbamates are used as household and agricultural insecticides. The incidence of toxicity, routes of exposure, modes of encountering the agents, and distribution are similar to those of the OPs. Most carbamates display low dermal toxicity, except for aldicarb, which is highly toxic by both oral ingestion and dermal contact.

The mechanism of toxicity, however, differs slightly from that of the OP compounds. Table 28.2 summarizes the chemical properties and toxic animal data of some popular carbamate insecticides. Figure 28.2 illustrates the functional substitutions of two carbamate compounds from the carbamate parent molecule. Substitution of the N-isopropylthiol group and the heterocyclic phenyl constituents for R′ on the parent carbamate molecule on aldicarb and carbaryl, respectively, accounts for the decrease in toxicity (Table 28.2) seen with these insecticides.

28.3.2 Mechanism of Toxicity

Carbamates *reversibly* inhibit ACh-Σ by carbamoylation of the esteratic site, allowing the carbamate–enzyme complex to quickly and spontaneously dissociate (within minutes to hours). Thus, toxicity is limited, as the enzyme is capable of rapid regeneration after binding.

28.3.3 Signs and Symptoms of Acute Toxicity

Because of the reversibility of the substrate–enzyme complex, the toxicity of the carbamates is proportionately of shorter duration and less intensity than that of the OPs. Effects are limited to activation of muscarinic and nicotinic sites. In addition, poor penetration of the blood–brain barrier limits the CNS toxicity of carbamates.

28.3.4 Clinical Management of Acute Poisoning

As with the OPs, treatment of carbamate poisoning is symptomatic. Atropine is indicated as an antidote. Since "aging" of the carbamoylated complex does not occur, pralidoxime is not indicated and may actually increase toxicity. Practically, when prior medical history does not allow distinguishing OP from carbamate exposure, pralidoxime administration should not be withheld.

28.4 ORGANOCHLORINE (OC) INSECTICIDES

28.4.1 Chemical Characteristics

OC insecticides are low–molecular weight, fat-soluble compounds with greater selectivity for lipid storage in insects. As a group, these polychlorinated cyclic compounds are structurally similar, especially the stereo isomers dieldrin and endrin. Table 28.3 summarizes the chemical properties and human and animal toxicological profiles

TABLE 28.2
Names, Chemical Properties, and Rodent Toxicity of Carbamate Insecticides Arranged According to Rodent Oral LD_{50}

Common Name	Chemical Name or Synonym	Physical Characteristics	Median Oral Rodent LD_{50} (mg/kg)
Aldicarb	Temik	Crystalline solid, sol in water	~1.0
Carbofuran	Furadan	White crystalline solid, sol in water	2.0
Methiocarb	Mesurol	White crystalline solid, insol in water	60–70
Carbaryl	Sevin	Crystalline solid, sol in water	250
Bufencarb	Bux	Yellow solid, sol in xylene and alc	61–97
Propoxur	Baygon	Crystalline solid, sol in alc and org liq	85

alc: alcohol; insol: insoluble; liq: liquid; org: organic; sol: soluble.

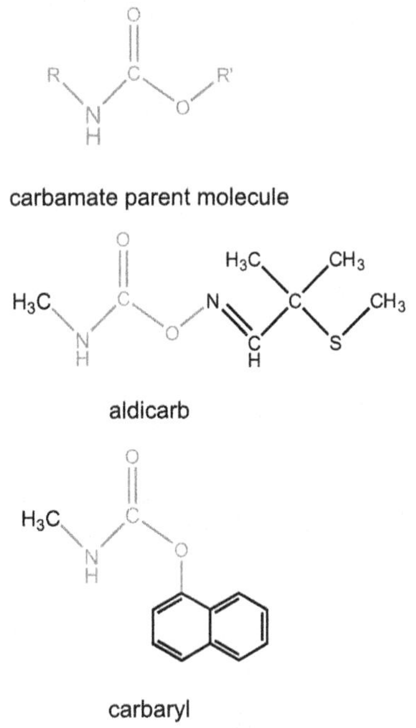

carbamate parent molecule

aldicarb

carbaryl

FIGURE 28.2 Structures of carbamate parent molecule and selective carbamate compounds are illustrated. Blue highlight indicates structural similarities retained from the parent molecule.

of some popular OC insecticides. Figure 28.3 illustrates the structural features of three representative OC compounds. The stereo configuration of endrin is noticeably different from the planar structures of lindane and DDT. The polychlorinated attachments on each of the compounds, however, are similar and are responsible for the insecticidal activity.

28.4.2 Occurrence and Uses

OC compounds have applications as insecticides in agriculture and as industrial and commercial pesticides, especially in mosquito abatement and red fire ant control. Because of their high lipid solubility, high carcinogenic potency, and potential for low-level accumulation with chronic exposure, some of these agents are no longer produced in the United States (endrin, aldrin, and dieldrin). DDT (dichlorodiphenyltrichloroethane; Table 28.3, Figure 28.3) was developed during World War II for the protection of armed forces against disease-carrying insects and is the best known of the chlorinated compounds. The persistence of DDT in the environment and the contribution of the insecticide to reduction of wildlife populations prompted the reevaluation of the utility of DDT and related chlorinated hydrocarbons. Heavy use of DDT in the United States before 1966 was probably responsible for a previously undetected epidemic of premature births. Epidemiological studies on humans since then have been suggestive of human reproductive toxicity of DDT. The pesticide is still widely used and highly effective in areas where mosquito-borne malaria is a major public health problem.

Gamma (γ)-benzene hexachloride (lindane; Figure 28.3), however, is currently used as a pediculocide, scabicide, and ectoparasiticide. It is available as a shampoo and lotion for the topical treatment of conditions due to itch mite (*Sarcoptes scabiei*), lice (*Pediculus humanus* sp.), and tick (dog and deer tick) infestations in humans and animals. The preparation is applied once for several minutes as a 1% lotion or shampoo, after which it is rinsed off. One reapplication may be required to eliminate the parasites.

28.4.3 Mechanism of Toxicity

From 1990 to 2000, over 30,000 poisonings resulting from OC exposure were reported in the United States. Most of the human toxicity data associated with OC is derived from clinical case studies of excessive inadvertent application, over-exposure, frequent reapplications, and accidental and intentional ingestion of lindane. Most exposures to lindane occur in infants, pregnant women, and nursing mothers. Overall, the OC compounds have proved to be selective neurotoxins. The toxic effects are greater in young victims since the compounds readily penetrate their immature skin. Prolonged contact with the agent, compromised skin, or application over a wide surface area also increases the chances for dermal absorption and systemic toxicity.

OCs alter membrane chloride ion permeability and interfere with normal GABA-ergic (γ-aminobutyric acid) neuronal transmission. The loss of this inhibitory neurotransmitter translates into the neurotoxic sequelae.

28.4.4 Signs and Symptoms of Acute Toxicity

Most of the toxic effects of OCs are due to CNS stimulation. Inhibition of GABA leads to motor, sensory, and behavioral changes, typically manifested as apprehension, irritability, confusion, sensory disturbances, dizziness, tremors, and seizures.

TABLE 28.3
Names, Chemical Properties, and Rodent Toxicity of Organochlorine Insecticides Arranged According to Rodent Oral LD_{50}

Common Name	Chemical Name or Synonym	Physical Characteristics	Toxicological Properties	Median Oral Rodent LD_{50} (mg/kg)
Endrin[a]	Mendrin, Endrex	Colorless to light tan solid, mild odor, sol in org liq	Stereoisomer of dieldrin	7.5–18
Aldrin[a]	Aldrite, Aldrex	Tan-dark brown solid, sol in org liq	Metabolized to dieldrin	39–60
Dieldrin[a]	Octalox, Dieldrex	Light tan solid, mild odor, mod sol in org liq	Toxicity similar to aldrin	40–50
Lindane	γ-benzene hexachloride; Kwell®	White crystalline solid, insol in water	Used therapeutically as a pediculocide; CNS toxicity	80–90
Heptachlor	Heptamul	White or light tan waxy solid; camphor odor	Blood dyscrasias, liver necrosis	100–160
DDT	Dichlorodiphenyl-trichloroethane	Colorless to white solid, mild odor, sol in org liq	CNS toxicity	110–120
Chlordane	Toxichlor, Velsicol 1068, Corodane, Belt	Colorless or amber viscous liquid, sol in org liq	Highly toxic orally, dermally, and by inhalation in humans	340
Methoxychlor	Methoxy-DDT	Crystalline solid	Low toxicity, lower than DDT	>5000

[a] Discontinued in the United States.
insol: insoluble; liq: liquid; mod: moderately; org: organic; sol: soluble.

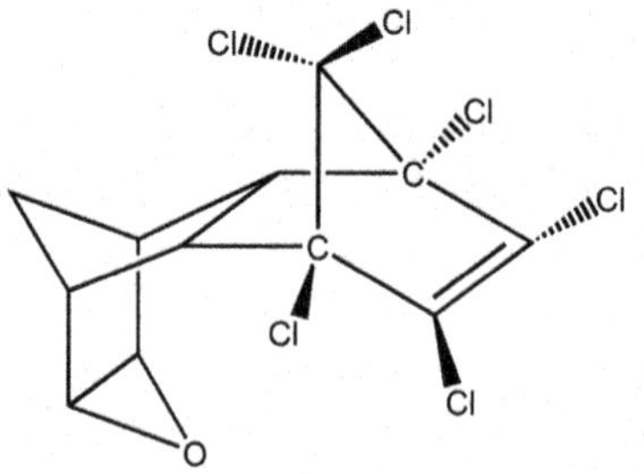

Endrin

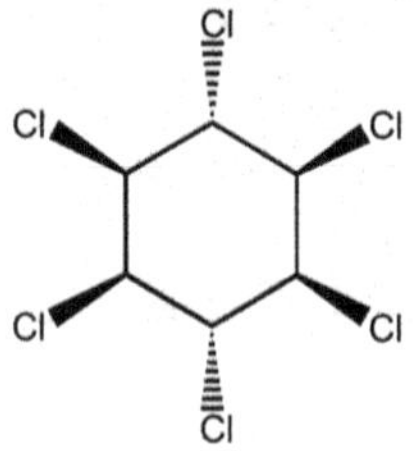

Lindane

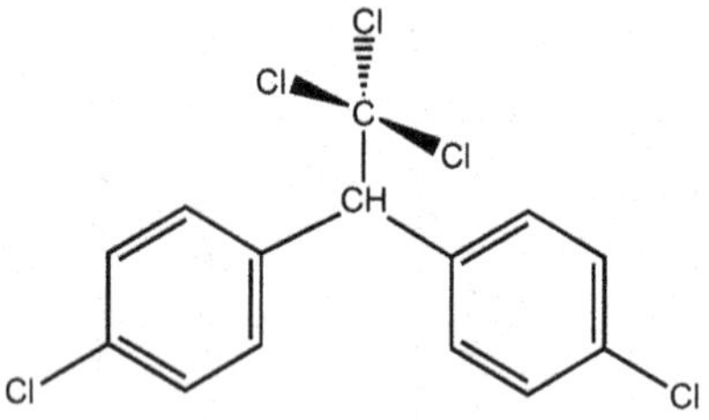

DDT

FIGURE 28.3 Structures of selected organochlorine insecticides. DDT = dichlorodiphenyltrichloroethane; lindane = γ-benzene hexachloride.

The chemicals sensitize involuntary autonomic muscle activity to catecholamines. In particular, myocardial arrhythmias are promptly precipitated. In addition, respiratory failure and hepatic and renal damage are consequences of ingesting a large quantity of lindane. As with the aliphatic and aromatic hydrocarbons, aspiration pneumonia is possible after inhalation of the organic vehicle delivering the insecticides.

28.4.5 Clinical Management of Acute Poisoning

Treatment involves symptomatic and supportive management of the condition. Decontamination, gastric lavage, and administration of activated charcoal reduce toxicity after oral ingestion. Washing of the affected area may reduce absorption after dermal exposure. Myocardial arrhythmias are managed with antiarrhythmics such as lidocaine, while benzodiazepines are indicated for preventing or reducing the development of seizures.

28.5 MISCELLANEOUS INSECTICIDES

28.5.1 Pyrethroid Esters

The pyrethroid insecticides are derived from the naturally occurring compound pyrethrum, from the dried flower heads of the yellow flower *Chrysanthemum cineriaefolium.** The synthetic acid-alcoholic esters are categorized into several classes of the active ingredients; namely, pyrethrins types I and II.† Pyrethrum flowers have been used as insecticides for centuries, particularly by Caucasian tribesmen and Armenians. The powdered form was introduced into the United States in 1855, after which its importation expanded tremendously. Type I pyrethrins include allethrine, permethrin, and cismethrin; type II pyrethrins include fenvalerate, deltamethrin, and cypermethrin. The compounds (0.17–0.33%) are combined with piperonyl butoxide or n-octyl-dicycloheptane dicarboximide (2–4%) in therapeutic nonprescription pediculocide preparations for the treatment of lice, tick, and mite infestation.‡ Pyrethrins are noted for their quick "knock-down" effect on flying insects, particularly flies and mosquitoes. The products are available in lotions, sprays, and shampoos for skin or scalp applications as well as for removal from furniture and bedding material.

Type I pyrethrins produce repetitive depolarization of axons by inhibiting inactivation of sodium channels. Type II pyrethrins have a similar mechanism but a longer duration of action and also affect GABA receptor–mediated chloride channels. Most poisonings, however, are benign and limited to contact irritant dermatitis, allergic dermatitis, and rhinitis. Pulmonary asthmatic reactions may occur in sensitive individuals. Some cases of severe poisoning have been reported from massive oral ingestion that resulted in coma, convulsions, and death.

Treatment is symptomatic and limited to alleviation of the inflammatory response. Oral or topical corticosteroids and H_1-antihistamine blockers may be of use as anti-inflammatory agents.

28.5.2 Nicotine

Nicotine is a pyridine alkaloid (1-methyl-2-[3-pyridyl] pyrrolidine) obtained from the cured and dried leaves of the tobacco plant, *Nicotiana* sp. The plant is indigenous to the southern United States and tropical South America (see Chapter 15 for a detailed description). Early American settlers in the nineteenth century recognized the insecticidal properties of nicotine, as they dusted their vegetable crops with finely powdered tobacco leaves. The leaves contain from 0.6 to 9% of the alkaloid along with other nicotinic derivatives. The demand for nicotine as an insecticide declined with the advent of the OP compounds.

---

* *Chrysanthemum* is from the ancient Greek name meaning golden flower; *cineriaefolium* is from two Greek words meaning ash-colored leaves.

† Cinerin I or II, and jasmolin I or II, are synonyms.

‡ These agents, found in commercial products RID® and NIX®, inhibit enzyme deactivation of the pyrethrins, thereby producing a synergistic effect with the insecticide.

The toxic effects of nicotine as a contact insecticide are identical to the effects of ingestion of large amounts of the compound in other commercial products. Nicotine stimulates nicotinic receptors of all sympathetic and parasympathetic ganglia, neuromuscular junction innervating skeletal muscle, and CNS pathways. Of the target organs affected by nicotine, the most significant is the cardiovascular system. Nicotine produces characteristic bradycardia or tachycardia. Nicotine's action on the CNS results in stimulation followed by predominance of parasympathetic overtone—salivation, lacrimation, urination, defecation, and vomiting. Continued exposure progresses to muscular weakness, tremors, hypotension, and dyspnea. Convulsions and respiratory paralysis are advanced complications of unattended nicotine toxicity. Treatment of poisoning is primarily symptomatic and involves decontamination, induction of emesis (depending on the time of onset of symptoms), and maintenance of vital signs.

28.5.3 Boric Acid (H_3BO_3)

Boric acid has numerous traditional therapeutic and commercial uses as a topical astringent (for minor swelling), as an ophthalmic bacteriostatic agent (in over-the-counter eye wash solutions), and as a topical antiseptic, herbicide, and insecticide. As a commercial, nontherapeutic product, the substance is an effective detergent, cleaning aid, flame retardant, herbicide, and insecticide, primarily useful as an "ant and roach killer." The undiluted powder is mixed in water with more palatable substances, such as flour or granular sugar, to attract the bugs. Consequently, its easy availability, accessibility, and similarity to products such as infant formulas have facilitated its ability to interchange with these substances. This ease of confusion with other products accounts for the persistent number of poisoning cases. It has also been inadvertently substituted in solution for Epsom salts (magnesium sulfate powder) and mistakenly used as a muscle relaxant.

Topically, concentrated solutions of boric acid precipitate the formation of a desquamating, erythematous rash (*boiled lobster* appearance), characterized by the production of severe redness, pain, and blisters. The chemical readily penetrates abraded skin and along with oral ingestion, has the potential for systemic toxicity. Symptoms of systemic toxicity include lethargy, fever, and muscular weakness with progression to development of tremors and convulsions. Individuals who develop hypotension, CNS depression, renal failure, jaundice, and cardiovascular collapse are at risk of death.

As with nicotine, treatment of boric acid poisoning is primarily symptomatic and involves decontamination, gastric lavage, and maintenance of vital signs. Washing the contaminated area with soap and water effectively neutralizes its acidic reaction. Induction of emesis, however, is not recommended, since regurgitation of acidic material contributes to further esophageal corrosion.

28.5.4 Rotenone

Rotenone (tubotoxin, derrin) is a colorless-to-red, odorless insecticidal principle derived from the *Derris* plant genus.* The plant is predominantly grown in the

* *Derris* roots were used as fish poisons for centuries by native tribes of Central and South America.

southern United States and tropical South America. Rotenone and the chemically related rotenoids (toxicarol, tephrosin, and sumatrol) are a series of naturally occurring cyclic aromatic hydrocarbons. The agents are widely employed in agriculture for controlling chewing and sucking insects and are dusted or sprayed on garden plants and crops as well.

The action of rotenone parallels that of the pyrethrins in providing quick *knock-down* of flying insects. Although it is considered to possess generally low toxicity, respiratory, dermal, oral, or ocular exposure results in symptoms that mimic the chemical irritants (Chapter 25). Pulmonary and systemic toxicity leading to respiratory depression, seizures, and coma have been reported. Management of poisoning is supportive and symptomatic.

28.5.5 Diethyltoluamide (DEET)

DEET (the *m*-isomer of *N,N*-diethyl-3-methylbenzamide) is a popular insect repellant available in topical preparations at 5% to 100% concentrations (OFF® Insect Repellant). About 5–10% of a dermal application is absorbed due to its high lipid solubility and stored in lipid compartments, resulting in a prolonged plasma half-life (2.5 hours). Although it is generally considered a chemical with low toxicity, dermal reactions reflect excessive topical application of sprays, creams, lotions, or alcohol-soaked DEET-containing towelettes. Ocular or dermal irritation is generally limited to allergic reactions and is the most frequent complaint. Prolonged dermal contact with significant absorption, or oral ingestion, has precipitated CNS toxicity. This is manifested by the development of headache, lethargy, confusion, and tremors. Hypotension, seizures, and coma are rare.

Treatment is largely symptomatic and supportive and involves decontamination, gastric lavage, and maintenance of vital signs, if necessary. Washing the contaminated area with soap and water and induction of emesis are recommended if needed.

28.6 METHODS OF DETECTION

As with the aliphatic and aromatic hydrocarbons, quantitative analysis of insecticides is usually performed on samples from ambient or contaminated areas, including air, water, and soil. Gas chromatographic (GC) techniques are sensitive and specific depending on the internal and external standards used. Although most of the insecticides have low vapor pressures, the volatility of the liquids, coupled with a suitable detector, renders them suitable for monitoring through a GC column (solid phase). Most of the solid insecticides are soluble in organic liquids, which also makes them amenable to GC analysis. Flame ionization, electron capture, mass spectrometry (GC-MS), and nitrogen-specific detectors have been employed for specific identification.

REFERENCES

SUGGESTED READINGS

Albertson, T.E. and Ferguson, T.J., Chapter 46. Insecticides: chlorinated hydrocarbons, pyrethrins, and DEET, in *Toxicology Secrets*, Ling, L.J., Clark, R.F., Erickson, T.B., and Trestrail, J.H. (Eds.), Hanley & Belfus, Philadelphia, PA, 2001.

Boobis, A.R., Ossendorp, B.C., Banasiak, U., Hamey, P.Y., Sebestyen, I., and Moretto, A., Cumulative risk assessment of pesticide residues in food. *Toxicol. Lett.* 180, 137, 2008.

Burns, C.J., McIntosh, L.J., Mink, P.J., Jurek, A.M., and Li, A.A., Pesticide exposure and neurodevelopmental outcomes: review of the epidemiologic and animal studies. *J. Toxicol. Environ. Health B Crit. Rev.* 16, 127, 2013.

Chomchai, S., Chapter 45. Insecticides: organophosphates and carbamates, in *Toxicology Secrets*, Ling, L.J., Clark, R.F., Erickson, T.B., and Trestrail, J.H. (Eds.), Hanley & Belfus, Philadelphia, PA, 2001.

Costa, L.G., Giordano, G., Guizzetti, M., and Vitalone, A., Neurotoxicity of pesticides: a brief review. *Front. Biosci.* 13, 1240, 2008.

CRC Handbook of Chemistry & Physics, W.M. Hayes (Ed.), 93rd ed., 2012.

Dawson, A.H., Eddleston, M., Senarathna, L., Mohamed, F., Gawarammana, I., Bowe, S.J., Manuweera, G., and Buckley, N.A., Acute human lethal toxicity of agricultural pesticides: a prospective cohort study. *PLoS Med.* 7, e1000357, 2010.

Gangemi, S., Miozzi, E., Teodoro, M., Briguglio, G., De Luca, A., Alibrando, C., Polito, I., and Libra, M., Occupational exposure to pesticides as a possible risk factor for the development of chronic diseases in humans. *Mol. Med. Rep.* 14, 4475, 2016.

Goldman, L.R., Managing pesticide chronic health risks: U.S. policies. *J. Agromedicine* 12, 67, 2007.

Hatcher, J.M., Pennell, K.D., and Miller, G.W., Parkinson's disease and pesticides: a toxicological perspective. *Trends Pharmacol. Sci.* 29, 322, 2008.

Katz, T.M., Miller, J.H., and Hebert, A.A., Insect repellents: historical perspectives and new developments. *J. Am. Acad. Dermatol.* 58, 865, 2008.

Leonard, P., Gianessi, L.P., and Marcelli, M.B., *Pesticide Use in U.S. Crop Production: 1997 National Summary Report*, National Center for Food and Agricultural Policy (NCFAP), Washington, DC, 2000.

London, L., Beseler, C., Bouchard, M.F., Bellinger, D.C., Colosio, C., Grandjean, P., Harari, R., Kootbodien, T., Kromhout, H., Little, F., Meijster, T., Moretto, A., Rohlman, D.S., and Stallones, L., Neurobehavioural and neurodevelopmental effects of pesticide exposures. *Neurotoxicology* 33, 887, 2012.

The Merck Index: An Encyclopedia of Chemicals, Drugs, and Biologicals, Maryadele, J. (Ed.), 15th ed., Merck, Whitehouse Station, NJ, 2013.

Moser, V.C., Animal models of chronic pesticide neurotoxicity. *Hum. Exp. Toxicol.* 26, 321, 2007

Narahashi, T., Zhao, X., Ikeda, T., Nagata, K., and Yeh, J.Z., Differential actions of insecticides on target sites: basis for selective toxicity. *Hum. Exp. Toxicol.* 26, 361, 2007.

Patnaik, P., *Comprehensive Guide to Hazardous Properties of Chemical Substances*, 3rd ed., Van Nostrand Reinhold, 2007.

Sadasivaiah, S., Tozan, Y., and Breman, J.G., Dichlorodiphenyltrichloroethane (DDT) for indoor residual spraying in Africa: how can it be used for malaria control? *Am. J. Trop. Med. Hyg.* 77, 249, 2007.

U.S. Environmental Protection Agency (U.S. EPA), Chapter 5. Organophosphates, in *Recognition and Management of Pesticide Poisonings*, 6th ed., Roberts, J.R. and Reigart, J.R. (Eds.), 2013.

REVIEW ARTICLES

Abu-Qare, A.W. and Abou-Donia, M.B., Combined exposure to DEET (N,N-diethyl-m-toluamide) and permethrin: pharmacokinetics and toxicological effects. *J. Toxicol. Environ. Health B. Crit. Rev.* 6, 41, 2003.

Ahlborg, U.G., Lipworth, L., Titus-Ernstoff, L., Hsieh, C.C., Hanberg, A., Baron, J., Trichopoulos, D., and Adami, H.O., Organochlorine compounds in relation to breast cancer, endometrial cancer, and endometriosis: an assessment of the biological and epidemiological evidence. *Crit. Rev. Toxicol.* 25, 463, 1995.

Banerjee, B.D., The influence of various factors on immune toxicity assessment of pesticide chemicals. *Toxicol. Lett.* 107, 21, 1999.

Damalas, C.A. and Eleftherohorinos, I.G., Pesticide exposure, safety issues, and risk assessment indicators. *Int. J. Environ. Res. Public Health* 8, 1402, 2011.

DeLorenzo, M.E., Scott, G.I., and Ross, P.E., Toxicity of pesticides to aquatic microorganisms: a review. *Environ. Toxicol. Chem.* 20, 84, 2001.

Gunasekara, A.S., Rubin, A.L., Goh, K.S., Spurlock, F.C., and Tjeerdema, R.S., Environmental fate and toxicology of carbaryl. *Rev. Environ. Contam. Toxicol.* 196, 95, 2008.

He, F., Neurotoxic effects of insecticides–current and future research: a review. *Neurotoxicology* 21, 829, 2000.

Karalliedde, L.D., Edwards, P., and Marrs, T.C., Variables influencing the toxic response to organophosphates in humans. *Food Chem. Toxicol.* 41, 1, 2003.

Kassa, J., Review of oximes in the antidotal treatment of poisoning by organophosphorus nerve agents. *J. Toxicol. Clin. Toxicol.* 40, 803, 2002.

Kwong, T.C., Organophosphate pesticides: biochemistry and clinical toxicology. *Ther. Drug Monit.* 24, 144, 2002.

Lotti, M., Low-level exposures to organophosphorus esters and peripheral nerve function. *Muscle Nerve* 25, 492, 2002.

Murphy, M.J., Rodenticides. *Vet. Clin. North Am. Small Anim. Pract.* 32, 469, 2002.

Nakajima, M. Biochemical toxicology of lindane and its analogs. *J. Environ. Sci. Health B.* 18, 147, 1983.

O'Malley, M., Clinical evaluation of pesticide exposure and poisonings. *Lancet* 349, 1161, 1997.

Ragoucy-Sengler, C., Tracqui, A., Chavonnet, A., Daijardin, J.B., Simonetti, M., Kintz, P., and Pileire, B., Aldicarb poisoning. *Hum. Exp. Toxicol.* 19, 657, 2000.

Ray, D.E. and Forshaw, P.J., Pyrethroid insecticides: poisoning syndromes, synergies, and therapy. *J. Toxicol. Clin. Toxicol.* 38, 95, 2000.

Reigart, J.R. and Roberts, J.R., Pesticides in children. *Pediatr. Clin. North Am.* 48, 1185, 2001.

Walker, B. Jr. and Nidiry, J., Current concepts: organophosphate toxicity. *Inhal. Toxicol.* 14, 975, 2002.

Weselak, M., Arbuckle, T.E., and Foster, W., Pesticide exposures and developmental outcomes: the epidemiological evidence. *J. Toxicol. Environ. Health B. Crit. Rev.* 10, 41, 2007.

Wolansky, M.J. and Harrill, J.A., Neurobehavioral toxicology of pyrethroid insecticides in adult animals: a critical review. *Neurotoxicol. Teratol.* 30, 55, 2008.

Yang, C.C. and Deng, J.F., Intermediate syndrome following organophosphate insecticide poisoning. *J. Chin. Med. Assoc.* 70, 467, 2007.

Ye, M., Beach, J., Martin, J.W., and Senthilselvan, A. Occupational pesticide exposures and respiratory health. *Int. J. Environ. Res. Public Health* 10, 6442, 2013.

29 Herbicides

29.1 INTRODUCTION

Herbicides constitute a wide variety of chemicals, whose main classes include:

1. Chlorphenoxy acids and their esters
2. Triazines
3. Substituted ureas
4. Dipyridyl derivatives
5. Mono- or dinitro aromatics

In the agricultural industry, the control of weeds and noxious vegetation is accomplished by the application of various contact herbicides that produce the rapid elimination of plant tissues for prolonged periods. Military use of defoliants, and commercial nonagricultural community and household use of herbicides along highways, commercial property, pedestrian walkways, parks, and lawns, accounted for 40% of the total herbicide consumption in the United States in 2001. Domestic maintenance of home lawns and gardens requires the use of selective herbicides that kill broad-leaf vegetation without serious damage to the preferred grasses.

Although the acute and chronic toxicity of the herbicides is generally low, acute contact dermatitis and chronic exposure account for a number of nonfatal poisonings from widespread application of the compounds. The number of cases average about several hundred annually. As with the insecticides, herbicide poisonings result from inadvertent ingestion of herbicide applicators and solutions for home use as well as dermal and respiratory exposure from industrial occupational exposure.

29.2 CHLORPHENOXY COMPOUNDS

29.2.1 Chemical Characteristics

Most of the chlorphenoxy compounds are odorless, colorless, crystalline compounds with low vapor pressures. The herbicides are effective against broad-leafed weeds and plants and some annual grasses. They are conveniently divided into selective and nonselective classes:

1. *Selective* herbicides eliminate undesirable plant species but produce little deleterious effects on other plants in the contact area.
2. *Nonselective* herbicides destroy all plant life within the zone of application.

Table 29.1 outlines the classes, chemical properties, and general human-toxic characteristics of the chlorphenoxy herbicides.

TABLE 29.1
Names, Chemical Properties, and Human Toxicity of Chlorphenoxy and Bipyridyl Contact Herbicides Arranged According to Oral Rodent LD_{50}

Common Name	Chemical Name	Physical Characteristics	Toxicological Effects	Median Oral Rodent LD_{50} (mg/kg)
		Chlorphenoxy Acids		
TCDD (dioxin)	2,3,7,8-tetrachloro-dibenzo-*p*-dioxin	Colorless to tan solid; banned by EPA	*Dermal exposure*: irritation, acne-like rash (chloracne, esp. with dioxin);*respiratory exposure*: NV, lethargy, stupor, weakness, somnolence, ataxia, hepatic and gastric disturbances, death from CV arrhythmias	0.22–0.45
2,4-D	2,4-dichlorophenoxy acetic acid	White to yellow crystalline powder		100–500
2,4,5-T	2,4,5-trichlorophenoxy acetic acid	Colorless to tan solid; banned by EPA (1985)		300–500
Silvex	2-(2,4,5-trichlorophenoxy) propionic acid	Crystalline solid, banned by EPA (1985)		650
		Bipyridyl Derivatives		
Paraquat	1,1′-dimethyl-4,4′-bipyridinium; methylviologen	Colorless to yellow crystals	*Ingestion*, *inhalation*: respiratory irritant, pulmonary edema, interstitial fibrosis	100–125
Diquat (aquacide)	1,1′-ethylene-2,2′-bipyridilium	Colorless to yellow crystals	Low toxicity in humans	125–230
Picloram (amodon)	4-amino-3,5,6-trichloropicolinic acid	White, crystalline powder	Mild toxicity in animals	>2500

29.2.2 Occurrence and Uses

Structurally, the chlorophenoxy herbicides consist of a simple aliphatic carboxylic acid moiety attached to a chlorine- or methyl-substituted aromatic ring through an ether bond. The mechanism of neurotoxicity is not known.

2,4-D (2,4-dichlorophenoxy acetic acid) was originally introduced in the 1940s as a plant growth stimulant (*auxin-like* compound) only later to retain the distinction of being the first organic herbicide. Its residues have been detected in soil, sediments, groundwater, and marine estuaries. Similarly, 2,4,5-T (2,4,5-trichlorophenoxy acetic acid) and its contaminant TCDD (2,3,4,8-tetrachloro-dibenzo-*p*-dioxin [dioxin]) are potent, highly toxic agents and more effective in eradicating woody plants.* The structural features of 2,4,5-T and TCDD (dioxin) are highlighted in Figure 29.1.

29.2.3 Signs and Symptoms and Mechanism of Acute Toxicity

Although the mechanism is not clear, the toxicity of these compounds increases with decreasing phenoxy side chain length. Symptoms are similar for all chlorphenoxy compounds and are usually mild after a single exposure. There is no apparent

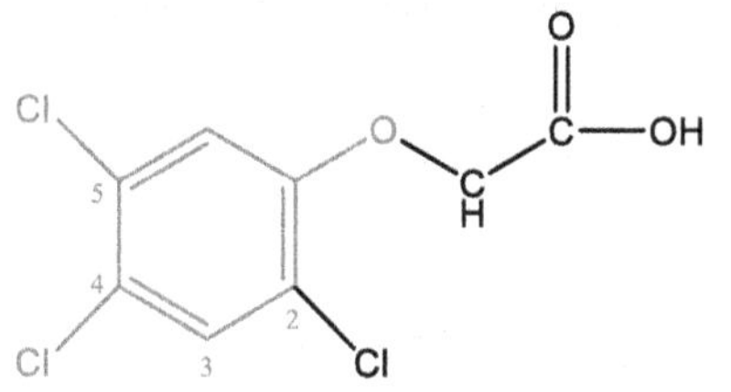

2,4,5-T

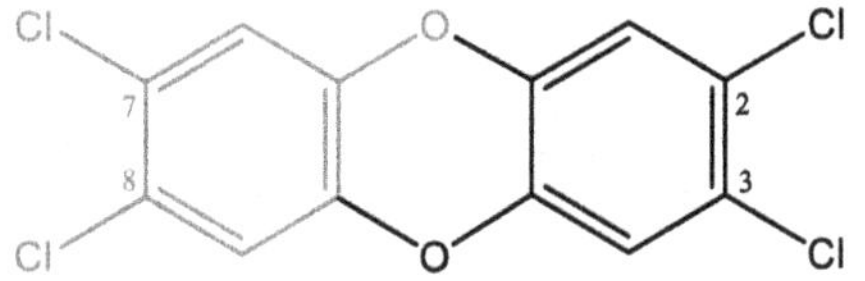

2,3,7,8-TCDD (dioxin)

FIGURE 29.1 Structures of the chlorphenoxy herbicides—2,4,5-T and TCDD. These representative compounds retain their structural similarities, the chlorphenoxy substituents, highlighted in light shade; 2,4,5-T = 2,4,5-trichlorophenoxy acetic acid.

* In the late 1960s, over 11 million gallons of "Agent Orange" was used as a defoliant during the Vietnam war. The substance contained a mixture of 2,4-D and 2,4,5-T and was inadvertently contaminated with trace amounts of dioxin. Exposure to dioxin during the defoliation missions of the war (Operation Ranchland) prompted veterans exposed to the defoliant to file class action lawsuits in the 1970s against the U.S. government. The chemical demonstrated carcinogenic and teratogenic properties in animals on chronic exposure. Epidemiological studies as to dioxin's chronic effects on human health are inconclusive.

compartmental accumulation. Toxic effects are related to their herbicidal action and depend on their translocation to all parts of the plants to which they are applied. Inorganic salts and aliphatic esters of 2,4-D ionize readily in aqueous solution and are not absorbed into the plant sap. Generally, they remain on the leaf surface (*contact* herbicides), thus ensuring sufficient amounts for local human and mammalian contact. Alcoholic esters, however, penetrate the waxy cuticle surface and are translocated to the plant's conducting system (*translocated* herbicides). These agents are considered more efficient herbicides and transfer less toxicity on contact with humans or animals.

Toxicity from commercial applications is a direct result of dermal absorption, inhalation, or oral ingestion. Acute toxic effects from single oral ingestions are listed in Table 29.1 (oral ingestion of 4 to 6 g is potentially fatal in humans). Nausea, vomiting, muscular weakness, and hypotension are the most notable symptoms. Chloracne, a severe form of dermatitis, is characterized by small, black follicular papules on the arms, face, and neck of exposed individuals in contact with dioxin contaminants. In addition, toxicity from the petroleum distillate solvent vehicles is associated with significant morbidity (see Chapter 25).

29.2.4 Clinical Management of Acute Poisoning

Decontamination and washing of exposed skin areas with mild soap neutralize the acidic properties. Gastric lavage is of unknown effectiveness. Eye rinsing is important for ocular exposure. Airway stabilization and activated charcoal are important initial supportive measures if necessary. Since the agents have generally low pKa (acidic) values, alkaline diuresis may enhance renal elimination. This is achieved by infusion of isotonic sodium bicarbonate to raise urine pH to 8, if possible, while frequently monitoring urine pH. Unfortunately, no trials have yet been performed to assess the effectiveness of this approach.

29.3 BIPYRIDYL HERBICIDES

29.3.1 Chemical Characteristics, Occurrence, and Uses

Paraquat (PQ) and diquat are nonselective bipyridyl herbicides used widely in agricultural and commercial residential applications (lawn maintenance) to eradicate broad-leaf plants and shrubs. The chemicals inhibit plant photosynthesis by interfering with $NADPH/NADP^+$ redox cycling. Their usefulness as contact herbicides resides in their ability to promote the reseeding of lawns and gardens within 24 hours after application. The two structurally similar compounds are often combined in commercial products. The acute toxicity of diquat, however, is lower than that of than PQ in humans (Figure 29.2).

29.3.2 Toxicokinetics

PQ is poorly absorbed through the skin and by inhalation, although both contribute to significant accumulation via systemic absorption. Most of the compound

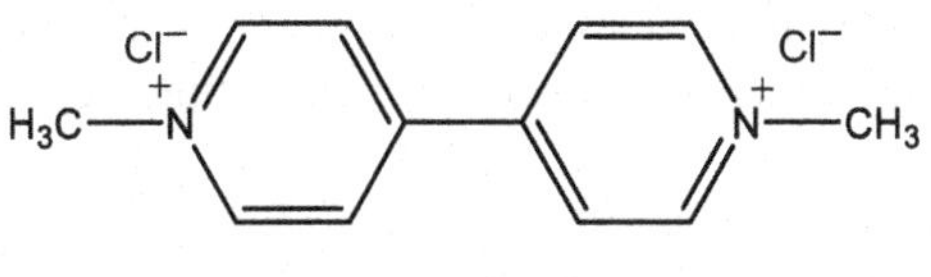

paraquat hydrochloride

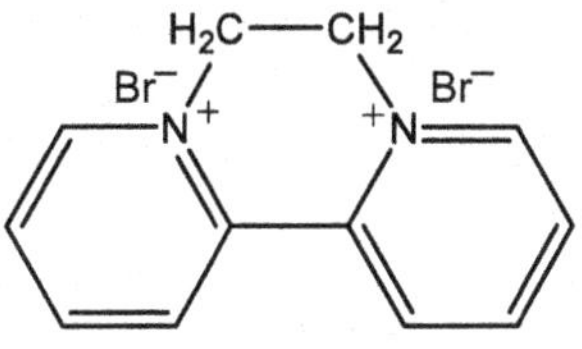

diquat dibromide

FIGURE 29.2 Structures of the bipyridal herbicides—paraquat and diquat.

remains at the site of exposure with dermal contact, whereas only about 10% is absorbed orally.

29.3.3 Mechanism of Toxicity

The toxicity of PQ was first recognized when U.S. Drug Enforcement Administration (DEA) programs were commissioned for the purpose of eradicating marijuana plants grown illicitly in the southwestern United States. Using aerial reconnaissance, the herbicide was sprayed over fields suspected of devoting large areas to marijuana cultivation. Undaunted by the antidrug program, individuals who smoked surviving marijuana plants inhaled significant amounts of the herbicide. The associated pulmonary manifestations were categorically identified with PQ toxicity.

PQ is extremely toxic to humans and laboratory animals; the mechanism of this toxicity is illustrated in the following reaction. The oxidized ($PQ_{[O]}$) compound accumulates in the lungs and kidneys and undergoes redox cycling reactions (single electron reduction). It is metabolized by nicotinamide adenine dinucleotide phosphate (NADPH)-cytochrome P-450-dependent reductase to the reduced ($PQ_{[H]}$) intermediate radical:

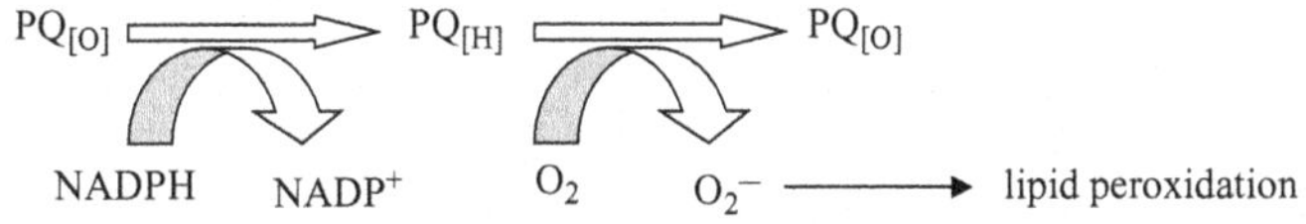

$$PQ_{[O]} \xrightarrow{NADPH \rightarrow NADP^+} PQ_{[H]} \xrightarrow{O_2 \rightarrow O_2^-} PQ_{[O]}$$

$O_2^- \longrightarrow$ lipid peroxidation

In turn, the reaction catalyzes the oxidation of NADPH to $NADP^+$, resulting in a depletion of the reducing equivalent. With continued exposure to the herbicide, there is an eventual decrease in the $NADPH/NADP^+$ ratio, thus depriving the cells of the protective role of reducing equivalents. The reduced intermediate ($PQ_{[H]}$) reacts with molecular oxygen to reform the oxidized PQ species (known as the *single electron reduction of oxygen*). The intracellular product of oxygen reduction is the superoxide anion radical (O_2^-). Through a subsequent series of catalyzed reactions, O_2^- may form hydroxyl radicals, which are capable of causing lipid peroxidation and damage to vital cell membranes.

Glutathione (GSH) and superoxide dismutase also play a role as cellular defense factors against reactive oxygen species generated in tissues. GSH and superoxide dismutase depletion selectively enhance PQ-induced oxidative stress by further inhibiting the cell's ability to prevent the damaging effects of free radical generation. In addition, dismutated oxygen ($O^{\bullet}$, singlet O_2) accumulates from O_2^-, producing an excess of hydroperoxides.* This precipitates a series of potentiating reactions undermining the stability of unsaturated lipids within cell membranes.

29.3.4 Signs and Symptoms of Acute Toxicity

Because of its interaction with pulmonary reactive tissue at the site of exposure, PQ selectively concentrates in alveolar type I and type II pneumocytes as well as in renal epithelial cells. The effect of the chemical is characterized as a "hit-and-run" event since the redox cycling and free radical formation occur after the toxin is eliminated. Thus, three phases of PQ poisoning are described based on the dose and route of exposure, outlined in Table 29.2. The initial clinical picture in patients who have ingested small doses and survive for 3–4 days is dominated by severe, painful ulceration of the lips, tongue, pharynx, and larynx, leading to dysphagia, cough, and inability to clear saliva and other secretions. Although death is typically delayed for 2 to 3 weeks in Phase II and up to 1 week in Phase III, the majority of cases that reach these phases result in death. Acute alveolitis and pulmonary fibrosis, characterized by cough, dyspnea, tachypnea, edema, atalectasis, collagen deposition, and fibrotic development, are eventual consequences of Phases II and III.

29.3.5 Clinical Management of Acute Poisoning

The development of pulmonary sequelae is associated with a poor prognosis. In general, PQ poisoning is managed aggressively with symptomatic and supportive care. Evaluation of esophageal erosion and determination of serum and gastric fluid levels of PQ are beneficial in predicting outcome. Activated charcoal, gastric lavage, and/or administration of a cathartic may prevent further absorption. Forced diuresis and hydration are effective only if intervention is attempted soon after ingestion and in the presence of lower doses of the herbicide. Although high-flow oxygen is generally used to counteract hypoxia, its use risks further acceleration of oxidative lung damage. Since in most cases, peak plasma concentrations of PQ are attained before the patient reaches the emergency department, treatment is unlikely to be effective. In general, whole bowel irrigation, forced diuresis, hemodialysis, hemoperfusion, surgical approaches, radiotherapy, and most pharmacological interventions do not appear to delay or ameliorate the consequences.

---

* Reactive oxygen species (ROS).

TABLE 29.2
Phases of Paraquat Toxicity and Associated Clinical Effects

Phases of Toxicity	Time Range (days)	Minimum Dose (mg/kg)	Toxic Effects: Oral Ingestion	Toxic Effects: Local Exposure
I. Asymptomatic or mild	1–5	20[a]	Nausea, vomiting, diarrhea; ulceration; intestinal hemorrhage, hemoptysis; oliguria	GI, dermal and ocular irritation
II. Moderate to severe	2–8	20–40	Vomiting, diarrhea; systemic accumulation, pulmonary fibrosis	Severe GI, dermal, and ocular irritation, inflammation, and ulceration of skin and mucous membranes
III. Severe: acute fulminant toxicity	3–14	>40	Liver, kidney, cardiac, pulmonary failure	Marked ulceration as with Phase II

[a] Doses as low as 4 mg/kg have resulted in death.
GI: gastrointestinal.

29.4 MISCELLANEOUS HERBICIDES

Table 29.3 summarizes the chemical and toxicological properties of selected miscellaneous herbicides. The triazines, substituted ureas, and nitroaromatic and chloroanilide classes are frequently used as contact, preemergence, and select herbicides. Their low to moderate toxicity to humans and animals makes them suitable for agricultural, industrial, and household utility. Their structural commonalities (highlighted in blue in Figure 29.3), such as the nitrogen moieties, explain their relative effectiveness and selective toxicology (Figure 29.3).

29.5 METHODS OF DETECTION

Gas chromatography-mass spectrometry (GC-MS), liquid chromatography (LC), high-performance liquid chromatography (HPLC), and immunoassay techniques are used to identify herbicides in air, soil, and biological samples. Most of the solids are soluble in organic liquids, which renders them suitable for monitoring through a packed GC column (solid phase). Flame ionization, electron capture, or mass spectrometry (MS) are suitable detectors with sensitivities in the micromolar range. Recently, several methods have been suggested for biomonitoring for exposure to herbicides or their metabolites, including methods incorporating conjugation or derivatization in conjunction with HPLC-electrochemical detection (HPLC-EC), HPLC-diode array detection (HPLC-DAD), HPLC/thermospray (TSP)-MS, and HPLC-MS/MS. The limits of detection of these methods are 25, 0.2, 200, and 0.5 ng/ml, respectively. Although the HPLC-MS/MS method has a slightly higher limit of detection (LOD) than the HPLC-DAD method, it offers a higher degree of selectivity.

TABLE 29.3
Names, Chemical Properties, and Human and Animal Toxicity of Triazines, Substituted Ureas, Nitroaromatic, and Miscellaneous Contact Herbicides

Common Name	Chemical Name	Physical Characteristics	Toxicological Effects	Median Oral Rodent LD_{50} (mg/kg)
Triazines				
Atrazine (Atranex, Primatol A)	2-chloro-4-ethylamino-6-isopropylamino-*s*-triazine	White, crystalline solid	Low toxicity in humans and animals: anemia, weakness, enteritis	1750
Substituted Ureas				
Tribunil (methabenz-thiazuron)	1,3-dimethyl-3-(2-benzo-thiazolyl) urea	White, crystalline solid	Pulmonary congestion in animals	>2500
Monuron (Monurex)	3-(4-chlorophenyl)-1,1-dimethyl urea	Crystalline solid, faint odor	Moderate toxicity, carcinogenic in laboratory animals	3700
Mono- and Dinitro Aromatics				
Nitrofen (TOK)	2,4-dichloro-1-(4-nitrophenoxy) benzene	White, crystalline powder	Highly toxic in laboratory animals, carcinogenic in laboratory animals	0.6
Dinoseb (Caldon)	2-sec-butyl-4,6-dinitrophenol	Orange-brown viscous liquid	Used as herbicide and insecticide, highly toxic in laboratory animals	10–30
Miscellaneous				
Alachlor (lazo)	2-chloro-2,6-diethyl-*N*-(methoxymethyl) acetanilide	Crystalline solid	Low toxicity in humans and animals, oncogenic in laboratory animals	1200

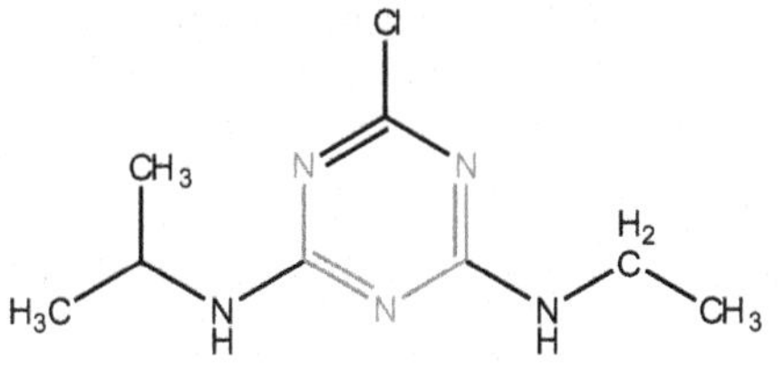

FIGURE 29.3 Structures of selected triazine, substituted urea, and nitro aromatic herbicides. Their structural similarities, highlighted in blue, such as the triazine, urea, and dinitro moieties (for atrazine, monuron, and dinoseb, respectively), account for the relative effectiveness and selective toxicity of nitrogen-containing molecules.

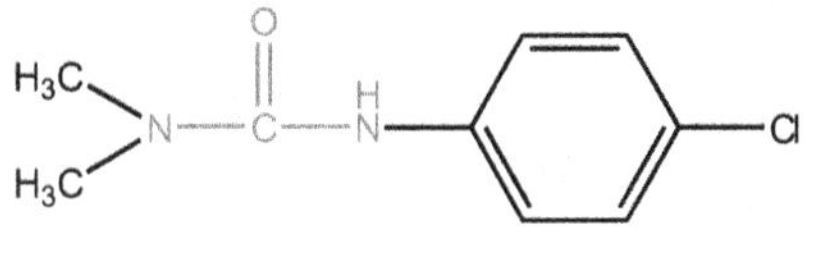

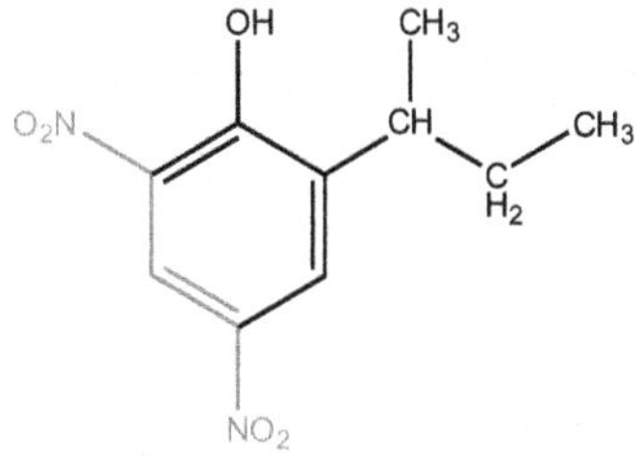

REFERENCES

SUGGESTED READINGS

Abdollahi, M., Ranjbar, A., Shadnia, S., Nikfar, S., and Rezaie, A., Pesticides and oxidative stress: a review. *Med. Sci. Monit.* 10, RA141, 2004.

Barr, D.B., Biomonitoring of exposure to pesticides. *J. Chem. Health Saf.* 231, 1, 2008.

Bradberry, S.M., Proudfoot, A.T., and Vale, J.A., Poisoning due to chlorophenoxy herbicides. *Toxicol. Rev.* 23, 65, 2004.

Dayan, F.E. and Duke, S.O., Natural compounds as next-generation herbicides. *Plant Physiol.* 166, 1090, 2014.

Eddleston, M. and Bateman, D.N., Pesticides. *Medicine* 35, 12, 2007.

Ferguson, T.J. and Albertson, T.E., Chapter 48. Herbicides, in *Toxicology Secrets*, Ling, L.J., Clark, R.F., Erickson, T.B., and Trestrail, J.H. (Eds.), Hanley & Belfus, Philadelphia, PA, 2001.

Li, S., Crooks, P.A., Wei, X., and de Leon, J., Toxicity of dipyridyl compounds and related compounds. *Crit. Rev. Toxicol.* 34, 447, 2004.

The Merck Index: An Encyclopedia of Chemicals, Drugs, and Biologicals, 15th ed., Maryadele, J. (Ed.), Merck, Whitehouse Station, NJ, 2013.

Roberts, D.M. and Buckley, N., Urinary alkalinization for acute chlorophenoxy herbicide poisoning, *Cochrane Database of Systematic Reviews* 2007, Issue 1. The Cochrane Collaboration, Wiley Publishers, New York.

REVIEW ARTICLES

Barr, D.B. and Needham, L.L., Analytical methods for biological monitoring of exposure to pesticides: a review. *J. Chromatogr. B* 778, 5, 2002.

Bismuth, C., Garnier, R., Baud, F.J., Muszynski, J., and Keyes, C., Paraquat poisoning. An overview of the current status. *Drug Saf.* 5, 243, 1990.

Bjørling-Poulsen, M., Andersen, H.R., and Grandjean, P., Potential developmental neurotoxicity of pesticides used in Europe. *Environ. Health* 7, 50, 2008.

Bradberry, S.M., Watt, B.E., Proudfoot, A.T., and Vale, J.A., Mechanisms of toxicity, clinical features, and management of acute chlorophenoxy herbicide poisoning: a review. *J. Toxicol. Clin. Toxicol.* 38, 111, 2000.

Costa, L.G., Basic toxicology of pesticides. *Occup. Med.* 12, 251, 1997.

Curtis, C.F., Should DDT continue to be recommended for malaria vector control? *Med. Vet. Entomol.* 8, 107, 1994.

Daniel, O., Meier, M.S., Schlatter, J., and Frischknecht, P., Selected phenolic compounds in cultivated plants: ecologic functions, health implications, and modulation by pesticides. *Environ. Health Perspect.* 107, 109, 1999.

de Almeida, R.M. and Yonamine, M., Gas chromatographic–mass spectrometric method for the determination of the herbicides paraquat and diquat in plasma and urine samples. *J. Chromatogr. B* 853, 260, 2007.

Eriksson, P. and Talts, U., Neonatal exposure to neurotoxic pesticides increases adult susceptibility: a review of current findings. *Neurotoxicology* 21, 37, 2000.

Garabrant, D.H. and Philbert, M.A., Review of 2,4-dichlorophenoxyacetic acid (2,4-D) epidemiology and toxicology. *Crit. Rev. Toxicol.* 32, 233, 2002.

Jones, G.M. and Vale, J.A., Mechanisms of toxicity, clinical features, and management of diquat poisoning: a review. *J. Toxicol. Clin. Toxicol.* 38, 123, 2000.

Kraehmer, H., Laber, B., Rosinger, C., and Schulz, A., Herbicides as weed control agents: state of the art: I. Weed control research and safener technology: the path to modern agriculture. *Plant Physiol.* 166, 1119, 2014.

Landrigan, P., Powell, K., James, L.M. and Taylor, P.R., Paraquat and marijuana: epidemiological assessment. *Am. J. Public Health* 73, 784, 1983.

Narahashi, T., Nerve membrane ion channels as the target site of insecticides. *Mini. Rev. Med. Chem.* 2, 419, 2002.

Roeder, R.A., Garber, M.J., and Schelling, G.T., Assessment of dioxins in foods from animal origins, *J. Anim. Sci.* 76, 142, 1998.

Suntres, Z.E., Role of antioxidants in paraquat toxicity. *Toxicology* 180, 65, 2002.

Vale, J.A., Meredith, T.J., and Buckley, B.M., Paraquat poisoning: clinical features and immediate general management. *Human Toxicol.* 6, 41, 1987.

30 Rodenticides

30.1 INTRODUCTION

Rodenticides are a diverse group of chemically and structurally *unrelated* compounds. Unlike the herbicides and insecticides, whose toxicity is uniformly categorized according to their chemical classification, the rodenticides are functionally organized according to their general category and their toxicity in rodents (LD_{50}), as outlined in Table 30.1. In addition, the substances differ in their mechanisms of action, clinical toxicity, effective doses, structural formulas, sources, and other uses.

Over $1.1 billion of commercial, industrial, and household rodenticide-containing products were produced and consumed in the United States in 2001, exceeding the production and consumption of all other countries combined. The American Association of Poison Control Centers (AAPCC) reported that of the 90,261 single-substance nonpharmaceutical exposures in 2007, over 12,000 exposure cases were due to rodenticides, in most cases the long-acting anticoagulant "superwarfarins" (Bronstein et al., 2008). Although these products are most familiar for their use in the extermination of mice and rats, they are also employed in the elimination of small mammals (squirrels, chipmunks), snakes, and frogs. Table 30.1 outlines some properties of popular rodenticides that are still commercially available. Except for fluoroacetate, which is found in one of the most poisonous plants on the planet and is not available in the United States, the substances will be further discussed individually.*

The acute and chronic toxic profiles of the rodenticides differ significantly from each other in severity and mechanism. Exposure due to the ready accessibility of the compounds accounts for a number of nonfatal poisonings, averaging about several hundred cases annually. As with the herbicides and insecticides, these cases result from inadvertent ingestion of commercial packages for home use as well as dermal and respiratory exposure as occupational hazards.

30.2 ANTICOAGULANTS

30.2.1 Chemical Characteristics

The warfarins (the carbon-3-substituted 4-hydroxycoumarin derivatives) and "super-warfarins" (brodifacoum, indanediones) (Figure 30.1) are popular rodenticides, whose properties were originally isolated from the sweet clover plant.† By 1950,

* The toxicity of strychnine, a popular rodent poison, is discussed extensively in Chapter 15. The toxicity of some metals, such as arsenic and zinc, is outlined in Chapter 26.

† Sweet clover (*Melilotus species*), long popular as food for grazing animals, contains various substances in the coumarin family. The plant has been used in the treatment of varicose veins and venous insufficiency. The name *Melilotus* originates from the Greek word for honey, *meli*, and a term for clover-like plants, *lotos*.

TABLE 30.1
Names, Chemical Properties, and Toxicity of Common Rodenticides

Common Name	Chemical Symbol, Name, or Synonym	Physical Characteristics	Toxicological Properties or Mechanism	Median Oral Rodent LD_{50} (mg/kg)
		Metals		
Arsenic	As	Brilliant gray metal, I	Human carcinogen; binds to sulfur-containing enzymes	15–40 (trivalent salts)
Barium	Ba	Yellow-white malleable metal, salts S	Hypokalemia, skeletal and cardiac muscle toxicity	20
Phosphorus (P)	Elemental, white P	White (also black or red) solid, SS	Nonspecific enzyme inhibitor; irritant	<5
Thallium	Tl	Blue-white soft metal, salts usually S	Inhibits sulfhydryl-containing enzymes	25 (sulfate)
Zinc Phosphide	Zn_3P_2	Grey crystals, I	Nonspecific enzyme inhibitor; irritant	40–45
		Anticoagulants		
Brodifacoum	Talon, Ratak+	Off-white powder, I	Inhibit Vitamin K synthesis	0.27
Norbromide	Raticate, Shoxin	White crystals, I		5.3
Warfarin (4-hydroxycoumarin derivative)	Coumadin® (sodium salt)	Crystals, salts are S		8–100
		Miscellaneous		
Strychnine	Strychnidin-10-one	White, crystalline powder, S	Inhibits glycine, an inhibitory amino acid neurotransmitter	5 (sulfate)
Red Squill (*Liliaceae* plant)	Scillaren A, B; sea onion	Bulbs of the red variety, granular, bitter powder, SS	Digoxin-like toxicity	1–10
Fluoroacetate (*Dichapitalum cymosum*)	Poisonous plant of South Africa (*gifblaar* poison); *Compound 1080*	Sodium salt, fine white powder, S; methyl, ethyl esters, S	Irritant, causes convulsions and ventricular fibrillation	2–5

Solubility in water: I: insoluble; S: soluble; SS: slightly soluble.

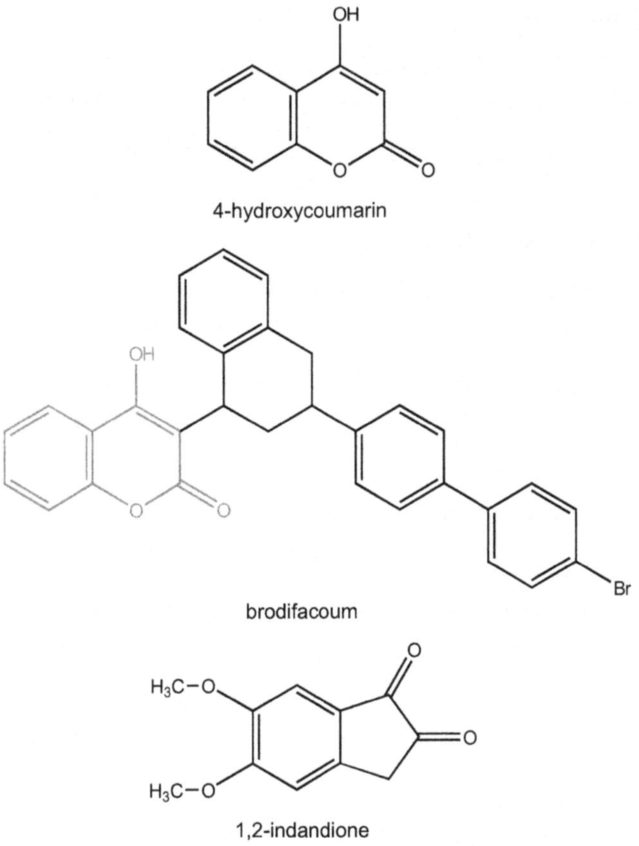

4-hydroxycoumarin

brodifacoum

1,2-indandione

FIGURE 30.1 Structures of 4-hydroxycoumarin (parent molecule), brodifacoum (long-acting coumarin derivative, with the parent molecule in blue), and 1,2-indandione.

the warfarins had achieved international notoriety as the most useful rodenticides. The clinical feasibility of oral anticoagulants in humans became known soon after with the realization of their relatively low toxicity. Rodent resistance to warfarin, however, became prevalent in the 1960s via autosomal dominant gene transmittance. By the 1970s, novel second-generation, very long-acting anticoagulants (superwarfarins) were synthesized to combat rodent resistance.

30.2.2 Commercial and Clinical Use

Today, warfarins are incorporated in rodent powder bait formulas and in rodent drinking water at concentrations from 0.025% to 0.5%. Clinically, the oral anticoagulants are used for the treatment and prevention of thromboembolic-related disorders, including myocardial infarction, cerebrovascular disease, venous thrombosis, pulmonary embolism, and other intravascular coagulation disorders. Whether directed for use as rodenticides or in the clinical management of coagulative disorders, the desirable or toxic effects of oral anticoagulants differ only quantitatively.

30.2.3 Toxicokinetics

Warfarins are rapidly and completely absorbed, reaching peak plasma concentrations within 1 hour. In circulation, warfarins are almost completely bound to plasma albumin (97–99%), localize to lipid and protein compartments (V_d=0.15 l/kg), and have a long half-life (about 35 hours).*

30.2.4 Mechanism of Toxicity

Warfarins are vitamin K antagonists that inhibit vitamin K–dependent newly synthesized precursor coagulation factor proteins S, C, and Z and clotting factors II, VII, IX, and X. The prothrombin effect is well understood and is the basis of detection of clinical poisoning. The effect of warfarin on prothrombin is shown in Figure 30.2 (W). Warfarins inhibit the carboxylation of glutamate residues (Glu) to γ-carboxyglutamate (Gla) in the conversion of descarboxyprothrombin (DCPT) to prothrombin (clotting factor II). Thus, the anticoagulant effect is not demonstrated until vitamin K plasma and liver storage are depleted (prothrombin levels must fall below 25% of baseline values). This criterion affects the kinetics of accumulation of the drug but not its biological effect. When vitamin K is not regenerated, clotting factors are not activated, and a coagulopathy results, involving both the extrinsic and intrinsic pathways. In addition, the ratio of vitamin K-epoxide to vitamin K1-reduced form is increased, because the inactive compound cannot be reduced to the active parent form. An increase in this ratio has also been documented in human overdoses. Thus, the earliest onset of toxicity or clinical activity is not obvious until at least 21 to 72 hours after exposure or therapeutic administration, respectively. In addition, to be effective rodenticides, the traditional warfarins require a minimum of 21 days of several feedings. The surperwarfarins are lethal after only one or two feedings, making them correspondingly more toxic.

30.2.5 Signs and Symptoms of Acute Toxicity

Accidental exposure from ingested rodent bait mostly involves children, whereas intentional ingestion is associated with attempted suicides or attempts to feign illness. Mild to moderate adverse effects are often seen during routine therapeutic management of coagulative disorders. Single exposures are rare but involve acute, high-dose ingestion. Consequently, signs and symptoms usually appear with repeated ingestion or following an initial therapeutic course.

Gingival bleeding, epistaxis, joint and muscle pain, easy bruising, and an abnormal PT time† are among the initial features of anticoagulant overdose. Intentional ingestion of high doses or chronic repeated administrations are associated with severe coagulopathies, including hematuria, bloody stools, intracranial hemorrhage, and shock.

---

* The racemic forms, levorotatory or S-(−)-enantiomorphs, are active warfarin isomers, while the more potent superwarfarins have longer half-lives (156 hours) and higher lipid solubility and are more selective for hepatic enzymes.

† Prothrombin.

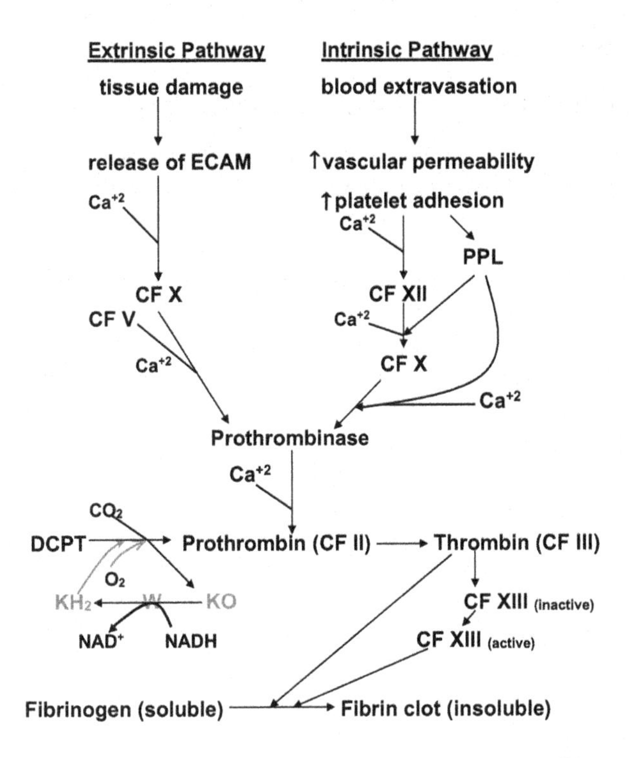

FIGURE 30.2 The intrinsic and extrinsic clotting factor cascades. The cascades illustrate the roles of tissue damage or blood extravasation (exposure to collagen fibers) in precipitating the formation of the fibrin clot. Warfarin (W) inhibits the enzymatic reduction of vitamin K-epoxide (KO) to the reduced hydroquinone (KH_2) derivative. CF: clotting factor; CF II: prothrombin; CF III: thrombin; CF V: proaccelerin; CF X: Stuart–Prower factor; CF XII: Hageman factor; CF XIII: fibrin stabilizing factor; DCPT: descarboxyprothrombin; KH_2: vitamin K hydroquinone (reduced form); KO: vitamin K epoxide (oxidized form).

30.2.6 Clinical Management of Acute Poisoning

Replacement of blood loss and reversal of the anticoagulant effects are the primary goals of therapy. A unit of fresh frozen plasma* replenishes lost multiple clotting factors and restores blood volume. Chronic anticoagulant management necessitates the administration of vitamin K_1 (phytonadione). Subcutaneous or intravenous administration of this active form of vitamin K† rapidly corrects PT within 24 hours. Maintenance with the oral dosage form may be continued for several weeks as needed.

* One unit of 300 ml of fresh frozen plasma (FFP) is prepared from whole blood within 6 hours of collection and stored at −18°C. FFP is also used for the treatment of immunoglobulin deficiencies, burn trauma, and complement dysfunction.

† Other vitamin K derivatives, K_3 (menadione) and K_4 (menadiol), are therapeutically ineffective.

30.3 PHOSPHORUS (P)

30.3.1 Chemical Characteristics, Occurrence, and Uses

A large class of irritating, corrosive, and asphyxiating phosphorus-containing compounds includes some broad categories of derivatives: 1. tri- and pentavalent halide, sulfide, and oxide derivatives; 2. phosphine gas; 3. organic phosphate, and phosphoric and phosphorus acid esters; and 4. organophosphorus insecticides. Only the more commonly encountered agents are discussed here (organophosphorus insecticides are thoroughly discussed in Chapter 28).

White (elemental) phosphorus is a white or colorless, spontaneously flammable, highly toxic solid. It readily combines with oxidizing agents to form explosive mixtures. It is used in inorganic analytical chemistry and as a rodenticide. Its practical commercial application allows it to be spread on food (especially cheese) to attract mice and rats. Red phosphorus is a less toxic and less reactive but flammable compound, yet its fumes are highly irritating on ignition. Red P is used to make safety matches and smoke bombs (for law enforcement) and as a reactant for the manufacturing of pyrotechnic substances, fertilizers, P halides, and rodenticides. Phosphine is a flammable, highly toxic colorless gas with a characteristic fishy odor. It is used as a fumigant rodenticide and insecticide and in the production of electronic components.

30.3.2 Toxicokinetics

Exposure to P and its derivatives occurs by oral ingestion or through inhalation of the flammable gases. Dermal, gastrointestinal, and respiratory irritation are hallmarks of P poisoning.

30.3.3 Mechanism and Signs and Symptoms of Acute Toxicity

Halides, oxides, and sulfides of P act as pulmonary irritants and dermal corrosive compounds. They are capable of spontaneous ignition, producing flames and fumes. They can combine with moist air or oxidizing agents to generate irritating or corrosive acidic conditions.

Phosphate esters (organic phosphates such as triethyl phosphate) inhibit acetyl cholinesterase. Oral ingestion, therefore, produces neurotoxicity, muscular weakness, and paralysis, not unlike the organophosphate insecticides (see Chapter 28).* Ingestion of elemental white P is associated with nausea, vomiting, diarrhea, and phosphorescent vomitus and stools (known as the *smoking stool syndrome*). Mucosal burning, abdominal pain, and a characteristic "garlic odor" to the breath are frequent complaints from gastrointestinal irritation. Tremors, convulsions, jaundice, liver and cardiovascular failure, and coma develop following severe intoxication. Mortality rates of approximately 25 to 75% are noted from complications of systemic toxicity.

* It is important to note that not all organic phosphate esters are responsible for the development of neurotoxicity. In fact, the toxicity of trimethyl phosphate is related to its carcinogenic effect, while triphenyl phosphate lacks neurotoxicity.

Phosphine gas is generated when solid rodenticides, such as zinc or aluminum phosphides, contact oxidizing agents or weak acids. The liberation and inhalation of the gas imparts a characteristic "fish odor" breath, especially when the powder mixes with stomach acid. Exposure produces signs and symptoms similar to those of pulmonary irritants and toxic products of combustion (see Chapter 25). Inhalation of phosphine vapors and fumes induces upper and lower respiratory tract injury (URT and LRT, respectively). Upper airway and ophthalmic injury is distinguished by local inflammation and irritation of ocular, oral, and nasal mucous membranes, including conjunctivitis, lacrimation, rhinitis, and pharyngitis. LRT symptoms include cough, wheezing, and tightness of chest with painful breathing. Early symptoms include nausea, fatigue, tremors, dizziness, and hypotension, followed by pulmonary edema, cardiogenic shock, central nervous system depression, convulsions, and coma.

Because of its proclivity for bone matrix, chronic, low-dose exposure to phosphates risks its accumulation in the skeleton. Peculiar symptoms develop, such as tooth pain, sore mandible (classic *phossy jaw*), and nonspecific nutritional imbalances.

30.3.4 Clinical Management of Acute Poisoning

Treatment of phosphate poisoning is primarily supportive and symptomatic, and necessitates the use of gastric lavage and administration of activated charcoal and sodium bicarbonate to reduce absorption and neutralize acidity, respectively. Treatment and decontamination should be instituted soon after exposure or on dermal contact. Induction of vomiting is generally not recommended due to phosphate's potentially corrosive nature. Cardiopulmonary support, renal perfusion, prevention of circulatory collapse, and oxygenation may be required.

30.4 RED SQUILL

30.4.1 Chemical Characteristics, Occurrence and Uses

Red squill* consists of the bulb of the red variety of *Urginea maritima* (Mediterranean squill, *Liliaceae* family), most of which is imported for use as a rodenticide. More recently, the less expensive importation of *Urginea indica* (Indian squill) has made this variety commercially more popular.

Squill contains many digoxin-like cardioactive glycosides, of which scillaren A, scillaren B, and proscillaridin A (Figure 30.3)† comprise most of the glycoside fraction. The compounds possess emetic, cardiotonic, and diuretic properties and were at one time ingredients of cough preparations (Cosanyl®). Scillaren and other cardioactive glycosides (strophanthin) were officially recognized for years and considered efficacious, but their unreliable therapeutic stability and variable effects caused them to lose their clinical utility to the digitalis derivatives. Poisoning due to accidental or intentional ingestion of rodenticides containing scillaren compounds is uncommon.

---

* *Scilla* is from the Greek meaning split, and refers to the separating scales of the plant. It is indigenous to the Mediterranean coasts of Spain, France, Italy, and Greece.

† Interestingly, current preclinical studies suggest that proscillaridin A possesses antiproliferative and anticancer activity.

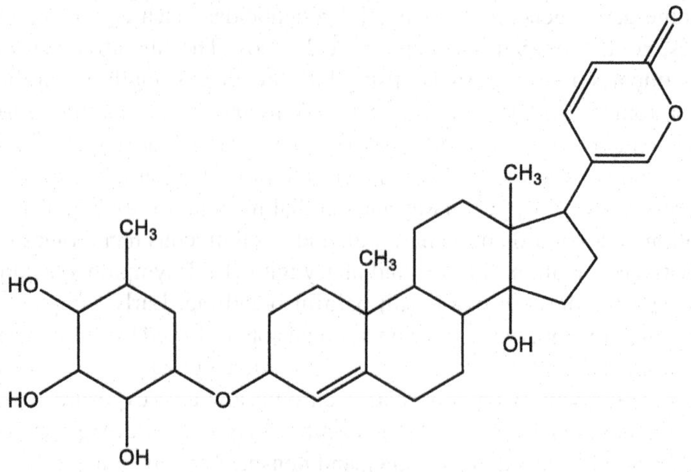

Proscillaridin A

FIGURE 30.3 Structure of proscillaridin A, a glycosidic constituent of red squill. Note the similarity of proscillaridin A, the active component of red squill, to the cardiac glycoside digoxin (Figure 21.3).

30.4.2 Mechanism and Signs and Symptoms of Acute Toxicity

The toxicity of red squill is similar to that of digoxin. Initial symptoms appear as blurred vision, arrhythmias, convulsions, and coma (interestingly, rodents do not possess a vomiting reflex and consequently, succumb to cardiac arrest, respiratory failure, and convulsions).

30.4.3 Clinical Management of Acute Poisoning

Clinical management requires intervention similar to that described for digoxin overdose (see Chapter 20).

30.5 METALS: THALLIUM, BARIUM*

30.5.1 Thallium (Tl)

As with many potent compounds of metallic or botanical origin whose toxicity was yet to be recognized, Tl was used among the ancient civilizations for the treatment of syphilis, dysentery, tuberculosis, and ringworm, and as a depilatory agent. Today, the odorless and tasteless metal is used in photoelectric cells, in semiconductor components, and rarely, as a rodenticide.

Tl inhibits oxidative phosphorylation through its ability to interfere with sulfhydryl-containing enzymes. Acute ingestion induces nausea, vomiting, diarrhea,

* Although arsenic is an active ingredient of some rodenticides, its toxicity is discussed in Chapter 26.

and tremors.* Neurologically, Tl precipitates a syndrome resembling Guillain–Barré—that is, acute febrile polyneuritis. The condition is characterized by agitation, confusion, pain, paresthesias, and weakness of the extremities radiating from the face and arms to the thorax and trunk. The syndrome is generally associated with repeated viral infections and eventually progresses to respiratory and cardiovascular collapse, convulsions, coma, and death.

Treatment of Tl intoxication is symptomatic and supportive. Chelation therapy is generally ineffective, although Prussian blue (potassium ferricyanoferrate) has shown some ability to decrease Tl absorption by forming insoluble complexes with the metal. Potassium chloride administration also prevents renal Tl reabsorption, thereby reducing blood Tl levels.

30.5.2 Barium (Ba)

Barium and its salts are used in the production of electronic components, paints, ceramics, lubricating oils, and textiles, as a contrast agent in radiology (sulfate salt), and in analytical chemistry. Ba interferes with K^+ efflux from cells, thereby causing a reduction in extracellular K^+. Consequently, signs and symptoms of Ba poisoning are related to the production of (1) hypokalemia and (2) skeletal and cardiac muscle abnormalities (Table 30.1). Myoclonus, muscular rigidity, ventricular arrhythmias, vomiting, and diarrhea are secondary to these effects.

Treatment of Ba exposure is symptomatic and supportive, and requires the correction of the K^+ imbalance. K^+ administration, therefore, remains the most effective means of counteracting Ba-induced toxicity.

30.6 METHODS OF DETECTION

Although suspected rodenticides can be measured by a variety of assays, including radioimmunoassay (RIA), enzyme-linked immunosorbent assay (ELISA), and HPLC, tests that address biomonitoring are more appropriate for diagnosing coagulopathies. For instance, prolongation of prothrombin time (PT) or INR (a PT ratio) is usually the first laboratory indication that a coagulopathy exists. In severe cases, the plasma thromboplastin time (PTT) can also be prolonged, representing involvement of both the intrinsic and extrinsic coagulation systems. Other bleeding disorders must be also be ruled out, such as disseminated intravascular coagulation (DIC) or hemophilias. Measurement of the specific clotting factors affected by warfarin confirms the diagnosis, such that all four vitamin K–dependent factors will be decreased (II, VII, IX, and X), while others may appear normal. Measurement of the vitamin K-epoxide/K1-reduced form ratio can help determine warfarin poisoning (normal ratio = 2–3). Finally, atomic absorption spectroscopy is used to identify thallium and barium rodenticides in air, soil, and biological samples.

* Alopecia is a classic pathognomonic sign of Tl toxicity. Unlike arsenic poisoning, however, Tl is not deposited in hair follicles.

REFERENCES

SUGGESTED READINGS

Barthold, C.L. and Schier, J.G., Organic phosphorus compounds—nerve agents. *Crit. Care Clin.* 21, 673, 2005.

Bentur, Y., Raikhlin-Eisenkraft, B., and Lavee, M., Toxicological features of deliberate self-poisonings. *Hum. Exp. Toxicol.* 23, 331, 2004.

Blumenthal, M. (Ed.), *The Complete German Commission E Monographs: Therapeutic Guide to Herbal Medicines*, American Botanical Council, Austin, TX, 1999, P218–P219.

Bronstein, A.C., Spyker, D.A., Cantilena, L.R. Jr, Green, J.L., Rumack, B.H., and Heard, S.E., 2007 Annual Report of the American Association of Poison Control Centers' National Poison Data System (NPDS): 25th Annual Report. *American Association of Poison Control Centers. Clin Toxicol (Phila).* 46, 927, 2008.

Caravati, E.M., Caravati, E.M., Erdman, A.R., Scharman, E.J., Woolf, A.D., Chyka, P.A., Cobaugh, D.J., Wax, P.M., Manoguerra, A.S., Christianson, G., Nelson, L.S., Olson, K.R., Booze, L.L., and Troutman, W.G.,Long-acting anticoagulant rodenticide poisoning: an evidence-based consensus guideline for out-of-hospital management. *Clin. Toxicol.* 45, 1, 2007.

Card, D.J., Francis, S., Deuchande, K., and Harrington, D.J., Superwarfarin poisoning and its management. *BMJ Case Rep.* bcr2014206360, 2014.

Das, A.K., Chakraborty, R., Cervera, M.L., and de la Guardia, M., Determination of thallium in biological samples. *Anal. Bioanal. Chem.* 385, 665, 2006.

Feinstein, D.L., Akpa. B.S., Ayee, M.A., Boullerne, A.I., Braun, D., Brodsky, S.V., Gidalevitz, D., Hauck, Z., Kalinin, S., Kowal, K., Kuzmenko, I., Lis, K., Marangoni, N., Martynowycz, M.W., Rubinstein, I., van Breemen, R., Ware, K., and Weinberg, G., The emerging threat of superwarfarins: history, detection, mechanisms, and countermeasures. *Ann. N.Y. Acad. Sci.* 1374, 111, 2016.

Goh, C.S., Hodgson, D.R., Fearnside, S.M., Heller, J., and Malikides, N., Sodium monofluoroacetate (Compound 1080) poisoning in dogs. *Aust. Vet. J.* 83, 474, 2005.

Hoffman, R.S., Thallium toxicity and the role of Prussian blue in therapy. *Toxicol. Rev.* 22, 29, 2003.

Ibrahim, D., Froberg, B., Wolf, A., and Rusyniak, D.E., Heavy metal poisoning: clinical presentations and pathophysiology. *Clin. Lab. Med.* 26, 67, 2006.

Kendrick, D.B., Mosquito repellents and superwarfarin rodenticides—are they really toxic in children? *Curr. Opin. Pediatr.* 18, 180, 2006.

Makris, M., Management of excessive anticoagulation or bleeding. *Semin. Vasc. Med.* 3, 279, 2003.

Peter, A.L. and Viraraghavan, T., Thallium: a review of public health and environmental concerns. *Environ. Int.* 31, 493, 2005.

Proudfoot, A.T., Bradberry, S.M., and Vale, J.A., Sodium fluoroacetate poisoning. *Toxicol. Rev.* 25, 213, 2006.

Riyaz, R., Pandalai, S.L., Schwartz, M., and Kazzi, Z.N., A fatal case of thallium toxicity: challenges in management. *J. Med. Toxicol.* 9, 75, 2013.

Thompson, D.F. and Callen, E.D., Soluble or insoluble prussian blue for radiocesium and thallium poisoning? *Ann. Pharmacother.* 38, 1509, 2004.

Thornton, S.L., Oller, L., and Coons, D.M., Annual report of the University of Kansas health system poison control center. *Kans. J. Med.* 11, 24, 2018.

United States Environmental Protection Agency (USEPA), *Controlling Rodents and Regulating Rodenticides.* www.epa.gov/rodenticides, last viewed March 2018.

Vindenes, V., Karinen, R., Hasvold, I., Bernard, J.P., Mørland, J.G., and Christophersen, A.S.,Bromadiolone poisoning: LC-MS method and pharmacokinetic data. *J. Forensic Sci.* 53, 993, 2008.

Whitlow, K.S., Belson, M., Barrueto, F., Nelson, L., and Henderson, A.K., Tetramethylenedisulfotetramine: old agent and new terror. *Ann. Emerg. Med.* 45, 609, 2005.

Winnicka, K., Bielawski, K., Bielawska, A., and Surazyński, A., Antiproliferative activity of derivatives of ouabain, digoxin and proscillaridin A in human MCF-7 and MDA-MB-231 breast cancer cells. *Biol. Pharm. Bull.* 31, 1131, 2008.

Yogendranatha, N. , Herath, H.M.M.T.B., Sivasundaram, T., Constantine, R., and Kulatunga, A., A case report of zinc phosphide poisoning: complicated by acute renal failure and tubulo interstitial nephritis. *BMC Pharmacol. Toxicol.* 18, 37, 2017.

REVIEW ARTICLES

Bosse, G.M. and Matyunas, N.J., Delayed toxidromes. *J. Emerg. Med.* 17, 679, 1999.

Costa, L.G., Basic toxicology of pesticides. *Occup. Med.* 12, 251, 1997.

Cruickshank, J., Ragg, M., and Eddey, D., Warfarin toxicity in the emergency department: recommendations for management. *Emerg. Med. (Fremantle)* 13, 91, 2001.

Dorman, D.C., Toxicology of selected pesticides, drugs, and chemicals. Anticoagulant, cholecalciferol, and bromethalin-based rodenticides. *Vet. Clin. North Am. Small Anim. Pract.* 20, 339, 1990.

El Bahri, L., Djegham, M., and Makhlouf, M., *Urginea maritima* L (Squill): a poisonous plant of North Africa. *Vet. Hum. Toxicol.* 42, 108, 2000.

Galvan-Arzate, S. and Santamaria, A., Thallium toxicity. *Toxicol. Lett.* 99, 1, 1998.

Leonard, A. and Gerber, G.B., Mutagenicity, carcinogenicity and teratogenicity of thallium compounds. *Mutat. Res.* 387, 47, 1997.

Moore, D., House, I., and Dixon, A., Thallium poisoning. Diagnosis may be elusive but alopecia is the clue. *BMJ* 306, 1527, 1993.

Mowry, J.B., Spyker, D.A., Cantilena, L.R. Jr, McMillan, N., and Ford, M.,2013 Annual Report of the American Association of Poison Control Centers' National Poison Data System (NPDS): 31st Annual Report. *Clin. Toxicol. (Phila.)* 52, 1032, 2015.

Murphy, M.J., Rodenticides. *Vet. Clin. North Am. Small Anim. Pract.* 32, 469, 2002.

Petterino, C. and Paolo, B., Toxicology of various anticoagulant rodenticides in animals. *Vet. Hum. Toxicol.* 43, 353, 2001.

Rodenberg, H.D., Chang, C.C., and Watson, W.A., Zinc phosphide ingestion: a case report and review. *Vet. Hum. Toxicol.* 31, 559, 1989.

Spahr, J.E., Maul, J.S., and Rodgers, G.M., Superwarfarin poisoning: a report of two cases and review of the literature. *Am. J. Hematol.* 82, 656, 2007.

Stone, W.B., Okoniewski, J.C., and Stedelin, J.R., Anticoagulant rodenticides and raptors: recent findings from New York, 1998-2001. *Bull. Environ. Contam. Toxicol.* 70, 34, 2003.

Sutcliffe, F.A., MacNicoll, A.D., and Gibson, G.G., Aspects of anticoagulant action: a review of the pharmacology, metabolism and toxicology of warfarin and congeners. *Rev. Drug Metab. Drug. Interact.* 5, 225, 1987.

Watt, B.E., Proudfoot, A.T., Bradberry, S.M., and Vale J.A., *Toxicol. Rev.* 24, 259, 2005.

Weiner, M.L., Salminen, W.F., Larson, P.R., Barter, R.A., Kranetz, J.L., and Simon, G.S., Toxicological review of inorganic phosphates. *Food Chem. Toxicol.* 39, 759, 2001.

31 Chemical Carcinogenesis and Mutagenesis

31.1 INTRODUCTION

Cancer is a classification of diseases in which there is an uncontrolled proliferation of cells that express varying degrees of fidelity to their precursor cell of origin. This cell proliferation occurs in almost all tissues throughout a lifespan and is influenced by a variety of circumstances. During the normal physiologic state, the delicate balance between cell proliferation and apoptosis (programmed cell death) is perpetuated to ensure the integrity and proper function of organs and tissues. Mutations in DNA that lead to cancer development interfere with this orderly process by disrupting its regulation. Consequently, the initiation of a *cancer* is due to an abnormal and uncontrolled progression of cell proliferation characterized by unregulated cell division and metastasis (spreading) of foci of cells to distant tissues.

Carcinogenesis represents the unwarranted appearance or increased incidence of abnormal cell proliferation occurring in an age-matched low-risk group whose propensity for development of cancer is statistically lower. A *carcinogen* is any chemical or viral agent that increases the frequency or distribution of new tumors, results in their appearance within a low-risk or otherwise early age group, or results in the introduction of new pathological growths otherwise absent in experimental controls.

Most chemical carcinogens require metabolic activation before demonstrating carcinogenic potential. As with most toxic phenomena, a minimum dosage is necessary to elicit a carcinogenic event. *Epigenetic* or *nongenotoxic* carcinogens enhance the growth of tumors by mechanisms other than through alteration of DNA. These chemicals influence absorption, biotransformation, alter post-translational modifications, or reduce elimination of the initiating agent. *Cocarcinogens* redirect hormonal action on cell proliferation and inhibit intercellular communication, thus allowing greater proliferative capacity. Alternatively, they induce immunosuppression of protective immunological pathways.

The carcinogenic process is, on occasion, a response to a mutation occurring within the genetic material of normal cells, resulting in uncontrolled cell division and transformation to the immortal phenotype. The unrestrained and often rapid proliferation of cells is further characterized as benign or malignant. *Benign* tumors do not metastasize, are usually confined to local target organ areas, are clinically susceptible to therapeutic intervention, and carry a more favorable prognosis. Alternatively, *malignant* tumors metastasize to distant organ locations, are not necessarily amenable to therapeutic intervention, and possess a less favorable prognosis. *Mutagenesis* refers to the ability of a virus or chemical agent to induce changes in the genetic sequence of mammalian or bacterial cells, thus altering the phenotypic expression of cell characteristics. *Genotoxicity* refers to the ability of an agent to induce heritable changes in genes that exercise homeostatic control in somatic cells while increasing the risk of influencing benign or malignant transformation. Genotoxic substances induce genotoxicity either by binding directly to DNA or by indirectly altering the DNA sequence, resulting in irreversible damage. It is also important to note, however, that genotoxic substances are not necessarily carcinogenic.

In addition, the interaction of a chemical, physical, or viral agent with nucleic acids results in the disruption of the transfer of genetic information and the development of genotypic and phenotypic consequences. Lastly, *mitogenesis* is the induction of cell division (mitosis) within eukaryotic or prokaryotic cells by stimulating transit through the cell cycle. Prolonged and continuous exposure to growth factors is required to commit cells to the cell cycle.

By definition, cancer must have at least six criteria:

1. Self-sufficiency in growth signals
2. Insensitivity to antigrowth signals
3. Evasion of apoptosis
4. Tissue invasion and metastasis
5. Sustained angiogenesis
6. Limitless replicative potential

The induction of a neoplasm is a multistage process that occurs over a long period of time, the stages of which include initiation, promotion, and progression (see section 31.3). The causes of most human cancers remain unidentified; however, considerable evidence suggests that environmental and lifestyle factors, and chemical agents, are important contributors. For example, tobacco smoking appears to be responsible for approximately 30% of all cancer deaths in developed countries. The influence of environmental factors on the incidence of cancer in a population is supported by epidemiological observations:

1. While overall cancer incidence is reasonably constant between countries, incidences of specific tumor types vary considerably.
2. Large differences in cancer incidence exist within populations of a single country.
3. Migrant populations assume the cancer incidence of their new environment within one to two generations.
4. Cancer incidence within a population changes rapidly.

Thus, understanding the molecular and cellular process underlying chemical carcinogenesis is of critical importance for carcinogenic risk assessment as well as the development of mechanistically based chemopreventive and therapeutic strategies for the management of human cancers.

31.2 MECHANISMS OF CHEMICAL CARCINOGENESIS

31.2.1 Metabolism

Carcinogenic agents initiate cancer progression by using one of two pathways:

1. Parent chemicals can cause cancers directly, such as with heavy metals.
2. Some chemicals, such as organic chemicals, require metabolic activation to reactive intermediates to affect the carcinogenic process; for example, benzo(a)pyrene is first converted to an electrophilic intermediate, which then covalently binds to cellular macromolecules. Accordingly, chemical carcinogens that require metabolic activation to exert their carcinogenic effects are termed *procarcinogens*, whereas their highly reactive metabolites are designated *ultimate* carcinogens. Metabolic intermediates between procarcinogens and ultimate carcinogens are called *proximate* carcinogens. Figure 31.1 illustrates the metabolic activation of benzo(a)pyrene.

Cytochrome P450 Phase 1 and Phase 2 enzymes are involved in the metabolism of carcinogens and usually result in the formation of reactive metabolites. Phase 2 enzyme-catalyzed reactions often lead to detoxification and elimination, which effectively protects against chemical carcinogenesis. Over time, the constant assault on cells by reactive metabolites that linger at the cellular level probably lead to interaction with cellular macromolecules and the carcinogenic process.

31.2.2 Chemistry

It is recognized that there is a relationship between the chemical structure of a suspected carcinogen and its carcinogenic activity. Molecules that contain aromatic moieties, such as aromatic nitro groups, N-oxides, mono- and dialkylamino groups, and aliphatic and aromatic epoxides, are the chemicals most associated with carcinogenic activity. These structural "alerts" are apparently useful for carcinogenic risk assessment.

31.2.3 Free Radicals and Reactive Oxygen Species

As with the chemical activation of procarcinogens to proximate and ultimate carcinogens, the bioactivation of chemicals also results in the formation of free radicals and reactive oxygen species (ROS)—that is, chemical species that cause oxidative damage to biomolecules, including nucleic acids (DNA), proteins, and lipids. These reactive species are generated by ionizing radiation, ultraviolet light, and a variety of exogenous and endogenous cellular sources (Figure 31.2).

ROS interaction with DNA induces the formation of adducts, which may lead to the activation of proto-oncogenes and/or the inactivation of tumor suppressor genes. Free radicals and ROS are also able to target cell signaling molecules, including transcription factors and protein kinase cascades, resulting in altered cell signal

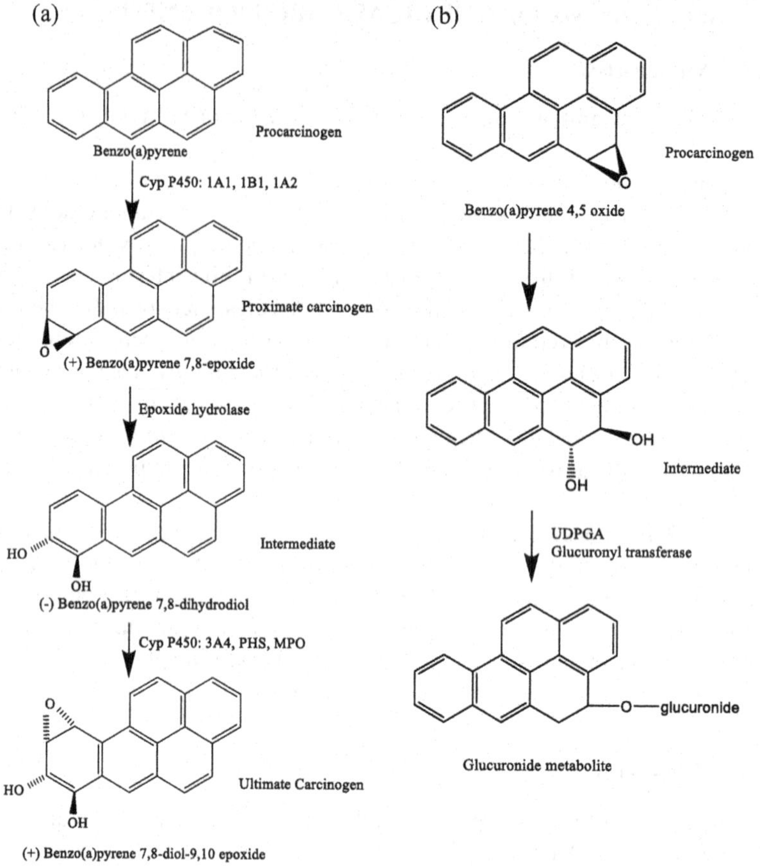

FIGURE 31.1 (a) Metabolic activation of benzo(a)pyrene to form the proximate and ultimate carcinogens from the action of phase I enzymes; (b) metabolic detoxification of benzo(a) pyrene 4,5-oxide to form the glucuronide metabolite from the action of phase II enzymes. MPO: myeloperoxidase; PHS: prostaglandin H synthase; UDPGA: uridine diphosphate glucuronic acid.

transduction and gene expression. The ability of free radicals and ROS to elicit DNA mutations and cause dysregulated cell signaling contributes to the multistage carcinogenesis pathway (Figure 31.2).

31.2.4 Mutagenesis

As mentioned previously, *mutagenesis* occurs when the mammalian or bacterial cell genetic sequence is altered such that transcription and translation are transformed. This induces a modification of the phenotypic expression of cell characteristics, which more often results in permanent pathologic changes rather than beneficial adjustments. In general, carcinogenics are separated into genotoxic and nongenotoxic classifications. Thus, a genotoxic carcinogen is able to interact with DNA directly,

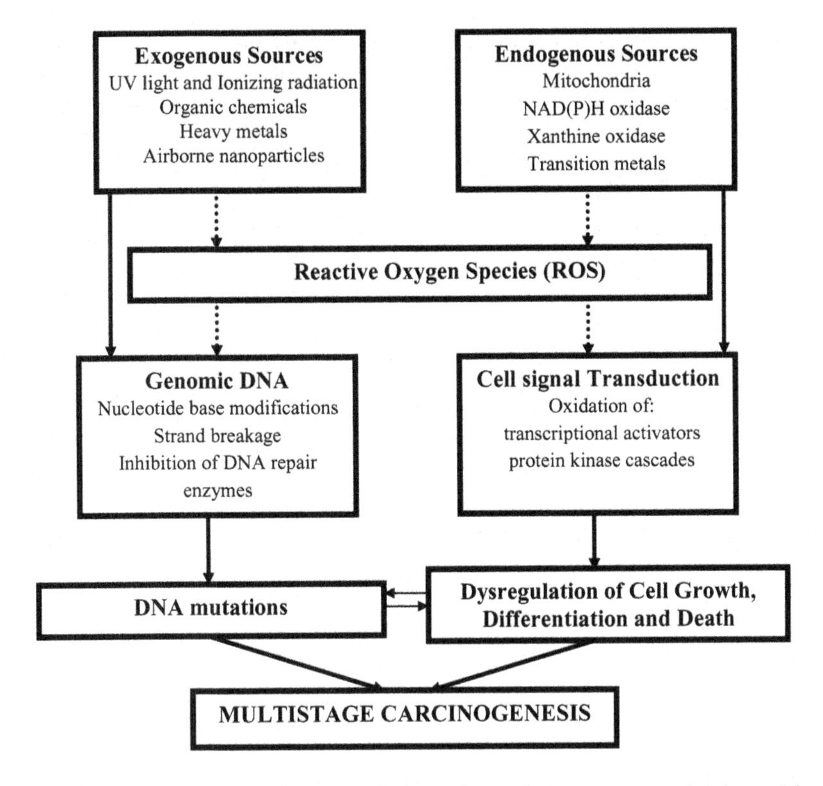

FIGURE 31.2 Involvement of free radicals and reactive oxygen species in multistage carcinogenesis.

leading to mutations. The production of mutations is due primarily to chemical or physical alterations in the structure of DNA that result in inaccurate replication of that gene. The process includes two major steps:

1. Structural DNA alteration and cell proliferation lead to permanent DNA alteration. The interaction of reactive intermediates derived from the bioactivation of carcinogens with DNA results in the formation of DNA adducts, DNA strand breaks, and DNA–protein cross-links. Carcinogens that result in the formation of bulky DNA adducts often specifically react with the sites in the purine ring, particularly the N7 position of guanine. This target is the most nucleophilic site in DNA, where many ultimate carcinogens form covalent adducts. Direct alkylation of DNA bases by several alkylating carcinogenic agents has also been demonstrated.
2. DNA modification resulting from hydroxylation of DNA nucleotide bases is often a manifestation of ROS reactivity. While changes occur in all four bases, the most understood step is the formation of 8-hydroxy-2′-deoxyguanosine.

The interaction of carcinogens with DNA produces adducts not only on DNA bases but also on the sugar and phosphate backbone. The covalent modifications of DNA by carcinogenic agents may ultimately lead to the development of cancers. Although the modifications of DNA by carcinogens are critically involved in the induction of DNA mutations, the mutagenesis process is also greatly affected by various DNA repair mechanisms.

31.2.5 DNA Repair

As the interaction of ultimate carcinogens with DNA results in the formation of various DNA adducts, the continued presence of the adduct prompts the cellular machinery to repair the structural alteration. It is estimated that more than 100 genes are dedicated to DNA repair, the purpose of which is to correct the structural damage induced by carcinogens. In this way, mutagenesis is prevented. Alternatively, the persistence of DNA–carcinogen adducts is indicative of the insufficiency of DNA repair.

31.2.6 Epigenetic Carcinogenesis

As discussed earlier, *genotoxic carcinogens* directly interact with DNA through the presence of reactive electrophilic metabolites in the chemical, leading to mutations. In contrast, *nongenotoxic (or epigenetic) carcinogens* do not directly affect DNA in the induction of cancer.* In fact, nongenotoxic carcinogens:

1. Are nonmutagenic;
2. Show no evidence of direct chemical reactivity with DNA;
3. Do not possess common chemical structural features within the class;
4. Exhibit a well-established dose–threshold effect;
5. Exhibit a lower carcinogenic potential than genotoxic carcinogens.

While DNA damage is described as a general mechanism of the carcinogenic effect of genotoxic carcinogens, nongenotoxic mechanisms are more heterogeneous. The proposed mechanisms of nongenotoxic carcinogenesis are summarized in Table 31.1.

31.3 MULTISTAGE CARCINOGENESIS

The process of carcinogenesis is divided into three experimentally defined stages: *tumor initiation, tumor promotion*, and *tumor progression*. This multistage development of carcinogenesis requires the conversion of benign hyperplastic cells to the malignant state, involving invasion and metastasis as manifestations of further genetic and epigenetic changes. Figure 31.3 correlates with the following discussion and illustrates the events that occur during the three stages of chemical carcinogenesis. The diagram also outlines the requirements necessary for the completion of the multistage process.

* The National Toxicology Program (NTP) report on chemical carcinogenicity estimates that 40% of chemicals classified as carcinogens are nongenotoxic (epigenetic) carcinogens, with the liver as the most common target organ (NTP, 2015); https://ntp.niehs.nih.gov/annualreport/2015/index.html.

TABLE 31.1
Mechanisms of Nongenotoxic Carcinogenesis

Chemical Effect	Indirect Effect on DNA	Cellular Consequences
Chronic cytotoxicity	Increased DNA replication	Prolonged stimulation of cell proliferation; increased secretion of trophic hormones
Chronic cytotoxicity	Preexisting DNA damage	Inhibition of apoptosis
Acute or chronic cytotoxicity	Impairment of DNA replication; impairment of DNA repair mechanisms	Prolonged stimulation of cell proliferation
Acute or chronic cytotoxicity	Impairment of DNA replication; impairment of DNA repair mechanisms	Disruption of gap-junction communication
Acute or chronic cytotoxicity	Receptor-mediated pathway	Dysregulation of cell signaling and gene expression
Acute or chronic cytotoxicity	Impairment of DNA repair mechanisms	Altered DNA methylation, acetylation, histone modification

31.3.1 Tumor Initiation

The early concept of tumor initiation indicated that the initial changes in chemical carcinogenesis involved irreversible genetic alterations. Recent data from molecular studies of preneoplastic human lung and colon tissues, however, implicate epigenetic changes as an early event in carcinogenesis. DNA methylation of promoter regions of genes can transcriptionally silence tumor-suppressor genes. Thus, carcinogen–DNA adduct formation is central to the theory of chemical carcinogenesis and may be a necessary, but not a sufficient, prerequisite for tumor initiation. The formation of DNA adducts, causing either the *activation* of a proto-oncogene or the *inactivation* of a tumor-suppressor gene, is the tumor-initiating event. One important characteristic of this stage is its irreversibility, such that the genotype/phenotype of the initiated cell is conferred during the process. Chemicals capable of initiating cells are referred to as *initiating agents.* Furthermore, initiation without the subsequent promotion and progression rarely yields malignant transformation.

31.3.2 Tumor Promotion

In multistage carcinogenesis, tumor promotion comprises the selective clonal expansion of initiated cells through a mechanism of gene activation. Because the accumulation rate of mutations is proportional to the rate of cell division, clonal expansion of initiated cells produces a larger population of cells that are at risk of further genetic changes and malignant conversion. Tumor promoters are generally nonmutagenic, are not solely carcinogenic, and often are able to mediate their biologic effects without metabolic activation. In addition, they do not directly interact with DNA. Interestingly, tumor promotion is reversible, although the continued presence of the

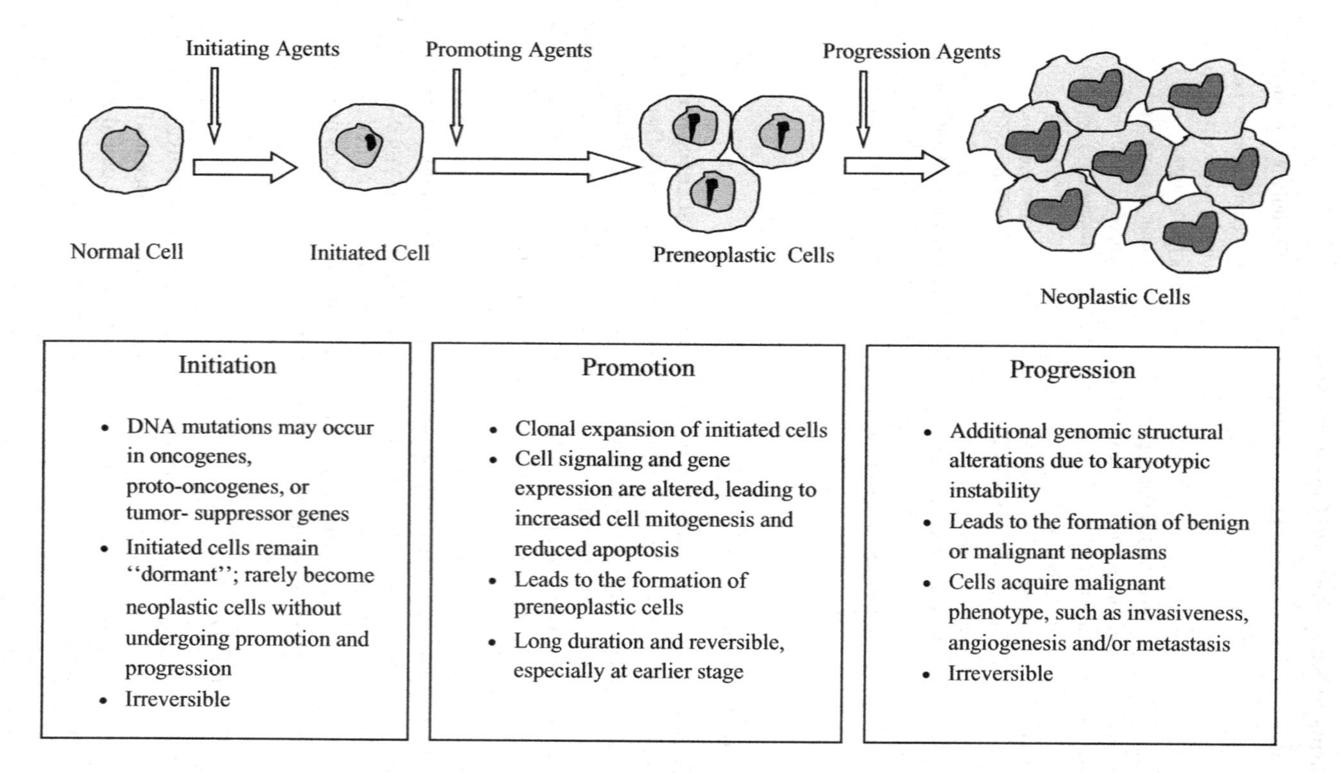

FIGURE 31.3 Schematic illustration of the three stages of chemical carcinogenesis.

promoting agent maintains the state of the promoted cell population (preneoplastic lesion). Thus, the stage appears to have a long duration and is a preferred target for experimental manipulation. Examples of typical tumor promoters include tetradecanoyl phorbol acetate (TPA), phenobarbital, and 2,3,7,8-tetrachlorodibenzo-*p*-dioxin (TCDD).

During tumor promotion, *malignant conversion* may occur, in which a preneoplastic cell is transformed into the malignant phenotype but requires further genetic changes. Frequent, repeated administration of the tumor promoter is more important than the total dose. In addition, if the tumor promoter is discontinued before malignant conversion has occurred, premalignant or benign lesions may regress. Tumor promotion contributes to the process of carcinogenesis by the expansion of a population of initiated cells that are then at risk for malignant conversion. The conversion of a fraction of these cells to malignancy is accelerated in proportion to the rate of cell division and the quantity of dividing cells in the benign tumor or preneoplastic lesion. In part, these genetic changes result from infidelity of DNA synthesis.

31.3.3 Tumor Progression

Tumor progression comprises the expression of the malignant phenotype and the tendency of malignant cells to acquire more aggressive characteristics over time. Also, metastasis may involve the ability of tumor cells to secrete proteases that allow invasion of secondary organs beyond the immediate primary tumor location. A prominent characteristic of the malignant phenotype is the propensity for genomic instability and uncontrolled growth. Further genetic and epigenetic changes occur, including the activation of proto-oncogenes and the functional loss of tumor-suppressor genes. Loss of function of tumor-suppressor genes usually occurs in a bimodal fashion and most frequently involves point mutations in one allele and loss of the second allele by a deletion, recombinational event, or chromosomal nondisjunction. These phenomena confer growth advantage on the cells as well as the capacity for regional invasion and ultimately, distant metastatic spread. Despite evidence for an apparent scheduling of certain mutational events, it is the accumulation of these mutations, and not the order or the stage of tumorigenesis in which they occur, that appears to be the determining factor.

31.4 CARCINOGENIC CHARACTERISTICS

Experimentally, a carcinogen is an agent that produces a statistically significant increased incidence of malignant neoplasms, compared with untreated control animals, when administered to previously untreated animals. Carcinogens may be chemicals, physical forces (such as ultraviolet light and gamma radiation), or biological catalysts (such as viruses). Chemical carcinogens are the most commonly investigated toxicologic mediators.

31.4.1 Classification

Chemical carcinogens are classified according to their chemical characteristics, such as:

1. Organic chemicals—benzo(a)pyrene, aflatoxin B1, and benzene
2. Inorganic chemicals—arsenic, cadmium, chromium, and nickel
3. Hormonal substances—estrogens, anabolic and androgenic steroids

Furthermore, chemical carcinogens are classified as genotoxic or nongenotoxic (epigenetic) carcinogens, based on reactivity with DNA, or as animal or human carcinogens, depending on evidence of carcinogenicity in animals and humans. The International Association for Research on Cancer (IARC) evaluates and categorizes human carcinogens accordingly:

Group 1—these agents have sufficient epidemiological evidence of carcinogenicity in humans; as such, they are labeled as *known human carcinogens*; for example, arsenic, aflatoxin B1, benzene, estrogens, vinyl chloride, nickel, and chromium belong to this group.

Group 2A—these substances possess sufficient evidence of carcinogenicity in animals but have limited evidence of carcinogenicity in humans and are therefore labeled as *probably carcinogenic* to humans; for example, benz(a) anthracene, polychlorinated biphenols, and styrene oxide.

Group 2B—these either possess limited evidence of carcinogenicity in humans or have sufficient evidence of carcinogenicity in animals and inadequate evidence of carcinogenicity in humans; they are labeled as *possibly carcinogenic* to humans; for example, styrene, urethane.

Group 3—this group is not classifiable as to carcinogenicity.

Group 4—these chemicals possess inadequate evidence of carcinogenicity in both animals and humans and thus, are *probably not carcinogenic* to humans.

31.4.2 Identification

The identification of carcinogenic agents is a critical step in carcinogenic risk assessment. While epidemiological studies provide the most definitive means of establishing carcinogenicity to humans from exposure to an agent, they are not protective of human health, since conclusions on carcinogenicity are typically derived after occurrence. In addition, epidemiological studies have limited sensitivity for identification of carcinogens. Thus, potential carcinogens are identified based on the execution of *in vitro* and animal experiments and are classified according to the following scheme:

1. Short-term bioassays—include mutagenesis assays and neoplastic transformation in cell culture. The duration of these assays is from several weeks to 3 months.

2. Medium-term bioassays—target qualitative and quantitative analysis of preneoplastic evidence. These assays are performed in animals for 2–8 months.
3. Long-term animal bioassays—the time frame is 18–24 months.

U.S. Food and Drug Administration (FDA) regulations require chronic 2-year bioassay rodent studies for identifying potential carcinogens. Further information on testing requirements can be found in the References.

31.4.3 Carcinogenic Potential

Not all carcinogenic chemicals are equally successful in inducing neoplasia; that is, they exhibit different carcinogenic potencies. The carcinogenic potential of a chemical is defined as the slope of the dose–response curve for the induction of neoplasms. This definition, however, is not sufficient for estimating carcinogenic potential based on data from chronic carcinogenesis bioassays. Historically, several methods have been developed to measure the carcinogenic potential, particularly the tumorigenesis dose rate 50 (TD_{50}). This value has been popular for the estimation of the carcinogenic potential for a number of chemicals. The TD_{50} is calculated as the dose rate (mg/kg body weight/day) of carcinogen, which when administered chronically for a standard period induces neoplasms in half of the test animals. The value is adjusted for spontaneous neoplasms and is an important component of carcinogenic risk assessment (Table 31.2).

31.4.4 Carcinogenic Risk Assessment

The process of carcinogenic risk assessment closely parallels general toxicologic risk assessment as outlined in Chapter 2. Specifically, the steps include:

1. Identification and evaluation of the carcinogen and its potential
2. Mechanism elucidation
3. Exposure assessment
4. Dose–response assessment
5. Qualitative and quantitative estimation of risk in humans

As previously noted, information and conclusions generated from epidemiological studies are tempered by the use of various laboratory experimental methods due to the nonprotective feature and limited sensitivity of the former. Several limitations, however, are encountered in the scientific and practical application of the information derived from laboratory bioassays to estimate human risk. For instance, approximating carcinogenic risk based on uncomplicated laboratory data requires major assumptions to extrapolate carcinogenic information from bioassays to humans, examples of which are outlined in Chapter 2. Another significant problem is the dose–response relationship. High doses, such as the maximum tolerated dose (MTD), of a test agent are used in the 2-year animal bioassay for identification of carcinogenicity. The actual levels of human

TABLE 31.2
Examples of Known Human Carcinogenic Chemicals Found in Various Workplaces

Chemical Agents	Commercial Uses	Cancer Induction
	Metals	
Arsenic	Copper mining and smelting	Cancer of skin, lung, or liver
Cadmium	Smelting, electronics, welding	Bronchial carcinoma
Chromium and chromates	Tanning, pigment manufacturing	Cancer of nasal sinus and bronchus
Nickel	Nickel refining	Cancer of nasal sinus and bronchus
	Acrylamides	
Benzidine and α-naphthylamine	Dye, textile, and rubber tire industry	Bladder cancer
Para-aminodiphenyl	Chemical industry	Bladder cancer
Polycyclic aromatic hydrocarbons (PAH)	Steel manufacturing, roofing	Skin or lung cancer
Benzene	Manufacturing, petroleum refining	Leukemia
Vinyl chloride	Chemical industry	Liver angiosarcoma

exposure to a particular potential carcinogenic chemical are generally much lower than those used in laboratory animal experiments. Therefore, applying the carcinogenic data obtained in animals treated with high doses of a test substance to the carcinogenic potential in humans, whose exposure to that agent is considerably lower, poses a significant hurdle in extrapolation statistics. In addition, extrapolation from high dose to low dose in carcinogenic risk assessment is further complicated by the assumption of the existence of a threshold for genotoxic carcinogens.

Qualitative and quantitative risk estimation presents another consideration in the determination of carcinogenic risk in humans. Qualitative risk estimation is easier to develop based on qualitative analysis of the information obtained from various bioassays. In contrast, quantitative risk estimation is more cumbersome and subject to greater variations. To validate quantitative risk analysis, the process should integrate biological data, toxicokinetic information, and toxicodynamic parameters of specific carcinogenic chemicals or mixtures. To satisfy this objective, various mathematical models and algorithms have been developed and used for the quantitative estimation of carcinogenic risk in humans. Among them are the one-hit (*linear*), multi-hit (*k-hit*), multistage, extreme value, log-probit, and MKV biomathematics models, as well as biologically based cancer modeling, some of which have been proposed by the U.S. Environmental Protection Agency (EPA) for quantitative analysis of carcinogenic risk.

31.4.5 Carcinogenic and Genotoxic Agents

Both collective and individual exposures often define environmental toxicology. Epidemiological research, supported by experimental investigations, has led to the determination of a "profile" of various risk factors. This profile is perpetually incomplete, since humans incessantly modify the environment, thus increasing the risk of interaction with agents against which there may be no protective mechanisms. The multiplicity of chemical or physical carcinogens does not render them amenable to study, as the development of carcinoma is, in all probability, due not to contact with a single carcinogenic factor but rather, to the sum of many possible causative agents.

Of the various environmental hazardous compounds, cigarette smoke enjoys the highest causal relationship with cancer risk in humans. Tobacco smoking plays a major role in the etiology of lung, oral cavity, and esophageal cancers as well as a variety of chronic degenerative diseases. Although cigarette smoke is a mixture of about 4000 chemicals, including more than 60 known human carcinogens, 4-methylnitrosamino-1-(3-pyridyl)-1-butanone (nicotine-derived nitrosamino ketone [NNK]) is the most carcinogenic tobacco-specific nitrosamine. NNK induces lung tumors in mice, rats, and hamsters, and the International Agency for Research on Cancer has designated NNK and NNN (N-nitrosonornicotine) as known human carcinogens. NNK is metabolically activated by CYP P-450 enzymes in the lung and generates O6-methylguanine in DNA. The reaction generates G:C to A:T mutations, with the subsequent activation of K-Ras proto-oncogene and the development of tumor initiation.

Radiation also contributes to the causative physical factors that induce human cancers. Radiation promotes double-strand breaks (DSBs) in DNA, which leads to chromosome aberrations and cell death and generates a variety of oxidative DNA damage. Because of the genotoxicity potential, radiation at high doses evidently results in the appearance of various tumors in humans. Even at low doses, residential exposure to radioactive radon and its decay products, for instance, may account for about 10–20% of all lung cancer deaths worldwide.

31.5 CANCER CHEMOPREVENTION

31.5.1 General Considerations

It is important to preface the topic of cancer chemoprevention by proposing the following—most of the evidence for the prevention of known or frequently occurring cancers is *á priori*. In fact, epidemiological data contributes to most of our understanding of the mechanisms of carcinogenic chemicals. Also, animal experimentation may support the discussion by expanding on the explanation of the factors that increase the risk for multistage carcinogenesis. Nevertheless, there are two complementary strategies for understanding how predilection for cancer, and the agents that predispose humans to cancer, can be avoided:

1. Avoidance of exposure
2. Modulation of behavior, particularly by enriching the host defense

In reference to the second strategy, therefore, cancer chemoprevention is defined as the use of natural or synthetic pharmacological or dietary agents to prevent, arrest, or reverse carcinogenesis at its earliest stages. Because carcinogenesis is a multistage process and often has a latency period of years, there is considerable opportunity for intervention via chemoprevention. Animal experiments and human clinical trials support this contention (see Review Articles).

31.5.2 Cancer Chemopreventive Agents

Due to our increased understanding of the molecular and cellular mechanisms underlying carcinogenesis, a number of potential targets for cancer chemoprevention have recently been identified. The chemopreventive agents most often cited include the:

1. Selective estrogen receptor modulators, such as tamoxifen and raloxifene
2. Anti-inflammatory agents, including aspirin and celecoxib
3. Antioxidants and phase 2 enzyme inducers, such as vitamin C, vitamin E, omega-3 fatty acids, isothiocyanates, and dithiolethiones
4. Agents that selectively modulate cellular receptors and signal transduction, such as retinoids and vitamin D analogs

31.5.3 Mechanisms Underlying Cancer Chemoprevention

The mechanisms that contribute to the protective effects against a particular carcinogenic agent may operate according to the following premises:

1. Mechanisms leading to alteration of the toxicokinetics of carcinogens
2. Mechanisms resulting in the inhibition of the multiple stages of carcinogenesis

Chemopreventive agents may alter carcinogen toxicokinetics by either inhibiting the absorption of carcinogens or increasing their detoxification. This is actuated primarily by inducing phase 2 and antioxidative enzymes. Alternatively, chemopreventive drugs inhibit the three stages of carcinogenesis (initiation, promotion, and progression) by modulating a number of cellular mechanisms and pathways, such as 1. stimulation of DNA repair, 2. induction of apoptosis and inhibition of cell proliferation, 3. stimulation of immune system, and 4. inhibition of neovascularization.

Several recent studies have elucidated the molecular events leading to the induction of antioxidative and phase 2 enzymes by chemopreventive agents. These reports are described in detail in the references.

REFERENCES

SUGGESTED READINGS

Atienzar, F.A. and Jha, A.N., The random amplified polymorphic DNA (RAPD) assay and related techniques applied to genotoxicity and carcinogenesis studies: a critical review. *Mutat. Res.* 613, 76, 2006.

Au, W.W., Usefulness of biomarkers in population studies: from exposure to susceptibility and to prediction of cancer. *Int. J. Hyg. Environ. Health* 210, 239, 2007.

da Costa, A.N. and Herceg, Z., Detection of cancer-specific epigenomic changes in biofluids: powerful tools in biomarker discovery and application. *Mol. Oncol.* 6, 704, 2012.

Gately, S. and Li, W.W., Multiple roles of COX-2 in tumor angiogenesis: a target for antiangiogenic therapy. *Semin. Oncol.* 31, 2, 2004.

Jacobson-Kram, D. and Contrera, J.F., Genetic toxicity assessment: employing the best science for human safety evaluation. Part I: Early screening for potential human mutagens. *Toxicol. Sci.* 96, 16, 2007.

Jakóbisiak, M., Lasek, W., and Gołab, J., Natural mechanisms protecting against cancer. *Immunol. Lett.* 90, 103, 2003. Erratum in: *Immunol. Lett.* 91, 255, 2004.

Kirkland, D.J., *In vitro* approaches to develop weight of evidence (WoE) and mode of action (MoA) discussions with positive *in vitro* genotoxicity results. *Mutagenesis* 22, 161, 2007.

Klaassen, C.D., *Casarett & Doull's Toxicology: The Basic Science of Poisons*, 8th ed., McGraw-Hill, New York, 2013, chapter 8.

Klaunig, J.E., Acrylamide carcinogenicity. *J. Agric. Food Chem.* 56, 5984, 2008.

Little, M.P., Heidenreich, W.F., Moolgavkar, S.H., Schöllnberger, H., and Thomas, D.C., Systems biological and mechanistic modelling of radiation-induced cancer. *Radiat. Environ. Biophys.* 47, 39, 2008.

Lorge, E., Gervais, V., Becourt-Lhote, N., Maisonneuve, C., Delongeas, J.L., and Claude, N., Genetic toxicity assessment: employing the best science for human safety evaluation part IV: A strategy in genotoxicity testing in drug development: some examples. *Toxicol. Sci.* 98, 39, 2007.

Marks, F., Fürstenberger, G., and Müller-Decker, K., Tumor promotion as a target of cancer prevention. *Recent Results Cancer Res.* 174, 37, 2007.

NTP (National Toxicology Program), Department of Health and Human Services, Report on Carcinogens (RoC), http://ntp.niehs.nih.gov, last viewed March 2009.

Nishigori, C., Hattori, Y., and Toyokuni, S., Role of reactive oxygen species in skin carcinogenesis. *Antioxid. Redox Signal.* 6, 561, 2004.

Nomura, T., Transgenerational effects from exposure to environmental toxic substances. *Mutat. Res.* 659, 185, 2008.

O'Brien, J., Renwick, A.G., Constable, A., Dybing, E., Müller, D.J., Schlatter, J., Slob, W., Tueting, W., van Benthem, J., Williams, G.M., and Wolfreys, A.,Approaches to the risk assessment of genotoxic carcinogens in food: a critical appraisal. *Food Chem. Toxicol.* 44, 1613, 2006.

Romani, M., Pistillo, M.P., and Banelli, B., Environmental epigenetics: crossroad between public health, lifestyle, and cancer prevention. *Biomed. Res. Int.* 2015, 587983, 2015.

Siddiqui, I.A., Afaq, F., Adhami, V.M., and Mukhtar, H., Prevention of prostate cancer through custom tailoring of chemopreventive regimen. *Chem. Biol. Interact.* 171, 122, 2008.

Trosko, J.E., Dietary modulation of the multistage, multimechanisms of human carcinogenesis: effects on initiated stem cells and cell-cell communication. *Nutr. Cancer* 54, 102, 2006.

Trosko, J.E., Chang, C.C., Upham, B.L., and Tai, M.H., Ignored hallmarks of carcinogenesis: stem cells and cell-cell communication. *Ann. N.Y. Acad. Sci.* 1028, 192, 2004.

Valavanidis, A., Fiotakis, K., and Vlachogianni, T., Airborne particulate matter and human health: toxicological assessment and importance of size and composition of particles for oxidative damage and carcinogenic mechanisms. *J. Environ. Sci. Health C Environ. Carcinog. Ecotoxicol. Rev.* 26, 339, 2008.

Wu, K.M., Ghantous, H., and Birnkrant, D.B., Current regulatory toxicology perspectives on the development of herbal medicines to prescription drug products in the United States. *Food Chem. Toxicol.* 46, 2606, 2008.

REVIEW ARTICLES

Bode, A.M. and Dong, Z., Molecular and cellular targets. *Mol. Carcinog.* 45, 422, 2006.

Carnero, A., Blanco-Aparicio, C., Kondoh, H., Lleonart, M.E., Martinez-Leal, J.F., Mondello, C., Scovassi, A., Bisson, W.H., Amedei, A., Roy, R., Woodrick, J., Colacci, A., Vaccari, M., Raju, J., Al-Mulla, F., Al-Temaimi, R., Salem, H.K., Memeo, L., Forte, S., Singh, N., Hamid, R.A., Ryan, E.P., Brown, D.G., Wise, J.P. Sr., Wise, S.S., and Yasaei, H., Disruptive chemicals, senescence and immortality. *Carcinogenesis* 36, S19, 2015.

Christofori, G., New signals from the invasive front. *Nature* 441, 444, 2006.

Claxton, L.D. and Woodall Jr., G.M., A review of the mutagenicity and rodent carcinogenicity of ambient air. *Mutat. Res.* 636, 36, 2007.

Ghanayem, B.I. and Hoffler, U., Investigation of xenobiotics metabolism, genotoxicity, and carcinogenicity using Cyp2el(-/-) mice. *Curr. Drug Metab.* 8, 728, 2007.

Gonzalez, F.J. and Kimura, S., Study of P450 function using gene knockout and transgenic mice. *Arch. Biochem. Biophys.* 409, 153, 2003.

Guengerich, F.P., Forging the links between metabolism and carcinogenesis. *Mutat. Res.* 488, 195, 2001.

Hengstler, J., Bogdanffy, M.S., Bolt, H.M., and Oesch, F., Challenging dogma: thresholds for genotoxic chemicals? The case of vinyl acetate. *Annu. Rev. Pharmacol. Toxicol.* 43, 485, 2003.

Kuper, H., Boffetta, P., and Adami, H.O., Tobacco use and cancer causation: association by tumour type. *J. Intern. Med.* 252, 206, 2002.

Laconi, E., Doratiotto, S., and Vineis, P., The microenvironments of multistage carcinogenesis. *Semin. Cancer Biol.* 18, 322, 2008.

Mohan, C.G., Gandhi, T., Garg, D., and Shinde, R., Computer-assisted methods in chemical toxicity prediction. *Mini. Rev. Med. Chem.* 7, 499, 2007.

Nguyen, T., Sherratt, P.J., and Pickett, C.B., Regulatory mechanisms controlling gene expression mediated by the antioxidant response element. *Annu. Rev. Pharmacol. Toxicol.* 43, 233, 2003.

Preston, R.J. and Williams, G.M., DNA-reactive carcinogens: mode of action and human cancer hazard. *Crit. Rev. Toxicol.* 35, 673, 2005.

Preston, R.J., Quantitation of molecular endpoints for the dose-response component of cancer risk assessment. *Toxicol. Pathol.* 30, 112, 2002.

Richardson, S.D., Plewa, M.J., Wagner, E.D., Schoeny, R., and Demarini, D.M.,Occurrence, genotoxicity, and carcinogenicity of regulated and emerging disinfection by-products in drinking water: a review and roadmap for research. *Mutat. Res.* 636, 178, 2007.

Saha, S.K., Lee, S.B., Won, J., Choi, H.Y., Kim, K., Yang, G.M., Dayem, A.A., and Cho, S.G., Correlation between oxidative stress, nutrition, and cancer initiation. *Int. J. Mol. Sci.* 18, 1544, 2017.

Shaughnessy, D.T., McAllister, K., Worth, L., Haugen, A.C., Meyer, J.N., Domann, F.E., Van Houten, B., Mostoslavsky, R., Bultman, S.J., Baccarelli, A.A., Begley, T.J., Sobol, R.W., Hirschey, M.D., Ideker, T., Santos, J.H., Copeland, W.C., Tice, R.R., Balshaw, D.M., and Tyson, F.L., Mitochondria, energetics, epigenetics, and cellular responses to stress. *Environ. Health Perspect.* 122, 1271, 2014.

Sporn, M.B. and Suh, N., Chemoprevention: an essential approach to controlling cancer. *Nat. Rev. Cancer* 2, 537, 2002.

32 Reproductive and Developmental Toxicity

Anirudh J. Chintalapati and Frank A. Barile

32.1 INTRODUCTION

Drugs or chemicals that induce structural malformations, physical dysfunction, behavioral alterations, or genetic deficiencies in the fetus, or impair fertility, are referred to as *teratogens*. The expression of the teratogenic response—that is, the embryotoxic effect of a compound on the growth and development of the fetus—is usually manifested at birth or in the immediate postnatal period. In addition, a chemical may interfere early in the reproductive cycle (first trimester) by impairing fertilization or interfering with implantation. These initial interactions between the prefertilized ovum and the xenobiotic may result in interference with ovum production, fertilization, implantation, or embryo development, or may initiate spontaneous termination of pregnancy.

Clinical and nontherapeutic drug use during the prenatal period has increased dramatically in recent years. About 90% of women have ingested one or more medications while pregnant (FDA, 2018). Excluding the commonly prescribed prenatal vitamins, iron supplements, and tocolytic* drugs, women under 35 years of age on average consume three prescriptions during the course of their pregnancies; for women over 35, this number rises to five. Some of the drugs used most extensively during pregnancy include antibiotics, analgesics, and narcotics, followed by antiepileptics, antihypertensives, antinauseants, psychotherapeutic agents, and respiratory medications. For most of these drugs, no well-controlled studies have been conducted in pregnant humans. Despite the current lack of information on the safety of drug products during pregnancy, there appears to be little reluctance to prescribe.

Although initially alarming, lack of sufficient toxicological investigation is evidence for the complacency and dearth of knowledge both in the public health community and within the medical profession, and highlights the insidious dangers of drug use during pregnancy. In addition, the well-intended practice of prescribing lower doses of drugs during pregnancy to minimize exposure to the fetus is misguided. The complicating factor lies in the ubiquitous availability and distribution of therapeutic drugs, as well as environmental chemicals and herbal products, that necessarily infiltrate the maternal/fetal environment. Although some therapeutic drugs are screened for their embryotoxic or teratogenic potential in animal reproductive studies, the majority of chemicals are not. In fact, the U.S. Food and Drug

* Any drug used to suppress premature labor.

Administration (FDA) and other federal agencies do not impose extensive requirements on pharmaceutical and chemical companies to provide such information. Also, most drug development programs do not routinely incorporate screening or follow-up of drugs for potential embryotoxic effects during preclinical or clinical phases or in the postmarketing period. Finally, chemicals not destined to be used clinically, but found in abundance throughout the environment (pesticides, solvents, and metals), generally do not undergo any teratogenic testing. This leaves a multitude of chemicals used in the marketplace without human embryotoxic data.

32.2 HISTORY AND DEVELOPMENT

The thalidomide disaster of 1961 initiated a significant effort to establish a program for teratogenicity testing. Developed as a sedative/hypnotic with no particular advantage over drugs of the same class, thalidomide was initially shown to lack teratogenic effects in all species tested *except* rabbits. Soon after its introduction to the European market, the drug was linked to the development of a relatively rare birth defect known as *phocomelia*. The epidemic proportions of the teratogenic effect of thalidomide prompted the passage of the Harris–Kefauver Amendment in 1962, one of many additions since to the Federal Pure Food and Drug Act. The Amendment requires extensive pharmacological and toxicological preclinical research before a therapeutic compound can be marketed.*

32.3 SUMMARY OF MATERNAL–FETAL PHYSIOLOGY

32.3.1 DEFINITIONS

Fetal development involves the gradual establishment and modification of anatomical structures from its beginning at fertilization (conception) to maturity. Figure 32.1 illustrates the phases of embryonic and fetal development as they occur within the trimesters. This figure denotes how the first 12 weeks roughly correspond to the first trimester. This phase is characterized as the period of time for *embryonic development*, when the fertilized ovum differentiates into precursor stem cells, which progress to the launching of fetal membranes and the embryonic disc (weeks 1–3, *embryo*). The remainder of the pregnancy (13 to 38 weeks) is dedicated to *fetal development*, when the established blueprints of organs and tissues undergo further growth and maturity. Together, the two processes occurring during the weeks of embryonic and fetal expansion are referred to as *prenatal development*. Thus, in humans, the gestational period is normally of 9 months' duration (38 weeks or 280 days from the first day of the last menstrual period, assuming a regular 28 day cycle). *Postnatal development* therefore begins at birth and progresses to maturity (i.e., for approximately 18 years). Particular events that distinguish the trimesters in Figure 32.1 are further described in the following subsection.

---

* Interestingly, almost 50 years later, thalidomide has resurfaced (Thalomid®) and is used in the United States for the treatment of erythema nodosum leprosum. The package is printed within and without with red warning labels alerting the prescriber, pharmacist, and patient to the precaution against its use in women of childbearing potential.

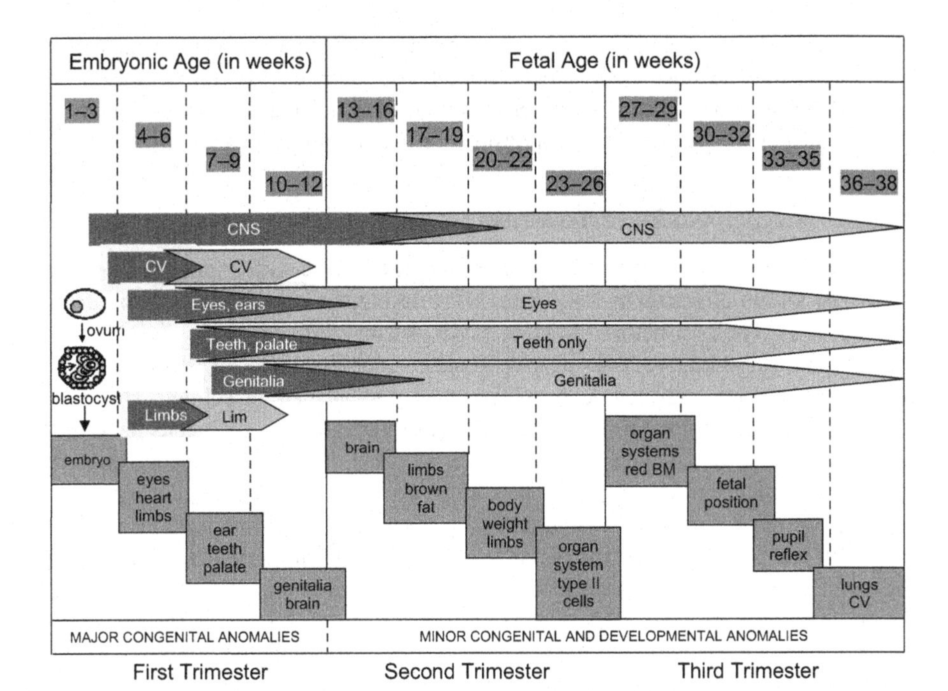

FIGURE 32.1 Timeline of Fetal Development and Teratogenic Susceptibility.

32.3.2 First Trimester

The *first trimester* is the period corresponding to embryonic and early fetal development (weeks 1 through 12). The main events occurring during the first trimester include:

1. *Cleavage* (days 1–6): a series of cell divisions following fertilization of the ovum by the sperm cell and terminating at the blastocyst stage (day 6, weeks 1–3 in figure).
2. *Implantation* (days 7–10, not shown): adhesion, attachment, and penetration of the uterine endometrial lining by the blastocyst; the trophoblast, amniotic cavity, and inner cell mass form and expand during this stage; further development of the yolk sac, chorion, allantois, amnion, and embryonic disc (from the inner cell mass) ensues during weeks 3 and 4 (beginning of organogenesis).
3. *Placentation* (from weeks 2 to 10): starting as early as week 2, blood vessels form around the periphery of the blastocyst to establish the presence of the placenta; the placenta is characterized by the formation of chorionic villi and the appearance of fetal vessels.
4. *Embryogenesis* (also during weeks 4–12)—the process of differentiation and folding of the embryonic disc, producing a physically and developmentally distinct and recognizable embryo. During these weeks, critical organs take shape and continue development, including the sensory organs, the central nervous system (CNS), and the cranial and facial skeleton and long bones.

Because of the critical events occurring during this phase, the first trimester represents the most sensitive period of fetal development to external stimuli. These structures are the most sensitive to drugs and chemicals (represented and labeled in red) and are responsible for major congenital anomalies.

32.3.3 Second and Third Trimesters

The *second trimester* corresponds to the third to the sixth month of gestation (weeks 13 to 26) and is characterized by rapid fetal development of organs and systems (Figure 32.1). These structures, such as CNS, sensory organs, teeth, gums, limbs, and external genitalia, are less sensitive to pharmacological and toxicological interactions with drugs and chemicals (represented and labeled in yellow). However, exposure to xenobiotics during the second trimester is also responsible for minor congenital anomalies, which are manifested as functional defects in the newborn (e.g., development of type II pulmonary alveolar epithelial cells during expansion of the lungs in preparation for secretion of surfactant). The *third trimester* displays rapid growth and maturation of established organ systems as the foundation for independent survival and function. For example, the development of critical organs during this phase includes formation of red bone marrow (Figure 32.1: BM, weeks 27–29), orientation of the fetal position (weeks 30–32), establishment of brain stem reflexes (weeks 32–35), and final preparation of the pulmonary and cardiovascular (CV, weeks 36–38) systems for birth.

32.4 MECHANISMS OF DEVELOPMENTAL TOXICITY

32.4.1 Susceptibility

As noted earlier, the first 8 to 12 weeks of embryogenesis qualify as a sensitive period for the fetus to the presence of exogenous substances. For instance, exposure to compounds 21 days after fertilization has the potential for greatest embryotoxicity, resulting in the death of the fetus, whereas exposure to compounds between the third and eighth weeks is potentially teratogenic—that is, these agents may interfere with organogenesis. These effects are distinguished by the development of gross anatomic, metabolic, or functional defects and the precipitation of spontaneous abortions.

Because the schematic for embryonic development is already established by the beginning of the second trimester, exposures to drugs or chemicals during the second or third trimester are less likely to be teratogenic. However, they still may interfere with the growth and maturation of critical organ systems, resulting in underdevelopment or alterations of normal organ function.

32.4.2 Dose–Response and Threshold

Exchange of nutrients, gases, and waste material between the maternal and fetal circulations starts from the fifth week after fertilization with the initiation of the formation of the placenta. Maternal and fetal circulations do not mix. Instead, maternal blood enters intervillous spaces (sinuses) of the placenta through ruptured maternal arteries and then drains into uterine veins for return to the maternal circulation. Fetal blood enters the placenta by a pair of umbilical arteries and returns to the fetal circulation by the umbilical vein. The umbilical arteries branch into capillaries surrounded by the syncytial trophoblast, forming the network of chorionic villi. Solutes, gases, and nutrients from the maternal circulation enter the sinuses, surround and bathe chorionic villi, and traverse the epithelium and connective tissues of the villi before penetrating the fetal capillary endothelial cells. Soluble substances are then carried toward the embryo through the umbilical vein.

Chemicals enter the fetal circulation as with any other membrane barrier—that is, the rate of exchange between the maternal and fetal circulations depends on the established equilibrium across the placenta and the rate of penetration of the chemical through the endothelial/epithelial barriers. Lipid-soluble drugs and chemicals, and dissolved gases, readily infiltrate placental membranes, establishing high concentrations of the substance in venous cord blood. Thus, compounds may freely circulate around the embryo within 32 to 60 minutes after maternal administration. In addition, exogenous substances can change the dynamic tone of placental blood vessels, altering nutrient, gas, and waste exchange. In addition, they restrict oxygen transport to fetal tissues, change the biochemical dynamics of the two circulations, and interfere with the endocrine function of the placenta.*

32.5 DRUGS AFFECTING EMBRYONIC AND FETAL DEVELOPMENT

32.5.1 Classification

The U.S. FDA has established five categories of drug safety during pregnancy. Table 32.1 outlines the requirements for safety of prescribed drugs according to these criteria. The regulations act as guidelines for benefit-to-risk decisions for the use of medications during pregnancy. Pharmacological preparations marketed in the United States bear a drug safety category designation.

32.5.2 Drug Classes

Tables 32.2 through 32.4 categorize a variety of drug classes, their therapeutic uses, and some of their major adverse effects on the human fetus. The tables further distinguish agents whose teratogenicity is most frequently seen when administered during the critical first trimester (Table 32.2) and during the second and third trimesters (Table 32.3). Finally, substances whose complications are associated with undesirable or hazardous social habits and illicit drug use are listed in Table 32.4.

It is important to remember that because embryonic tissues grow rapidly and have a high DNA turnover rate, they are vulnerable to many drugs and chemicals that may not necessarily promote adverse reactions in children or adults. As explained earlier,

* In its endocrine capacity, the placenta secretes human chorionic gonadotropin, placental prolactin, relaxin, estrogen, and progesterone. Also, the degree of chemical penetration of placental barriers relies less on the number and thickness of the layers and more on the dynamic physicochemical interactions between the chemical and the placenta.

TABLE 32.1
FDA-Established Drug Safety Categories during Pregnancy

Category	Description
A	Adequate, well-controlled studies in pregnant women have not shown an increased risk of fetal abnormalities
B	Animal studies have revealed no evidence of harm to the fetus; however, there are no adequate and well-controlled studies in pregnant women; *or* Animal studies have shown an adverse effect, but adequate and well-controlled studies in pregnant women have failed to demonstrate a risk to the fetus.
C	Animal studies have shown an adverse effect, and there are no adequate and well-controlled studies in pregnant women; *or* No animal studies have been conducted, and there are no adequate and well-controlled studies in pregnant women.
D	Adequate well-controlled or observational studies in pregnant women have demonstrated a risk to the fetus. However, the benefits of therapy may outweigh the potential risk.
X	Adequate well-controlled or observational studies in animals or pregnant women have demonstrated positive evidence of fetal abnormalities. The use of the product is contraindicated in women who are, or may become, pregnant.

the most sensitive period for embryonic development is within the first trimester of pregnancy, when the fetus is in the early stages of gestation. Some classes of agents require repeated, continuous, or high-dose exposure to elicit an untoward reaction. Other more potent drugs, such as antineoplastic agents and synthetic retinoids, target rapidly proliferating cells and tissues. Their effects are significant and occur with higher frequency. In fact, the latter category carries a 25% risk of birth defects and anomalies. Some of the teratogenic risks estimated with specific classes of drugs are based on previous observations of effects with known teratogens.* Other teratogenic effects are based on clinical reports of women undergoing treatment for chronic conditions with anticonvulsants, thyroid medications, psychoactive drugs, and bacterial and fungal antibiotics. Lastly, complications surrounding multiple and illicit drug use (Table 32.4) and their effects on the human fetus are difficult to assign to individual compounds. Intravenous drug users (IDU) are at greater risk of developing anemia, bacteremia, hepatitis, endocarditis, venereal disease, and AIDS.† Thus, children born to drug-addicted mothers are more likely to suffer the consequences of such complications than to succumb to the effects of the indirect action of the compound.

* The effects of sex hormones are based on the 1950s experience with the routine administration of diethylstilbestrol (DES) during pregnancy for the treatment of threatened or habitual abortions. Daughters of the mothers exposed to DES displayed *in utero* and post-natal an increased incidence of clear cell adenocarcinoma of the vagina, menstrual dysfunction, spontaneous abortions, incompetent cervix, and preterm labor. In males, exposure to DES resulted in meatal stenosis and hypospadias both *in utero* and post-natal.

† Acquired immune deficiency syndrome.

TABLE 32.2
Drug Classes That Affect Embryogenesis and Organogenesis in the Human Fetus (Primarily the First Trimester)

Drug Class[a]	Representative Drugs within This Class	Therapeutic Use	Some Adverse Effects on the Human Fetus
Antineoplastic agents	Methotrexate, 6-MP, cyclophosphamide, chlorambucil, busulfan	Treatment of cancers and neoplasms	Growth retardation, craniofacial hypoplasia, clubfoot
Synthetic retinoids (vitamin A)	Isotretinoin	Acne vulgaris	Cardiac defects, hydrocephalus, isolated mental retardation, spontaneous abortions
Androgenic, anabolic, and progestational hormones	Synthetic androgens (testosterone derivatives), progestins, estrogens	Various indications (see Chapter 17)	Masculinization of female external genitalia, increased risk of reproductive developmental consequences later in life (see text footnotes on DES)
Anticonvulsants	Phenytoin, phenobarbital, carbamazepine	Epilepsy and related disorders	Cleft palate; cardiac, craniofacial, visceral abnormalities; mental retardation (effects are based on severity of condition, dose, and concurrent use of anticonvulsants)
Vaccines	Rubella (other vaccines may be given depending on risk of infection)	Prevention of bacterial and viral infections	Placental and fetal infections
Thyroid drugs	I^{131}, T_3, methimazole, PTU	Hyper- and hypothyroidism	Fetal hypothyroidism, goiter, scalp defects (methimazole only)
Psychoactive agents	Lithium carbonate	Manic depression	Cardiovascular malformations (19% frequency)
Anticoagulants	Coumarins	Hypercoagulation disorders	"Fetal warfarin syndrome"[b]

[a] The drugs listed here have been shown to cause, or increase the risk for, human fetal abnormalities (pregnancy categories D and X). Compounds that have demonstrated animal teratogenicity, but not human fetal toxicity, are not included in this table (pregnancy categories A, B, and C). The toxic profiles of most of these agents are discussed in other chapters. The teratogenicity of these agents is generally associated with exposure early in pregnancy (i.e., during the first 12 weeks of gestation).

[b] "Fetal warfarin syndrome" is seen in 25% of fetuses exposed to coumarin anticoagulants during the first trimester and displays as mental retardation, bone stippling, and optic atrophy. Exposure during the second and third trimesters is manifest as optic atrophy, cataracts, mental retardation, microcephaly, and microophthalmia.

6-MP: 6-mercatopurine; I^{131}: radioactive isotope iodine-131; PTU: propylthiouracil; T_3: triiodothyronin.

TABLE 32.3
Drug Classes That Affect Human Fetal Development, Labor, or Delivery or are Associated with Postnatal or Neonatal Complications

Drug Class[a]	Therapeutic Use	Representative Drugs within This Class	Some Adverse Effects in the Human Fetus
Opioids and related narcotics; salicylates	Analgesia	Any of the opioids	Narcotic withdrawal syndrome
		Salicylates, ASA	Fetal kernicterus, delayed onset of labor (large doses), neonatal bleeding
Neuroleptics, sedative/hypnotics	Psychoses	Diazepam (when given near term or during labor and delivery)	*Perinatal*: depression, irritability, tremors, hyperreflexia; *delivery*: low Apgar scores, neurologic depression (no significant fetal risk associated with phenothiazines, tricyclics, and most anxiolytics)
		Lithium carbonate	*Perinatal effects*: lethargy, hypotonia, goiter, poor feeding
Antibiotics	Bacterial, viral, and fungal infections	Tetracyclines	Retardation of bone growth, permanent yellowish staining of teeth, ↓ resistance to dental caries
		Aminoglycosides	Ototoxicity
		Chloramphenicol	"Gray baby syndrome" (when administered near term)[b]
		Sulfonamides	Jaundice (when administered near term)
Cardiovascular drugs and diuretics	Cardiac disorders, hypertension	Propranolol	Bradycardia, hypoglycemia, growth retardation
		Thiazide diuretics	↓ Fetal oxygenation, hyponatremia, hypokalemia, thrombocytopenia
		ACE inhibitors	Fetal renal failure, oligohydramnios sequence[c]
Oral hypoglycemics	Diabetes mellitus	Any of the oral hypoglycemics	Profound hypoglycemia in newborns
Drugs used during labor and delivery	Local anesthesia	Local anesthetics (lidocaine)	*Maternal effects*: CNS depression, bradycardia
	Caesarean delivery	Thiopental (barbiturates)	Concentration in fetal liver
	To arrest premature labor	Magnesium sulfate	Lethargy, hypotonia, respiratory depression

[a] Teratogenicity with these agents is associated with exposure later in pregnancy (i.e., during the second and third trimesters).

[b] "Gray baby syndrome" is associated with the inability of the newborn to metabolize the drug, resulting in high blood levels. The syndrome is characterized by cardiovascular and circulatory collapse.

[c] Characterized by craniofacial deformities, limb contractures, and poor lung development when administered primarily during the second and third semesters.

ACE: angiotensin-converting enzyme; ASA: acetylsalicylic acid (aspirin).

TABLE 32.4
Adverse Effects of Social Habits and Illicit Drugs on the Human Fetus

Social Habits and Illicit Drug Use[a]	Some Adverse Effects on the Human Fetus
Cigarette smoking[b]	Reduction in birth weight; ↑ risk for spontaneous abortions, placenta previa, premature rupture of membranes, congenital heart defects, orofacial clefts, SIDS; ↓ physical, intellectual, and behavioral development[b]
Consumption of alcoholic beverages	Fetal alcohol syndrome: effects include spontaneous abortions, impairment of physical (CNS dysfunction, CV defects, growth retardation), intellectual (mental retardation), and behavioral development
Marijuana, opioids, amphetamines	Effects of individual agents are complicated by the unstable social, economic, and emotional environment of illicit and multiple drug use
Cocaine	Spontaneous abortions, fetal tissue hypoxia, intrauterine growth retardation, circulatory compromise, skeletal defects, hyperactivity, tremors
Caffeine	No apparent risk with moderate coffee drinking

[a] Teratogenicity with these agents is associated with exposure during any stage of pregnancy. Cigarette smoking and alcohol consumption are the most common drug/behavioral addictions; alcohol is the leading teratogenic substance known among pregnant women.

[b] Effects of cigarette smoking on fetal development are mediated by the chronic hypoxic consequences of carbon monoxide and by the vasoconstrictor/stimulating effects of nicotine on fetal tissues.

SIDS: sudden infant death syndrome.

32.6 ENDOCRINE DISRUPTING CHEMICALS (EDCS)

An EDC is defined as "an exogenous (non-natural) chemical, or mixture of chemicals, that interferes with endogenous hormonal action." Naturally occurring EDCs, such as the phytoestrogens genistein and coumestrol, are occasionally found in human and animal foods. EDCs target and modulate the endocrine system (ES), which comprises a series of secretory glands and their hormonal secretions. The release of glandular secretions into the systemic circulation is responsible for regulation of a variety of biological functions essential for metabolism, growth, and reproduction. Table 32.5 summarizes a list of endocrine glands and their hormonal effects. Consequently, EDCs interfere with and disrupt hormonally dependent homeostasis of physiological systems, which may result in lifelong toxic manifestations.

EDCs commonly appear as additives in commercial products such as pesticides, herbicides, and plasticizers, and as contaminants in processed foods, potable water, and personal care goods. Recent estimates suggest that the total annual health cost from EDC exposure in the United States, including expenditures for treatment of associated disorders accompanied by loss of productivity, is $340 million, which accounts for 2.3% of GDP. A list of commonly encountered EDCs and their probable disease links is summarized in Table 32.6.

TABLE 32.5
List of Endocrine Glands, Hormones, and Physiological Effects

Endocrine Gland	Hormone(s) Secreted	General Effect(s)
Adipose tissue	Peptides, cytokines, and steroids	Metabolic homeostasis
Adrenal	Glucocorticoids—cortisol	Stress and immune response
	Mineralocorticoids—aldosterone	Blood pressure regulation
	Sex steroids—DHEA	Musculoskeletal growth
Hypothalamus	GHRH	Growth
	TRH	Metabolism
	CRH	Stress and immune response
	GnRH	Reproduction
	Dopamine	Inhibits lactation
Ovary	Estrogens	Reproduction
	Progesterones	
Pancreas	Insulin	Regulate blood sugar levels
	Glucagon	
Parathyroid	Parathyroid hormone	Calcium homeostasis
Pituitary	GH	Promotes growth, and metabolism
	TSH	Metabolism
	ACTH	Stimulates adrenals to produce cortisol
	LH	Initiates and maintains reproductive cycle
	FSH	Initiates and maintains reproductive cycle
	Prolactin	Milk production
	Oxytocin	Uterine contraction
	Vasopressin	Electrolyte and blood pressure balance
Testes	Androgens—Testosterone	Reproduction
Thyroid	Thyroid hormone	Metabolism
	Calcitonin	Calcium balance

32.6.1 Mechanism of Toxicity

EDC-mediated toxicity is a multiple mechanistic paradigm. Despite the absence of an established toxic pathway and profile, the biological adverse effects of EDCs are mediated by selective manipulation of the ES in two distinct yet interlinked pathways. Initially, EDCs mimic the effect of natural endogenous hormones; moreover, EDCs then interfere with their synthesis, transport, metabolism, and elimination, thereby altering hormonal plasma concentrations. The former process, however, requires the EDC to bear significant structural similarity to the endogenous molecules, rendering them a remarkable affinity to endogenous receptors to evoke or antagonize a biological response. Alternatively, toxicity caused by EDCs can be described as genomic or may follow nongenomic mechanisms, as outlined in Figure 32.2. For instance, BPA (Figure 32.3), used in the manufacture of plastic and epoxy resins, exhibits genomic toxicity through its ability to bind to both membrane and nuclear estrogen receptors (ERs), triggering changes in the expression of estrogen-responsive genes. In contrast,

TABLE 32.6
List of Common EDCs, Applications, and Probable Disease Links

EDC	Applications	Disease Links
Alkylphenols	Detergents, additives in fuels, lubricants, and phenolic resins	Breast cancer
Atrazine	Herbicide	Delayed puberty, carcinogenesis
Bisphenol A (BPA)	Plastic manufacture, epoxy resins in food contact materials	Breast and ovarian cancers, neurobehavioral anomalies, altered pubertal development
Cadmium	Cigarette smoke, batteries, photovoltaic cells, fertilizers	Reproductive organ damage causing testicular and ovarian injury, placental damage
DDT[a]	Pesticides	Breast cancer, and other cancers
Dioxin	By-products of fuel combustion	Cancer, testicular damage, diminished sperm quality, oligospermia, infertility, neurobehavioral abnormalities
Ethinyl estradiol	Synthetic steroid for oral contraception	Endometrial cancer, hepatotoxicity
Lead	Paint, drinking water, gasoline	Preterm birth, renal toxicity, oxidative stress, neurobehavioral abnormalities including diminished cognitive and learning skills
Mercury	Seafood, burning coal	Pancreatic damage, metabolic abnormalities, diabetes
Phthalates	Plastics, fragrances, medical tubing	Metabolic abnormalities, birth defects, oligospermia, asthma, neurobehavioral abnormalities
PCBs[b]	Electrical coolant, in building materials	Cancer, immunosuppression, thyroid atrophy, porphyria

[a] Dichlorodiphenyltrichloroethane.
[b] Polychlorinated biphenyls.

genistein (Figure 32.4), a phytoestrogen, exerts its cellular toxicity via a nongenomic pathway by inhibiting tyrosine kinase and protein kinase-B cell signaling pathways, producing cellular apoptosis. EDCs can also stimulate non-receptor-mediated endocrine disturbances through inhibition of the synthesis and metabolism of endogenous hormones. This phenomenon is typical of toxicity caused by polychlorinated biphenyls (PCBs), which inhibit estrogen sulfotransferases, raising the plasma concentration and half-life of circulating 17β-estradiol (Figure 32.5). Lastly, EDCs may act directly on the glucocorticoid or mineralocorticoid receptors or on steroidogenic pathways. Examples of these EDCs include dioxin, lindane, and PCBs, which ultimately affect the synthesis of adrenal steroids.

32.6.2 Effects of EDCs on Female Reproductive System

Exposure to EDCs has a significant impact on the organogenesis, development, and function of both male and female reproductive systems. However, the extent of impact is dependent on the developmental time frame of exposure (prenatal,

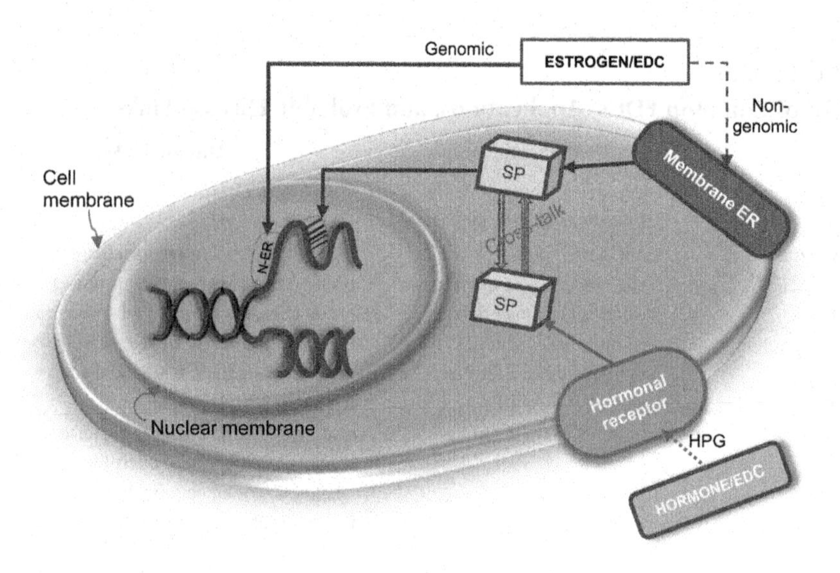

FIGURE 32.2 EDC toxicity pathways. HPG: hypothalamic–pituitary–gonadal axis; N-ER: nuclear estrogen receptor; SP: signaling protein to transcription activating site in nucleus.

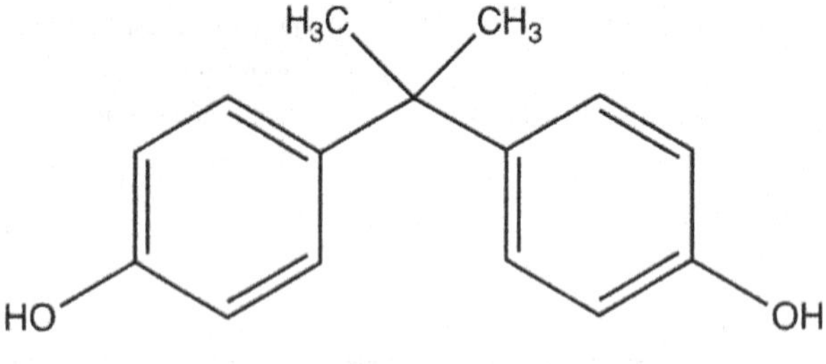

FIGURE 32.3 Chemical structure of bisphenol A (BPA).

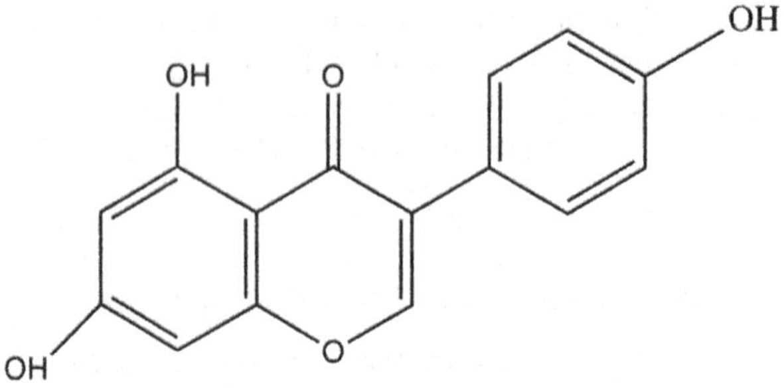

FIGURE 32.4 Chemical structure of genistein.

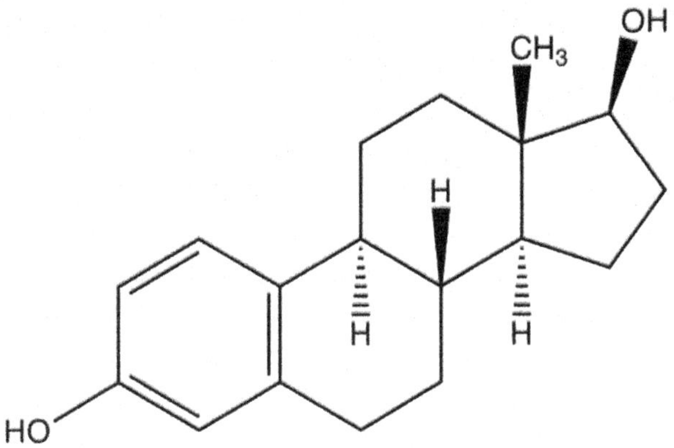

FIGURE 32.5 Chemical structure of 17β-estradiol.

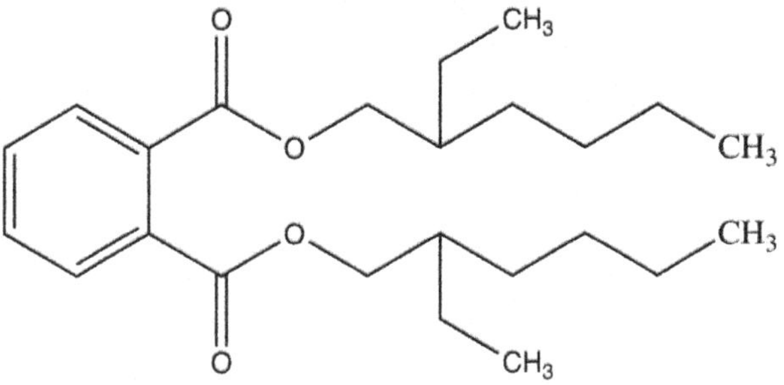

FIGURE 32.6 Chemical structure of di-(2-ethylhexyl) phthalate (DEHP).

perinatal, adolescence, adulthood), the route of exposure, and the pharmacokinetic and pharmacodynamic properties of the EDCs. Mammalian female reproductive system development is characterized by hormone-dependent (HD) and hormone-independent (HI) phases, of which the latter extends from fetal to prepubertal age and the former starts at puberty. Despite hormone independence, the fetal phase exhibits enormous sensitivity to EDC insult, principally from agents that mimic endogenous steroid hormones. For instance, BPA has estrogenic properties, which when present *in utero*, instigate irregular estrous cycles and early vaginal opening in female offspring.

The HD phase is characterized by the production and secretion of endogenous steroid hormones in response to gonadotropins secreted by the anterior pituitary. Estrogens are the primary steroid hormones responsible for maturation, developmental progression of the reproductive tract, initiation of the ovulation cycle, and development of secondary sexual characteristics. Following ovulation, the corpus luteum of the ovary produces and secretes progesterone, which maintains the endometrium for blastocyst development. EDCs affect the HD phase by either potentiating or inhibiting the effects of estrogens and progesterone to a variable extent, thus causing premature or delayed puberty, disruption of ovulation, and early onset of reproductive senescence. For example, di-(2-ethylhexyl) phthalate (DEHP, Figure 32.6), a chemical commonly added to plastics with a suspected human carcinogen profile, has shown evidence of reproductive, developmental, and neuroendocrine toxicity (ATSDR, 2001).

The hypothalamic–pituitary–gonadal (HPG) axis is the fundamental regulator of the HD phase of the female reproductive system. The hypothalamus secretes gonadotropin-releasing hormone (GnRH), which stimulates the anterior pituitary to produce the gonadotropins (GTs)—follicle stimulating hormone (FSH) and luteinizing hormone (LH). FSH and LH are shuttled to the ovaries via the systemic circulation and fuel the production of endogenous sex steroids—estrogens and progesterone. Figure 32.7 outlines the schematic of the HPG. The steroids serve as both positive and negative feedback signals to promote or inhibit the production of GnRH and GTs. Although knowledge of the specific effects of EDCs on the development and function of the HPG axis is lacking, their interference with this pathway is understood to result in partial to complete infertility.

32.6.3 Effects of EDCs on Male Reproductive System

The effect of EDCs on the development and function of the male reproductive system is hypothesized to occur by multiple mechanisms involving alterations in steroidogenesis, oxidative stress, and epigenetic (EG) pathways. As noted earlier, EDCs interfere with steroidogenesis by meddling with the HPG axis (Figure 32.8). For instance, exposure to phthalates such as di-(2-ethylhexyl) phthalate and BPA causes decreased testosterone, FSH, and LH levels. *In vivo* studies suggest that exposure to phthalates is linked to ROS induction, lipid peroxidation, and initiation of apoptosis in

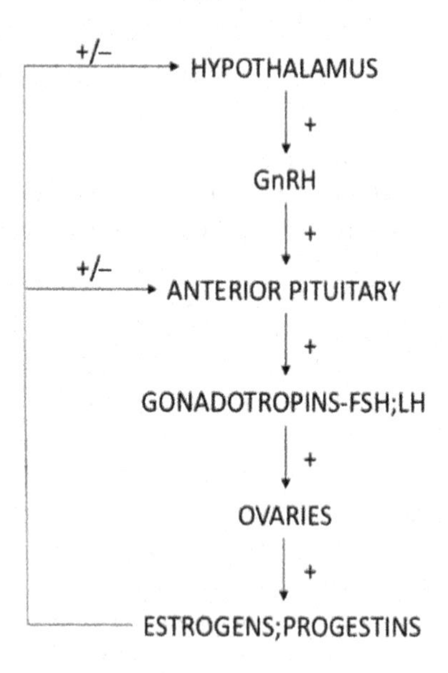

FIGURE 32.7 Diagram of hypothalamic–pituitary–gonadal axis in female. FSH: follicle stimulating hormone; GnRH: gonadotropin releasing hormone; LH: luteinizing hormone; –: negative feedback; +: positive feedback.

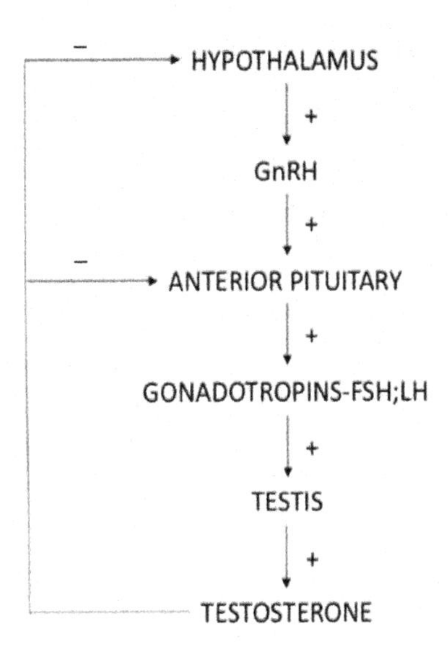

FIGURE 32.8 Diagram outlining schematic of hypothalamic–pituitary–gonadal axis in male. FSH: follicle stimulating hormone; GnRH: gonadotropin releasing hormone; LH: luteinizing hormone; –: negative feedback; +: positive feedback.

spermatocytes. EG mechanistic changes include DNA methylation, posttranslational modifications of histone proteins, and formation of microRNAs. DNA methylation is a transcriptionally repressive, transgenerational EG change. Neonatal exposure of rats to BPA led to aberrant DNA methylation in testis, indicating methylation-mediated EG changes as a possible mechanism of BPA-induced adverse effects on spermatogenesis and fertility.

32.6.4 Management of EDC Exposure

It is scientifically and toxicologically accepted that EDC exposure precipitates gonadotoxicity, cancer, and transgenerational reproductive anomalies. More detailed studies are necessary to propose well-documented therapeutic protocols to clarify and understand EDC toxicity. Based on existing evidence, screening programs can efficiently detect neoplasms and other endocrine consequences of EDC exposure. However, therapeutic testing and screening cannot proceed without extensive documentation of environmental and occupational exposure. Regulatory agencies should promote further toxicology research on reproductive health hazards pertaining to EDC exposure and their consequences over multiple generations.

REFERENCES

SUGGESTED READINGS

Agency for Toxic Substances and Disease Registry (ATSDR), Center for Disease Control and Prevention (CDC), *Toxicological Profiles*, DEHP; 2001. From: https://www.atsdr.cdc.gov/toxprofiles/tp.asp?id=684&tid=65; last viewed March 2019.

Barile, F.A., Chapter 10, Toxicology testing for fertility and reproduction, in *Principles of Toxicology Testing*, 2nd ed., CRC Press, Boca Raton FL, 2013.

Burns-Naas, L.A., Hastings, K.L., Ladics, G.S., Makris, S.L., Parker, G.A., and Holsapple, M.P.,What's so special about the developing immune system? *Int. J. Toxicol.* 27, 223, 2008.

Cho, E. and Li, W.J., Human stem cells, chromatin, and tissue engineering: boosting relevancy in developmental toxicity testing. *Birth Defects Res. C Embryo Today* 81, 20, 2007.

Gore, A.C., Developmental programming and endocrine disruptor effects on reproductive neuroendocrine systems. *Front. Neuroendocrinol.* 29, 358, 2008.

Guthrie, O.W., Aminoglycoside induced ototoxicity. *Toxicology* 30, 249, 2008.

Koren, G., Hutson, J., and Gareri, J., Novel methods for the detection of drug and alcohol exposure during pregnancy: implications for maternal and child health. *Clin. Pharmacol. Ther.* 83, 631, 2008.

Myllynen, P., Pasanen, M., and Vähäkangas, K., The fate and effects of xenobiotics in human placenta. *Expert Opin. Drug Metab. Toxicol.* 3, 331, 2007.

Piersma, A.H., Alternative methods for developmental toxicity testing. *Basic Clin. Pharmacol. Toxicol.* 98, 427, 2006.

Porter, R.S. and Kaplan, J.L. (Eds.), *The Merck Manual of Diagnosis and Therapy*, 19th ed., Merck Research Laboratories, NJ, 2011, chapter 18.

Rippon, H.J., Lane, S., Qin, M., Ismail, N.S., Wilson, M.R., Takata, M., and Bishop, A.E.,Embryonic stem cells as a source of pulmonary epithelium *in vitro* and *in vivo*. *Proc. Am. Thorac. Soc.* 5, 717, 2008.

Rogers, J.M., Tobacco and pregnancy: overview of exposures and effects. *Birth Defects Res. C Embryo Today* 84, 1, 2008.

Shankar, K., Ronis, M.J., and Badger, T.M., Effects of pregnancy and nutritional status on alcohol metabolism. *Alcohol Res. Health* 30, 55, 2007.

U.S. Food and Drug Administration (U.S. FDA), Natinal Center for Toxicological Research (NCTR), Women's Health Topics, updated 2014. From: https://www.fda.gov/ForConsumers/ByAudience/ForWomen/WomensHealthTopics/ucm117976.htm.

Weston, A.D., Ozolins, T.R., and Brown, N.A., Thoracic skeletal defects and cardiac malformations: a common epigenetic link? *Birth Defects Res. C Embryo Today* 78, 354, 2006.

Women's Health Initiatives, Office on Women's Health, Dept. of Health and Human Services (DHHS). www.womenshealth.gov/30-achievements/25; last updated March 2018, last viewed March 2019.

Younglai, E.V., Wu, Y.J., and Foster, W.G., Reproductive toxicology of environmental toxicants: emerging issues and concerns. *Curr. Pharm. Des.* 13, 3005, 2007.

REVIEW ARTICLES

Allegaert, K., van den Anker, J.N., Naulaers, G., and de Hoon, J., Determinants of drug metabolism in early neonatal life. *Curr. Clin. Pharmacol.* 2, 23, 2007.

Augustine-Rauch, K., Alternative experimental approaches for interpreting skeletal findings in safety studies. *Birth Defects Res. B Dev. Reprod. Toxicol.* 80, 497, 2007.

Barrow, P.C., Reproductive toxicology studies and immunotherapeutics. *Toxicology* 185, 205, 2003.

Dencker, L. and Eriksson, P., Susceptibility *in utero* and upon neonatal exposure. *Food Addit. Contam.* 15, 37, 1998.

Dietert, R.R. and Dietert, J.M., Possible role for early-life immune insult including developmental immunotoxicity in chronic fatigue syndrome (CFS) or myalgic encephalomyelitis (ME). *Toxicology* 247, 61, 2008.

Kelce, W.R. and Wilson, E.M., Environmental antiandrogens: developmental effects, molecular mechanisms, and clinical implications. *J. Mol. Med.* 75, 198, 1997.

Mantovani, A. and Calamandrei, G., Delayed developmental effects following prenatal exposure to drugs. *Curr. Pharm. Des.* 7, 859, 2001.

Mittendorf, R., Teratogen update: carcinogenesis and teratogenesis associated with exposure to diethylstilbestrol (DES) *in utero. Teratology* 51, 435, 1995.

Newbold, R.R., Prenatal exposure to diethylstilbestrol (DES). *Fertil. Steril.* 89, e55, 2008.

Newbold, R.R Perinatal exposure to environmental estrogens and the development of obesity. *Mol. Nutr. Food Res.* 51, 912, 2007.

Riecke, K. and Stahlmann, R., Test systems to identify reproductive toxicants. *Andrologia* 32, 209, 2000.

Swan, S.H., Intrauterine exposure to diethylstilbestrol: long-term effects in humans. *APMIS*, 108, 793, 2000.

Ulbrich, B. and Palmer, A.K., Neurobehavioral aspects of developmental toxicity testing. *Environ. Health Perspect.* 104, 407, 1996.

Woodruff, T.J. , Zeise, L., Axelrad, D.A., Guyton, K.Z., Janssen, S., Miller, M., Miller, G.G., Schwartz, J.M., Alexeeff, G., Anderson, H., Birnbaum, L., Bois, F., Cogliano, V.J., Crofton, K., Euling, S.Y., Foster, P.M., Germolec, D.R., Gray, E., Hattis, D.B., Kyle, A.D., Luebke, R.W., Luster, M.I., Portier, C., Rice, D.C., Solomon, G., Vandenberg, J., and Zoeller, R.T., Meeting report: moving upstream—evaluating adverse upstream end points for improved risk assessment and decision-making. *Environ. Health Perspect.* 116, 1568, 2008.

33 Radiation Toxicity

33.1 PRINCIPLES OF RADIOACTIVITY

Radiation is a form of energy whose sources are both synthetic and naturally occurring. Small quantities of radioactive materials occur naturally in the environment (atmosphere, water, and food) and cause what is referred to as *internal* exposure. *External* exposure results from sunlight radiation and from synthetic and other naturally occurring radioactive materials.

The observation by Becquerel in 1896 of the fogging of photographic plates by uranium salts prompted an intense search for an understanding of the phenomenon of radioactivity. The scientific experimental persistence of the Curies, Schmidt, Debierne, Rutherford, Bohr, and Soddy led to the discoveries of radioactive isotopes and the identification of the types of radiation emitted by these elements. The landmark publication of Bohr's theory of atomic structure (1913) soon followed.* In 1942, Fermi prepared the schematics for the first nuclear reactor 5 years after Rutherford's death. Rutherford's realization of the development of nuclear power until then remained unfulfilled. Since then, constructive research on the atomic nucleus has resulted in the means to harness this energy, not only for the production of electricity and nuclear weapons but also for its application to the medical, pharmaceutical, chemical, and agricultural industries.

The electromagnetic spectrum consists of a wide range of wave-propagated energy, expressed as radiation energy. The spectrum is based on the photon energy (ρ) emitted from atomic sources, corresponding to frequency (in cycles per second [cps; Hz]), which is inversely proportional to its wavelength λ (in meters) (Figure 33.1) and is represented as

$$\text{Frequency (Hz)} = 1/\lambda \text{ (meters)}$$

A proton is a particle of radiation energy with a mass of 1836 m (mass units)† and possesses a charge of +1. Electrons and neutrons constitute the other basic components of atoms, both stable and radioactive (their charges equal −1 and 0, respectively). Radiation energy above the UV range (i.e., higher frequencies, to the left of UV in Figure 33.1) is called *ionizing radiation* because of the ability of photon energy to displace electrons from their atomic nuclei. UV, X-rays, and γ-rays represent the most commonly encountered high-energy rays capable of inducing cellular damage. Nonionizing radiation occurs at frequencies below the visible spectrum (lower frequencies and to the right of VIS in Figure 33.1).

* Bohr's theory of atomic structure was based on Rutherford's nuclear theory and Planck's quantum theory.

† For simplicity, an electron registers at 1 m and, being the smallest of the three particles, inconveniently weighs in at 9.1019×10^{-28} g.

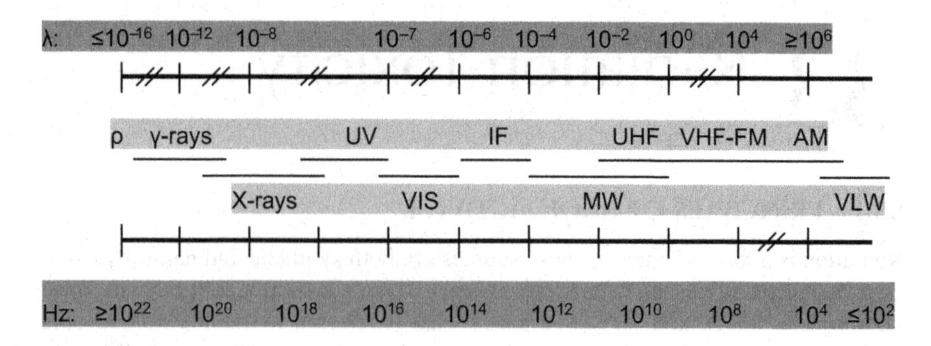

FIGURE 33.1 The electromagnetic spectrum.

33.2 IONIZING RADIATION

33.2.1 Biological Effects of Ionizing Radiation

Ionizing radiation induces somatic changes in cells and tissues by displacing electrons from their atomic nuclei, resulting in the intracellular ionization of molecules. Depending on the dose and length of exposure, the effects can be immediate, chronic, or delayed. Thus, reversible or irreversible DNA changes are induced, initiating a series of events that culminate in the production of a mutagenic response, a carcinogenic response, the inhibition of cell replication, or cell death. In addition, it is reasonable to expect that a change in DNA due to radiation exposure is inheritable and may result in genetic defects in the offspring (for the mechanisms of carcinogenesis, see Chapter 31).

33.2.2 Sources

Undesirable exposure to ionizing radiation emanates from a variety of isotope sources. High-energy diagnostic or therapeutic X-rays, used in the treatment of cancer, result in localized radiation injury (exposure to whole-body therapeutic irradiation is rare). Occupational exposure involves variable amounts of radioactivity from nuclear reactors, linear accelerators, and sealed cesium, americium, and cobalt sources used in therapeutic instruments and detectors. Radium and radon gas are naturally occurring hazardous isotopes embedded in the earth's crust and are often encountered in communities and households close to concentrated sources. Large amounts of radioactivity have also escaped from nuclear reactors, such as Three Mile Island in Pennsylvania (1979) and Chernobyl, Ukraine (1986).

The Chernobyl reactor accident of April 1986 provides the best-documented example of a massive radionuclide release in which large numbers of people across a broad geographical area were exposed acutely to radioiodines released into the atmosphere. The reactor accident resulted in massive releases of ^{131}I and other radioiodines. Approximately 4 years after the accident, a sharp increase in the incidence of thyroid cancer among children and adolescents in Belarus and Ukraine (the areas covered by the radioactive plume) was observed. The most comprehensive and reliable data available describing the relationship between thyroid radiation dose and risk for thyroid cancer following an environmental release of ^{131}I were documented following this accident.*

33.2.3 Clinical Manifestations

Total dose- and rate-dependent effects of radiation—that is, radiation dose per unit of time—result most often in acute cell injury. A single, rapid high dose of radiation may be less injurious if exposure occurs over weeks or months. Injury resulting from exposure also depends on the cumulative body surface area exposed, where acute exposure to high whole-body doses is fatal. In addition, rapidly proliferating cells are more susceptible to the effects of ionizing radiation than cells with a lower turnover rate. Lymphoid, hematopoietic, gonadal, endothelial, osteogenic, and gastrointestinal and pulmonary epithelial cells are most sensitive. Consequently, it is not unreasonable to expect that the initial and persistent adverse reactions associated with cancer radiation therapy will include hair loss, nausea, vomiting, diarrhea, bone marrow suppression, coughing, dyspnea, and dermal changes. Patients with bone marrow suppression have increased susceptibility to infection and bleeding disorders. Sloughing of the epithelial lining of the GI tract, resulting in vomiting and diarrhea, facilitates excessive fluid loss and precipitates electrolyte imbalances. In addition, acute or chronic occupational or accidental exposures to significant doses of ionizing radiation predictably produce skin cancers, leukemias, osteogenic sarcomas, and lung carcinomas.

Cell membrane disruption, resulting in swelling and erythema, occurs after initial exposure to low doses of radiation. At higher levels, acute reversible necrosis is possible. More persistent damage and chronic ischemia occur with continuous radiation of slowly regenerating cells. Chronic effects are characterized by fibrosis and scarring of irradiated tissues. Irreversible necrosis, impaired wound healing, and inflammation of affected organs (dermatitis, cystitis, and enteritis) are significant consequences of chronic radiation exposure. Table 33.1 categorizes and summarizes the clinical effects associated with acute and prolonged ionizing radiation, starting from the lowest-dose exposure. Regeneration of cells and tissues depends on the dose and rate of irradiation and the commencement of supportive therapy.

The most useful and rapid method for clinical assessment of the degree of radiation exposure, especially ionizing radiation, is determination of the patient's total blood lymphocyte count. Serial determinations are performed every 6 h for at least 48 h. A 50% fall in total lymphocytes every 24 h for 2 days is indicative of a potentially lethal injury.

---

* It is noteworthy that the thyroid radiation exposures after Chernobyl were virtually all *internal*, from radioiodines. Despite some degree of uncertainty in the doses received, it is reasonable to conclude that the contribution of external radiation was negligible for most individuals. Thus, the increase in thyroid cancer seen after Chernobyl is attributable to ingested or inhaled radioiodines. A comparable burden of excess thyroid cancers could conceivably accrue should U.S. populations be similarly exposed in the event of a nuclear accident.

TABLE 33.1
Signs and Symptoms of Acute Radiation Syndromes

Syndrome	Radiation Dose[a]	Phase, Time-Dependent Reaction, or Target Organ	Toxic Effects
Acute radiation sickness	Following radiation therapy ≤0.05 Gy	Initial reaction	NVD, anorexia, headache, malaise, tachycardia; usually subsides
Intermediate delayed effects	≤0.05 Gy	Prolonged, repeated exposure	Decreased fertility, libido; anemia, leukopenia
		Higher doses or high localized exposure	Skin ulceration, atrophy, keratosis; alopecia; delayed (years) appearance of carcinomas, sarcomas
Extensive radiation therapy	≥0.05 Gy (prolonged or cumulative exposure to selected organs)	Kidneys	Proteinuria, renal insufficiency, HT; ≥20 Gy <5 weeks: fibrosis, renal failure
		Muscles	Painful myopathy, atrophy
		Lungs	Pulmonary fibrosis, pneumonitis (>30 Gy)
		Mediastinum	Pericarditis, myocarditis
		Abdominal lymph nodes	Ulceration, fibrosis, perforation of bowel
Hematopoietic syndrome	2–10 Gy	Initial reaction	Anorexia, apathy, NV; may subside
		Continuous or high-dose exposure	Atrophy of lymph nodes, spleen, BM, immunosuppression; pancytopenia, bacterial infections, hemorrhage
GI syndrome	≥4 Gy	Initial reaction	NVD, dehydration, vascular collapse
		Continuous or high-dose exposure	Progressive GI atrophy and necrosis, bacteremia, loss of plasma volume
Cerebral syndrome	>30 Gy	Prodromal phase	Nausea, vomiting
		Listless phase	Apathy, drowsiness, prostration
		Tremulous phase	Ataxia, convulsions, death (within hours)
Late somatic–genetic defects	Variable	Somatic cells	Leukemia, thyroid, skin, bone cancers
		Germ cells	↑ Risk of mutations and genetic defects

[a] The gray (Gy) is the amount of energy absorbed by a tissue and applies to all types of radiation; the roentgen (R) is the amount of X- or γ-radiation in air; for X- or γ-radiation, the sievert (Sv) equals the Gy after factoring for its biological effect (1 sievert = 100 rem).

BM: bone marrow; GI: gastrointestinal; HT: hypertension; NVD: nausea, vomiting, diarrhea.

33.2.4 Nuclear Terrorism and Health Effects

Although the U.S. Centers for Disease Control (CDC) has not been able to assess the level of threat of a terrorist nuclear attack, it has regularly participated in emergency response drills. In cooperation with other U.S. federal, state, and local agencies, the CDC has developed, tested, and implemented extensive national radiological emergency response plans. The adverse health consequences of a terrorist nuclear attack vary according to the type of attack and the distance a person is from the radioactive emission. Potential terrorist attacks may include a small radioactive source with a limited range of impact, a nuclear detonation involving a wide area of impact, or an attack on a nuclear power plant. In any of these events, injury or death may occur as a result of the blast itself or as a result of debris thrown from the blast. In addition, individuals may experience external and/or internal exposure. Depending on the dose, type, route, and length of time of the exposure, the signs and symptoms are not unlike those of ionizing radiation.

The U.S. Food and Drug Administration (FDA) has provided guidance documents and regulations in the event of a terrorist nuclear attack (FDA, 2018(a)). A national emergency response plan would be activated and would include federal, state, and local agencies. In general, individuals can reduce the potential exposure and subsequent health consequences by limiting the time, increasing the distance, or keeping a physical barrier between them and the source. In addition, the FDA has determined that administration of potassium iodide (KI) is a safe and effective means of blocking uptake of radioiodines by the thyroid gland in a radiation emergency under certain specified conditions of use (U.S. FDA, 2018(b)).*

The effectiveness of KI as a specific blocker of thyroid radioiodine uptake is well established. The recommended dose is 130 mg per day for adults and children above 1 year of age and 65 mg per day for children below 1 year of age. When administered at this level, KI is effective in reducing the risk of thyroid cancer in individuals or populations at risk for inhalation or ingestion of radioiodines. KI competes with radioactive iodine for uptake into the thyroid, thus preventing incorporation of the radioactive molecules into the gland. The radioactive compounds are subsequently excreted in the urine.

33.3 ULTRAVIOLET (UV) RADIATION

33.3.1 Biological Effects of Ultraviolet Radiation

UV rays are of lower frequency and longer wavelength than ionizing radiation (about 10^{-8} to 10^{-7} meters; Figure 33.1). Thus, the effects of UV radiation are less penetrating and benign. Unlike ionizing radiation, skin damage induced by UV rays (e.g., from sun exposure) is mediated principally by the generation of reactive oxygen species (ROS) and the interruption of melanin production. As with ionizing radiation,

* The recommendations were formulated after reviewing studies relating radiation dose to thyroid disease risk. The FDA's evaluations relied on estimates of *external* thyroid irradiation after the nuclear detonations at Hiroshima and Nagasaki. It was concluded that the risks of short-term use of small quantities of KI were outweighed by the benefits of suppressing radioiodine-induced thyroid cancer.

however, cumulative or intense exposure to UV rays precipitates DNA mutations—that is, base pair insertions, deletions, single-strand breaks, and DNA–protein cross-links. DNA repair mechanisms play an important role in correcting UV-induced DNA damage and preventing further consequences of excessive sunburn. Melanin production by melanocytes increases, and the epidermis thickens in an attempt to prevent future damaging effects. The protective ability of antioxidant enzymes and DNA repair pathways diminishes with age, thus setting the conditions for the development of skin neoplasms later in life.

33.3.2 Sources

Radiation burn (*sunburn*) is commonly caused by acute intense or prolonged exposure to the sun's UV rays. The recent fashionable phenomenon of obtaining a "tan" in tanning salons has resulted in excess exposure to tanning lamps, although these lamps produce more UVA than UVB.* Despite the availability of "sun blocking" products containing 5% PABA (para-aminobenzoic acid) that protect against the damaging effects of UV radiation, complications from sunburn still abound. The dissemination of public knowledge concerning the harmful effects of excessive exposure to the sun has not abated the development of its serious clinical effects, including the production of acute reactions (sunburn), chronic changes (skin cancer), or photosensitivity.

33.3.3 Clinical Manifestations

The extent of solar injury depends on the type of UV radiation, the duration and intensity of exposure, clothing, season, altitude and latitude, and the amount of melanin pigment present in the skin. In fact, individuals differ greatly in their response to sun. Fair-skinned persons have fewer melanin-producing cells. Consequently, they are more sensitive to UV rays than people of dark-skinned races, although the skin of the latter is still reactive and can become sunburned with prolonged exposure. Also, the harmful effects of UV rays are filtered out by glass, smog, and smoke but enhanced by reflecting off snow and sand. The indiscriminate use of chlorofluorocarbons in aerosol propellants depletes the UV-blocking properties of ozone in the stratosphere, thus allowing a greater intensity of UV rays to penetrate through the protective upper atmospheric layers.

As with thermal, chemical, or electrical burns, severe solar radiation exposure is classified according to the severity and depth of the burn. *First-degree* burns are generally red, sensitive, and moist. The absence of blisters and blanching of the skin with application of light pressure are characteristic features. *Second-degree* burns are classified as superficial, intermediate, or deep, with partial skin loss. The presence of erythematous blisters with exudate is typical of *second-degree* burns. *Third-degree* burns involve deep dermal, whole skin loss. The skin appears black, charred, and leathery. Although subdermal vessels do not blanch with applied pressure, the

* The shorter wavelengths of the sunburn-producing UVB radiation (280–320 nm λ) are more damaging than UVA rays (320–400 nm λ).

areas exposed are generally anesthetic or insensitive to pain stimuli. Although generally not associated with sunburn, *fourth-degree* burns involve deep tissue and structure loss. Hypertrophic scars and chronic granulations develop unless skin grafting treatment is instituted.

Tap water compresses are effective in relieving symptoms of sunburn. Topical corticosteroids or astringent gels (such as aloe) may relieve the discomfort associated with extensive severe sunburn. Occlusive ointments or creams should not be applied to ruptured blisters or wounds.

Table 33.2 summarizes the chronic effects of sunlight and photosensitivity reactions. For instance, general dermatologic reactions, such as dermatoheliosis and elastosis, occur with greater frequency with chronic sunlight exposure (Table 33.2). Actinic keratoses, described as precancerous keratotic lesions, result from years of sun exposure, especially in fair-skinned individuals. The lesions present as pink, poorly marginated, scaly or crusted superficial growths on skin. Depending on the number of lesions present, they are usually treated with cryotherapy (freezing), curettage (surgical scraping), or topical application of 5-fluorouracil (5-FU).

Squamous cell carcinoma, basal cell carcinoma, and malignant melanomas are high-risk tumors associated with prolonged exposure to UV rays (Table 33.2).

Squamous cell carcinoma (SCC) is a precancerous lesion that appears as a papule and often develops on damaged fair-skinned individuals following prolonged UV exposure. Some SCCs develop from actinic keratosis lesions on common areas such as the rim of the ears, face, and lips. Although the cure rate is 95% with proper treatment, it nevertheless accounts for 2300 deaths per year, particularly because of its ability to metastasize (about 3%) to the lymphatic vessels. Treatment for SCC depends on its type, location, and risk for metastasis. Most frequently, SCC is

TABLE 33.2
Chronic Effects of Sunlight and Photosensitivity Reactions

Reactions to Excessive Sunlight	Clinical Effects[a]
General dermatologic	Dermatoheliosis—aging of the skin due to chronic exposure to sunlight
	Elastosis—yellow discoloration of skin with accompanying small nodules
	Wrinkling, hyperpigmentation, atrophy, and dermatitis
Actinic keratoses	Precancerous keratotic lesions, appear after many years of exposure to UV rays
Squamous/basal cell carcinoma	Occurs more commonly in light-skinned individuals exposed to extensive UV rays during adolescence
Malignant melanomas	Associated with increased, intense, prolonged exposure to UV light
Photosensitive reactions	Erythema and erythema multiforme lesions; urticaria, dermatitis, bullae; thickened, scaling patches

[a] See text for description of terms.

removed by complete surgical excision, cryosurgery, curettage, or electrodesiccation. Chemoradiation therapy follows high-risk SCC. Persons who develop an SCC are at higher risk for future appearance of lesions. Consequently, regular dermal screening examinations, avoidance of sunlight, and early treatment prevent complications and improve prognosis.

Basal cell carcinoma (BCC) is the most common form of cancer. The malignant tumors arise from the deepest layer of the epidermis, and most are located on the face, hands, and other sun-exposed parts of fair-skinned persons. Early diagnosis and successful treatment are usually accompanied by a favorable prognosis. The appearance of the tumor varies from a red, irritated, patchy papule to a smooth, reddish-yellowish macule. BCC treatment modalities depend on factors similar to SCC—that is, BCC is removed by complete surgical excision, cryosurgery, curettage, or electrodesiccation.* Chemoradiation therapy also follows high-risk BCC. In contrast to an SCC, persons who develop BCC have a 30% greater frequency of recurrence within 5 years.

Malignant melanoma (MM), or cutaneous melanoma, grows from melanocytes and is associated with a high frequency of metastasis. As the fifth and sixth most common disease in men and women, respectively, MM spreads through local lymphatic or hematogenous routes. It appears as a nest of nevus cells (altered melanocytes), usually presenting as tan to brown macules or papules with well-defined rounded borders. The lesion typically changes in size, shape, impression, or color, with the appearance of bleeding. MM develops on sun-exposed areas of the body such as the shoulders, head or neck in men, and arms or legs in women. In the early appearance of MM, abnormal cells originate only on the outer epidermis. As the cells differentiate, the tumor advances to deeper layers of the skin (about 4 mm), shows additional growth around the original tumor cells (satellite tumor) or metastasizes to the lymphatics. Consequently, MM is "staged" according to these invasive characteristics. Treatment involves removal of the tumor by complete surgical excision, followed by grafting, chemotherapy, and radiation therapy. The 5-year survival rate for treated advanced (stage III) melanoma patients is about 60%, while the disease and the treatment protocol compromise quality of life.

Photosensitive reactions involve the development of unusual responses to sunlight, sometimes exaggerated in the presence of a variety of precipitating, seemingly unrelated circumstances. Erythema (redness) or dermatitis (inflammation of the skin or mucous membranes) is an acute response to direct UV light. It is mediated by dilation and congestion of superficial capillaries and type IV hypersensitivity cell-mediated reactions. In addition, macules, papules, nodules, and target ("bull's eye"-shaped) lesions are seen in the multiforme variation of erythematous reactions. As with erythematous reactions, solar urticaria is a pruritic skin eruption characterized by transient wheals of varying shapes and sizes. The lesions have well-defined erythematous margins and pale centers. The development of urticaria is mediated by capillary dilation in the dermis in response to the release of vasoactive amines (histamine and kinins) on sun exposure. More severe photosensitivity reactions lead to the development of bullae, thin-walled blisters on the skin or mucous membranes

* Involves scraping of the tumor tissue and destroying the thin surrounding layer with heat.

greater than 1 cm in diameter and containing clear, serous fluid. Dehydration, scaling, scarring, fibrosis, and necrosis develop as the exposed areas heal.

Numerous factors contribute to the development of photosensitivity reactions. Xeroderma pigmentosum,* lupus erythematosus,† and porphyrias‡ are among the pathologic conditions that are accompanied by photosensitivity reactions. The ingestion or application of antibacterial and antifungal antibiotics (sulfonamides, tetracyclines; griseofulvin, respectively) and thiazide diuretics is responsible for occasional photosensitive reactions. Contact with dermal products containing coal tar, salicylic acid, plant derivatives, and ingredients of colognes, perfumes, cosmetics, and soaps are also sources of the problem.

Preventive treatment and avoidance of sunlight (use of protective clothing, sunscreens, and shaded areas) are generally helpful in averting skin reactions. Other therapies include the use of topical or oral corticosteroids, H_1-blockers (antihistamines), antimalarial drugs (hydroxychloroquine), topical sunscreens with moisturizers, and psoralen UV light (PUVA).

33.4 NONIONIZING RADIATION

33.4.1 Sources

Nonionizing radiation emanates from sources with lower energy (low frequency) and longer wavelengths (to the right of the visible spectrum; Figure 33.1). Low-frequency radiation, such as infrared light (IF), ultrasound, microwaves (MW), and laser energy, is commonly encountered in the home and in the workplace. Televisions, radios (UHF, ultrahigh-frequency TV waves; VHF-FM/AM, very high-frequency TV and radio FM/AM), cellular phones, wireless devices, and microwave ovens are among the household sources. Plastic, wood, metal, and chemical industries account for occupational exposures. Medical applications (diathermy) are another source of exposure.

33.4.2 Biological Effects and Clinical Manifestations

In contrast to ionizing radiation, the lower energy of nonionizing radiation induces vibrational and rotational movements of intracellular atoms and molecules. Accordingly, burns and thermal injury are the most common consequences of exposure to direct low-energy rays. Deep penetration of dermal and subcutaneous tissue is observed when tissue intercepts the path of infrared or microwave rays. The latter are also capable of disrupting the normal function of electronic medical devices such as subcutaneously implanted cardiac pacemakers and monitors.

---

* A genetic disorder characterized by extreme photosensitivity and increased risk of skin cancer in sun-exposed skin. It is usually due to the lack of DNA repair enzymes in damaged skin.

† A systemic, inflammatory, multi-organ, autoimmune disease with characteristic development of rashes and joint pain.

‡ Porphyria cutanea tarda is the most common form of porphyria. The syndrome is a sporadic, autosomal dominant condition characterized by skin fragility on the dorsal hands and forearms, and facial hyperpigmentation.

REFERENCES

SUGGESTED READINGS

Gavrilin, Y.I., Khrouch, V.T., Shinkarev, S.M., Krysenko, N.A., Skryabin, A.M., Bouville, A., and Anspaugh, L.R., Chernobyl accident: reconstruction of thyroid dose for inhabitants of the republic of Belarus. *Health Phys.* 76, 105, 1999.

Likhtarev, I.A., Sobolev, B.G., Kairo, I.A., Tronko, N.D., Bogdanova, T.I., Oleinic, V.A., Epshtein, E.V., and Beral, V., Thyroid cancer in the Ukraine. *Nature* 375, 365, 1995.

Porter, R.S. and Kaplan, J.L. (Eds), *The Merck Manual of Diagnosis and Therapy*, 19th ed., Merck Research Laboratories, Whitehouse Station, NJ, 2011, chapter 22.

Robbins, J. and Schneider, A.B., Thyroid cancer following exposure to radioactive iodine. *Rev. Endoc. Metab. Dis.* 1, 197, 2000.

U.S. Food and Drug Administration (U.S. FDA), National Center for Toxicological Research (NCTR). From:. www.fda.gov/oc/ocm/radplan.html, last viewed March 2019(a).

U.S. Food and Drug Administration (U.S. FDA), National Center for Toxicological Research (NCTR). From: https://www.fda.gov/Drugs/EmergencyPreparedness/Bioterrorism andDrugPreparedness/ucm072248.htm, last viewed March 2019(b).

Zvonova, I.A. and Balonov, M.I., Radioiodine dosimetry and prediction of consequences of thyroid exposure of the Russian population following the Chernobyl accident, in *The Chernobyl Papers. Doses to the Soviet Population and Early Health Effects Studies.* Volume I, Mervin, S.E. and Balonov, M.I. (Eds.), Research Enterprises, Richland, Washington, 1993, p. 71.

REVIEW ARTICLES

Baselet, B., Rombouts, C., Benotmane, A.M., Baatout, S., and Aerts, A., Cardiovascular diseases related to ionizing radiation: the risk of low-dose exposure (Review). *Int. J. Mol. Med.* 38, 1623, 2016.

Baudouin, C., Charveron, M., Tarroux, R., and Gall, Y., Environmental pollutants and skin cancer. *Cell Biol. Toxicol.* 18, 341, 2002.

Betlazar, C., Middleton, R.J., Banati, R.B., and Liu, G.J., The impact of high and low dose ionising radiation on the central nervous system. *Redox Biol.* 9, 144, 2016.

Brent, R.L., Utilization of developmental basic science principles in the evaluation of reproductive risks from pre- and post-conception environmental radiation exposures. *Teratology* 59, 182, 1999.

Cardis, E., Richardson, D., and Kesminiene, A., Radiation risk estimates in the beginning of the 21st century. *Health Phys.* 80, 349, 2001.

Dainiak, N., Hematologic consequences of exposure to ionizing radiation. *Exp. Hematol.* 30, 513, 2002.

Hogan, D.E. and Kellison, T., Nuclear terrorism. *Am. J. Med. Sci.* 323, 341, 2002.

Kadhim, M. Salomaa, S., Wright, E., Hildebrandt, G., Belyakov, O.V., Prise, K.M., and Little, M.P., Non-targeted effects of ionizing radiation—implications for low dose risk. *Mutat. Res.* 752, 84, 2013.

Ko, S. Chung, H.H., Cho, S.B., Jin, Y.W., Kim, K.P., Ha, M., Bang, Y.J., Ha, Y.W., and Lee, W.J., Occupational radiation exposure and its health effects on interventional medical workers: study protocol for a prospective cohort study. *BMJ Open* 7, e018333, 2017.

Lomax, M.E., Gulston, M.K., and O'Neill, P., Chemical aspects of clustered DNA damage induction by ionising radiation. *Radiat. Prot. Dosimetry* 99, 63, 2002.

Maisin, J.R., Chemical protection against ionizing radiation. *Adv. Space Res.* 9, 205, 1989.

Mettler, F.A. Jr. and Voelz, G.L., Major radiation exposure—what to expect and how to respond. *N. Engl. J. Med.* 346, 1554, 2002.

Moulder, J.E., Radiobiology of nuclear terrorism: report on an interagency workshop (Bethesda, MD, December 17–18, 2001). *Int. J. Radiat. Oncol. Biol. Phys.* 54, 327, 2002.

Moulder, J.E., Pharmacological intervention to prevent or ameliorate chronic radiation injuries. *Semin. Radiat. Oncol.* 13, 73, 2003.

Moysich, K.B., Menezes, R.J., and Michalek, A.M., Chernobyl-related ionising radiation exposure and cancer risk: an epidemiological review. *Lancet Oncol.* 3, 269, 2002.

Nagataki, S., and Nyström, E., Epidemiology and primary prevention of thyroid cancer. *Thyroid* 12, 889, 2002.

Nohynek, G.J. and Schaefer, H., Benefit and risk of organic ultraviolet filters. *Regul. Toxicol. Pharmacol.* 33, 285, 2001.

Pinnell, S.R., Cutaneous photodamage, oxidative stress, and topical antioxidant protection. *J. Am. Acad. Dermatol.* 48, 1, 2003.

Rahu, M., Health effects of the Chernobyl accident: fears, rumours and the truth. *Eur. J. Cancer* 39, 295, 2003.

Rosen, E.M., Fan, S., Goldberg, I.D., and Rockwell, S., Biological basis of radiation sensitivity. Part 1: Factors governing radiation tolerance. *Oncology (Huntington)* 14, 543, 2000.

Skorga, P. Persell, D.J., Arangie, P., Gilbert-Palmer, D., Winters, R., Stokes, E.N., and Young, C.,Caring for victims of nuclear and radiological terrorism. *Nurse Pract.* 28, 24, 2003.

Sridhara, D.M., Asaithamby, A., Blattnig, S.R., Costes, S.V., Doetsch, P.W., Dynan, W.S., Hahnfeldt, P., Hlatky, L., Kidane, Y., Kronenberg, A., Naidu, M.D., Peterson, L.E., Plante, I., Ponomarev, A.L., Saha, J., Snijders, A.M., Srinivasan, K., Tang, J., Werner, E., and Pluth, J.M., Evaluating biomarkers to model cancer risk post cosmic ray exposure. *Life Sci. Space Res. (Amst.)* 9, 19, 2016.

Urbach, F., Forbes, P.D., and Davies, R.E., Modification of photocarcinogenesis by chemical agents. *J. Natl. Cancer Inst.* 69, 229, 1982.

34 Chemical and Biological Threats to Public Safety

34.1 INTRODUCTION

Terrorism is the threat or implementation of violent means to undermine, destabilize, inflict harm, or cause panic in a society. The threat of terrorism in the world, at any time and within any nation, is real. At the least, the induction of fear is a destabilizing force in a nation or state. The United States has only recently experienced serious terrorist threats with the events of September 11, 2001. Since then, the public's awareness of a variety of potential means that could be used as terrorist threats has prompted unprecedented political, military, biomedical, industrial, and community responses. Terrorist methods comprise the use of radiological, chemical, and biological agents capable of widespread, mass casualties and destruction. The ease of availability, the low cost of production, and the facility for wide dissemination of these means of destruction make them very attractive weapons.

Chemical and radiological toxicity have already been discussed throughout several chapters in this book (nerve gases, radiological toxicity). Therefore, only the applicable classes of chemical and radiological agents are outlined below. Consequently, this chapter elaborates on the mechanisms of previously unclassified pathogenic agents and their toxic biological products, some of which only recently have been identified as potential toxic terrorist threats.

The U.S. Food and Drug Administration (FDA), together with other federal agencies, has prepared a series of working guidelines and information to help prepare against the threat of bioterrorism. The Office of Pediatric Drug Development and Program Initiatives (OPDDPI) also identifies and facilitates the development of drug products that may be used in the treatment of conditions caused by agents released in the event of a terrorist attack. These agents are of a pathogenic, radiological, or chemical nature. Table 34.1 compiles high-priority chemical and biological agents, and the syndrome resulting from their exposure, according to their potential as a bioterrorist threat.

Category A includes high-priority agents and pathogens rarely seen in the United States that pose a risk to national security. They are highly infectious and easy to disseminate, and the clinical effects from exposure result in high mortality rates. By their nature, Category A agents are capable of inciting public panic and disruption. In addition, they require special action for public health preparedness. The agents or diseases in Category A include anthrax, botulism, plague, smallpox, tularemia, and organisms that induce viral hemorrhagic fevers.

Category B lists the second-highest-priority agents. These organisms are moderately easy to disseminate, result in moderate morbidity and low mortality rates, and require specific enhancements of the Centers for Disease Control (CDC)'s diagnostic capacity and enhanced disease surveillance. Other less identifiable substances in

TABLE 34.1
Chemical and Biological Agents with High Risk to National Security

Category	Category Description	Toxin or Disease	Causative Agent	Classification
A	Highly infectious, easily disseminated, high mortality rate	Anthrax	*Bacillus anthracis*	Spore-forming bacteria
		Botulism	*Clostridium botulinum*	Spore-forming bacteria
		Plague	*Yersinia pestis*	Enterobacteria
		Smallpox	Variola, vaccinia[a]	Poxvirus
		Tularemia	*Francisella tularensis*	Coccobacillus
		Hemorrhagic fever viruses	Lassa, Machupo, Junin	Arenavirus
			Ebola, Marburg	Filovirus
B	Moderate rates of infection, dissemination, morbidity and mortality	Brucellosis	*Brucella* sp.	Coccobacillus
		Epsilon toxin	*Clostridium perfringens*	Spore-forming bacteria
		Acute gastroenteritis	*Salmonella* sp., *Shigella* sp., *E. coli*	Enterobacteria
		Q fever	*Coxiella burnetii*	Obligate intracellular bacillus
		Ricin toxin	*Ricinus communis*	Castor beans
C	Highly infectious, ease of dissemination unknown	Hemorrhagic fever, ARDS	Hantavirus	Bunyaviridae
		Encephalitis	Bunyavirus	
		Encephalitis	Nipah virus	Paramyxoviridae

[a] Variola is the extinct human virus; vaccinia is the laboratory product used for smallpox vaccinations.
ARDS: adult respiratory distress syndrome.

Category B (not discussed in this chapter) include glanders *(Burkholderia mallei)*, melioidosis (*B. pseudomallei*), psittacosis (*Chlamydia psittaci*), Staphylococcal enterotoxin B, typhus fever (*Rickettsia prowazekii*), and viral encephalitis (alphaviruses, such as equine encephalitis).

The third category of highest-priority agents, *Category C*, includes pathogens that could be engineered for mass dissemination. Emerging threats, such as the Nipah virus and hantavirus, are available and easily produced and disseminated provided there exists some technical knowledge of microbiology. Thus, they have the potential for high morbidity and mortality.

34.2 BIOLOGICAL PATHOGENIC TOXINS AS THREATS TO PUBLIC SAFETY

Naturally occurring or laboratory-derived biological pathogens and their products, released intentionally or accidentally, can result in disease or death. Human exposure to these agents may occur through inhalation, skin (cutaneous) exposure, or ingestion of contaminated food or water. Following exposure, physical symptoms may be delayed and sometimes confused with naturally occurring illnesses, thus contributing to the possible postponement in response. In addition, biological warfare agents may persist in the environment and cause problems long after their release.

The agents discussed here fall into three major classes of microorganisms: bacteria,* rickettsia, and viruses. In some cases, hazardous bacterial toxins are also produced as by-products of their pathogenic metabolism. To aid in understanding the pathogenic features of infectious agents, several tables are included. Table 34.2 describes the biological features of Category A agents. Table 34.3 outlines the clinical aspects of the same Category A organisms, and Table 34.4 defines and describes signs, symptoms, and syndromes associated with the infectious agents. Incubation periods, duration of illnesses, means of transmission, treatment, and prognosis are also noted for each of the organisms.

34.2.1 Anthrax (*Bacillus anthracis*)

Anthrax is an infectious disease caused by the spore-forming bacterium *Bacillus anthracis*. The organism is an obligate aerobe and facultative anaerobe. The highly resistant, prominent polypeptide capsule of the endospore renders *B. anthracis* immune to most methods of disinfection or natural processes of inactivation.† Thus the organism may be present in the soil for decades, occasionally infecting grazing goats, sheep, and cattle. When ingested, the hibernating, dehydrated, protected spores release viable bacteria on contact with gastrointestinal (GI) fluids. Human infection occurs by three routes of exposure to anthrax spores: cutaneous, GI, and inhalation. Although human cases of anthrax are infrequent in North America, the U.S. military views anthrax as a potential biological terrorist threat because of its high resistance and ease of communicability through the air.

The development of anthrax as a biological weapon by several foreign countries has been documented. Prior to the recent deliberate release of anthrax-laden letters throughout the United States in the fall of 2001, much of what is known about inhalation anthrax was learned from the accidental discharge of spores in 1979 in Sverdlovsk in the former Soviet Union. This accident resulted in 79 cases of anthrax and 68 deaths.

Cutaneous infection is the most common manifestation of anthrax in humans, accounting for 95% of cases. Inoculation through exposed skin results from contact with contaminated soil or animal products such as hair, hides, and wool. A painless papule at the site of inoculation progresses rapidly to an ulcer surrounded by vesicles and then to a necrotic eschar. Systemic infection, complicated by massive edema and painful lymphadenopathy, is fatal in 20% of patients.‡

---

* Bacterial toxins are products of bacterial metabolism or components of their structures that stimulate immunologic responses or alter normal physiologic function.

† Interestingly, the capsule is observed in clinical specimens and is not synthesized *in vitro* unless special growth conditions are employed.

‡ Since person-to-person transmission does not occur, there is no need to immunize or treat contacts, such as household family, friends, or coworkers, unless they also were also exposed to the same source of infection.

TABLE 34.2
Biological Features of Category A Agents

Disease	Causative Agent	Bacterial/ViralFeatures	Biological Characteristics	Mode of Transmission
Anthrax	*Bacillus anthracis*	Spore-forming bacteria	Obligate aerobe; facultative anaerobe	Cutaneous; GI; inhalation
Botulism	*Clostridium botulinum*	Spore-forming bacteria	Gram-positive anaerobic rod	GI
Plague	*Yersinia pestis*	Enterobacteria	Bubonic plague	Bite of infected flea
			Pneumonic plague	Inhalation
Smallpox	Variola, vaccinia[a]	Poxvirus	Largest viral family	Inhalation, salivary route (person-to-person)
Tularemia	*Francisella tularensis*	Coccobacillus	Gram-negative spherical/rod	GI, inhalation; bite of arthropods
Hemorrhagic fever	Lassa, Junin, Machupo	Arenavirus	RNA viruses	Inhalation (person-to-person)
	Ebola, Marburg	Filovirus		

[a]Variola is the extinct human virus; vaccinia is the laboratory product used for smallpox vaccinations.
ARDS: adult respiratory distress syndrome.

TABLE 34.3
Clinical Aspects of Category A Agents

Causative Agent	Occurrence	Incubation Period	Unique Clinical Features[a]	Current Available Treatment
Bacillus anthracis	Cutaneous—often; GI—rare; inhalation— occasional	2–5 days; shorter with inhalation	Resistant endospores; wool sorter's disease	Fluoroquinolones, doxycycline; AVA; supportive
Clostridium botulinum	Common	12–36 hours	Botulinum neurotoxin	Botulinum antitoxin
Yersinia pestis	Bite of fleas—uncommon	7 days	Buboes: fever, lymphadenopathy	Streptomycin, tetracycline, chloramphenicol, gentamicin
	Inhalation—uncommon	3 days	Fever, pneumonia-like S&S	
Variola, vaccinia[a]	Eradicated in 1980	7–12 days	Pocks: rash, fever, dissemination	Vaccination
Francisella tularensis	100 cases per year	3–6 days	Glandular, typhoidal syndrome	Vaccination; streptomycin, fluoroquinolones, doxycycline, gentamicin
HFV	Endemic in sub-Saharan Africa	2–21 days	Hemorrhagic fevers, DIC, encephalitis	Ribavirin (arenavirus only)

[a] See text for description.

AVA: anthrax vaccine adsorbed; DIC: disseminated intravascular coagulation; GI: gastrointestinal; HFV: hemorrhagic fever viruses; S&S: signs and symptoms.

TABLE 34.4
Description of Signs, Symptoms, and Syndromes Associated with Infectious Agents

Syndrome	Description
Cellulitis	Acute, diffuse, edematous suppurative inflammation of the subcutaneous tissue and muscle; sometimes associated with abscess formation; caused by introduction of bacteria into a trauma or surgical wound or burn, or in an immunocompromised patient
Conjunctivitis	Inflammation of the conjunctiva or inner lining of the eyelid consisting of hyperemia and discharge; usually a result of acute bacterial infection
Cutaneous	Pertaining to the skin; dermal
Eschar	A slough resulting from scar formation associated with thermal burn, corrosive application, or gangrene
Gas gangrene	Death of tissue in considerable mass and with loss of vascularity; putrefaction associated with release of bacterial sulfide gases containing greenish-black metabolites
Myalgia	Pain in a muscle or muscles
Myonecrosis	Destruction of muscle fibers
Myositis	Inflammation of voluntary muscles (inflammatory myopathy)
Necrotizing enteritis	Destruction of membrane and muscle fibers of the lining of the GI tract
Pseudomembranous	A layer resembling an organized membrane but actually composed of coagulated fibrin with bacteria and leukocytes, especially on mucous membranes
Suppurative	Purulent or pus-forming; due to infiltration and sequestration of bacteria

Ingestion of undercooked or raw, infected meat causes the rare GI infection. This is characterized by an acute inflammation of the intestinal tract. Initial signs of nausea, anorexia, vomiting, and fever are followed by abdominal pain, bloody vomitus, and severe diarrhea. As with the cutaneous form, systemic disease can progress rapidly.

Respiration of dried, airborne spores leads to inhalation anthrax (*wool-sorters' disease*). Initial symptoms of inhalation anthrax follow a prolonged, asymptomatic latent period (from 1 week up to 2 months) and present as fever, dyspnea, cough, headache, vomiting, chills, and chest and abdominal pain. After several days, a progressive, dramatic worsening of the infection occurs, leading to fever, pulmonary edema, and lymphadenopathy. Shock and death occur within 3 to 7 days of initial signs and symptoms.

The mortality rates from anthrax vary, depending on exposure and age, and are approximately 20% for cutaneous anthrax without antibiotics and 25–75% for GI anthrax. Inhalation anthrax has a fatality rate that is 80% or higher. Anthrax is susceptible to an early course of antibiotic treatment with penicillin, doxycycline, and fluoroquinolones (an efficacy supplement for ciprofloxacin was approved by the FDA on August 30, 2000 for postexposure inhalation anthrax).

The anthrax vaccine is an effective control measure and was developed from an attenuated strain of *B. anthracis*. Clinical studies have calculated the efficacy level of the vaccine at about 92.5%. The vaccine derives from cell-free culture filtrates of this strain and in its final formulation, is adsorbed onto an aluminum salt. The licensed anthrax vaccine, anthrax vaccine adsorbed (AVA),* is recommended for individuals at risk for occupational exposure and persons involved with diagnostic, clinical, or investigational activities with *B. anthracis* spores. Vaccination is not available to the general public and is not recommended to prevent disease.

34.2.2 Botulism

Botulism is caused by *Clostridium botulinum*, a gram-positive, spore-forming anaerobic rod. Like *B. anthracis*, the bacterium lives in soil and the intestines of animals. Besides botulism, the production of protein toxins by a variety of clostridial species is associated with a range of diseases, including tetanus, gas gangrene, food poisoning, diarrhea, and pseudomembranous colitis.

Botulism is caused by the release of preformed botulinum toxin secreted by the organism in contaminated food. Interestingly, the organism's presence in the food or existence in the victim's GI tract is not necessary for virulence. Anaerobic conditions favor the growth of the spores in contaminated meats, vegetables, and fish. The heat-resistant spores survive food processing and canning in sufficient numbers to cause toxicity on release of the toxin.

The botulinum toxin is a neurotoxin capable of preventing the release of acetylcholine at peripheral cholinergic synapses (see Chapter 17 for a description of the cholinergic system). Clinically, the disease has an onset of 12–36 hours. Signs and symptoms are related to inhibition of skeletal muscle innervation and include flaccid muscular paralysis, blurred vision, difficulty swallowing, and respiratory paralysis. Striated muscle groups weaken as a result of the descending paralysis, eventually affecting the neck and extremities.

Supportive care and maintenance of vital functions, especially respiration, is of the utmost importance in treatment. Specific antitoxin is available for some of the neurotoxin types (A, B, and E). With good supportive care and antitoxin administration, the mortality rate is reduced to 25%. Prolonged or permanent muscle paralysis may persist indefinitely.

34.2.3 Plague (*Yersinia pestis*)

Plague is an infectious disease of animals and humans caused by the bacterium *Yersinia pestis*. All yersinia infections are zoonotic, capable of spreading from rodents and their fleas (*urban plague*) as well as from squirrels, rabbits, field rats, and cats (*sylvatic plague*). Historically, pandemics resulting from yersinia infections have devastated human populations. The first of three urban plagues started in Egypt (541 AD) and spread through the Middle East, North Africa, Asia, and Europe, killing over 100 million persons. The Middle Ages (1340s) recorded a second pandemic that probably originated from Asia and spread through Europe, resulting in 25 million

* To date, there is only a single anthrax vaccine licensed and manufactured in the United States (BioPort, Lansing, MI).

deaths in Europe. Recent history (1895) recorded a pandemic that began in Hong Kong and spread to Africa, India, Europe, and the Americas, leaving 10 million deaths in its wake over 20 years. The recognition of public health and maintenance of hygienic standards has essentially eradicated urban plague from most communities, although some cases are reported in the United States annually. Clinically, *Y. pestis* infections are manifested as bubonic plague and pneumonic plague.

The bite of an infected flea starts the incubation period for *Y. pestis* of about 7 days. The resulting clinical signs and symptoms of *bubonic plague* include the development of buboes (lymphadenopathy or swelling of the axillary or inguinal lymph nodes) and fever. In the absence of antibiotics, fatal bacteremia develops rapidly.

Aerosolized transmission of *Y. pestis* characterizes the highly infectious *pneumonic plague.** A shorter incubation period results in fever, headache, malaise, and a cough accompanied by blood and mucus. Without early antibiotic intervention, the pneumonia progresses rapidly over 2 to 4 days to septic shock and death.

Early treatment with streptomycin, tetracycline, and chloramphenicol is effective in treating pneumonic plague. Prophylactic antibiotic treatment for 7 days will protect persons at risk for close (face-to-face) contact with infected patients. There is no vaccine against plague.†

34.2.4 Brucellosis (*Brucella suis*)

Brucellosis is caused by Gram-negative coccobacillus bacteria of the genus *Brucella*. Infections are generally zoonotic, with animal reservoirs maintained in sheep, goats, cattle, deer, elk, pigs, and dogs. Human infections are acquired by contact with infected animals, consumption of contaminated unpasteurized dairy products, and improper laboratory handling (inhalation and cutaneous exposure).‡ Over 500,000 cases are reported worldwide annually. The disease is frequently a problem in countries that do not have good standardized and effective public health and domestic animal health programs. Areas currently listed as high risk are countries in the Mediterranean basin (Portugal, Spain, Southern France, Italy, Greece, Turkey, and North Africa), South and Central America, Eastern Europe, Asia, Africa, the Caribbean, and the Middle East. Unpasteurized cheeses ("village cheeses") from these areas may present a particular risk for tourists. Close person-to-person transmission is rare. The incidence is lower in the United States (100 to 200 cases annually), and it is most commonly reported from California and Texas and in residents and visitors from Mexico.

Depending on the species, human brucellosis causes a range of symptoms. Acute disease has an onset of 2 months. Initial symptoms are nonspecific and consist of fever,§ sweating, headache, back pain, and malaise. Cutaneous, neurologic, and car-

---

* Person-to-person transmission occurs through close inhalation of infected respiratory droplets.

† Although gentamicin is not FDA approved for treatment of pneumonic plague, the Center for Civilian Biodefense Studies Working Group on Civilian Biodefense has recommended it along with streptomycin as a preferred therapy.

‡ Owners of dogs infected with *B. canis* (the species common in dogs) are not considered to be at risk of acquiring brucellosis, provided they are not exposed to infected secretions from untreated animals.

§ In untreated patients, the fever is at times intermittent, from which is derived the name *undulant fever*.

diovascular complications characterize severe infections. Chronic complications involve recurrent fevers, joint pain, and fatigue.

Because *Brucella* species are slow growing on initial laboratory isolation, diagnosis depends on serological identification in blood or bone marrow isolates. Treatment with doxycycline and rifampin in combination, for 6 weeks, prevents recurring infections. The timing of treatment and severity of the illness dictate the length of recovery, from several weeks to several months. Although the mortality rate is low (<2%) and is usually associated with the development of cardiovascular complications (endocarditis), the lack of an effective human vaccine is of concern as a bioterrorist threat. Adherence to recommended public hygiene guidelines is the most effective preventive measure against brucellosis.

34.2.5 Salmonellosis (*Salmonella* Species)

Salmonellosis is caused by infection with *Salmonella* species, a Gram-negative bacillus. It is a member of the *Enterobacteriaceae* family that colonizes the GI tracts of many species of animals, including chickens, cattle, and reptiles. Because of the large number of reactions that *Salmonella* display against human antibodies, over 2400 known serogroups have been classified into three pathological categories. These include:

1. *Salmonella* that are highly adapted to human hosts, such as *S. typhi* (Group D salmonella) and *S. paratyphi* (Group A), which produce typhoid fever and paratyphoid fever, respectively
2. *Salmonella* that are adapted to nonhuman hosts and that cause disease almost exclusively in animals (with a few exceptions)
3. *Salmonella* that are not adapted to specific hosts, which comprise the majority of the 2200 serotypes, designated *S. enteritidis* (Group D non-typhoidal *Salmonella*).

Infections with species of *S. enteritidis* account for 85% of all *Salmonella* infections in the United States (typhoid fever infections caused by *S. typhi* are rare, with only 400 to 500 cases reported annually). About 50% of salmonellosis cases are caused by *S. enteritidis* and *S. typhimurium*. Although the proportion of salmonellosis caused by these serotypes has decreased since 1996, an increasing number of isolates show resistance to multiple antimicrobial drugs. For instance, the incidence of a previously lesser-known species, *S. newport*, has increased by 34% from 1996 to 2001. It has thus become the third most frequent serotype, with many isolates resistant to more than nine antimicrobial drugs.

Approximately 1.4 million cases of salmonellosis infections occur in the United States annually, most of which are diagnosed as *gastroenteritis*. The disease results primarily from infection with group D non-typhoidal *Salmonella*. Gastroenteritis usually starts 12 to 48 hours after ingestion of the organisms with nausea and mild to severe abdominal pain, followed by watery diarrhea, sudden fever, and sometimes vomiting and dehydration. Non-typhoidal salmonellosis is usually mild to moderate, is self-limiting, lasts 4 to 7 days, and has a mortality of less than 1%. A chronic carrier state is possible in non-typhoidal salmonellosis but is more common with typhoid fever. In any event, chronic carriers shed organisms in the stools for at least a month, sometimes more than a year, after disease. In addition, persistent shedding of organisms in the stool for more than 1 year occurs in 0.2 to 0.6% of patients with non-typhoidal salmonellosis. Therefore, food handlers pose a serious epidemic risk.*

The mechanism of contraction of infection with group D non-typhoidal *Salmonella* is similar to but more complicated than that for typhoid fever. Fecal–oral transmission occurs in individuals by direct or indirect contact with animals, foodstuffs, and excretions from infected animals or humans. As few as 100,000 organisms are required for human inoculation. Once ingested, the organisms that successfully escape the acidic stomach contents penetrate and pass through to the ileum and colon. The organism penetrates intact mucosal membrane barrier cells of the ileum or colon, multiplies within the mucosal cells, and activates an inflammatory response. The resulting pathology is mostly due to damage within the GI tract.

Occasionally, more severe, protracted illness and long-term consequences result from *Salmonella* poisoning, including (1) enteric fever, (2) focal (localized) infections, and (3) bacteremia.

1. *Enteric fever* is a systemic syndrome manifested by fever, prostration, and bacteremia. Enteric fever is attributable primarily to group B, *S. typhi*, and milder group A *S. paratyphi*, referred to as *paratyphoid fever* (see later). B. *Focal infections* of infected organs start in the GI tract and disseminate to the liver, gall bladder, and appendix. The organisms can proliferate to cause cardiac murmurs, pericarditis, pneumonia, rheumatoid-like arthritis, and osteomyelitis. C. *Bacteremia*, characterized by sustained septicemia with *S. typhi*, *S. paratyphi*, *S. choleraesuis*, and *S. enteritidis*, is relatively uncommon in patients with gastroenteritis. However, *Salmonella* from groups B and C1 can cause bacteremia lasting about 1 week.

Other routes of contagion include ingestion of poorly cooked meat and handling of raw infected meat. Unsuspecting ingestion of contaminated poorly cooked poultry, raw milk or eggs, and egg products is often the perpetrator (the eggs may be contaminated both on their surface and within). Outbreaks are more common in summer months and are often related to contaminated egg or chicken salads.

Patients with non-typhoidal salmonellosis are at risk of developing a less common but significantly adverse disorder known as *Reiter's syndrome*. The postinfection sequelae are initiated by a bacterial-induced, aberrant, hyperactive immune response. The disease is characterized by arthritis, particularly affecting the knees, ankles, and feet. About 50% of patients experience prolonged recurrence of arthritis over several years.

Isolation of the organism from the stools, blood, pus, vomitus, and urine aids in the presumptive diagnosis. Identification is based on initial growth of the organism on appropriate culture media, followed by biochemical and serological confirmation.

* Gastroenteritis and bacteremia are common and more virulent among infants and children, in the elderly, and in immunocompromised individuals. Salmonellosis also occurs 20 times more often in patients with AIDS, sickle cell anemia, cirrhosis, and leukemia.

There is no acceptable antibiotic cure for uncomplicated non-typhoidal salmonellosis. Gastroenteritis is treated symptomatically with fluids, electrolytes, and a bland diet (for dehydration and continuous fever). Antibiotics prolong excretion of the organism and are unwarranted in uncomplicated cases. Antibiotic resistance is more common with non-typhoidal salmonella than with *S. typhi*. In addition, antibiotics can prolong the shedding of organisms in the stools after the drug has been discontinued. Infections with *S. typhi* and *S. paratyphi* are treated with fluoroquinolones (e.g., ciprofloxacin), chloramphenicol, gentamicin, trimethoprim/sulfamethoxazole, or broad-spectrum cephalosporins.

As with brucellosis, the lack of an effective vaccine warrants concern that *Salmonella* can be used as a possible bioterrorist tool. Epidemiologically, preventing contamination of foodstuffs by infected humans is paramount. Contaminated raw eggs may pass unrecognized in some foods, such as homemade hollandaise sauce, caesar and other salad dressings, homemade ice cream, mayonnaise, cookie dough, and frostings. Ready-to-eat food, raw food, fruits, vegetables, or prepared desserts must be properly cooked, handled, stored, and/or refrigerated. Hand washing is essential before handling any food and between different food items. It is easy to observe how direct deliberate or inadvertent contamination can occur in the absence of proper hygienic food handling measures.

34.2.6 Typhoid Fever

Typhoid fever is a less common but life-threatening infection caused by *Salmonella typhi*. About 400 cases occur each year in the United States, of which 70% are acquired through international travel. In the developing world, it affects about 12.5 million persons each year. The organism is responsible for *enteric fever*. As with other *Salmonella*, *S. typhi* penetrates intact intestinal mucous membranes after oral ingestion and remains viable within phagocytic macrophages. The immune cells migrate to the liver, spleen, and bone marrow, transporting the organism to critical destinations. *S. typhi* eventually reactivates 10 to 14 days after ingestion. Gradually increasing nonspecific fever, headache, myalgias, malaise, and anorexia ensue and persist for about 1 week. GI symptoms resume. The cycle continues with bacteremia and colonization of the gallbladder and with reinfection of the intestinal tract.

Humans are the natural host for *S. typhi*. Persons with typhoid fever, and about 1% to 5% of those who recover from the infection (carriers), maintain chronic colonization of the organism. The latter can carry the organism for more than 1 year after symptomatic disease, with the gallbladder acting as the physiologic incubator. Both are capable of shedding *S. typhi* in their feces.

As with other forms of salmonellosis, *S. typhi* is acquired through fecal–oral transmission. Typhoid fever is preventable by avoiding ingestion of poorly cooked, raw, or unwashed foods. Drinking bottled water or boiling potable water removes the risk of contamination. The disease is usually treated with antibiotics (ampicillin, trimethoprim-sulfamethoxazole, and fluoroquinolones), but vaccination is recommended especially for travelers to endemic areas. The vaccine is available for parenteral administration (ViCPS) and in capsule form (Ty21a) and requires 1 and 2 weeks, respectively, for induction of adequate antibody titers. The complications of untreated typhoid fever are associated with a 20% mortality rate.

34.2.7 Shigellosis (Shigella Species)

Shigellosis is caused by infection with *Shigella* species, a Gram-negative bacillus. As with the other members of the *Enterobacteria* family, the organisms colonize the GI tracts of many species of animals. Several species of *Shigella* are further divided into 45 serogroups: *Shigella sonnei* (group D) accounts for over 65% of shigellosis in the United States; *S. flexneri* (group B) accounts for the rest (about 450,000 total unconfirmed cases in the United States each year). *S. dysenteriae* type 1 and *S. boydii* are rare, although they continue to be important causes of disease in the developing world. Over 150 million cases of shigellosis occur annually worldwide.* As with salmonella infections, fecal–oral transmission of contaminated food is the major route of transmission.

Children account for 70% of shigellosis cases. The syndrome is characterized by abdominal cramps, diarrhea (often containing abundant blood and mucus), fever, and stomach cramps. The onset is 1 to 2 days, and duration is self-limiting to 5 to 7 days. Severe infections with high fever are associated with seizures in children less than 2 years old. Asymptomatic carriers are contagious by shedding organisms in the feces.

Diagnosis and pathogenicity are similar to those of salmonellosis. Although the disease is self-limiting, antibiotic treatment shortens the duration of illness and prevents the spread of the organism. Ampicillin, trimethoprim/sulfamethoxazole, nalidixic acid, or fluoroquinolones are indicated. As with salmonellosis, about 3% of patients with S. *flexneri* infections develop Reiter's syndrome, lasting for months or years. Vaccination and preventive measures are similar to those recommended for salmonellosis.

34.2.8 *Escherichia Coli* O157:H7

This Gram-negative enteropathogenic bacterium is an emerging cause of foodborne illness. The *Escherichia* genus consists of five, species of which *E. coli* is the most common and clinically important. Approximately 73,000 cases of infection occur annually in the United States. *E. coli* possesses a broad range of virulence factors, exotoxins, and adhesion molecules allowing the organisms to attach to the GI and urinary tracts.

A large number of enteropathogenic groups of *E. coli* inhabit the small and large intestines of healthy cattle and are further classified according to their serologic serotypes. Most *E. coli* strains are part of the normal human bacterial flora. *E. coli* serotype O157:H7, however, is not. In fact, it is one of the most virulent strains, responsible for producing fatal enterotoxins.†

E. coli is the most common microorganism responsible for sepsis and urinary tract infections. It is also a prominent cause of neonatal meningitis and gastroenteritis in

* Epidemics of *S. dysenteriae* type 1 have occurred in Africa and Central America with case fatality rates of 5–15%. In addition, *S. dysenteriae* produces an exotoxin (Shiga toxin) that damages intestinal epithelium and glomerular endothelial cells.

† The serotype designation (O:, H:) refers to the expression of surface and fimbrial antigens, respectively, that uniquely characterize this genus.

developing nations. Strains of *E. coli* responsible for gastroenteritis are subdivided into six groups: enterotoxigenic, enteropathogenic, enteroinvasive, enterohemorrhagic, enteroaggregative, and diffuse aggregative. Each is responsible for a variety of diseases including traveler's and infant diarrhea, dysentery, intra-abdominal infections, hemolytic uremic syndrome (HUS),* and hemorrhagic colitis.

The organism's ability to invade intestinal epithelial cells, its capacity to release exotoxins, and its ability to express adhesion molecules confer the properties necessary for producing gastroenteritis. The disease is characterized by acute dysentery accompanied by abdominal cramps with little or no fever.

Infection is acquired through ingestion of poorly cooked ground beef; consumption of unpasteurized milk and juice, sprouts, lettuce, and salami; and contact with cattle. Waterborne transmission occurs through swimming in contaminated lakes or pools or by drinking inadequately chlorinated water. The organism is easily transmitted from person to person and has been difficult to control in child day-care centers.

Stool cultures and serology confirm the presence of *E. coli* O157:H7 in suspected infections. With the exception of severe complications, most infections are self-limiting to 5 to 10 days. Antibiotic treatment is unwarranted and may actually precipitate kidney infections. Antidiarrheal agents, such as loperamide, should also be avoided.

Thorough cooking and proper handling of ground beef and hamburgers are effective in preventing *E. coli* infections. Raw meat should be kept separate from ready-to-eat foods. Hands, counters, and utensils should be washed with hot soapy water after contact with raw meat. Drinking unpasteurized milk, juice, or cider should be avoided. Fruits and vegetables should be washed thoroughly to avoid exposure.

34.2.9 Cholera (*Vibrio cholerae*)

Cholera is an acute diarrheal illness caused by intestinal infection with the Gram-negative, facultative, anaerobic bacterium *Vibrio cholerae*. Many species of this curved bacillus are also implicated in human infections (*V. parahaemolyticus* and *V. vulnificus*), although *V. cholerae* is the best-known member of the genus. The species are further subdivided into over 200 serogroups on the basis of their somatic O antigens. *V. cholerae* serogroup O1 or O139 is responsible for classic epidemic cholera and produces cholera toxin.

Although the infection is often mild or asymptomatic, approximately 1 in 20 infected persons has severe disease, which is primarily due to the secretion of the toxin. The cholera toxin is similar to the heat-labile enterotoxin of *E. coli*. During infection, the toxin binds to receptors on intestinal epithelial cells, stimulating hypersecretion of water and electrolytes. Within 2–3 days of ingestion, copious watery diarrhea, vomiting, rapid loss of body fluids, and leg cramps develop. The symptoms eventually progress to dehydration, metabolic acidosis, shock, and cardiovascular collapse. The condition is fatal (25–50%) if untreated (interestingly, the organism is not flushed from the intestinal tract during a diarrheal episode but adheres to the mucosa via attachment pili).

* Chronic kidney failure and neurologic impairment are frequent complications of HUS.

As with other enterobacteria, cholera is acquired by ingesting contaminated water or food. Large epidemic outbreaks are related to fecal contamination of water supplies or street-vended foods. The organism is occasionally transmitted through eating raw or undercooked shellfish that are naturally contaminated. While 0–5 cases per year are reported in the United States, there are no identifiable risk groups. Risk is extremely low (1 per million), even in travelers. It is a major cause of epidemic diarrhea, however, throughout the developing world. It is responsible for a global pandemic in Asia, Africa, and Latin America for the last four decades. A modest increase in imported cases since 1991 is related to an ongoing epidemic that began in 1991 in South America. The disease is acquired, or carried back in contaminated food, by travelers to endemic countries. A few cases have occurred in the United States among persons who traveled to South America or ate contaminated food brought back by travelers.

Fluid and electrolyte replacement is effective in preventing death, but the logistics of delivery in remote areas during pandemics renders this option difficult to achieve. Doxycycline, trimethoprim-sulfamethoxazole, and furazolidone reduce the bacterial burden and toxin production. Eradication of *V. cholerae* is very unlikely since the organism is found in natural estuarine and marine environments and is associated with chitinous shellfish.

Because of advanced water and sanitation systems, cholera is easily prevented and treated and is not a major threat in industrialized nations. Currently available killed parenteral vaccines offer incomplete protection of relatively short duration and have been discontinued; no multivalent vaccines are available for O139 infections. Newer killed, whole cell, oral vaccines for the O1 strain may offer some protection, but neither is available in the United States (Dukoral® from Biotec AB and Mutacol® manufactured by Berna).

34.2.10 Smallpox

Naturally occurring smallpox (*variola major*) is a member of the *orthopoxvirus* family, the largest and most complex family of viruses. It is easily transmissible via close person-to-person contact. The virus accounted for 7% to 12% of all deaths in eighteenth-century England. The last case of smallpox in the United States occurred in 1949 and the last case in the world, in Somalia, in 1977. Routine vaccination in the United States was stopped in 1971; the World Health Organization (WHO) determined in 1980 that smallpox had been successfully eradicated from the world.

Smallpox virus was inhaled and replicated in the respiratory tract. Dissemination occurred via lymphatics, resulting in viremia. The lymphatic vessels provided the pathways for the virus to spread to the spleen, bone marrow, liver, and skin (characteristic rash). A second viremia caused the development of additional lesions. Depending on which of the two variants of smallpox was contracted, mortality ranged from 15% to 40% (for variola major) to 1% (variola minor).

The incubation period for variola was about 7 to 17 days. Initial symptoms included high fever, fatigue, and head and back aches. The characteristic lesions (*pocks*) appeared in 2–3 days on the face, arms, and legs. Lesions became pus-filled and began to crust early in the second week. Eschar developed, separated, and sloughed off after 3–4 weeks.

Smallpox was very contagious through the respiratory/salivary route and infected humans only through accidental or occupational exposure. Individuals were most contagious during the first week of signs and symptoms, corresponding to high-viremic periods.

Although routine vaccination against smallpox ended in 1972, the level of continuous immunity among persons vaccinated before 1972 is uncertain. Recently, the threat that smallpox could be used as a weapon of bioterrorism has prompted the development of new vaccination strategies for both military personnel and civilians. Vaccinia, a form of cowpox, is used for the production of smallpox vaccine. The procedure consists of scratching live virus into the patient's skin and observing for the formation of vesicle and pustules. In October 2002, the FDA approved a new 100-dose kit for smallpox vaccine (Dryvax®, Wyeth), the only currently licensed smallpox vaccine. All supplies of Dryvax® for civilian use are under CDC control. Complications and potential risks related to vaccination occur primarily in immunocompromised individuals and, more recently, in persons with preexisting cardiovascular disorders. The most serious complications are encephalitis, progressive infection (vaccinia necrosum), and myocardial infarction.

34.2.11 Tularemia (*Francisella tularensis*), Q Fever (*Coxiella burnetti*), and Viral Hemorrhagic Fevers (VHF)

Tularemia is caused by the Gram-negative coccobacillus *Francisella tularensis.* Humans acquire the zoonotic infection through ticks (arthropod vectors), rabbits, and domestic pets. About 100 cases annually are reported in the United States. Clinically, tularemia is a plague-like, glandular, typhoidal disease. GI and pneumonic syndromes are characterized by fever, chills, headache, myalgia, and malaise. Streptomycin and gentamicin are the antibiotics of choice. A live-attenuated vaccine is available but rarely used.

Q fever is caused by an obligate intracellular bacterium, *Coxiella burnetti*, closely related to *Francisella.* Most human infections are acquired through inhalation from contact with animal reservoirs (arthropods are not important vectors with Q fever). The infection is characterized by acute onset of an influenza-like syndrome (fever, headache, and malaise), followed by pneumonia, hepatitis, endocarditis, and pulmonary disease. Tetracyclines are the drugs of choice for acute infections; rifampin plus doxycycline is used for chronic infections. Vaccines are protective if administered before *C. burnetti* is contracted.

Viral hemorrhagic fevers (VHF) are a clinically related group of viral diseases with a diverse etiology. The family of *Filoviridae* includes the Marburg and Ebola viruses. Similarly, the family of *Bunyaviridae* includes the hantavirus and bunyavirus (which primarily causes encephalitis in humans). All are RNA viruses endemic in Africa and with the exception of the bunyavirus, cause severe or fatal hemorrhagic fevers. The condition is characterized by acute onset of fever, headache, generalized myalgia, conjunctivitis, and severe prostration, followed by various hemorrhagic symptoms. The organism facilitates the destruction of endothelial cells, leading to vascular injury and increased capillary permeability, leukopenia, and thrombocytopenia. Antiviral therapy has not been shown to be clinically useful.

Arenaviruses, including the Lassa, Junin, and Machupo viruses, are endemic in Africa and South America. The organisms stimulate cell-mediated inflammatory responses following an incubation period of 10–14 days. The syndrome is manifested by fever, petechiae, and cardiac, hepatic, and splenic damage. No vasculitis or neuronal damage is observed. The recently approved antiviral drug, Ribavirin®, has demonstrated some activity against Lassa fever, which has a 50% mortality rate if untreated.

34.3 CHEMICAL AGENTS AS THREATS TO PUBLIC SAFETY

34.3.1 Nerve Gases

The nerve gases were developed during World War II for possible use as chemical warfare agents; the first compound to be developed was tetraethyl pyrophosphate (TEPP). The biological action of the nerve gases, such as sarin (GB),* tabun (GA), and soman (GD), is similar to, but more toxic than, the organophosphate (OP) insecticides (see Chapter 28). The clear, colorless, tasteless liquids inhibit the action of acetylcholine esterase (ACh-∑) by forming an *irreversible* OP–ACh-∑ complex, rendering it incapable of hydrolyzing acetylcholine (ACh). Inhibition of the enzyme results in accumulation and over-stimulation of ACh at autonomic and somatic receptors. Excessive stimulation of nicotinic receptors is followed by skeletal muscle paralysis. These circumstances account for the toxic manifestations of OP insecticides as well as the nerve gases.

Soman, a methylphosphonofluoridic acid ester, is miscible with organic solvents and water. It is readily absorbed through the skin. Soman is the most toxic of the three "G" agents and one of the most toxic compounds ever synthesized, with fatalities occurring with an oral dose of 10 μg/kg in humans. *Sarin*, a methylphosphonofluoridic acid ester, and *tabun*, a dimethylphosphonofluoridic acid ester, are also miscible with organic solvents and water and are readily absorbed through the skin (tabun possesses a bitter almond smell). Recently, however, *VX* (methylphosphonothioic acid ester) has emerged as a more toxic nerve agent. Its chemical and biological properties satisfy the requirements for VX as a chemical bioterrorist threat. The odorless, tasteless, oily liquid was originally developed in the United Kingdom in the 1950s. Unlike more volatile aromatic and aliphatic hydrocarbons, heating VX liquid renders it suitable for inhalation, dermal, or ocular contact with the airborne vapors. Contamination of food and water supplies is also possible. Of concern is its ability to accumulate in physiological compartments, its slow metabolic degradation, and its high density, enabling it to spread throughout low-lying areas.

The onset of symptoms is within seconds for inhalation of vapors and hours for dermal contact or oral ingestion. In addition, they all decompose in the presence of bleach (or base). As with all the nerve agents, VX produces a severe cholinergic syndrome, terminating in convulsions, respiratory failure, and death with high doses. Recovery from mild or moderate exposure to a nerve agent is possible. Treatment relies on removal from exposure, supportive measures, decontamination, and rapid

* The deliberate release of sarin in a Tokyo subway in 1995 by a terrorist organization was responsible for 12 deaths and 5000 casualties.

administration of atropine (antimuscarinic) and pralidoxime (ACh-∑ reactivator). The FDA has recently approved pyridostigmine bromide, an anticholinesterase agent, as a prophylactic drug for U.S. military personnel to increase survival after exposure to soman poisoning during combat.

Nerve agents are metabolized quickly in air and soil and after several hours in water, incurring a low incidence of environmental accumulation. The risk for accidental exposure to the general population is low, except for military personnel or individuals working in or around storage plants.

34.3.2 Vesicants, Chemical Asphyxiants, and Pulmonary Irritants

Vesicants, such as nitrogen mustard compounds, are capable of causing tissue necrosis on extravasation. These alkylating agents have a delayed onset of action (about 6–8 hours), although cellular necrosis ensues immediately. Inhalation produces a typical pulmonary irritation. The toxicity of vesicating agents is more severe than that of lacrimating agents, the latter of which were subsequently developed for domestic law enforcement. Sulfur mustard compounds (H, HD, and HT; "mustard gas") possess a garlic odor, although they are sometimes odorless. The compounds are generally colorless or yellowish liquids, do not accumulate in the environment, and are not a threat for accidental public exposure.

Lacrimating agents (benzyl bromide), simple asphyxiants (inert gases), chemical asphyxiants (carbon monoxide, cyanide), and pulmonary irritants (carbon disulfide, phosgene) are now recognized as potential chemical bioterrorist threats. Most of the clinical toxicity of these compounds is discussed in detail in Chapter 25. Some of their applications as bioterrorist weapons will be further considered here.

Cyanide has risen on the high-priority list for bioterrorist threats, although its toxicity is recognized historically. Also known by the military designations AN (for hydrogen cyanide) and CK (for cyanogen chloride), cyanide may have been used during the Iran–Iraq war in the 1980s. It was probably used along with other chemical agents against the inhabitants of the Kurdish city of Halabja in northern Iraq. The compounds are released in air, soil, or water as industrial waste products in the electroplating and metallurgical industries. Additional uses of cyanides include photographic development, fumigating ships, and in the mining industry. Cyanides are also metabolic products of bacteria, fungi, algae, and plants.

As a terrorist threat, AN gas is most toxic. The colorless gas has a faint bitter almond odor. The Environmental Protection Agency (EPA) allows a maximum contaminant level of 0.2 mg/l of cyanide in drinking water. It requires the reporting of spills or accidental releases into the environment of 1 lb or more of cyanide compounds. The clinical toxicity and treatment of cyanide poisoning are discussed in Chapter 23.

34.3.3 Ricin (*Ricinus communis*)

Ricin is derived from the processing of the castor bean and its seeds (*Euphorbiaceae*) in the extraction of castor oil.* The seeds, plant, fruit, and constituents are stable in

* *Ricinus* is Latin for tick; the seed resembles some arthropods in shape and markings.

extremely hot or cold temperatures. The plant is indigenous to temperate and tropical India, Africa, and South America. The seeds have been found in Egyptian tombs. The oil appears to have had only technical uses until the eighteenth century, when its medicinal application was explored.

Ricin is composed of two lectins found in the seeds, ricins I and II. The compounds, especially ricin II, bind to and inactivate the 60S ribosomal subunit in somatic cells, thus blocking protein synthesis. The potent cholinergic properties of the oil have rendered it useful for decades as an "over the counter" laxative. It is one of the most potent of plant toxins.

Ingestion of intact castor bean seeds is unlikely to cause deleterious effects for several days, although the ingestion of chewed castor beans rarely results in significant morbidity. Gradually, however, it produces nausea, vomiting, dyspnea, and diarrhea. Gastroenteritis follows and is characterized by severe bloody diarrhea, vomiting, and dehydration. Mental confusion, seizures, and hyperthermia complicate the scenario.

Inhalation of ricin powder is likely to produce a cough, dyspnea, nausea, and vomiting within a few hours. Pulmonary congestion and cyanosis could soon follow. *Injection* of a lethal amount of ricin (estimated to be about 500 μg) at first would cause local muscle paralysis and lymph node necrosis near the injection site. Massive stomach and intestinal hemorrhaging would ensue, followed by multiple organ failure. Death occurs within 36–48 hours and is due to focal necrosis of the liver, spleen, lymph nodes, intestine, and stomach.

Ricin is otherwise not an environmental metabolic product, and unintentional ricin poisoning is highly unlikely. Its presence, therefore, suggests deliberate contamination; hence the bioterrorist concern.* The manufactured material can be dissolved in water, vaporized, or injected, thus facilitating exposure through oral, respiratory, or parenteral routes.

Antidotes are not available for ricin poisoning. Treatment necessitates supportive emergency measures, including maintenance of respiration and renal perfusion, and gastric decontamination.

REFERENCES

SUGGESTED READINGS

Borio, L. Inglesby, T., Peters, C.J., Schmaljohn, A.L., Hughes, J.M., Jahrling, P.B., Ksiazek, T., Johnson, K.M., Meyerhoff, A., O'Toole, T., Ascher, M.S., Bartlett, J., Breman, J.G., Eitzen, E.M. Jr., Hamburg, M., Hauer, J., Henderson, D.A., Johnson, R.T., Kwik, G., Layton, M., Lillibridge, S., Nabel, G.J., Osterholm, M.T., Perl, T.M., Russell, P., and Tonat, K., Hemorrhagic fever viruses as biological weapons: medical and public health management. *JAMA*, 287, 2391, 2002.

Broad, W.J. and Miller, J., Threats and Responses: The Bioterror Threat; Health Data Monitored for Bioterror Warning, *New York Times*, National Desk, January 27, 2003.

* In 1978, Georgi Markov, a Bulgarian writer and journalist, died of severe gastroenteritis and multi-organ failure after he was attacked by a man in London with an umbrella suspected of containing a ricin pellet in the tip. Other reports have implicated the use of ricin during the Iran–Iraq war in the 1980s.

Center for Biologics Evaluation and Research (CBER), U.S. Food and Drug Administration, *Countering Bioterrorism and Emerging Infectious Diseases*, last updated 2018, last viewed March 2019.

Center for Biologics Evaluation and Research (CBER), U.S. Food and Drug Administration, *Emergency Preparedness and Response, Counterterrorism and Emerging Threats, Medical Countermeasures Initiative*, last updated 2018, last viewed March 2019.

Drazen, J.M., Perspective: smallpox and bioterrorism. *N. Engl. J. Med.*, 346, 1262, 2002.

Emergency Preparedness and Response, Centers for Disease Control and Prevention, *Crisis & Emergency Risk Communication (CERC)*. https://emergency.cdc.gov/cerc/index.asp, last updated May 2018, last viewed March 2019.

Kman, N.E. and Nelson, R.N., Infectious agents of bioterrorism: a review for emergency physicians. *Emerg. Med. Clin. North Am.* 26, 517, 2008.

Prokop, L.D., Weapons of mass destruction: overview of the CBRNEs (Chemical, Biological, Radiological, Nuclear, and Explosives). *J. Neurol. Sci.* 249, 50, 2006.

Recommendations of the Advisory Committee on Immunization Practices: Use of Anthrax Vaccine in the United States. *MMWR Recomm. Rep.* 59, 1, 2010.

U.S. Army Medical Research Institute of Infectious Diseases, *USAMRIID's Medical Management of Biological Casualties Handbook*, Zembek, Z.F. (lead editor), 7th ed., 2011.

REVIEW ARTICLES

Agency for Toxic Substances and Disease Registry (ATSDR), *Toxicological profiles*. U.S. Department of Health and Human Services, Public Health Service. www.atsdr.cdc.gov/toxprofiledocs/index.html, last updated 2018, last viewed March 2019.

Bozeman, W.P., Dilbero, D. and Schauben, J.L., Biologic and chemical weapons of mass destruction. *Emerg. Med. Clin. North Am.* 20, 975, 2002.

Dixon, T.C., Meselson, M., Guillemin, J., and Hanna, P.C., Anthrax. *N. Engl. J. Med.* 341, 815, 1999.

Franz, D.R. and Zajtchuk, R., Biological terrorism: understanding the threat, preparation, and medical response, *Dis. Mon.* 48, 493, 2002.

Grundmann, O., The current state of bioterrorist attack surveillance and preparedness in the US. *Risk Manag. Healthc. Policy.* 7, 177, 2014.

Hamburg, M.A., Bioterrorism: responding to an emerging threat. *Trends Biotechnol.* 20, 296, 2002.

Henretig, F.M., Cieslak, T.J., and Eitzen, E.M. Jr., Biological and chemical terrorism. *J. Pediatr.* 141, 311, 2002.

Jefferson, C., Lentzos, F., and Marris, C., Synthetic biology and biosecurity: challenging the "Myths". *Front Public Health* 2, 115, 2014.

Karwa, M., Bronzert, P., and Kvetan, V., Bioterrorism and critical care. *Crit. Care Clin.* 19, 279, 2003.

Kimmel, S.R., Mahoney, M.C., and Zimmerman, R.K., Vaccines and bioterrorism: smallpox and anthrax. *J. Fam. Pract.* 52, S56, 2003.

Lashley, F.R., Factors contributing to the occurrence of emerging infectious diseases. *Biol. Res. Nurs.* 4, 258, 2003.

Moran, G.J., Talan, D.A., and Abrahamian, F.M., Biological terrorism. *Infect. Dis. Clin. North Am.* 22, 145, 2008.

McQuiston, J.H., Childs, J.E., and Thompson, H.A., Q fever. *J. Am. Vet. Med. Assoc.* 221, 796, 2002.

Menrath, A.,Tomuzia, K., Frentzel, H., Braeunig, J., and Appel, B.,Survey of systems for comparative ranking of agents that pose a bioterroristic threat. *Zoonoses Public Health* 61, 157, 2014.

Nyamathi, A.M., Fahey, J.L., Sands, H., and Casillas, A.M., Ebola virus: immune mechanisms of protection and vaccine development. *Biol. Res. Nurs.* 4, 276, 2003.

Olsnes, S. and Kozlov, J.V., Ricin. *Toxicon* 39, 1723, 2001.

Orloski, K.A. and Lathrop, S.L., Plague: a veterinary perspective. *J. Am. Vet. Med. Assoc.*, 222, 444, 2003.

Rosenbloom, M., Leikin, J.B., Vogel, S.N., and Chaudry, Z.A., Biological and chemical agents: a brief synopsis, *Am. J. Ther.* 9, 5, 2002.

Sandvig, K. and van Deurs, B., Entry of ricin and Shiga toxin into cells: molecular mechanisms and medical perspectives. *EMBO J.* 19, 5943, 2000.

Slater, L.N. and Greenfield, R.A., Biological toxins as potential agents of bioterrorism. *J. Okla. State Med. Assoc.* 96, 73, 2003.

Slater, M.S. and Trunkey, D.D., Terrorism in America. An evolving threat. *Arch. Surg.* 132, 1059, 1997.

Terriff, C.M., Schwartz, M.D., and Lomaestro, B.M., Bioterrorism: pivotal clinical issues. Consensus review of the Society of Infectious Diseases Pharmacists. *Pharmacotherapy* 23, 274, 2003.

Wenzel, R.P., Recognizing the real threat of biological terror. *Trans. Am. Clin. Climatol. Assoc.* 113, 42, 2002.

Index

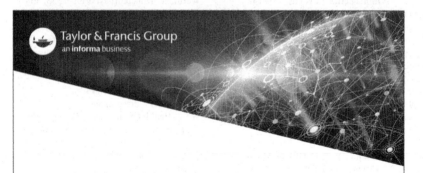